Seed Fate

Predation, Dispersal and Seedling Establishment

Dedication

We dedicate this volume to Daniel Janzen, whose many ideas
germinated, took root and bore fruit, including this book

Seed Fate

Predation, Dispersal and Seedling Establishment

———————————

Edited by

Pierre-Michel Forget

Département Ecologie et Gestion de la Biodiversité, Muséum National d'Histoire Naturelle, Brunoy, France

Joanna E. Lambert

Department of Anthropology, University of Wisconsin, Madison, USA

Philip E. Hulme

NERC Centre for Ecology and Hydrology, Banchory, UK

and

Stephen B. Vander Wall

Department of Biology, University of Nevada, Reno, USA

CABI Publishing

CABI Publishing is a division of CAB International

CABI Publishing
CAB International
Wallingford
Oxfordshire OX10 8DE
UK

CABI Publishing
875 Massachusetts Avenue
7th Floor
Cambridge, MA 02139
USA

Tel: +44 (0)1491 832111
Fax: +44 (0)1491 833508
E-mail: cabi@cabi.org
Website: www.cabi-publishing.org

Tel: +1 617 395 4056
Fax: +1 617 354 6875
E-mail: cabi-nao@cabi.org

A catalogue record for this book is available from the British Library, London, UK.

Library of Congress Cataloging-in-Publication Data
Symposium on "Post-Primary Seed Fate: Predation and Secondary Dispersal" (2002 : Panama, Panama)
  Seed fate: predation, dispersal, and seedling establishment / edited by Pierre-Michel Forget . . . [et al.].
    p. cm.
  Some papers were presented at the Symposium on "Post-Primary Seed Fate: Predation and Secondary Dispersal" held in Panama City, Panama, 29 July–2 August, 2002.
  Includes bibliographical references and index.
  ISBN 0-85199-806-2 (alk. paper)
  1. Seeds--Congresses. I. Forget, Pierre-Michel. II. Title.

SB113.3.S96 2002
631.5′21--dc22   2004006553

ISBN 0 85199 806 2

Typeset by AMA DataSet Ltd, UK.
Printed and bound in the UK by Biddles Ltd, King's Lynn.

Contents

Contributors

Bruce Anderson, *Botany Department, University of Cape Town, P. Bag Rondebosch 7701, South Africa*

Ellen Andresen, *Centro de Investigaciones en Ecosistemas, Universidad Nacional Autónoma de México, A.P. 27-3, Morelia, C.P. 58089, Michoacán, México (E-mail: andresen@ate.oikos.unam.mx)*

Harald Beck, *Duke University, Center for Tropical Conservation, PO Box 90381, Durham, NC 27708-0381, USA (E-mail: harald@duke.edu)*

Colin A. Chapman, *Department of Zoology, University of Florida, Gainesville, FL 32611, USA and Wildlife Conservation Society, 2300 Southern Boulevard, Bronx, NY 10460, USA*

Eleanor A. Collins, *CSIRO Sustainable Ecosystems and the Rainforest Cooperative Research Centre, PO Box 780, Atherton, Qld 4883, Australia, and Department of Tropical Biology, James Cook University of North Queensland, Townsville, Qld 4810, Australia*

James W. Dalling, *Department of Plant Biology, University of Illinois, Urbana-Champaign, 265 Morrill Hall, 505 S Goodwin Ave., Urbana, IL 61801, USA, and Smithsonian Tropical Research Institute, Apartado 2072, Balboa, Republic of Panama (E-mail: dallingj@life.uiuc.edu)*

Jan den Ouden, *Centre for Ecosystem Studies, Wageningen University, PO Box 47, 6700 AA, Wageningen, The Netherlands (E-mail: Jan.denOuden@wur.nl)*

Andrew J. Dennis, *CSIRO Sustainable Ecosystems and the Rainforest Cooperative Research Centre, PO Box 780, Atherton, Qld 4883, Australia (E-mail: Andrew. Dennis@csiro.au)*

François Feer, *Département Ecologie et Gestion de la Biodiversité, Muséum National d'Histoire Naturelle, UMR 5176 CNRS-MNHN, 4 avenue du Petit Château, F-91800 Brunoy, France*

Renate C. Fischer, *Institute of Botany, Department of Morphology and Reproduction Ecology, University of Vienna, Rennweg 14, A-1030 Vienna, Austria*

Pierre-Michel Forget, *Département Ecologie et Gestion de la Biodiversité, Muséum National d'Histoire Naturelle, UMR 5176 CNRS-MNHN, 4 avenue du Petit Château, F-91800 Brunoy, France (E-mail: pmf@mnhn.fr)*

Graham N. Harrington, *CSIRO Sustainable Ecosystems, PO Box 780, Atherton, Qld 4883, Australia*

Fumio Hayashi, *Department of Biology, Tokyo Metropolitan University, Minamiosawa 1-1, Hachioji, Tokyo 192-0397, Japan*

Kazuhiko Hoshizaki, *Department of Biological Environment, Faculty of Bioresource Sciences, Akita Prefectural University, Akita 010-0195, Japan (E-mail: khoshiz@akita-pu.ac.jp)*

Philip E. Hulme, *NERC Centre for Ecology and Hydrology, Hill of Brathens, Banchory AB31 4BW, UK (E-mail: pehu@ceh.ac.uk)*

Nor Azman Hussein, *Forest Research Institute Malaysia, Kepong, Kuala Lumpur, Malaysia*

Nobuo Ishii, *Japan Wildlife Research Center, Shitaya, Taito, Tokyo, Japan*

Patrick A. Jansen, *Centre for Ecosystem Studies, Wageningen University, PO Box 47, 6700 AA, Wageningen, The Netherlands; Present address: Community and Conservation Ecology Group, University of Groningen, PO Box 14, 9750 AA, Haren, The Netherlands (E-mail: p.a.jansen@biol.rug.nl)*

Toshio Katsuki, *Tama Forest Science Garden, Forestry and Forest Product Research Institute, Todori, Hachioji, Tokyo 193-0843, Japan*

Johannes Kollmann, *Department of Ecology, Royal Veterinary and Agricultural University, Rolighedsvej 21, 1958 Frederiksberg C, Denmark (E-mail: jok@kvl.dk)*

Joanna E. Lambert, *Department of Anthropology, University of Wisconsin, Madison, WI 53706, USA (E-mail: jelambert@wisc.edu)*

Karl W. Larsen, *Department of Natural Resource Sciences, University College of the Cariboo, Kamloops, British Columbia V2C 5N3, Canada*

Hong-Jun Li, *State Key Laboratory of Integrated Management of Pest Insects and Rodents in Agriculture, Institute of Zoology, Chinese Academy of Sciences, Beijing 100080, P.R. China*

Geoffrey J. Lipsett-Moore, *Department of Tropical Biology, James Cook University of North Queensland, Townsville, Qld 4810, Australia*

William S. Longland, *USDA Agricultural Research Service, 920 Valley Road, and The Program in Ecology, Evolution and Conservation Biology, University of Nevada, Reno, NV 89557, USA*

Veronika Mayer, *Institute of Botany, University of Vienna, Rennweg 14, A-1030 Vienna, Austria (E-mail: veronika.mayer@univie.ac.at)*

Jeremy J. Midgley, *Botany Department, University of Cape Town, P. Bag Rondebosch 7701, South Africa (E-mail: midgleyj@botzoo.uct.ac.za)*

Hideo Miguchi, *Faculty of Agriculture, Niigata University, Niigata 950-2181, Japan*

Shingo Miura, *Forestry and Forest Products Research Institute, 1 Matsunosato, Tsukuba, Ibaraki 305-8687, Japan*

Evan M. Notman, *Organization for Tropical Studies, Apartado 676-2050 San Pedro, Costa Rica; Present address: US Senate Committee on Agriculture, Nutrition and Forestry, Washington DC, USA (E-mail: evannotman@comcast.net)*

Toshinori Okuda, *National Institute for Environmental Studies, Onogawa, Tsukuba, Ibaraki, Japan*

Paulo S. Oliveira, *Departamento de Zoologia, C.P. 6109, Universidade Estadual de Campinas, 13083-970 Campinas SP, Brazil*

Silvester Ölzant, *Institute of Botany, Department of Morphology and Reproduction Ecology, University of Vienna, Rennweg 14, A-1030 Vienna, Austria*

Luciana Passos, *Departamento de Botânica, C.P. 6109, Universidade Estadual de Campinas, 13083-970 Campinas SP, Brazil*

Marco A. Pizo, *Departamento de Botânica, C.P. 199, Universidade Estadual Paulista, 13506-900, Rio Claro SP, Brazil (E-mail: pizo@rc.unesp.br)*

Kirsten M. Silvius, *Environmental Center, Krauss Annex 19, 2500 Dole St, University of Hawaii, Honolulu, HI 96822-2303, USA (E-mail: silvius@hawaii.edu)*

Ruben Smit, *Centre for Ecosystem Studies, Wageningen University, PO Box 47, 6700 AA, Wageningen, The Netherlands*

Michael Steele, *Department of Biology, Wilkes University, Wilkes-Barre, PA 18766, USA (E-mail: msteele@wilkes.edu)*

Noriko Tamura, *Tama Forest Science Garden, Forestry and Forest Product Research Institute, Todori, Hachioji, Tokyo 193-0843, Japan (E-mail: haya@ffpri.affrc.go.jp)*

Tad C. Theimer, *Department of Biological Sciences, Northern Arizona University, Flagstaff, AZ 86011, USA (E-mail: tad.theimer@nau.edu)*

Stephen B. Vander Wall, *Department of Biology-314 and The Program in Ecology, Evolution and Conservation Biology, University of Nevada, Reno, NV 89557, USA (E-mail: sv@med.unr.edu)*

Ana Cristina Villegas, *US State Department, Bureau of Oceans, Environmental and Scientific Affairs, Washington, DC, USA*

Luc A. Wauters, *Department of Structural and Functional Biology, University of Insubria, Varese, I-21100 Varese, Italy*

Dan G. Wenny, *Illinois Natural History Survey, Lost Mound Field Station, 3159 Crim Drive, Savanna, IL 61074, USA (E-mail: dwenny@inhs.uiuc.edu)*

David A. Westcott, *CSIRO Sustainable Ecosystems and the Rainforest Cooperative Research Centre, PO Box 780, Atherton, Qld 4883, Australia*

Zhi-Shu Xiao, *State Key Laboratory of Integrated Management of Pest Insects and Rodents in Agriculture, Institute of Zoology, Chinese Academy of Sciences, Beijing 100080, P.R. China*

Masatoshi Yasuda, *Forestry and Forest Products Research Institute, 1 Matsunosato, Tsukuba, Ibaraki 305-8687, Japan (E-mail: yasuda@mammalogist.jp)*

Zhi-Bin Zhang, *State Key Laboratory of Integrated Management of Pest Insects and Rodents in Agriculture, Institute of Zoology, Chinese Academy of Sciences, Beijing 100080, P.R. China (E-mail: zhangzb@ioz.ac.cn)*

Preface

Since the pioneering work of Janzen (1969, 1970, 1971), a large number of ecologists have become interested in analysing seed fate, whether seeds are preyed on or secondarily dispersed by either invertebrates or vertebrates. This growing interest in seed ecology was particularly obvious for both tropical and temperate habitats during the Third Symposium-Workshop on Frugivory and Seed Dispersal that took place in São Pedro, Brazil (see Silva *et al.*, 2000; Levey *et al.*, 2002) and where the four editors and many of the authors of this book met and interacted, especially during a workshop on 'Methods Used to Study Seed Fate and Removal' chaired by P.-M. Forget. During this meeting, it was evident that there is continued interest in the topic of post-dispersal seed fate among the international community, with new findings in the tropics (Forget and Vander Wall, 2001). A number of the papers included in this book were presented during another Symposium on 'Post-Primary Seed Fate: Predation and Secondary Dispersal' that took place at the annual conference of the Association for Tropical Biology and Conservation (ATBC) in Panama City, a meeting jointly organized by the Smithsonian Tropical Research Institute and the ATBC. During this symposium, our goal was to move a step forward beyond previous symposia on frugivores and seed dispersal (Estrada and Fleming, 1986; Fleming and Estrada, 1993), with original talks presenting and evaluating recently collected data on the roles that tropical animals, both invertebrate and vertebrate, play while preying on and/or secondarily moving seeds that have been transported by other means, mostly primary seed dispersers. None the less, the philosophy of this book was comparable to that of previous cited ones related to frugivory and seed dispersal; that is to offer students and scholars in the field an updated review of our current knowledge of seed ecology focusing on the destiny of seeds before or after being released or removed from the parent plant.

The purpose of this book is thus to present and to evaluate the most recent data on seed fate in diverse geographical regions around the world. This has been achieved by inviting leading scientists involved in research on seed ecology and, most particularly, animal–plant interactions from the perspective of predation and primary and secondary seed dispersal. The goal was also to evaluate questions relating to seed fate at several scales: temperate/tropical, continental, regional and smaller, within-site comparisons. In addition to a broad geographic assessment, we aimed to evaluate the impact of a variety of animal taxa on seed fate: from small invertebrates to medium- and large-bodied mammals. Finally, we wanted to evaluate these interactions and their influence on plant and animal strategies at both ecological and evolutionary scales. We aimed for a worldwide review of this topic with chapters dealing with organisms from a diversity of habitats in the tropics, i.e. from savanna to lowland tropical, montane cloud and subtropical forests in Brazil, Costa Rica, French

Guiana, Mexico, Panama, Peru, Venezuela, and the North American continent as a whole, central Africa and eastern Africa, Malaysia, China and Queensland in Australia, and from a diversity of habitat types in temperate and boreal ecosystems from mountain to arid desert habitats in northern China, Europe, Japan, North America and South Africa. The preface of Levey *et al.* (2002) ended with: 'We hope the ideas expressed in this book will disperse, take root and grow!'. Today, we may add that not only these ideas established and developed in the community, but they are now fruiting on its branches. This book is one of these fruit, and we hope its seeds will disperse, again and again, and successfully establish in many fields. We also hope those reviews will help students in many regions to start research and will continue to fill that gap of knowledge regarding seed fate and the events occurring between plant and offspring.

Pierre-Michel Forget
Joanna Lambert
Phil Hulme
Stephen Vander Wall

References

Estrada, R. and Fleming, T.H. (1986) *Frugivores and Seed Dispersal.* Dr W. Junk Publishers, Dordrecht, The Netherlands, 392 pp.

Fleming, T.H. and Estrada, R. (1993) *Frugivory and Seed Dispersal: Ecological and Evolutionary Aspects*, Vol. 107/108. Kluwer Academic Publishers, Dordrecht, The Netherlands, 392 pp.

Forget, P.-M. and Vander Wall, S.B. (2001) Scatter-hoarding rodents and marsupials: convergent evolution on diverging continents. *Trends in Ecology & Evolution* 16, 65–67.

Janzen, D.H. (1969) Seed-eaters versus seed size, number, toxicity and dispersal. *Evolution* 23, 1–27.

Janzen, D.H. (1970) Herbivores and the number of tree species in tropical forests. *American Naturalist* 104, 501–528.

Janzen, D.H. (1971) Seed predation by animals. *Annual Review of Ecology and Systematics* 2, 465–492.

Levey, D.J., Silva, W.R. and Galetti, M. (2002) *Seed Dispersal and Frugivory: Ecology, Evolution and Conservation.* CAB International, Wallingford, UK, 511 pp.

Silva, W., Galetti, M., Pizo, M.A., Levey, D. and Green, R. (2000) *3rd Symposium-Workshop on Frugivores and Seed Dispersal. Biodiversity and Conservation Perspectives. Programs and abstracts.* Universidade Estadual de Campinas and Universidade Estadual Paulista, São Pedro, Brazil, 294 pp.

Acknowledgements

The idea for this book first germinated during the Third International Symposium-Workshop on Frugivores and Seed Dispersal held in São Pedro (SP), Brazil, 6–11 August 2000. It later matured during a Symposium on 'Post-Primary Seed Fate: Predation and Secondary Dispersal' held during the annual conference of the Association for Tropical Biology and Conservation in Panama City, Panama, 29 July–2 August 2002. After this second meeting, the list of invited contributors was expanded in order to cover a greater scope of topics and geographic regions. We thank the organizers, especially Doug Levey and Joe Wright, and the various sponsors of these two conferences both for the invitation and the organization of workshop and symposia, which were crucial steps in the elaboration of this book. For financial support we especially thank the French Foreign Office, the Muséum National d'Histoire Naturelle and the Centre National de la Recherche Scientifique, the Association for Tropical Biology and Conservation (ATBC) in Washington, DC and the Smithsonian Tropical Research Institute (STRI) in Balboa, Panama, for travel grants and invitations to P.-M. Forget to Brazil and Panama. Each manuscript was reviewed by at least two editors and one outside reviewer. Reviews were provided by A. Beattie, H. Beck, W. Bond, R. Boulay, S. Brewer, S. Chauvet, J.R. Corlett, J. den Ouden, A. Dennis, J. Fragoso, A. Gautier-Hion, J. Gomez, M. Guariguata, K. Harms, K. Hoshizaki, P. Jansen, B. Kaplin, W.S. Longland, V. Mayer, J. Midgley, A. Moles, G. Murray, M. Norconk, E. Notman, J.-F. Ponge, M. Price, J. Refisch, E. Schupp, K. Silvius, P. Smallwood, T. Theimer, K. Vulinec, L. Wauters and D. Wenny. To all contributors and reviewers, we express our gratitude for their hard work and for adhering to the schedule. We are also thankful to the book publisher Tim Hardwick at CABI Publishing for his interest and for granting the use of an MSN website, which facilitated exchanges of documents and comments among editors and contributors during the review process. The use of the World Wide Web was very helpful during the editing of the book, often at the expense of our private lives. Indeed, with authors located around the planet, from eastern Australia, Japan and China to the western USA and southern Brazil, from Denmark to South Africa, there is no doubt that Internet communication improved and accelerated book editing. Because we often worked around the clock, thanks to the 19-h time interval between East and West, we thank our children and partners, Raphaël, Cristina Poletto-Forget, Jerry Jacka and Kathie Vander Wall for patience while our familial time was monopolized during the final step of book editing.

1   Seed Fate Pathways: Filling the Gap between Parent and Offspring

Stephen B. Vander Wall,[1] Pierre-Michel Forget,[2] Joanna E. Lambert[3] and Philip E. Hulme[4]

[1]Department of Biology and the Program in Ecology, Evolution and Conservation Biology, University of Nevada, Reno, NV 89557, USA; [2]Muséum National d'Histoire Naturelle, Département Ecologie et Gestion de la Biodiversité, UMR 5176 CNRS-MNHN, 4 avenue du Petit Château, F-91800 Brunoy, France; [3]Department of Anthropology,University of Wisconsin, Madison, WI 53706, USA; [4]NERC Centre for Ecology and Hydrology, Hill of Brathens, Banchory AB31 4BW, UK

Introduction

Seeds are mature ovules containing an embryo and stored nutrients inside a protective seed coat. Seeds are the products of sexual reproduction in most vascular plants (i.e. gymnosperms and angiosperms) and are the means by which plants produce genetically diverse offspring capable of surviving in variable and changing environments. Mature seeds remain dormant (*sensu lato*) for days to many years, and in this state they are able to tolerate adverse conditions and stressful environments (e.g. intense cold, heat, drought, darkness) that could not be endured by most plants (Baskin and Baskin, 1998). The particulate nature and small size of seeds relative to mature, seed-bearing plants lend mobility and also make it possible for a plant to produce large numbers of seeds during its life. Seeds are the principal means by which plants move across landscapes. The biotic and abiotic factors that affect seeds have unparalleled importance in the demography and evolution of plants.

The problem of seedling recruitment from the plant's perspective is that the environment is hostile and most points on the landscape are unsuitable for seedling establishment. Many organisms attempt to obtain the nutrients within seeds for their own purposes and kill the embryo in the process (Janzen, 1971; Hulme and Benkman, 2002). These risks are often intensified near the parent (Janzen, 1970; Connell, 1971), so effective dispersal usually must move seeds well beyond the influence of the parent plant (Howe and Smallwood, 1982; Herrera, 2002). Common modes of dispersal include abiotic forces (e.g. wind or rain splash), ballistic actions (e.g. as projectiles from exploding capsules), inadvertently by animals (e.g. within vertebrate guts, inside the cheek pouches of primates, or attached to the outer surface of animals), or purposely by animals (i.e. by seed-storing corvids, rodents or ants). These varied means of dispersal are associated with adaptations of fruit (the mature ovary wall and associated structures of angiosperms) that facilitate dispersal. Despite the diversity of dispersal mechanisms, most seeds move at random with respect to suitable establishment sites, and few seeds arrive at microsites conducive to seedling establishment. Janzen (1986) likened the task of a plant having its seeds reach a safe site for seedling establishment to a blind hunter trying to shoot a moving buffalo on an undulating plain. The hunter's success is likely to be low but depends on the amount of ammunition, size of the ammunition and the number of buffalo. Like the hunter's bullets, most seeds miss their target and fall on inhospitable ground (Schupp *et al.*, 1989; Schupp, 1993). What has not been appreciated until recently is that some seeds get to take two or more shots at the target (i.e. if they miss a suitable habitat or microsite after primary dispersal, secondary dispersal may eventually move them to a favourable site).

Interest in the ecology and function of seeds has a long history (Ridley, 1930). However, it is only since the mid-1970s that scientists have begun to tease apart the dynamic role seeds play in plant demography. Historically, interest focused primarily on seeds on or beneath plants. Most attention was directed towards pre-dispersal seed mortality, how seeds are morphologically adapted for dispersal, and the behaviour of animals that removed seeds and fruits from plants. When dispersal was studied, the focus was usually on the initial stage of dispersal such as the consumption of fruits by frugivores (Sutton, 1951; Rick and Bowman, 1961) or the frequency of occurrence of wind-dispersed seeds in seed traps located at varying distance from a seed source (Cremer, 1965; Greene and Johnson, 1989). Moreover, many important seed dispersing taxa (e.g. primates and caviomorph rodents; Forget, 1990; Lambert, 1999) received comparatively little attention relative to other, more conspicuous frugivores such as birds (Snow, 1971). Although seed dispersal has always been viewed as an important process (e.g. Ridley, 1930; van der Pijl, 1969; Thoreau, 1993), it was seldom studied from start to finish (e.g. Wang and Smith, 2002). Instead, the general patterns of dispersal were surmised from traits of propagules and, where relevant, the behaviour of animals that handled seeds. The latter stages of dispersal

were inferred from the patterns of seedling establishment. The occurrence of plants on oceanic islands, the establishment of seedlings outside the local distribution of conspecifics, and evidence for plant migration from fossils and pollen records all bore testament to the efficacy of seed dispersal mechanisms. But nearly all of the early studies on seed dispersal were incomplete in the sense that they failed to provide a full accounting of the diverse fates and various pathways that a population of seeds might follow between production and germination.

The lack of empirical studies of the intimate details regarding seed dispersal processes is a consequence of at least four factors. First, most seeds are small, and once they leave the parent plant they can be exceedingly difficult to find, let alone follow. Second, the relatively long dormancy of most seeds (ranging from several weeks to many years), especially in the temperate zone, greatly complicates the problem of following seeds until they germinate. Third, many scientists believed that it was not necessary to understand seed fate dynamics. Plants produced so many seeds over their lifespan and so few seeds survived to produce seedlings that it was generally believed that survival of seeds and the arrival of seeds at safe sites was essentially a stochastic process. The realization in the 1960s (Harper, 1967; Janzen, 1969) that natural selection is a factor during all stages of a plant's life cycle, including post-dispersal seed predation and microsite conditions at germination sites, increased our need to understand better the fate of seeds during the interval between seed production and germination. In addition, conceptual advances (e.g. Price and Jenkins, 1986) helped to point out the need for more detailed studies of seed fates. Fourth, and perhaps most importantly, there were few techniques for following seeds after they left the parent plant. The study of seed movement was limited to direct observation (e.g. Cahalane, 1942; Howe, 1977; Darley-Hill and Johnson, 1981; Forget and Wenny, Chapter 23, this volume). The technical difficulty of following small seeds over long periods of time contributed to an impression that seeds became nearly invisible between the time they left the parent and when the seeds germinated.

Things began to change with the emergence of methods that permitted researchers to follow seeds. These methods include the use of gamma-emitting radionuclides to label seeds and detect them with Geiger counters (Lawrence and Rediske, 1959; Radvanyi, 1966; Abbott and Quink, 1970), the use of metal objects in or on large seeds allowing them to be located using a metal detector (Stapanian and Smith, 1978; Sork, 1984), the use of string with an attached label or spool-and-line methods (Hallwachs, 1986; Forget, 1990; Yasuda et al., 1991), fluorescent dyes (Longland and Clements, 1995), and, for some very large seeds, radiotransmitters implanted inside seeds (Tamura, 1994; Soné and Kohno, 1996). Following these pioneering studies (discussed in more detail in Forget and Wenny, Chapter 23, this volume), detailed investigations of seed fates have proliferated. Coupled with these new methods for following seeds, it was realized that more rapid progress can be made in understanding seed demography if seeds are individually marked (e.g. numbered) for repeated observation. It is apparent from these studies that the pathways that seeds follow are often longer and more complex than previously assumed (Chambers and MacMahon, 1994; Vander Wall and Longland, Chapter 18, this volume). Prior to the availability of these methods, it was generally assumed that the dispersal of most seeds was a rather simple, direct process. Seeds move away from the parent plant, land or get deposited at a point on the landscape, and then either die or germinate at some later time. The view that has emerged since the early 1990s is that seed dispersal is a far more dynamic process, often involving multiple steps by two or more distinctly different mechanisms (e.g. Forget and Milleron, 1991; Kaufmann et al., 1991; Bohning-Gaese et al., 1999; Vander Wall, 2003). In earlier studies, when dispersed seeds disappeared it was often assumed that the seeds had been found by a predator and eaten. We have learned that this is often not the case. Instead, disappearance of a seed may simply indicate that a new stage in the dispersal process has begun.

With a growing realization that the pathways that seeds follow to favourable germination sites are diverse and often complicated, it is becoming increasingly important to have a detailed understanding of seed fates. A central element of any study that attempts to account for seed fates is a means of tracking seeds as they move from site to site. Additionally, a number of related studies need to be conducted in conjunction with studies of seed movement. Such studies might include, for example, measures of seed production, observational studies of the behaviour and effects of seed predators and dispersers, measurements of seed removal rates, the dynamics of soil seed banks, germination experiments under a range of environmental conditions that simulate possible deposition sites, and studies of post-germination seedling survivorship. Studies that have done this effectively include Holthuijzen et al. (1987), Kjellsson (1991), Hughes and Westoby (1992), Vander Wall (1994), Feer et al. (2001) and Lambert (2002). In addition, one might incorporate modelling studies and simulations into analyses of seed fates (e.g. Alvarez-Buylla and Martinez-Ramos, 1990; Alvarez-Buylla and Garcia-Barrios, 1991; Kalisz and McPeek, 1992; Charles-Dominique et al., 2003). By understanding questions such as when and where seed mortality occurs, how animals assist and/or hinder seed survival, and where germination is occurring but seedling establishment is failing, we will gain a fuller understanding of the selective forces that shape fruiting phenology, seed and fruit morphology, seed chemistry, dormancy schedules and other characteristics of seeds. The goal of this series of essays is to take stock of what we have learned about seed fates from studies throughout the world and to point out directions for future research.

Potential Seed Fates

Figure 1.1 is a generalized seed fate diagram, showing the most likely alternative

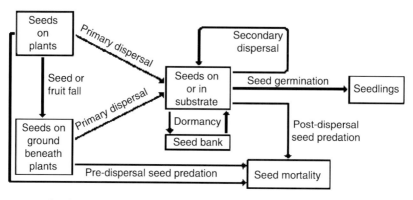

Fig. 1.1. A generalized seed fate diagram. Rectangles represent seed states and arrows between rectangles indicate movement or transition between states. See text for an explanation.

pathways that a seed might follow. Plants produce seeds within fruits, but here we focus on the seeds. Seeds on plants can be dispersed via a variety of primary dispersal mechanisms, including abiotic factors, like wind or ballistic projection, or biotic factors, like fruit-eating birds. Because of the diversity of fruit and seed types and the variety of dispersal processes, defining what constitutes primary dispersal can be complicated. Some seeds fall to the ground surface beneath the parent plant. Whether this should be viewed as primary seed dispersal depends on the structure of the fruit and how the seeds are typically dispersed. Consider two situations: first, some seeds are adapted for dispersal by ground-foraging animals, and falling to the ground is part of the adaptive syndrome of plants to get seeds to dispersers. The clearest example of this is nut fall by trees (e.g. walnuts, hickories, chestnuts, oaks, Brazil nuts, Carapa) that are dispersed by nut-caching animals (Boucher and Sork, 1979; Lewis, 1982; Forget, 1996; Peres and Baider, 1997). In these cases, nut fall beneath the tree is best viewed as nut presentation, not dispersal. Similarly, overripe fruits adapted for dispersal by ground-foraging vertebrates can fall beneath the parent plant. As long as the fruits remain below the parent, we view fruit fall as non-dispersal. However, if topography (e.g. steep slopes) assists fruits or nuts to move beyond the canopy of the parent

(e.g. Grinnell, 1936; Forget, 1992), then these longer movements should be viewed as primary seed dispersal. For animal-dispersed diaspores, whether the seed is taken from a plant canopy or the ground beneath a plant, we consider the animal-mediated movement of the seed to be primary dispersal. Second, if the fruit or seed is adapted for some means of dispersal (e.g. ballistic projection, wind, or fruit-eating animal) that typically carries seeds beyond the canopy of the parent, but the seed falls below the parent because of a failure of the dispersal mechanism, then we view this as primary dispersal. For example, a ballistically dispersed seed may hit an obstruction and fall below the parent (Hanzawa *et al.*, 1988; Forget, 1989; Espadaler and Gomez, 1996), a wind-dispersed maple (*Acer*) samara or pine (*Pinus*) seed could fall below the parent on a windless day (Guries and Nordheim, 1984; Nathan *et al.*, 2000), or animal-dispersed tropical seed may be spat out underneath a parent tree by cercopithecine primates (Lambert, 1999, 2001). In these cases, primary seed dispersal has occurred despite the poor quality of that dispersal.

Whether on a plant or on the ground beneath a plant, many seeds die before they can be dispersed. Animals that attack seeds before dispersal can occur are known as pre-dispersal seed predators (Janzen, 1971; Hulme and Benkman, 2002). These animals are often specialized seed predators that are attracted to chemical cues emanating from the parent plant and/or fruit. The most prominent of these organisms include bruchid beetles, weevils, moth larvae, seed bugs and wasps (Janzen, 1971; Hulme and Benkman, 2002). Birds and mammals, including some potential dispersers like seed-caching rodents, pitheciine primates and corvids, also kill numerous seeds before they are dispersed. Yet other seeds are killed or made unacceptable to dispersers by microbes that attack fruit.

Those seeds that do get dispersed arrive on (or sometimes in) a substrate. This substrate is usually the ground (e.g. soil, rock, plant litter), but it could also be a plant surface (e.g. faeces deposited on leaves or branches for epiphytes, plant parasites or other plants). When the dispersal mechanism moves the seed at random relative to favourable establishment sites, the initial point of deposition may or may not be advantageous to seedling establishment. At this point, the seed can experience one of several fates. First, the seed could be discovered by an animal and eaten. This post-dispersal seed predation is treated separately from pre-dispersal seed predation because both the phase and timing at which the seed is consumed and the organisms that consume it are different (Hulme and Benkman, 2002). These animals, including ants, rodents and birds, often specialize on seeds but forage opportunistically for a wide variety of seed species. Second, the seed could be moved to another site. Any significant movement of seeds after initial dispersal from plants is known as secondary dispersal (e.g. Greene and Johnson, 1997; Bohning-Gaese et al., 1999). In recent years, the importance of secondary dispersal has become increasingly evident (this volume). Secondary seed dispersal falls into two distinct categories: (i) the seed could be moved either by the same general mechanism as that involved in primary dispersal (e.g. dispersal of seeds over snow by wind; Matlack, 1989), or (ii) by a completely different mechanism (e.g. burial of seeds in faeces by dung beetles; Andresen and Feer, Chapter 20, this volume). We refer to the latter form of secondary dispersal as diplochory. We

feel that it is important to distinguish between these two possibilities because they have different implications for seeds. Seeds that are moved by a sequence of two or more distinctly different means of dispersal have a much greater chance of arriving at favourable germination sites, a benefit that is not shared with many forms of secondary dispersal involving only one means of dispersal (Vander Wall and Longland, 2004, and Chapter 18, this volume).

If a seed is not consumed or attacked by microbes, and if it lands in a suitable microsite, either as the result of primary or secondary dispersal, it could germinate. The delay from the time that a seed arrives at a substrate until it germinates can vary from a few days to many years (Garwood, 1983). Seeds that do not germinate when conditions appear to be favourable are said to be dormant, and dormant seeds make up the seed bank (Harper, 1977; Bulow-Olsen, 1984; Leck et al., 1989). Most large seeds and nuts germinate within 1 year triggered by seasonal cues (e.g. warm soil in spring, soil moisture in the rainy season). Other seeds can reside in the seed bank until the occurrence of some more unpredictable event (e.g. fire or a change in light levels because of a tree fall). During dormancy, seeds remain susceptible to predators and microbes (see Dalling, Chapter 3, this volume). Most seeds that remain in the seed bank for more than a year or two are small and generally overlooked by potential seed predators.

Our interest in seeds does not end when they germinate. This is because the seed often continues to exert an effect on the seedling for some time. For example, when seeds in a cache germinate, rodents will often excavate the soil around the seedlings to find ungerminated seeds at their base, killing or damaging seedlings in the process. Seedlings often serve as conspicuous 'flags' for buried seeds (Vander Wall, 1994; Pyare and Longland, 2000). In species with large nuts, the seed often continues to supply nutrients to the seedling for weeks to months. For woody plants (shrubs and trees), it is usually desirable to follow the fates of seedlings for several years in order to gauge the suitability of establishment sites (e.g. whether shrubs can serve as nurse plants for seedling trees). More generally, it is necessary to evaluate early seedling survival to judge the efficacy of seed dispersal because safe sites for seeds are not always suitable establishment sites for seedlings (Schupp, 1995).

The remainder of this book is divided into three sections (Fig. 1.1). The first section focuses on the sources and consequences of seed predation, including both pre-dispersal and post-dispersal seed losses. The second section covers primary dispersal from the plant and, where relevant, the ground beneath plants to substrates and the subsequent fate of those seeds, including seed germination and seedling establishment. The final section of the book examines the mechanisms and consequences of secondary seed dispersal and how it can influence seed fate at later stages of seed fate pathways. This organization is admittedly artificial, and all chapters contain information that pertains to two or more of these sections.

References

Abbott, H.G. and Quink, T.F. (1970) Ecology of eastern white pine seed caches made by small forest mammals. *Ecology* 51, 271–278.

Alvarez-Buylla, E.R. and Garcia-Barrios, R. (1991) Seed and forest dynamics: a theoretical framework and an example from the neotropics. *American Naturalist* 137, 133–154.

Alvarez-Buylla, E.R. and Martinez-Ramos, M. (1990) Seed bank versus seed rain in the regeneration of a tropical pioneer tree. *Oecologia* 84, 314–325.

Baskin, C.C. and Baskin, J.M. (1998) *Seeds: Ecology, Biogeography and Evolution of Dormancy and Germination.* Academic Press, San Diego, California, 666 pp.

Bohning-Gaese, K., Gaese, B.H. and Rabemanantsoa, S.B. (1999) Importance of primary and secondary seed dispersal in the Malagasy tree *Commiphora guillaumini. Ecology* 80, 821–832.

Boucher, D.H. and Sork, V.L. (1979) Early drop of nuts in response to insect infestation. *Oikos* 33, 440–443.

Bulow-Olsen, A. (1984) Diplochory in *Viola*: a possible relation between seed dispersal and soil seed bank. *American Midland Naturalist* 112, 251–260.

Cahalane, V. (1942) Caching and recovery of food by the western fox squirrel. *Journal of Wildlife Management* 6, 338–352.

Chambers, J.C. and MacMahon, J.A. (1994) A day in the life of a seed: movements and fates of seeds and their implications for natural and managed systems. *Annual Review of Ecology and Systematics* 25, 263–292.

Charles-Dominique, P., Chave, J., Dubois, M.-A., de Granville, J.-J., Riéra, B. and Vezzoli, C. (2003) Evidence of a colonization front of the palm *Astrocaryum sciophilum*, a possible indicator of paleoenvironmental changes in French Guiana. *Global Ecology and Biogeography* 12, 237–248.

Connell, J.H. (1971) On the role of natural enemies in preventing competitive exclusion in some marine animals and rain forest trees. In: Den Boer, P.J. and Gradwell, G. (eds) *Dynamics of Populations.* Centre for Agricultural Publishing and Documentation, Wageningen, The Netherlands, pp. 298–312.

Cremer, K.W. (1965) Dissemination of seed from *Eucalyptus regnans. Australian Forestry* 30, 33–37.

Darley-Hill, S. and Johnson, W.C. (1981) Acorn dispersal by blue jays (*Cyanocitta cristata*). *Oecologia* 50, 231–232.

Espadaler, X. and Gomez, C. (1996) Seed production, predation and dispersal in the Mediterranean myrmecochore *Euphorbia characias* (Euphorbiaceae). *Ecography* 19, 7–15.

Feer, F., Julliot, C., Simmen, B., Forget, P.-M., Bayart, F. and Chauvet, S. (2001) Recruitment, a multi-stage process with unpredictable result: the case of a Sapotaceae in French Guiana forest. *Revue d'Ecologie (La Terre et La Vie)* 56, 119–145.

Forget, P.-M. (1989) La régénération naturelle d'une espèce autochore de la forêt guyanaise *Eperua falcata* Aublet (Caesalpiniaceae). *Biotropica* 21, 115–125.

Forget, P.-M. (1990) Seed-dispersal of *Vouacapoua americana* (Caesalpiniaceae) by caviomorph rodents in French Guiana. *Journal of Tropical Ecology* 6, 459–468.

Forget, P.-M. (1992) Regeneration ecology of *Eperua grandiflora* (Caesalpiniaceae), a large-seeded tree in French Guiana. *Biotropica* 24, 146–156.

Forget, P.-M. (1996) Removal of seeds of *Carapa procera* (Meliaceae) by rodents and their fate in rainforest in French Guiana. *Journal of Tropical Ecology* 12, 751–761.

Forget, P.-M. and Milleron, T. (1991) Evidence for secondary seed dispersal in Panama. *Oecologia* 87, 596–599.

Garwood, N.C. (1983) Seed germination in a seasonal tropical forest in Panama: a community study. *Ecology* 53, 159–181.

Greene, D.F. and Johnson, E.A. (1989) A model of wind dispersal of winged or plumed seeds. *Ecology* 70, 339–347.

Greene, D.F. and Johnson, E.A. (1997) Secondary dispersal of tree seeds on snow. *Journal of Ecology* 85, 329–340.

Grinnell, J. (1936) Up-hill planters. *Condor* 38, 80–82.

Guries, R.P. and Nordheim, E.V. (1984) Flight characteristics and dispersal potential of maple samaras. *Forest Science* 30, 434–440.

Hallwachs, W. (1986) Agoutis *Dasyprocta punctata*: the inheritors of guapinol *Hymenaea courbaril* (Leguminosae). In: Estrada, R. and Fleming, T.H. (eds) *Frugivores and Seed Dispersal*. Dr W. Junk Publishers, The Hague, pp. 119–135.

Hanzawa, F.M., Beattie, A.J. and Culver, D.C. (1988) Directed dispersal: demographic analysis of an ant-seed mutualism. *American Naturalist* 131, 1–13.

Harper, J.L. (1967) A Darwinian approach to plant ecology. *Journal of Ecology* 55, 247–270.

Harper, J.L. (1977) *Population Biology of Plants*. Academic Press, New York, 892 pp.

Herrera, C.M. (2002) Seed dispersal by vertebrates. In: Herrera, C.M. and Pellmyr, O. (eds) *Plant–Animal Interaction: an Evolutionary Approach*. Blackwell Science, Padstow, UK, pp. 185–208.

Holthuijzen, A.M.A., Sharik, T.L. and Fraser, J.D. (1987) Dispersal of eastern red cedar (*Juniperus virginiana*) into pastures: an overview. *Canadian Journal of Botany* 65, 1092–1095.

Howe, H.F. (1977) Bird activity and seed dispersal of a tropical wet forest tree. *Ecology* 58, 539–550.

Howe, H.F. and Smallwood, P.D. (1982) Ecology of seed dispersal. *Annual Review of Ecology and Systematics* 13, 201–228.

Hughes, L. and Westoby, M. (1992) Fate of seeds adapted for dispersal by ants in Australian sclerophyll vegetation. *Ecology* 73, 1285–1299.

Hulme, P.E. and Benkman, C.W. (2002) Granivory. In: Herrera, C.M. and Pellmyr, O. (eds) *Plant–Animal Interaction: an Evolutionary*

Approach. Blackwell Science, Padstow, UK, pp. 132–154.

Janzen, D.H. (1969) Seed-eaters versus seed size, number, toxicity and dispersal. *Evolution* 23, 1–27.

Janzen, D.H. (1970) Herbivores and the number of tree species in tropical forests. *American Naturalist* 104, 501–528.

Janzen, D.H. (1971) Seed predation by animals. *Annual Review of Ecology and Systematics* 2, 465–492.

Janzen, D. (1986) Seeds as products. *Oikos* 46, 1–2.

Kalisz, S. and McPeek, M.A. (1992) Demography of an age-structured annual: resampled projection matrices, elasticity analyses and seed bank effects. *Ecology* 73, 1082–1093.

Kaufmann, S., McKey, D.B., Hossaert-McKey, M. and Horvitz, C.C. (1991) Adaptations for a two-phase seed dispersal system involving vertebrates and ants in a hemiepiphytic fig (*Ficus microcarpa*: Moraceae). *American Journal of Botany* 78, 971–977.

Kjellsson, G. (1991) Seed fate in an ant-dispersed sedge, *Carex pilulifera* L.: recruitment and seedling survival in tests of models for spatial dispersion. *Oecologia* 88, 435–443.

Lambert, J.E. (1999) Seed handling in chimpanzees (*Pan troglodytes*) and redtail monkeys (*Cercopithecus ascanius*): implications for understanding hominoid and cercopithecine fruit processing strategies and seed dispersal. *American Journal of Physical Anthropology* 109, 365–386.

Lambert, J.E. (2001) Red-tailed guenons (*Cercopithecus ascanius*) and *Strychnos mitis*: evidence for plant benefits beyond seed dispersal. *International Journal of Primatology* 22, 189–201.

Lambert, J.E. (2002) Exploring the link between animal frugivory and plant strategies: the case of primate fruit-processing and post-dispersal seed fate. In: Levey, D.J., Silva, W.R. and Galetti, M. (eds) *Seed Dispersal and Frugivory: Ecology, Evolution and Conservation*. CAB International, Wallingford, UK, pp. 365–379.

Lawrence, W.H. and Rediske, J.H. (1959) Radio-tracer technique for determining the fate of broadcast Douglas-fir seed. *Proceedings of the Society of American Forestry* 1959, 99–101.

Leck, M.A., Parker, V.T. and Simpson, R.L. (1989) *Ecology of Soil Seed Banks*. Academic Press, New York, 462 pp.

Lewis, A.R. (1982) Selection of nuts by gray squirrels and optimal foraging theory. *American Midland Naturalist* 107, 250–257.

Longland, W.S. and Clements, C. (1995) Use of fluorescent pigments in studies of seed caching by rodents. *Journal of Mammalogy* 76, 1260–1266.

Matlack, G.R. (1989) Secondary dispersal of seed across snow in *Betula lenta*, a gap-colonizing tree species. *Journal of Ecology* 77, 853–869.

Nathan, R., Safriel, U.N., Noy-Meir, I. and Schiller, G. (2000) Spatio-temporal variation in seed dispersal and recruitment near and far from *Pinus halepensis* trees. *Ecology* 81, 2156–2169.

Peres, C.A. and Baider, C. (1997) Seed dispersal, spatial distribution and population structure of Brazilnut trees (*Bertholletia excelsa*) in southeastern Amazonia. *Journal of Tropical Ecology* 13, 595–616.

Price, M.V. and Jenkins, S.H. (1986) Rodents as seed consumers and dispersers. In: Murray, D.R. (ed.) *Seed Dispersal.* Academic Press, Sydney, pp. 191–235.

Pyare, S. and Longland, W.S. (2000) Seedling-aided cache detection by heteromyid rodents. *Oecologia* 122, 66–71.

Radvanyi, A. (1966) Destruction of radio-tagged seeds of white spruce by small mammals during summer months. *Forest Science* 12, 307–315.

Rick, C.M. and Bowman, R.I. (1961) Galapagos tomatoes and tortoises. *Evolution* 15, 407–417.

Ridley, H.N. (1930) *The Dispersal of Plants Throughout the World.* Reeve, Ashford, UK, 745 pp.

Schupp, E.W. (1993) Quantity, quality and the effectiveness of seed dispersal by animals. *Vegetatio* 107/108, 15–29.

Schupp, E.W. (1995) Seed-seedling conflicts, habitat choice and patterns of plant recruitment. *American Journal of Botany* 82, 399–409.

Schupp, E.W., Howe, H.F., Augspurger, C.K. and Levey, D.J. (1989) Arrival and survival in tropical treefall gaps. *Ecology* 70, 562–564.

Snow, D.W. (1971) Evolutionary aspects of fruit-eating in birds. *Ibis* 113, 194–202.

Soné, K. and Kohno, A. (1996) Application of radiotelemetry to the survey of acorn dispersal by *Apodemus* mice. *Ecological Research* 11, 187–192.

Sork, V.L. (1984) Examination of seed dispersal and survival in red oak, *Quercus rubra* (Fagaceae), using metal-tagged acorns. *Ecology* 65, 1020–1022.

Stapanian, M.A. and Smith, C.C. (1978) A model for seed scatterhoarding: coevolution of fox squirrels and black walnuts. *Ecology* 59, 884–896.

Sutton, G.M. (1951) Dispersal of mistletoe by birds. *Wilson Bulletin* 63, 235–237.

Tamura, N. (1994) Application of radio-transmitter for studying seed dispersal by animals. *Journal of the Japanese Forestry Society* 76, 607–610.

Thoreau, H.D. (1993) *Faith in a Seed.* Island Press, Covelo, California, 283 pp.

van der Pijl, L. (1969) *Principles of Dispersal in Higher Plants.* Springer-Verlag, New York, 153 pp.

Vander Wall, S.B. (1994) Seed fate pathways of antelope bitterbrush: dispersal by seed-caching yellow pine chipmunks. *Ecology* 75, 1911–1926.

Vander Wall, S.B. (2003) Masting in pines alters the use of cached seeds by rodents and causes increased seed survival. *Ecology* 84, 3508–3516.

Vander Wall, S.B. and Longland, W.S. (2004) Diplochory: are two seed dispersers better than one? *Trends in Ecology and Evolution* 19, 155–161.

Wang, B.C. and Smith, T.B. (2002) Closing the seed dispersal loop. *Trends in Ecology and Evolution* 17, 379–385.

Yasuda, M., Nagagoshi, N. and Takahashi, F. (1991) Examination of the spool-and-line method as a quantitative technique to investigate seed dispersal by rodents. *Japanese Journal of Ecology* 41, 257–262.

2 Seed Predator Guilds, Spatial Variation in Post-dispersal Seed Predation and Potential Effects on Plant Demography: a Temperate Perspective

Philip E. Hulme[1] and Johannes Kollmann[2]

[1]NERC Centre for Ecology and Hydrology, Hill of Brathens, Banchory AB31 4BW, UK;
[2]Department of Ecology, Royal Veterinary and Agricultural University,
Rolighedsvej 21, 1958 Frederiksberg C, Denmark

Introduction

Although the importance of spatial scale in ecology has long been recognized (e.g. Greig-Smith, 1952), and in the last decade 'multi-scaled analyses' have been increasingly advocated (e.g. O'Neill, 1989; Wiens, 1989; Hoekstra et al., 1991; Levin, 1992; Schneider, 1994; Schupp and Fuentes, 1995), knowledge about the scaling of most ecological processes is still limited. This is particularly true for the interactions between plants and animals that often play a key role in ecosystem dynamics. Scaling matters because it affects the potential for co-evolutionary processes, for example during pollination and seed dispersal (Thompson, 1994). If plant–animal interactions are determined by specific habitat attributes at a local scale with little variation across different landscapes the potential for natural selection might be higher than in a case where the characteristics of the landscape have strong additional effects. The scaling of biotic interactions also has a bearing on species conservation, since interactions which are exclusively determined at the habitat or microhabitat level might be less vulnerable than those which also depend on the overall landscape structure.

Although post-dispersal seed predation has received considerable attention during the past 30 years, starting with the seminal reviews by Janzen (1969, 1971), it is still far from clear at which biotic scale, i.e. microhabitat, habitat or landscape units, levels of seed predation are most strongly differentiated (see for example Dennis et al., Chapter 7, this volume). Part of the attractiveness of the scaling of seed predation lies in its links to major theoretical concepts in ecology and evolution, among them density-dependent predation and species diversity as suggested by the Janzen–Connell hypothesis (Wright, 2002). Hundreds of studies on post-dispersal seed predation have been published from almost all biogeographical zones, and these results have been compiled in a number of reviews (Hulme, 1993, 1998; Crawley, 2000; Hulme and Benkman, 2002; Moles and Westoby, 2003). Variation in overall levels of predation seems to be extremely high as a result of not accounting for different spatial units (microhabitat, habitat), failure to distinguish between different seed predator guilds (vertebrates,

invertebrates) and differences in methodological approaches (seed density presented)
(e.g. Crawley, 2000). However, in some cases
the importance of spatial variation has been
reviewed but usually with specific reference
to single taxonomic groups (e.g. rodents;
Hulme, 1993), biomes (e.g. deserts; Hulme
and Benkman, 2002), or latitude (Moles and
Westoby, 2003). To date, no review has yet
attempted to integrate these approaches
across several spatial scales to assess
differences between biomes, continents or
habitats for different taxa.

The extensive literature on post-dispersal seed predation in the temperate zone
should allow us to answer the following
questions:

1. At which spatial scale(s) are patterns
of post-dispersal seed predation most
predictable?
2. Does the importance of spatial variation
differ across larger geographical units
(habitats, landscapes, biomes)?

Some Basic Definitions

Seed predator guilds

A 'guild' is a group of several species that
exploit the same class of environmental
resources in a similar way. Seed predator
guilds therefore group together species,
without regard to taxonomic position, that
overlap significantly in their niche requirements (Root, 1967; Simberloff and Dayan,
1991). A distinction is often made between
pre-dispersal guilds that feed on seeds on
the parent plant prior to seed dispersal (e.g.
crossbills, squirrels, weevils), and post-
dispersal guilds that scavenge for seeds
after they have been dispersed (e.g. pheasants, pigs, earthworms). However, many
granivores act as both pre- and post-
dispersal predators such that this guild
distinction becomes rather blurred. Due
to different methodological approaches,
the taxa involved and both evolutionary
and demographic consequences (Hulme
and Benkman, 2002), we have not

attempted to address pre-dispersal seed
predation.

A great variety of animals are post-
dispersal seed predators (Hulme, 1998;
Crawley, 2000): these include insects (especially ants and carabid beetles), some mol-
lusc, crab and fish species, granivorous birds
and small rodents, but also large mammals
such as deer or wild boar (see den Ouden
et al., Chapter 13, this volume). Post-
dispersal seed predators feed on diverse
and spatially heterogeneous resources and
thus most taxa share a requirement for generalist feeding habits. However, it is possible
to distinguish four major guilds of post-
dispersal seed predators: microtine and
murid rodents (in the following 'small
rodents'), passerine birds and two groups
of insects, i.e. seed harvesters (e.g. some
ant species) and seed parasites (e.g. weevils,
lygaeid bugs). These four guilds may all
share the same ecosystem and may even
overlap in the seed species predated but they
differ in the spatial grain of their foraging,
habitat requirements, degree of specialization and ability to cope with seed defences.
Distinguishing between these guilds is
essential to understand the ecosystem consequences of seed predation. Not only is the
assemblage of post-dispersal seed predators
diverse, but also its composition varies
considerably among different ecosystems. In
cool-temperate woodlands, the majority of
post-dispersal seed removal is attributable
to one or two species of small rodents (e.g.
Hulme, 1997; Kollmann et al., 1998). In temperate deserts harvester ants are significant
post-dispersal seed predators, whereas in
more humid ecosystems they act mainly
as seed dispersers (Hulme, 1998). However,
these generalizations regarding ecosystem
trends should be interpreted with caution.
Experimental studies in semi-arid ecosystems reveal marked intercontinental
differences in both the overall magnitude
of post-dispersal seed predation and the
relative importance of different guilds of
seed predator (Hulme and Benkman, 2002).
Rodents play a major role in northern
hemisphere deserts whereas ants appear
more important in the southern hemisphere,
where overall rates of post-dispersal seed

predation are considerably lower. Further studies are required to assess how granivory varies across a particular plant species' geographic range.

Spatial scales in seed predation

In the temperate zone, 'microhabitat', 'habitat', 'landscape', 'region' and 'zonobiome' are the most commonly recognized environmental spatial units (Table 2.1; see Kollmann, 2000). For a definition of these units it is convenient to delineate them on the basis of vegetation characteristics, since plants are structurally dominant in most terrestrial ecosystems and determine to a large extent the biotic and abiotic interactions within ecosystems. Additionally, one needs a rough estimate of the size of the five spatial categories, although the exact areas of each depend on the particular vegetation type. Several factors determine post-dispersal seed predation at each of these five spatial scales (Table 2.1) as described in more detail in the following section that highlights frequent overlap of the factors between scales.

The smallest scale used in our analysis is the 'microhabitat', which represents, for example, gaps in heathland, patches of lichens or dead wood on the forest floor. Microhabitats can range from a fraction of a square metre in grasslands to 100 m² in

forests. Studies that focus, for example, on a comparison of seed predation under different tree species, or examine differences between edge and interior forest, are choosing microhabitats as the unit of observation. Large forest gaps, on the other hand, usually develop a distinct vegetation which is different from the surrounding forest and can be characterized as a habitat and not as a microhabitat.

A 'habitat' can be defined as a spatial unit of a specific plant community, e.g. acid grassland, oakwood or foredune. A frequent comparison in seed predation studies is between woodland and neighbouring grassland habitats. Microhabitats are necessarily nested within habitats. 'Landscapes' are mosaics of plant communities and abiotic habitat features which are repeated in similar form over a kilometres-wide area (Forman, 1995). Within a landscape, several sets of attributes tend to be similar across the whole area, including the landforms, disturbance regimes, soils, vegetation types and local faunas, e.g. agricultural, montane or coastal landscapes. The next level in the scale hierarchy is the biogeographical 'region', e.g. the Danish islands in the Baltic Sea or the Scottish Highlands, which are determined by common topographic and climatic traits, and a more or less distinctive set of species and ecosystems. The same is true on an even larger scale for the 'zonobiome', e.g. the cold-temperate ('boreal') zone. In this review we follow the definition

Table 2.1. Characterization of the spatial scales on which the review on post-disperal seed predation in the temperate zone is based, and examples of key factors influencing post-dispersal seed predation (modified after Kollmann, 2000).

Scale	Area	Vegetation traits	Factors influencing seed predation
Microhabitat	0.01–10 (100) m²	Individual plants	Microtopography, soil characteristics, rocks, dead wood, ground cover, fruiting plants
Habitat	10–10,000 m²	Plant community	Vegetation height and cover, seed availability, predator avoidance, microclimate
Landscape	0.01–1,000 km²	Community mosaic	Habitat mosaic, fragmentation and connectivity, granivore abundance and movements
Region	> 10^3 km²	Regional vegetation	Plant and granivore species pool, phenology of seed production, dispersal and germination
Biome	> 10^4 km²	Zonobiome	Plant and granivore species pool, phenology of seed dynamics, history of human management

of zonobiomes as introduced by Heinrich Walter (Walter, 1962/1968); for a recent edition see Breckle (2002).

Ecological setting of the temperate zone

Since this chapter is focused on post-dispersal seed predation in the temperate zone *sensu lato*, some essential background information on the ecological setting seems to be necessary to understand differences in seed predator guilds and potential effects on plant community dynamics compared with tropical and subtropical habitats. The description of the temperate zone (about 23–66° latitude) follows Breckle (2002) and Körner (2002), and it distinguishes five temperate units: the Mediterranean zonobiome IV, the warm-temperate zonobiome V, the cool-temperate zonobiome VI, temperate steppe and desert vegetation (zonobiome VII) and the cold-temperate zonobiome VIII. A brief overview about the biogeographical characteristics of these zonobiomes is given in Table 2.2. For further information we refer to Breckle (2002).

Post-dispersal seed predation has been studied in most spatial units used in this review, but the boreal zonobiome, the alpine belt and coastal habitats seem to be under-investigated although seed predation can be significant in these areas (e.g. Maron and Simms, 1997; Nystrand and Granström, 2000; Nilson and Hjältén, 2003; Rudgers and Maron, 2003).

Spatial Heterogeneity and Scale-dependence in Post-dispersal Seed Predation

Differences between microhabitats, habitats and landscapes

Most studies from the temperate zone show that seed predators are patchily distributed and thus predation intensity varies spatially; similar results have been reported from tropical and subtropical ecosystems. Seed predation varies across a hierarchy of

spatial scales including: along topographic gradients and across a species' range, between habitats (e.g. woodland vs. grassland), among microhabitats within a single habitat (forest understorey vs. forest edge), and at an even finer scale within a single microhabitat (Louda, 1989; Hulme, 1998; Hulme and Benkman, 2002). As might be expected, spatial variation arises because some habitats, irrespective of seed availability, are more suitable for certain granivores than others (Janzen, 1971) or because seed defences might vary spatially (Louda *et al.*, 1987). The probability that a seed will be located and consumed by a granivore depends, among other factors, on food availability (Angelstam *et al.*, 1987; Schluter and Repasky, 1991; Fortier and Tamarin, 1998), predation risk (Brown, 1988), microclimate and substrate type (Andersen, 1978; Price and Heinz, 1984; Nystrand and Granström, 1997), and both inter- and intraspecific interactions.

Many studies in the temperate zone have detected variation in seed predation between microhabitats and some have found fine-scale differences in predation rates even between closely related plant species (e.g. Fig. 2.1). Considerable attention has focused on the role of vegetation cover in explaining the intensity of seed predation in different microhabitats. Frequently, fewer seeds are removed from open microhabitats (O'Dowd and Hay, 1980; Hay and Fuller, 1981; Mittelbach and Gross, 1984; Gill and Marks, 1991; Myster and Pickett, 1993; Reader, 1993; Wada, 1993; Hulme, 1994, 1996, 1997; Kollmann, 1995; Kollmann and Pirl, 1995; Alcantara *et al.*, 2000; Schreiner *et al.*, 2000; Garcia, 2001; Kollmann and Bassin, 2001; Smit *et al.*, 2001; see den Ouden *et al.*, Chapter 13, this volume; but see also Russell and Schupp, 1998; Castro *et al.*, 1999; Rey *et al.*, 2002; Obeso and Fernández-Calvo, 2003). This appears to occur when small rodents are the principal granivores since their abundance tends to be positively associated with vegetative cover that provides them with a screen from avian predators (Hulme, 1993). In contrast, harvester ants appear to forage preferentially in open areas and avoid dense vegetation

Table 2.2. Description of the eight 'zonobiomes' of the temperate zone *sensu lato* used in the present review on post-dispersal seed predation, based on Breckle (2002) and Körner (2002).

Zonobiomes	Climate	Soil types	Vegetation; typical plant families	Distribution
IV Arid-humid winter rain biome	Precipitation 400–1100 mm, winter rain, summer drought; summer < 35°C, winter > –6°C	Cambisol, luvisol, rendzina	Evergreen sclerophyllic forest and shrubland (degraded by grazing and fire); Ericaceae, Lamiaceae, Myrtaceae	California, Chile, Mediterranean region, S Africa, SW Australia
V Warm-temperate humid biome	Precipitation 1000–2000 (6000) mm; (almost) no frost	Humus-rich soils	Tall evergreen 'laurel' forest; Aquifoliaceae, Lauraceae, Magnoliaceae	Relictic and fragmented on all continents
VI.1 Cool-temperate biome (lowland)	Precipitation 500–1000 mm, max. in summer; annual temperature 5–15°C, frost to –25°C; growth period 5–8 months	Slightly acidic brown earth	Deciduous forest with arable field, grassland or heathland as secondary vegetation; Fagaceae, Juglandaceae, Hamamelidaceae	Central and eastern North America, Europe, East Asia
VI.2 Cool-temperate biome (montane)	Precipitation > VI.1, winter snow; growth period 3–6 months (7–12°C), 6–12 months with frost	Brown earth, podsol	Mixed deciduous–coniferous forest; Pinaceae	Northern hemisphere, Chile, S Africa, Australia, New Zealand
VI.3 Cool-temperate biome (alpine)	Precipitation and temperature < VI.2 but strongly dependent on micro-climate; growth period 6–16 weeks	Rendzina, ranker, podsol	Grassland, heathland or low scrub; Caryophyllaceae, Cyperaceae, Poaceae	Northern hemisphere, Chile, S Africa, Australia, New Zealand
VII.1 Arid-temperate biome (steppe)	Precipitation 250–500 mm; temperatures high in summer (< 40°C), low in winter (down to –50°C)	Chernozem	Shortgrass prairie, longgrass prairie, *Artemisia* steppe and vegetation dominated by thorny cushions, arable fields as secondary vegetation; Asteraceae, Poaceae	North America, Asia, Argentina
VII.2 Arid-temperate biome (desert)	Temperature regime similar to VII.1 but precipitation < 250 mm; summer drought, winter frost	(Often) saline soils	Halophytic vegetation; Chenopodiaceae, Zygophyllaceae, Polygonaceae	Northern and southern hemisphere
VIII Cold-temperate biome	Warm summer, cold winter (down to –70°C); growth period 3–5 months	Humus-rich brown earth, podsol	Coniferous forest, *Betula* forest; Ericaceae, Pinaceae, Salicaceae	Northern hemisphere

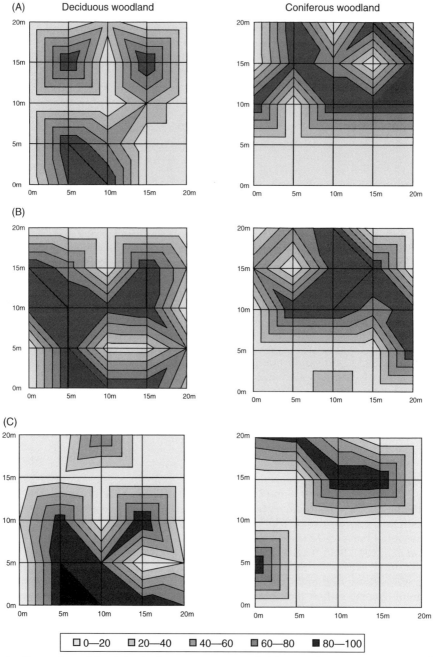

Fig. 2.1. Fine-scale spatial variation in the percentage removal of three seed taxa in two contrasting forest habitats in north-eastern England. Each plot depicts the intensity of seed removal interpolated across a 5 × 5 grid of seed depots spaced at 5 m intervals. The two sites represent a deciduous woodland with dense *Rubus fruticosus* cover and a neighbouring conifer plantation without any understorey vegetation. Three related legume taxa were studied in May 1996: (A) *Phaseolus aureus* Roxb, (B) *Phaseolus vulgaris* L. var. *haricot*, and (C) *Phaseolus vulgaris* L. var. *cannelloni*. Seed predation was due to small mammals (mostly *Apodemus sylvaticus*) and the patterns highlight that variation in seed predation between microhabitats can differ considerably among taxa even within the same habitat (P.E. Hulme, unpublished data).

(O'Dowd and Hay, 1980; Wilson, 1989; Crist and Wiens, 1994; Hulme, 1997). The few studies that have specifically examined post-dispersal seed removal by passerine birds also highlight a preference for open vegetation (Hulme, 1992; Nystrand and Granström, 1997; Robinson and Sutherland, 1999).

Similarly, variation in granivory between habitats has also been attributed to differences in vegetation structure, and particularly differences in plant cover (Watt, 1923; Harvey and Meredith, 1981; Mittelbach and Gross, 1984; Webb and Willson, 1985; Holmes, 1990). The principal determinant of rates of seed encounter within a habitat may actually be the spatial distribution of rodents and not their overall density (Hulme, 1994). Thus differences among habitats may reflect the abundance and spatial distribution of microhabitats preferred by rodents, rather than overall habitat attributes (Bowers and Dooley, 1993; Hulme, 1994; Hulme and Borelli, 1999). Seed predators may therefore significantly modify the seed shadows of plants, both within and between habitats.

The balance of evidence suggests that microhabitat effects for small rodents are a combined result of predator avoidance and resource availability (Lima and Dill, 1990; Hulme, 1993). Independently of the mode of seed dispersal, for many plant species most seeds fall close to the parent plant (Portnoy and Willson, 1993). Thus highly vegetated microhabitats (e.g. shrubs) could indeed reflect resource hotspots for seed predators (Garcia et al., 2001). Both Janzen (1970) and Connell (1971) suggested that granivores might preferentially feed on seeds beneath the parent plant either because they responded to the increased seed density (density-responsive granivores) or they were specialist granivores whose foraging was limited to within a certain distance of the parent (distance-responsive granivores). Numerous assessments of distance and density responsiveness of granivores suggest invertebrate granivores are more likely to feed in a distance- and/or density-responsive manner than vertebrates (Hammond and Brown, 1998). Although

small rodent species differ in their requirement for vegetation cover (Hulme, 1993), the weight of field evidence from the temperate zone suggests that removal of proportionally fewer seeds in open habitats may be a general characteristic of rodent seed predation. In addition, there is evidence that rodent granivores trade off food abundance and shelter, with the result that small rodents may feed where resources are relatively poor but where sites are safe from predators (Hansson, 1978; Lima and Dill, 1990; Hulme, 1996; Manson and Stiles, 1998). Small rodents may be more selective in terms of the species preyed upon when at risk from carnivorous predators (Hay and Fuller, 1981) and may only venture into open microhabitats when these contain highly preferred seed species (O'Dowd and Hay, 1980; Harris, 1984).

Carnivorous predators are not the only taxa that influence spatial variation in seed removal. For example, birds and beetles rarely feed on seeds close to anthills due to the antagonistic behaviour of ants (Niemelä et al., 1992; Nystrand, 1998). Alien invasive ants may reduce the foraging intensity and abundance of native granivorous ants (Carney et al., 2003). Interspecific competition among rodents in North American deserts has led some heteromyid rodents to feed primarily in open microhabitats (Price, 1978). Even within a single habitat, the distribution of a rodent species may change over time with unknown effects of microhabitat, resource or predation pressures (Fig. 2.2). Spacing behaviour as a result of demographic changes in population size or reproductive behaviour may dramatically alter any correlations between seed predation intensity and microhabitat (Hulme, 1994).

A further factor that will influence spatial patterns in rates of seed predation is the structure of the landscape (Angelstam et al., 1987; Verdú and García-Fayos, 1996). The ecotones between habitats may have very different rates of seed removal than either of the two neighbouring habitats (Fig. 2.3). In old fields, seed predation by small mammals appears to be higher close to forest edges for some tree species (Myster and Pickett, 1993; Ostfeld et al., 1997; Manson

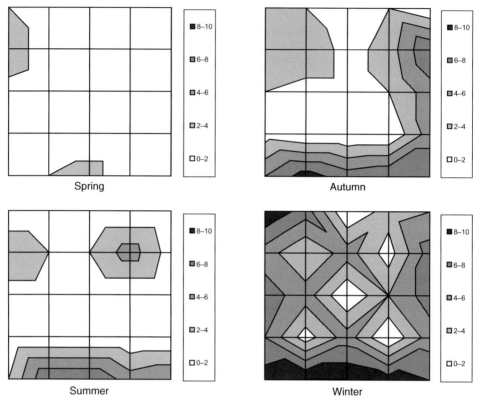

Fig. 2.2. Changes over 1 year in the number of captures of the rodent *Apodemus sylvaticus* at 25 trap points within grassland in southern England. Values represent the total number of captures summed across the three monthly trapping periods (4 trap nights each) covered within each season (e.g. maximum of 12 captures per point) (P.E. Hulme, unpublished data).

and Stiles, 1998; Jules and Rathcke, 1999; Kollmann and Buschor, 2002; Meiners and LoGiudice, 2003). However, other tree species show no edge responses in seed predation (Myster and Pickett, 1993). Similar trends have been found in agricultural habitats (Marino *et al.*, 1997) and roadsides (Manson *et al.*, 1999). However, trends at the ecotone not only reflect the vegetation composition of the neighbouring habitats but also landscape attributes such as the size of the habitats themselves. Habitat fragmentation may influence habitat choice of the seed predators (Andreassen *et al.*, 1998) and is known to increase rodent population densities (Collins and Barrett, 1997). Thus, small isolated woodlots may have higher seed predation rates than larger fragments (Telleria *et al.*, 1991) and this in turn could influence any edge effects. This may explain

why predation in forest fragments may be lower at the edge (Sork, 1983). Similarly, forest gaps may influence seed predation rates as a function of their size and vegetation structure. The fact that these two factors have rarely been controlled for in the field possibly explains the lack of generic insights from studies that have found higher rates of predation in gaps (Boman and Casper, 1995; Kollmann, 1997), while others have found the converse (Herrera *et al.*, 1994; Diaz *et al.*, 1999; Schreiner *et al.*, 2000) or no difference (Plucinski and Hunter, 2001).

The complexity of factors influencing spatial variation in seed predation may explain the lack of consistent trends across the numerous studies published to date. As one major source of variation in seed predation seems to be vegetation structure, structural differences within the same type

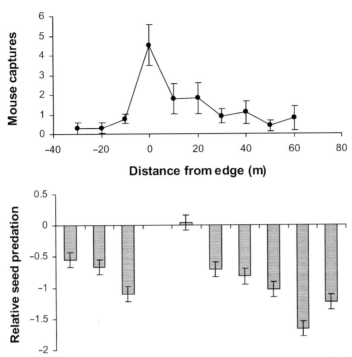

Fig. 2.3. Variation in rodent captures and seed predation with distance from a forest edge–oldfield ecotone. Values presented are estimates from Cox regressions compared to distance 0, i.e. the edge (means ± 1 SE; redrawn from Meiners and LoGiudice, 2003).

of habitat or microhabitat may explain at least some of the contradictions in the literature. A forest edge, for example, is not a homogeneous spatial unit: it might be densely vegetated or devoid of shrubs and herbs after recent management that strongly affects seed predation rates (Kollmann and Buschor, 2002). The magnitude of these sources of variation and the absence of consistent directional trends has made it difficult to construct accurate predictive models of post-dispersal seed predation and gauge its impact on plant species whose seeds are preyed upon (Brown and Venable, 1991). A new methodological synthesis is clearly required.

To date, the vast majority of post-dispersal seed predation studies have responded to natural environment gradients, e.g. natural gaps, fragments, etc. (see for example Theimer, Chapter 17, this volume), rather than attempting to manipulate the environment. Increasingly large-scale manipulations of ecosystems, such as deliberate

establishment of replicated forest (Laurance and Bierregaard, 1997) and grassland fragments (Holt et al., 1995) of different sizes and isolation, and the converse of creating artificial gaps within a continuous forest matrix (Tewksbury et al., 2002), are being used to answer fundamental ecological questions. These large-scale approaches need to be extended to a wider range of ecosystems and explicitly explore how such manipulations influence the dynamics of seed predation (Debinski and Holt, 2000). Within such experiments, assessments of seed predation should use hierarchical experimental designs nested within two or more spatial scales in order to explore issues such as the degree of spatial autocorrelation and the influence of local and regional processes on seed loss. Such a proposal may appear rather idealistic, yet without such large-scale collaborative ventures, progress in the understanding of the spatial ecology of seed predation will be slow if it continues to rely on the accumulation of numerous case studies reporting significant spatial variation in seed losses.

Differences between continents and zonobiomes

Differences in seed predation between zonobiomes are largely under-investigated, and there is no study that has specifically investigated predation rates in different biomes with identical methods. However, Hulme (1998) reviewed the literature on continental differences of seed predation in semi-arid habitats. He found that predation rates are lower in Australia, Argentina and Chile compared with North America and South Africa. Rodent seed predation seems to be particularly strong in warm and cold deserts of the northern hemisphere and in cool- and cold-temperate ecosystems, whereas ants act mostly as seed dispersers in the Mediterranean and in the warm-temperate zonobiome.

If one compares some case studies there seem to emerge some trends that might be used to generate future hypotheses. The more intense predation under dense vegetation in the cool-temperate zone is mostly caused by small rodents that get shelter and food under this vegetation type and forage mostly at night. This pattern can be changed in Mediterranean habitats where ants are the most important seed predators in open areas with highest activity in the daytime, whereas rodents are prominent predators under shrub vegetation at night (Hulme, 1997).

A more systematic review on large-scale variation in seed predation has been published by Moles and Westoby (2003). These authors standardized predation rates at 24-h periods after the start of the observation period, and used a species-centred approach by excluding all studies where species had been pooled. In addition, phylogenetic corrections were applied. Moles and Westoby (2003) found no significant trend in seed predation with latitude, although there was a negative correlation between latitude and seed mass. However, a different result emerges from a discontinuous approach

where the latitudinal gradient is broken up in major biomes and habitat type. A re-analysis of this data set revealed the following results; a non-parametric test was used, because the data were not normally distributed after arcsine (square root) transformation. There were significant differences between zonobiomes (Fig. 2.4A) with lowest seed predation rates in the tropical zonobiome ($8.2 \pm 1.7\%$ seed losses after 24 h, mean $\pm$ SE; $n = 95$ plant species), and higher rates in the warm-temperate zonobiome ($32.4 \pm 8.0\%$; $n = 21$) and in the cool-temperate zonobiome ($20.3 \pm 2.5\%$; $n = 59$; Kruskal–Wallis test, $\chi^2 = 35.4$, $P < 0.0001$). Further, the number of studies was sufficiently high for a comparison between cool-temperate forests ($28.8 \pm 5.0\%$ seed predation) vs. grasslands/oldfields ($13.0 \pm 1.3\%$) where significant differences also emerged (Fig. 2.4B, $\chi^2 = 4.8$, $P = 0.029$). However, sample size was too small for a similar analysis of data from the Mediterranean and cold-temperate zone.

The higher rates of granivory in cool-temperate forests compared with grasslands/oldfields in the same biome might be related to the habitat preferences of small rodents, which are the most important seed predators in the cool-temperate zonobiome. However, the higher rates of temperate granivory compared with the tropical zonobiome are difficult to explain as they contradict the global trend of more intense herbivory in the tropics (Coley and Barone, 1996), albeit seeds and leaves are somehow different resources. Possible explanations for this result include the following scenarios: (i) temperate ecosystems are more fragmented than higher or lower latitude ecosystems and hence predation rates are higher; (ii) losses of native carnivorous predators in temperate ecosystems are proportionally greater and thus herbivores tend to be food- rather than predator-limited; (iii) (a methodological point) 24-h removal rates, being higher in the temperate than in the tropical zone, may be similar after about a month depending on season. There is indeed a strong seasonal effect in the tropics as well as in the temperate zone (see Stapanian, 1986), with lower instantaneous

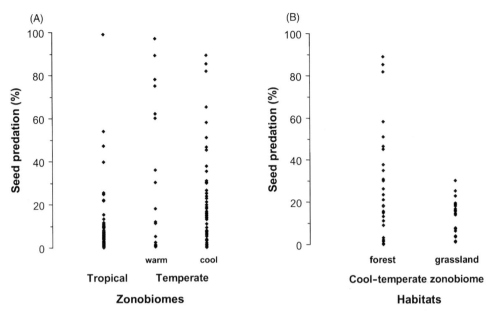

Fig. 2.4. Large-scale patterns in post-dispersal seed predation. (A) Differences between zonobiomes, i.e. tropical (I), warm-temperate (V) and cool-temperate (VI) *sensu* Breckle (2002), and (B) between two major habitat types in the cool-temperate zonobiome ('forest' vs. 'grassland' included grassland and oldfields). Re-analysis of the data set by Moles and Westoby (2003), i.e. seed predation rates per individual species after 24 h; for statistical results see the main text.

(24-h) removal during the fruit peak than during the lean season (see also Forget *et al.*, 2002). And (iv) (another methodological problem) seed predation studies in temperate systems are somewhat biased towards bird-dispersed plant species which tend to be large-seeded. If these species have a higher than average predation rate, then temperate studies may overestimate the mean levels of seed predation. Seed predation in temperate systems may strongly reflect characteristics of a few plant families, especially Rosaceae, while the Tropics reflect a wider taxonomic diversity.

Demographic Consequences of Seed Predation: Death or Dispersal?

There is still little information on the demographic consequences of seed predation. Most difficult for methodological reasons is actually the balance between predation and secondary dispersal. More frequent are experimental studies which compare plant

recruitment rates with and without seed predation.

When does seed predation matter?

High rates of seed predation do not always result in dramatic reductions in plant density. For example, for the desert annual *Erodium circutarium*, consumption of over 95% of seeds by heteromyid rodents was predicted to reduce plant density by only 30% (Soholt, 1973). Average rates of pre- and post-dispersal seed predation are often much lower than 95% (Hulme, 1998, 2002; Hulme and Benkman, 2002) and it is possible that even moderately high seed losses may have little impact on plant demography. The circumstances under which seed predators may influence the dynamics of plant populations have recently received detailed attention (Hulme, 1998, 2002; Crawley, 2000; Hulme and Benkman, 2002). In general, seed predation may play only a minor role in the demography of plants where: (i) plants regenerate primarily by vegetative means; (ii) seed losses to predators are buffered by the presence of a large persistent seed bank; (iii) regeneration is microsite- rather than seed-limited; and (iv) seed predators are satiated by large seed crops.

Can we use this information to assess in what habitats post-dispersal seed predators are likely to have significant impacts on plant demography in the temperate zone? The most straightforward approach is to identify the range of regenerative strategies exhibited by plant species in any one habitat and subsequently to assess their vulnerability to seed predation. Three major, though not mutually exclusive, regenerative strategies have been identified for flowering plants in temperate biomes (Hodgson et al., 1994): (i) vegetative expansion through the formation of persistent rhizomes, stolons or suckers; (ii) regeneration by seeds from a persistent seed bank; and (iii) regeneration by seeds that do not form a persistent seed bank. A working hypothesis is that habitats become ever more susceptible to the impacts of seed predators as the proportion of species reliant on regeneration strategy (iii) increases.

Using a compendium of species traits for the vascular flora of the British Isles, the likely importance of seed predation in plant demography can be assessed with reference to major north-western European habitats (Hodgson et al., 1994). In all but two of the dominant habitats of the British Isles (mires and arable), plant species that reproduce exclusively by short-lived seeds are better represented than species that rely exclusively on vegetative reproduction or regeneration from a persistent soil seed bank (Hulme and Benkman, 2002). In most habitats, and particularly those where disturbance is frequent (e.g. wastelands, quarries and riparian), species reproducing exclusively by short-lived seeds account for between 30 and 40% of the flora. Even among those plant species that adopt more than one regenerative strategy, species relying to some extent on non-persistent seeds are more common than those that do not. However, even in cases where seeds are the

only means for regeneration, the importance of seed predators will depend on the extent of microsite limitation.

Assessing whether plant recruitment is seed- or microsite-limited is fraught with difficulty because microsite availability will vary both spatially and temporally. Generalizations from experimental sowing of additional seeds into natural plant communities are constrained by the limited spatial and temporal scales they address. Nevertheless, it appears that seed-limited recruitment is common in many plant communities, particularly ruderal habitats, although within any community significant variation occurs among plant species (Turnbull et al., 2000). The few field studies that have simultaneously combined seed addition, disturbance and exclosure of seed predators in a factorial design indicate that the failure of many species to establish in dense vegetation may sometimes be due to higher rates of post-dispersal seed predation rather than to increased interference from established vegetation (Reader, 1993; Edwards and Crawley, 1999). These studies lend support to the view that the importance of seed limitation in communities of perennial plants may currently be underestimated (Turnbull et al., 2000).

Through the synchronous production of large seed crops at irregular time intervals (often described as masting) plants are thought to satiate seed predators and enhance their regenerative capacity (Silvertown, 1980; Kelly and Sork, 2002; see also Hoshizaki and Miguchi, Chapter 15, this volume). Large, irregular seed crops (masts) are probably more successful at limiting the impacts of specialist pre-dispersal seed predators than generalist post-dispersal seed predators (Gardner, 1977; Nilsson and Wästljung, 1987; Crawley and Long, 1995). This reflects the generalist feeding habit of small rodents, which enables them to persist in non-mast years by feeding on seeds of other species. In years of abundant food their numbers increase (Jensen, 1982) and they also have the ability to exploit superabundant food supplies through storing food in caches (Vander Wall, 1990). Densities of seedlings are often higher after mast than non-mast years (Gardner, 1977; Crawley and Long, 1995), since more seeds escape predation in mast years. Although the proportion of seeds destroyed by post-dispersal seed predators remains little changed, overall more seed escape predation in mast years. This is not the equivalent of stating that granivores have no effect during mast years, since the number of seedlings recruiting in the absence of predators is not known and could potentially be much greater. These findings for masting trees also suggest that recruitment is seed-limited for these species, since more seed production results in more seedlings.

The foregoing highlights the complexity of trying to state categorically that seed predators do or do not play an important role in plant communities. Simple comparisons of mean rates of seed predation across habitats (e.g. Hulme, 1993) are unlikely to provide an indication of the impact on plant population dynamics. What can be concluded is that, in the majority of habitats, there is a high probability that post-dispersal seed predators influence the dynamics of one or more plant species. Can we identify functional groups of plant species that might be most susceptible?

Are certain plant functional groups more susceptible to seed predation?

Once again using the compendium of species traits for the vascular flora of the British Isles (Hodgson et al., 1994), insights can be gained as to the relationship between seed dispersal syndrome and seed predation. Analysis suggests that seed predators may influence both the demography and evolution of animal-dispersed species more than they influence those processes in plant species with predominantly abiotic or unspecialized dispersal mechanisms (Hulme, 2002). In general, animal-dispersed species tend to rely proportionally more on regeneration by seeds, often lack a persistent seed bank and are more limited in recruitment by seed abundance than by microsite abundance. These trends appear strongest for

seeds actively dispersed by vertebrates. However, the relationships between seed persistence and plant demography (see above) and between seed dispersal syndrome, seed predation and regeneration strategy might have developed secondarily through other, unstudied variables, and could be explained through seed mass, i.e. vertebrate-dispersed seeds tend to be larger than abiotically dispersed seeds, and species with large seeds produce fewer seeds per adult plant per year and thus are more likely to be seed-limited than small-seeded species.

Seed production is usually less variable in plants dispersed by mutualistic frugivores than either among those that are synzoochorous or are dispersed by abiotic means (Herrera et al., 1998; Kelly and Sork, 2002). This pattern is consistent with the assumption that species dispersed by frugivores should rarely produce excessively large fruit crops since this might result in the frugivore population becoming satiated and less efficient in dispersal (Herrera et al., 1998). In contrast, synzoochorous species may irregularly produce large seed crops and potentially may satiate post-dispersal seed predators. Overall we may expect endozoochorous species to be most prone to the deleterious impacts of seed predators. A further implication of these findings is that current knowledge of the impacts of granivores on plant demography may be somewhat biased, with an over-representation of studies on animal-dispersed species (Hulme, 1993, 1998). Future research needs to address this imbalance and, more specifically, to focus on seed predation in plant species that differ in dispersal strategy.

Can we distinguish between dispersal and predation?

Many seed predation studies, especially those examining post-dispersal granivory, assume that removal of seeds by granivores ultimately results in the consumption of the seed. However, when seed predators cache seeds for later consumption and seed recovery is incomplete, the end result of 'granivory' may be seed dispersal. Because few studies have followed the fate of seeds removed by vertebrate granivores in the temperate zone (see other chapters in this volume), it is possible that the negative impact of seed predators on plant demography may be overestimated. However, it is equally dangerous to assume that all granivores act as mutualists (see Theimer, Chapter 17, this volume). If granivores act as mutualists, caching should facilitate dispersal away from the parent (enhancing both escape from natural enemies and opportunities for colonization) and/or dispersal into specific microsites.

The distance over which seeds might be successfully cached by granivores is often underestimated due to methodological problems of locating seeds transported more than several metres. Indeed, the studies that have attempted to estimate dispersal distances report relatively short mean dispersal distances: 4.1 m for *Fagus sylvatica* (Jensen, 1985), 5.4–19.9 m for *Pasania edulis* (Sone *et al.*, 2002), 7.8–9.5 m for *Purshia tridentata* (Vander Wall, 1994), 9.6 m for *Simplocarpus renifolius* (Wada and Uemura, 1994), 10.0 m for *Pinus strobus* (Abbott and Quink, 1970), 12.2–44.7 m for *Aesculus turbinata* (Hoshizaki and Hulme, 2002) and 15.3 m for *Quercus robur* (Jensen and Nielsen, 1986). While a small proportion of seeds may be dispersed over further distances, these mean dispersal distances are similar to those identified for bird-dispersed fleshy-fruited species, but significantly less than mean distances frequently achieved via synzoochorous birds (e.g. jays; Kollmann and Schill, 1996; see den Ouden *et al.*, Chapter 13 this volume) or wind dispersal (Willson, 1993). Thus, at least for microtine rodents, caching may not always enhance opportunities for colonization and may not move seeds far enough from parent trees to escape distance-responsive seed predators. Although small rodents in the temperate zone often bury seeds in separate caches, burial of several seeds within a cache will increase sibling competition among seedlings and may also attract density-dependent granivores. However, at least in some species and in

some habitats grouping of seedlings might actually be beneficial, as reported for *Pinus cembra*, a conifer dispersed by the European nutcracker at the alpine timberline (Kratochwil and Schwabe, 1994). Here, seedlings from the same cache in the soil can develop a common root system and later also fuse their aboveground stems. Thus, multi-stemmed trees may actually be genetic mosaics, which might be advantageous in a rather rough environment.

Seed-dispersing granivores often select particular cache sites, presumably to increase their chance of seed recovery (Bossema, 1979; Vander Wall, 1990). Rodents often use physical cues in the habitat to aid cache relocation and their caches are frequently made in the vicinity of fallen logs or at the base of trees (Vander Wall, 1990). While the selection of caches is certainly non-random, it is unclear whether these sites represent optimum microsites for plant establishment.

One of the key differences between caching and other forms of seed dispersal is the process of seed burial. For many large-seeded forest species, burial is a necessary requirement for germination, since otherwise seeds will often desiccate and die (e.g. Watt, 1919; Vander Wall, 1994; Kollmann and Schill, 1996). This benefit of burial declines with decreasing seed size and, for small seeds, especially herbaceous species, burial at the depths normally found for caches can significantly reduce germination rates (Froud-Williams *et al.*, 1984). Nevertheless, burial also protects seeds from post-dispersal seed predation (Kollmann and Schill, 1996; Hulme, 1998; Hulme and Borelli, 1999; Schreiner *et al.*, 2000). Thus for large seeds the benefit of caching may simply be burial, independent of the distance from the parent plant.

Successful seed dispersal is more likely through scatterhoarding than larderhoarding, because seeds are buried in many shallow caches and distributed among a variety of microhabitats. The large number of caches often results in incomplete seed recovery (Vander Wall, 1990). In contrast, larderhoards are often buried more deeply (frequently within animal burrows) and the single location makes recovery of seeds highly probable (Smith and Reichman, 1984). Even where recovery is incomplete, the depth of burial may prevent successful germination from larderhoards. Detailed studies have shown that in most cases survival and germination of seeds from scatterhoarded caches is low (Hulme and Benkman, 2002; Clark and Wilson, 2003). Germination rates reported for seeds cached by murid rodents in the temperate zone include: 0.02% for *Oryzopsis hymenoides* (McAdoo *et al.*, 1983), 0.8% for *Pinus strobus* (Abbott and Quink, 1970), 1.5–2.5% for *Quercus robur* (Jensen and Nielsen, 1986), 0–4% for *Fagus sylvatica* (Jensen, 1982) and 5–8.5% for *Purshia tridentata* (Vander Wall, 1994). However, germination rates can be underestimated when researchers lose track of seeds during repeated bouts of re-caching, and seedlings may emerge from secondary or even higher order caches (Vander Wall, 2002). The limited data on seedling emergence rates for seeds dispersed by other means show higher rates of seedling emergence. A comparable estimate of seedling emergence for *Phillyrea latifolia*, a bird-dispersed species, is 8% (Herrera *et al.*, 1994) and for two wind-dispersed *Acer* spp. 19–24% (Houle, 1995). It is clear that plants dispersed by granivores bear a high cost in terms of seeds consumed for a relatively low return in terms of seedling regeneration. However, even relatively low germination rates may translate into a substantial number of seedlings if seed abundance is high. Yet the few studies that have monitored the subsequent survival of seedlings, record high seedling mortality (Vander Wall, 1994), reflecting that cache locations may not necessarily be suitable for establishment.

Few studies in temperate ecosystems have shown unequivocally that seed dispersal by granivores is essential for plant regeneration (but see corresponding studies for tropical forests reviewed by Jansen and Forget, 2001). Nevertheless, caution must be applied when equating seed removal and seed predation; although most seeds removed are consumed, a small fraction may be dispersed to suitable microsites. Interpreting the consequences of seed removal will require knowledge of the granivores involved, their foraging behaviour and the reproductive traits of the plant. Thus, rodent scatterhoarding of large-seeded, mast-fruiting trees may be essential for regeneration of those trees. However, murid rodents that larderhoard the small seeds of a bird-dispersed shrub may significantly reduce plant recruitment. While it is impossible to know the fate of all seeds removed by granivores, application of basic ecological knowledge can shed light on the probabilities of seed dispersal versus predation.

Perspectives for Future Research

The majority of post-dispersal seed predation studies in the temperate zone have not quantified the relative contributions made by different granivore guilds. The separation of the impact of vertebrates and invertebrates on seed survival is important since these guilds of animals are likely to differ in the temporal and spatial scales of their effects, their functional responses, their preferences for plant species, and the consequences of their foraging ('predation' or 'secondary dispersal'). Studies that ignore these differences may misrepresent factors thought to be important in plant demography (Hulme, 1998). Thus, the absence of consistencies in habitat or microhabitat trends may reflect variation in the spatial distribution of different guilds of post-dispersal granivores. For example, in contrast to cool-temperate communities, microhabitat did not appear to determine the probability of predation in Mediterranean habitats (Verdú and García Fayos, 1996; Castro *et al.*, 1999; Rey *et al.*, 2002; Obeso and Fernández-Calvo, 2003). This reflects the increasing importance of ants in the early-successional stages of Mediterranean habitats (Diaz, 1992; Hulme, 1997). However, while seed removal may appear more homogeneous, seed predation may still be heterogeneous where losses to rodents in later seres result in seed predation while ants act primarily as secondary seed dispersal agents in open habitats (Hulme,

1997). We suggest that it is essential for future studies to distinguish the separate impacts of different seed predator guilds (e.g. Tamura *et al.*, Chapter 14, this volume). A diversity of methods exists to separate the impacts of rodents, invertebrates and birds (e.g. Kollmann and Bassin, 2001; Kollmann and Buschor, 2002) and these should be adopted as 'standard' when assessing the dynamics of seed predation. While this is increasingly the case, such studies still represent only a minority of cases. This is especially important where feeding by different guilds has different consequences (predation/dispersal) and/or seed preferences. Failure to address these differences may result in incorrect conclusions being drawn from the result of seed losses.

There are statistical shortcomings in many studies of post-dispersal seed predation, e.g. lack of independence of sample points when assessing edge effects along transects or pseudoreplication of microhabitats within a single habitat location. However, while these statistical issues are increasingly being addressed, attempts to compare results from many scattered studies (as done by Hulme, 1998; Hulme and Benkman, 2002; Moles and Westoby, 2003) continue to suffer from a bias in the experimental design of many seed predation studies. The standard approach when assessing microhabitat variation within a given habitat is often to select similar numbers of replicates for all microhabitats, thus ensuring a balanced experimental design. Summary statistics for the habitat, often used in comparative studies, will use mean levels of seed predation as representative of that habitat type. However, such summary statistics do not take into account the relative abundance of the microhabitats. The situation is illustrated by the following somewhat extreme example. Imagine comparing a shrubland and a grassland habitat, within each of which are found only two microhabitats, shrub and grass. The difference between the two habitats is the relative proportion of the two microhabitats, with the cover of the shrubland being composed of 70% shrubs while the grassland has only 20% shrub (the

remaining microhabitat is grass in both cases). Seed predation experiments sample each microhabitat with the same degree of replication. The results highlight for both habitats that an order of magnitude higher seed predation rates are found in shrub than in grass microhabitats. However, because the microhabitats are sampled equally, differences between grasslands and shrublands are non-significant. Contrast this with a study that assesses seed predation at random points in each habitat. These results are likely to show a more realistic value of the mean level of seed predation in each habitat (higher in shrublands than grasslands), the drawback being that the analysis of microhabitat effects may be limited by the unbalanced experimental design. There is therefore a danger that, unless habitats have been sampled at random, many means quoted for particular habitats are nothing of the sort and individual microhabitat values should be weighted by their relative abundance. This suggests, unfortunately, that many comparative studies published to date may not represent generic patterns. Thus, some of the points raised within this review should be treated with caution and emphasis placed on trends supported by several independent studies.

In summary, much progress has been made in furthering our understanding of seed predation since Janzen's (1969) seminal publication. Understandably, the more we learn, the more we realize we do not fully understand. In the most part, straightforward quantification of seed predation is insufficient, especially in temperate woodlands and grasslands. Nevertheless, further studies are needed in boreal, alpine and coastal ecosystems. In general, greater emphasis should be placed on assessing the relative impact of different seed predator guilds on plant species with distinct dispersal syndromes. Such studies should also manipulate the environment to assess the response of different guilds to spatial parameters. Fantastic opportunities exist for cross-continental experiments bringing together the science community. These recommendations are undoubtedly a major challenge but will be essential steps if the next 30 years of research are to be as illuminating as the last.

Acknowledgements

Earlier drafts of this chapter benefited greatly from comments and suggestions by Angela Moles, Luc Wauters, Patrick Jansen and Pierre-Michel Forget.

References

Abbott, H.G. and Quink, T.F. (1970) Ecology of eastern white pine seed caches made by small forest mammals. *Ecology* 51, 271–278.

Alcantara, J.M., Rey, P.J., Valera, F. and Sanchez-Lafuente, A.M. (2000) Factors shaping the seedfall pattern of a bird-dispersed plant. *Ecology* 81, 1937–1950.

Andersen, J. (1978) The influence of the substratum on the habitat selection of Bembidiini (Col., Carabidae). *Norwegian Journal of Entomology* 25, 119–138.

Andreassen, H.P., Hertzberg, K. and Ims, R.A. (1998) Space-use responses to habitat fragmentation and connectivity in the root vole *Microtus oeconomus*. *Ecology* 79, 1223–1235.

Angelstam, P., Hansson, L. and Pehrsson, S. (1987) Distribution of field mice *Apodemus*: the importance of seed abundance and landscape composition. *Oikos* 50, 123–130.

Boman, J.S. and Casper, B.B. (1995) Differential postdispersal seed predation in disturbed and intact temperate forest. *American Midland Naturalist* 134, 107–116.

Bossema, I. (1979) Jays and oaks: an eco-ethological study of symbiosis. *Behaviour* 70, 1–117.

Bowers, M.A. and Dooley, J.L. Jr (1993) Predation hazard and seed removal by small mammals: microhabitat versus patch scale effects. *Oecologia* 94, 247–254.

Breckle, S.W. (2002) *Walter's Vegetation of the Earth*, 4th edn. Springer Verlag, Berlin, 527 pp.

Brown, J.S. (1988) Patch use as an indicator of habitat preference, predation risk, and competition. *Behavioral Ecology and Sociobiology* 22, 37–47.

Brown, J.S. and Venable, D.L. (1991) Life-history evolution of seed-bank annuals in response to seed predation. *Evolutionary Ecology* 5, 12–29.

Carney, S.E., Byerley, M.B. and Holway D.A. (2003) Invasive Argentine ants (*Linepithema humile*) do not replace native ants as seed dispersers of *Dendromecon rigida* (Papaveraceae) in California, USA. *Oecologia* 135, 576–582.

Castro, J., Gomez, J.M., Garcia, D., Zamora, R. and Hodar, J.A. (1999) Seed predation and dispersal in relict Scots pine forests in southern Spain. *Plant Ecology* 145, 115–123.

Clark, D.L. and Wilson, M.V. (2003) Post-dispersal seed fates of four prairie species. *American Journal of Botany* 90, 730–735.

Coley, P.D. and Barone, J.A. (1996) Herbivory and plant defenses in tropical forests. *Annual Review of Ecology and Systematics* 27, 305–335.

Collins, R.J. and Barrett, G.W. (1997) Effects of habitat fragmentation on meadow vole (*Microtus pennsylvanicus*) population dynamics in experiment landscape patches. *Landscape Ecology* 12, 63–76.

Connell, J.H. (1971) On the role of natural enemies in preventing competitive exclusion in some marine animals and in tropical rainforest trees. In: den Boer, P.J. and Gradwell, G.R. (eds) *Dynamics of Populations*. Centre for Agricultural Publishing and Documentation, Wageningen, The Netherlands, pp. 298–310.

Crawley, M.J. (2000) Seed predators and plant population dynamics. In: Fenner, M. (ed.) *Seeds. The Ecology of Regeneration in Plant Communities*, 2nd edn. CAB International, Wallingford, UK, pp. 167–182.

Crawley, M.J. and Long, C.R. (1995) Alternate bearing, predator satiation and seedling recruitment in *Quercus robur* L. *Journal of Ecology* 83, 683–696.

Crist, T.O. and Wiens, J.A. (1994) Scale effects of vegetation on forager movement and seed harvesting by ants. *Oikos* 69, 37–46.

Debinski, D.M. and Holt, R.D. (2000) A survey and overview of habitat fragmentation experiments. *Conservation Biology* 14, 342–355.

Diaz, I., Papic, C. and Armesto, J.J. (1999) An assessment of post-dispersal seed predation in temperate rain forest fragments in Chiloe Island, Chile. *Oikos* 87, 228–238.

Diaz, M. (1992) Spatial and temporal patterns of granivorous ant seed predation in patchy cereal crop areas of central Spain. *Oecologia* 91, 561–568.

Edwards, G.R. and Crawley, M.J. (1999) Rodent seed predation and seedling recruitment in mesic grassland. *Oecologia* 118, 288–296.

Forget, P.M., Hammond, D.S., Milleron, T. and Thomas, R. (2002) Seasonality of fruiting and food hoarding by rodents in Neotropical forests: consequences for seed dispersal and seedling recruitment. In: Levey, D.J., Silva, W.R. and Galetti, M. (eds) *Seed Dispersal and Frugivory: Ecology, Evolution and Conservation.* CAB International, Wallingford, UK, pp. 241–256.

Forman, R.T.T. (1995) *Land Mosaics: the Ecology of Landscapes and Regions.* Cambridge University Press, Cambridge, 632 pp.

Fortier, G.M. and Tamarin, R.H. (1998) Movement of meadow voles in response to food and density manipulations: a test of the food-defense and pup-defense hypotheses. *Journal of Mammalogy* 79, 337–345.

Froud-Williams, R.J., Drennan, D.S.H. and Chancellor, R.J. (1984) The influence of burial and dry-storage upon cyclic changes in dormancy, germination and response to light in seeds of various arable weeds. *New Phytologist* 96, 473–481.

Garcia, D. (2001) Effects of seed dispersal on *Juniperus communis* recruitment on a Mediterranean mountain. *Journal of Vegetation Science* 12, 839–848.

Garcia, D., Zamora, R., Gomez, J.M. and Hodar, J.A. (2001) Frugivory at *Juniperus communis* depends more on population characteristics than on individual attributes. *Journal of Ecology* 89, 639–647.

Gardner, G. (1977) The reproductive capacity of *Fraxinus excelsior* on the Derbyshire limestone. *Journal of Ecology* 65, 107–118.

Gill, D.S. and Marks, P.L. (1991) Tree and shrub seedling colonization of old fields in central New York. *Ecological Monographs* 61, 183–205.

Greig-Smith, P. (1952) The use of random and contiguous quadrats in the study of the structure of plant communities. *Annals of Botany* 16, 293–316.

Hammond, D.S. and Brown, V.K. (1998) Disturbance, phenology and life-history characteristics: factors influencing distance/density-dependent attack on tropical seeds and seedlings. In: Newbery, D.M., Prins, H.H.T. and Brown, N.D. (eds) *Dynamics of Tropical Communities.* Blackwell Science, Oxford, pp. 401–474.

Hansson, L. (1978) Small mammal abundance in relation to environmental variables in three Swedish forest phases. *Studia Forestalia Suedica* 147, 1–40.

Harris, J.H. (1984) An experimental analysis of desert rodent foraging ecology. *Ecology* 65, 1579–1584.

Harvey, H.J. and Meredith, T.C. (1981) Ecological studies of *Peucedanum palustre* and their implications for conservation management at Wicken Fen, Cambridgeshire. In: Synge, H. (ed.) *The Biological Aspects of Rare Plant Conservation.* John Wiley & Sons, Chichester, UK, pp. 365–378.

Hay, M.E. and Fuller, P.J. (1981) Seed escape from heteromyid rodents: the importance of microhabitat and seed preference. *Ecology* 62, 1395–1399.

Herrera, C.M., Jordano, P., López-Soria, L. and Amat, J.A. (1994) Recruitment of a mast-fruiting, bird-dispersed tree: bridging frugivore activity and seedling establishment. *Ecological Monographs* 64, 315–344.

Herrera, C.M., Jordano, P., Guitián, J. and Traveset, A. (1998) Annual variability in seed production by woody plants and the masting concept: reassessment of principles and relationship to pollination and seed dispersal. *The American Naturalist* 152, 576–594.

Hodgson, J.G., Grime, J.P., Hunt, R. and Thompson, K. (1994) *The Electronic Comparative Plant Ecology.* Chapman and Hall, London.

Hoekstra, T.W., Allen, T.F.H. and Flather, L.H. (1991) Implicit scaling in ecological research. *BioScience* 41, 148–154.

Holmes, P.M. (1990) Dispersal and predation in alien *Acacia. Oecologia* 82, 288–290.

Holt, R.D., Robinson, G.R. and Gaines, M.S. (1995) Vegetation dynamics in an experimentally fragmented landscape. *Ecology* 76, 1610–1624.

Hoshizaki, K. and Hulme, P.E. (2002) Mast seeding and predator-mediated indirect interactions in a forest community: evidence from post-dispersal fate of rodent-generated caches. In: Levey, D.J., Silva, W.R. and Galetti, M. (eds) *Seed Dispersal and Frugivory: Ecology, Evolution and Conservation.* CAB International, Wallingford, UK, pp. 227–239.

Houle, G. (1995) Seed dispersal and seedling recruitment: the missing links. *Ecoscience* 2, 238–244.

Hulme, P.E. (1992) The ecology of a temperate plant in a mediterranean environment: post-dispersal seed predation of *Daphne laureola.* In: Thanos, C.A. (ed.) *Plant–Animal Interactions in Mediterranean Type Ecosystems.* Athens University Press, Athens, Greece, pp. 281–286.

Hulme, P.E. (1993) Post-dispersal seed predation by small mammals. *Symposium of the Zoological Society of London* 65, 269–287.

Hulme, P.E. (1994) Post-dispersal seed predation in grassland: its magnitude and sources of variation. *Journal of Ecology* 82, 645–652.

Hulme, P.E. (1996) Natural regeneration of yew (*Taxus baccata* L.): microsite, seed or herbivore limitation? *Journal of Ecology* 84, 853–861.

Hulme, P.E. (1997) Post-dispersal seed predation and the establishment of vertebrate dispersed plants in Mediterranean scrublands. *Oecologia* 111, 91–98.

Hulme, P.E (1998) Post-dispersal seed predation: consequences for plant demography and evolution. *Perspectives in Plant Ecology, Evolution and Systematics* 1, 32–46.

Hulme, P.E. (2002) Seed-eaters: seed dispersal, destruction and demography. In: Levey, D.J., Silva, W.R. and Galetti, M. (eds) *Seed Dispersal and Frugivory: Ecology, Evolution and Conservation*. CAB International, Wallingford, UK, pp. 257–273.

Hulme, P.E. and Benkman, C.W. (2002) Granivory. In: Herrera, C.M. and Pellmyr, O. (eds) *Plant–Animal Interactions. An Evolutionary Approach*. Blackwell Science, Oxford, pp. 132–154.

Hulme, P.E. and Borelli, T. (1999) Variability in post-dispersal seed predation in deciduous woodland: relative importance of location, seed species, burial and density. *Plant Ecology* 145, 149–156.

Jansen, P.A. and Forget, P.-M. (2001) Scatterhoarding by rodents and tree regeneration in French Guiana. In: Bongers, F., Charles-Dominique, P., Forget, P.-M. and Théry, M. (eds) *Nouragues: Dynamics and Plant–Animal Interactions in a Neotropical Rainforest*. Kluwer Academic Publishers, Dordrecht, The Netherlands, pp. 275–288.

Janzen, D.H. (1969) Seed-eaters versus seed size, number, toxicity and dispersal. *Evolution* 23, 1–27.

Janzen, D.H. (1970) Herbivores and the number of tree species in tropical forests. *The American Naturalist* 104, 501–508.

Janzen, D. (1971) Seed predation by animals. *Annual Review of Ecology and Systematics* 2, 465–492.

Jensen, T.S. (1982) Seed production and outbreaks of non-cyclic rodent populations in deciduous forests. *Oecologia* 54, 184–192.

Jensen, T.S. (1985) Seed–seed predator interactions of European beech, *Fagus silvatica* and forest rodents, *Clethrionomys glareolus* and *Apodemus flavicollis*. *Oikos* 44, 149–156.

Jensen, T.S. and Nielsen, O.F. (1986) Rodents as seed dispersers in a heath–oak wood succession. *Oecologia* 70, 214–221.

Jules, E.S. and Rathcke, B.J. (1999) Mechanisms of reduced *Trillium* recruitment along edges of old-growth forest fragments. *Conservation Biology* 13, 784–793.

Kelly, D. and Sork, V.L. (2002) Mast seeding in perennial plants: why, how, where? *Annual Review of Ecology and Systematics* 33, 427–447.

Kollmann, J. (1995) Regeneration window for fleshy-fruited plants during scrub development on abandoned grassland. *Ecoscience* 2, 213–222.

Kollmann, J. (1997) Schadfraß an Gehölzsamen auf Waldlichtungen und im Wald. *Forstwissenschaftliches Centralblatt* 116, 113–123.

Kollmann, J. (2000) Dispersal of fleshy-fruited species: a matter of spatial scale? *Perspectives in Plant Ecology, Evolution and Systematics* 3, 29–51.

Kollmann, J. and Bassin, S. (2001) Effects of management on seed predation in wildflower strips in northern Switzerland. *Agriculture, Ecosystems and Environment* 83, 285–296.

Kollmann, J. and Buschor, M. (2002) Edges effects on seed predation by rodents in deciduous forests of northern Switzerland. *Plant Ecology* 164, 249–261.

Kollmann, J. and Pirl, M. (1995) Spatial pattern of seed rain of fleshy-fruited plants in a scrubland–grassland transition. *Acta Oecologica* 16, 313–329.

Kollmann, J. and Schill, H.P. (1996) Spatial patterns of dispersal, seed predation and germination during colonization of abandoned grasslands by *Quercus petraea* and *Corylus avellana*. *Vegetatio* 125, 193–205.

Kollmann, J., Coomes, D.A. and White, S.M. (1998) Consistencies in post-dispersal seed predation of temperate fleshy-fruited species among seasons, years and sites. *Functional Ecology* 12, 683–690.

Körner, C. (2002) Die Vegetation der Erde. In: Sitte, P., Weiler, E.W., Kadereit, J.W., Bresinsky, A. and Körner, C. (eds) *Strasburger. Lehrbuch der Botanik*. Spektrum Akademischer Verlag, Heidelberg, Germany, pp. 1003–1043.

Kratochwil, A. and Schwabe, A. (1994) Coincidences between different landscape ecological zones and growth forms of Cembran pine (*Pinus cembra* L.) in subalpine habitats of the Central Alps. *Landscape Ecology* 9, 175–190.

Laurance, W.F. and Bierregaard, R.O. Jr (1997) *Tropical Forest Remnants: Ecology, Management, and Conservation of Fragmented*

Communities. University of Chicago Press, Chicago, Illinois, 616 pp.

Levin, S. (1992) The problem of pattern and scale in ecology. *Ecology* 73, 1943–1967.

Lima, S.L. and Dill, L.M. (1990) Behavioral decisions made under the risk of predation: a review and prospectus. *Canadian Journal of Zoology* 68, 619–640.

Louda, S.M. (1989) Predation in the dynamics of seed regeneration. In: Leck, M.A., Parker, V.T. and Simpson, R.L. (eds) *Ecology of Soil Seed Banks.* Academic Press, San Diego, California, pp. 25–51.

Louda, S.M., Farris, M.A. and Blua, M.J. (1987) Variation in methylglucosinolate and insect damage to *Cleome serrulata* (Capparaceae) along a natural soil moisture gradient. *Journal of Chemical Ecology* 13, 569–581.

Manson, R.H. and Stiles, E.W. (1998) Links between microhabitat preferences and seed predation by small mammals in old fields. *Oikos* 82, 37–50.

Manson, R.H., Ostfeld, R.S. and Canham, C.D. (1999) Responses of a small mammal community to heterogeneity along forest–old-field edges. *Landscape Ecology* 14, 355–367.

Marino, P.C., Gross, K.L. and Landis, D.A. (1997) Weed seed loss due to predation in Michigan maize fields. *Agriculture, Ecosystems and Environment* 66, 189–196.

Maron, J.L. and Simms, E.L. (1997) Effects of seed predation on seed bank size and seedling recruitment of bush lupine *Lupinus arboreus. Oecologia* 111, 76–83.

McAdoo, J.K., Evans, C.C., Roundy, B.A., Young, J.A. and Evans, R.A. (1983) Influence of heteromyid rodents on *Oryzopsis hymenoides* germination. *Journal of Range Management* 36, 61–64.

Meiners, S.J. and LoGiudice, K. (2003) Temporal consistency in the spatial pattern of seed predation across a forest–old field edge. *Plant Ecology* 168, 45–55.

Mittelbach, G.G. and Gross, K.L. (1984) Experimental studies of seed predation in old-fields. *Oecologia* 65, 7–13.

Moles, A.T. and Westoby, M. (2003) Latitude, seed predation and seed mass. *Journal of Biogeography* 30, 105–128.

Myster, R.W. and Pickett, S.T.A. (1993) Effects of litter, distance, density and vegetation patch type on postdispersal tree seed predation in old fields. *Oikos* 66, 381–388.

Niemelä, J., Haila, Y., Halme, E., Pajunen, T. and Punttila, P. (1992) Small-scale heterogeneity in the spatial distribution of carabid beetles in the southern Finnish taiga. *Journal of Biogeography* 19, 173–181.

Nilson, M.E. and Hjältén, J. (2003) Covering pine-seeds immediately after seeding: effects on seedling emergence and on mortality through seed-predation. *Forest Ecology and Management* 176, 449–457.

Nilsson, S.G. and Wästljung, U. (1987) Seed predation and cross pollination in mast-seeding beech (*Fagus sylvatica*) patches. *Ecology* 68, 260–265.

Nystrand, O. (1998) Post-dispersal predation on conifer seeds and juvenile seedlings in boreal forest. PhD thesis, Swedish University of Agricultural Sciences, Umeå, Sweden.

Nystrand, O. and Granström, A. (1997) Post-dispersal predation on *Pinus sylvestris* seeds by *Fringilla* spp: ground substrate affects selection for seed color. *Oecologia* 110, 353–359.

Nystrand, O. and Granström, A. (2000) Predation on *Pinus sylvestris* seeds and juvenile seedlings in Swedish boreal forest in relation to stand disturbance by logging. *Journal of Applied Ecology* 37, 449–463.

Obeso, J.R. and Fernández-Calvo, I.C. (2003) Fruit removal, pyrene dispersal, post-dispersal predation and seedling establishment of a bird-dispersed tree. *Plant Ecology* 165, 223–233.

O'Dowd, D.J. and Hay, M.E. (1980) Mutualism between harvester ants and a desert ephemeral: seed escape from rodents. *Ecology* 61, 531–540.

O'Neill, R.V. (1989) Perspectives in hierarchy and scale. In: Roughgarden, J., May, R.M. and Levin, S.A. (eds) *Perspectives in Ecological Theory.* Princeton University Press, Princeton, New Jersey, pp. 140–156.

Ostfeld, R.S., Manson, R.H. and Canham, C.D. (1997) Effects of rodents on survival of tree seeds and seedlings invading old fields. *Ecology* 78, 1531–1542.

Plucinski, K.E. and Hunter, M.L. (2001) Spatial and temporal patterns of seed predation on three tree species in an oak–pine forest. *Ecography* 24, 309–317.

Portnoy, S. and Willson, M.F. (1993) Seed dispersal curves, behavior of the tail of the distribution. *Evolutionary Ecology* 7, 25–44.

Price, M.V. (1978) The role of microhabitat in structuring desert rodent communities. *Ecology* 59, 910–921.

Price, M.V. and Heinz, K.M. (1984) Effects of body size, seed density, and soil characteristics on rates of seed harvest by heteromyid rodents. *Oecologia* 61, 420–425.

Reader, R.J. (1993) Control of seedling emergence by ground cover and seed predation in relation to seed size for some old-field species. *Journal of Ecology* 81, 169–175.

Rey, P.J., Garrido, J.L., Alcántara, J.M., Ramírez, J.M., Aguilera, A., García, L., Manzaneda, A.J. and Fernández, R. (2002) Spatial variation in ant and rodent post-dispersal predation of vertebrate-seeds. *Functional Ecology* 16, 773–781.

Robinson, R.A. and Sutherland, W.J. (1999) The winter distribution of seed-eating birds: habitat structure, seed density and seasonal depletion. *Ecography* 22, 447–454.

Root, R.B. (1967) The niche exploitation pattern of the blue-gray gnat-catcher. *Ecological Monographs* 37, 317–350.

Rudgers, J.A. and Maron, J.L. (2003) Facilitation between coastal dune shrubs: a non-nitrogen fixing shrub facilitates establishment of a nitrogen-fixer. *Oikos* 102, 75–84.

Russell, S.K. and Schupp, E.W. (1998) Effects of microhabitat patchiness on patterns of seed dispersal and seed predation of *Cercocarpus ledifolius* (Rosaceae). *Oikos* 91, 434–443.

Schluter, D. and Repasky, R.R. (1991) Wordwide limitation of finch densities by food and other factors. *Ecology* 72, 1763–1774.

Schneider, D.C. (1994) *Quantitative Ecology: Spatial and Temporal Scaling.* Academic Press, San Diego, California, 395 pp.

Schreiner, M., Bauer, E.M. and Kollmann, J. (2000) Reducing predation of conifer seeds by clear-cutting *Rubus fructicosus* agg. in two montane forest stands. *Forest Ecology and Management* 126, 281–290.

Schupp, E.W. and Fuentes, M. (1995) Spatial patterns of seed dispersal and the unification of plant population ecology. *Ecoscience* 2, 267–275.

Silvertown, J.W. (1980) The evolutionary ecology of mast seeding in trees. *Biological Journal of the Linnean Society* 14, 235–250.

Simberloff, D. and Dayan, T. (1991) The guild concept and the structure of ecological communities. *Annual Review of Ecology and Systematics* 22, 115–143.

Smit, R., Bokdam, J., den Ouden, J., Olff, H., Schot-Opschoor, H. and Schrijvers, M. (2001) Effects of introduction and exclusion of large herbivores on small rodent communities. *Plant Ecology* 155, 119–127.

Smith, C.C. and Reichman, O.J. (1984) The evolution of food caching by birds and mammals. *Annual Review of Ecology and Systematics* 15, 329–351.

Soholt, L.F. (1973) Consumption of primary production by a population of kangaroo rats (*Dipodomys merriami*) in the Mojave desert. *Ecological Monographs* 43, 357–376.

Sone, K., Hiroi, S., Nagahama, D., Ohkubo, C., Nakano, E., Murao, S. and Hata, K. (2002) Hoarding of acorns by granivorous mice and its role in the population processes of *Pasania edulis* (Makino) Makino. *Ecological Research* 17, 553–564.

Sork, V.L. (1983) Distribution of pignut hickory (*Carya glabra*) along a forest to edge transect and factors affecting seedling recruitment. *Bulletin of the Torrey Botanical Club* 110, 494–506.

Stapanian, M.A. (1986) Seed dispersal by birds and squirrels in the deciduous forests of the United States. In: Estrada, A. and Fleming, T.H. (eds) *Frugivores and Seed Dispersal.* Dr W. Junk, Dordrecht, The Netherlands, pp. 225–236.

Telleria, J.L., Santos, T. and Alcantara, M. (1991) Abundance and food-searching intensity of wood mice (*Apodemus sylvaticus*) in fragmented forests. *Journal of Mammalogy* 72, 183–187.

Tewksbury, J.J., Levey, D.J., Haddad, N.M., Sargent, S., Orrock, J.L., Weldon, A., Danielson, B.J., Brinkerhoff, J., Damschen, E.I. and Townsend, P. (2002) Corridors affect plants, animals, and their interactions in fragmented landscapes. *Proceedings of the National Academy of Sciences USA* 99, 12923–12926.

Thompson, J.N. (1994) *The Coevolutionary Process.* University of Chicago Press, Chicago, Illinois, 376 pp.

Turnbull, L.A., Crawley, M.J. and Rees, M. (2000) Are plant populations seed limited? A review of seed sowing experiments. *Oikos* 88, 225–238.

Vander Wall, S.B. (1990) *Food Hoarding in Animals.* University of Chicago Press, Chicago, Illinois, 445 pp.

Vander Wall, S.B. (1994) Seed fate pathways of antelope bitterbrush: dispersal by seed-caching yellow pine chipmunks. *Ecology* 75, 1911–1926.

Vander Wall, S.B. (2002) Secondary dispersal of Jeffrey pine seeds by rodent scatter-hoarders: the roles of pilfering, recaching and a variable environment. In: Levey, D.J., Silva, W.R. and Galetti, M. (eds) *Seed Dispersal and Frugivory: Ecology, Evolution and Conservation.*

CAB International, Wallingford, UK, pp. 193–208.

Verdú, M. and García-Fayos, P. (1996) Post-dispersal seed predation in a Mediterranean patchy landscape. *Acta Oecologica* 17, 379–391.

Wada, N. (1993) Dwarf bamboos affect the regeneration of zoochorous trees by providing habitats to acorn-feeding rodents. *Oecologia* 94, 403–407.

Wada, N. and Uemura, S. (1994) Seed dispersal and predation by small rodents on the herbaceous understory plant *Symplocarpus renifolius*. *American Midland Naturalist* 132, 320–327.

Walter, H. (1962/1968) *Die Vegetation der Erde in öko-physiologischer Betrachtung.* Gustav Fischer, Jena, 538/1001 pp.

Watt, A.S. (1919) On the causes of failure of natural regeneration in British oak-woods. *Journal of Ecology* 7, 173–203.

Watt, A.S. (1923) On the ecology of British beechwoods, with special reference to their regeneration: I. The causes of failure of natural regeneration of the beech (*Fagus sylvatica*). *Journal of Ecology* 11, 1–48.

Webb, S.L. and Willson, M.F. (1985) Spatial heterogeneity in post-dispersal predation on *Prunus* and *Uvularia* seeds. *Oecologia* 67, 150–153.

Wiens, J.A. (1989) Spatial scaling in ecology. *Functional Ecology* 3, 385–397.

Willson, M.F. (1993) Dispersal mode, seed shadows, and colonization patterns. *Vegetatio* 107/108, 261–280.

Wilson, C.G. (1989) Postdispersal seed predation of an exotic weed, *Mimosa pigra* L., in the Northern Territory. *Australian Journal of Ecology* 14, 235–240.

Wright, S.J. (2002) Plant diversity in tropical forests: a review of mechanisms of species coexistence. *Oecologia* 130, 1–14.

3   The Fate of Seed Banks: Factors Influencing Seed Survival for Light-demanding Species in Moist Tropical Forests

James W. Dalling

Department of Plant Biology, University of Illinois, Urbana-Champaign, 265 Morrill Hall, 505 S Goodwin Ave., Urbana, Illinois 61801, USA and Smithsonian Tropical Research Institute, Apartado 2072, Balboa, Republic of Panama

Introduction

Seed mass in species-rich moist tropical forests often varies over six or more orders of magnitude (Foster, 1982; Foster and Janson, 1985; Hammond and Brown, 1995). Most of these species, including some with the minutest seeds (Metcalfe, 1996; Metcalfe and Grubb, 1997), can be classified as 'shade-tolerant' with the ability to establish beneath a closed canopy. Seeds of the shade-tolerators vary widely in morphology and physiology, but many are recalcitrant, and most germinate within a few months of dispersal (Hall and Swaine, 1980; Ng, 1980; Garwood, 1983; Hopkins and Graham, 1987).

This chapter concerns the approximately 10–20% of tree species in moist tropical forests that can be classified as 'light-demanding' or 'gap-dependent' (Dalling *et al.*, 1998a; Molino and Sabatier, 2001). These species require higher levels of irradiance than are found in closed forest understoreys for successful seedling establishment. Variation in seed mass, morphology and physiology among these species can be as great as or greater than that of shade tolerators (Hammond and Brown,

1995). Establishment of light-demanding species occurs when seeds are dispersed directly to canopy openings, or perhaps more frequently, when disturbance cues the germination of seeds present in the soil seed bank (e.g. Putz and Appanah, 1987; Lawton and Putz, 1988; Dalling and Hubbell, 2002). Successful recruitment from a persistent seed bank depends on surviving an array of predators on the soil surface, remaining buried or being returned to shallow soil depths that permit seedling emergence, and avoiding predation or infection by soil invertebrates and pathogenic microbes.

The additional interactions that seeds encounter during their stay in the soil set apart studies of seed fate for light-demanding species from those of rapidly germinating shade-tolerant species in tropical forests. Greater duration of the seed stage of the regeneration cycle might also imply that sources of seed mortality play a more important role as a demographic filter influencing population growth and niche partitioning for these species. Despite this, however, seed fate in the soil remains poorly explored, with most studies restricted to relatively simple experimental assays of seed survival (e.g. Hopkins and Graham,

1987; Dalling *et al.*, 1998b; Murray and Garcia, 2002).

This chapter reviews the potential fates of small-seeded (< 100 mg) light-demanding species from dispersal to seedling emergence. It emphasizes the experimental and observational approaches necessary to uncover the interactions between seeds and soil macrofauna, and highlights the near absence of studies examining microbial interactions with seeds. Finally, studies of seed fate underscore the need to take an integrated life-history approach to the interpretation of seed traits. Light-demanding species exhibit remarkable diversity in seed characteristics, including seed mass variation over four orders of magnitude, adoption of varied dispersal mechanisms, and capacities for seed persistence ranging from weeks to decades (Dalling *et al.*, 1997, 2002; Murray and Garcia, 2002). Different combinations of seed traits may contribute to species coexistence by influencing the size of canopy gaps that can be detected by the seed and the seasonality of seedling emergence (Dalling *et al.*, 1997; Pearson *et al.*, 2002). However, many potential combinations of seed traits may be essentially equivalent, resulting in similar recruitment success despite the manifold fates that seeds can experience.

Effects of Dispersal Mode on Post-dispersal Seed Fate

Dispersal mode is one of the strongest determinants of the subsequent fate of seeds. Seeds dispersed by vertebrate frugivores may retain residual fruit pulp, or aril tissue, or faecal matter that attracts secondary seed dispersers and predators to the site of seed deposition. Experimental arrays of small seeds placed on the soil surface as intact drupes (Dalling *et al.*, 1998b), intact arillate seeds (Horvitz and Schemske, 1986, 1994; see Pizo *et al.*, Chapter 19, this volume), bird droppings (e.g. Byrne and Levey, 1993; Kaspari, 1993), and primate defecations (Estrada and Coates-Estrada, 1991) are typically removed by ants and beetles (see

Andresen and Feer, Chapter 20, this volume) within a few hours. Similarly, many ballistically dispersed seeds frequently carry elaiosomes (food bodies attached to the seed), which attract ants that subsequently carry seeds (see Mayer *et al.*, Chapter 10, this volume) to the nest site (Passos and Ferreira, 1996). In contrast, wind-dispersed seeds lack these attractants.

In a comparative study of secondary removal rates of seeds of six light-demanding species in Panama, Fornara and Dalling (2004, unpublished results) placed arrays of seeds of four vertebrate-dispersed, and two wind-dispersed species on microscope slides on the soil surface in five lowland and lower montane forests in Panama. Removal rates were remarkably similar among sites. Overall, 45% of seeds of the four vertebrate-dispersed species (*Miconia argentea*, Melastomataceae, 0.08 mg; *Cecropia peltata*, Cecropiaceae, 0.5 mg; *Trema micrantha*, Celtidaceae, 1.3 mg; *Apeiba aspera*, Tiliaceae, 14.2 mg) were removed over 2 days, primarily by ants. In contrast, only 2% of seeds of the wind-dispersed species (*Luehea seemanii*, Tiliaceae, 1.9 mg; *Jacaranda copaia*, Bignoniaceae, 4.7 mg) were removed. The explanation for the differential attractiveness of wind- versus vertebrate-dispersed seeds remains unclear. Residual fruit pulp on vertebrate-dispersed seeds that might have attracted ants was washed off prior to the experiment, while removal of the large samara attached to *Luehea* seeds did not influence their removal rate. Future work with a larger number of species will be needed to determine whether primary seed dispersal mode is indeed a predictor of subsequent seed fate. One possibility is that invertebrate seed predators may discriminate against wind-dispersed seeds based on their lower seed moisture content (Augspurger, 1988).

Dispersal mode may also indirectly affect seed fate by influencing the spatial pattern of seed dispersal. For large-seeded species, the number and density of seeds deposited, and the distance seeds are moved from reproductive conspecifics can affect the probability that seeds are encountered by predators and that predators will be satiated by seed resources (e.g. Augspurger and Kitajima, 1992; Burkey, 1994); similar patterns might also be expected for small-seeded species. For vertebrate-dispersed seeds, the number of seeds deposited in each dropping is likely to scale with body size. Large-bodied primates such as howler monkeys and Capuchin monkeys and ungulates, such as tapirs in the neotropics may deposit several thousand small seeds in a single defecation (Janzen, 1982; Andresen, 1999; Wehncke et al., 2003), contributing to locally aggregated dispersal patterns. Furthermore, physiology and behaviour of vertebrate frugivores may also determine seed fate. For example, in neotropical forests, howler and spider monkeys defecate at infrequent intervals, depositing seeds in large clumps at feeding roosts, latrines and sleeping roosts (Milton, 1984; Fragoso, 1997; Julliot, 1997). In contrast, Capuchin monkeys defecate frequently through the day, producing small dung piles that result in significantly lower rodent seed predation rates than dung piles left by howler monkeys (Zhang and Wang, 1995; Wehncke et al., 2004).

Mechanisms of Seed Incorporation into the Soil

In temperate forest and grassland communities seed size and morphology are important determinants of the rate of incorporation of seeds into the soil (Peart, 1984; Thompson et al., 1993). Larger seeds, and seeds with appendages for wind dispersal, are less likely to become embedded in soil particles or fall between soil aggregates and, as a consequence, are more conspicuous to seed predators that are primarily active on the soil surface (Thompson, 1987). Biophysical constraints on seed burial rates can therefore strongly influence the survival of dispersed seeds and may provide an explanation for the negative relationships found between seed size or shape (variance in seed dimensions), and seed persistence (Thompson and Grime, 1979; Rees, 1993; Bekker et al., 1998; Funes et al., 1999; but see Moles et al., 2000).

Similar relationships between size and persistence may exist for wind-dispersed species in tropical forests. In wet forests, passive seed burial may occur when rain splash covers seeds with fine soil particles, but frequent rain storms also block the spaces between soil aggregates, limiting opportunities for small seeds to percolate through soil pores (Pearson et al., 2003). In more seasonal forests, subjected to repeated wetting and drying cycles, soil particles form larger aggregates of up to a few centimetres in size with larger pores between them (Marshall et al., 1996). In strongly seasonal forests, soil cracks may appear in the dry season that extend tens of centimetres into the soil, permitting seeds to be deeply buried during the dry season (Garwood, 1989).

For seeds dispersed by frugivores, however, secondary dispersal and predation seem more likely fates than passive incorporation into the soil. In lowland forests, an array of ant species are attracted to defecated, regurgitated or fallen seeds. These seed-harvesting ants are remarkably abundant and diverse (e.g. 44 seed-harvesting ant species, and > 300 Pheidole ants/m^2 at La Selva, Costa Rica; Levey and Byrne, 1993; Kaspari, 1996), and remove seeds varying widely in size (0.5–15 mg; Kaspari, 1996; Fornara and Dalling, 2003, unpublished results). Ant removal of seeds in the litter is less important in montane forests, however, as ant abundance declines sharply above about 1500 m (McCoy, 1990; Olson, 1994; Samson et al., 1997; Brühl et al., 1999).

The fate of ant-harvested seeds remains somewhat unclear, but a detailed study by Levey and Byrne (1993) of Pheidole ants at La Selva has begun to reveal the complexity of the ant–seed interaction. Pheidole are small ants (minor workers < 2 mm) with small colony sizes (< 100 individuals) that forage in the litter and on the soil surface removing seeds from frugivore faeces and caching them in nests consisting of partially decomposed twigs. Most small seeds of Melastomataceae removed by Pheidole are consumed in the nest, but a small fraction (6%) of seeds are deposited in refuse piles or abandoned as nests are periodically relocated to new twigs (Levey and Byrne, 1993). Thus, removal by litter ants permits some seeds to become incorporated in the litter/soil seed bank, and may result in some dispersal of seeds to nutrient-rich micro-habitats (refuse piles) more favourable for seedling growth than the original site of deposition (see Pizo *et al.*, Chapter 19, this volume).

However, not all seeds harvested by ants remain close to the soil surface. Larger seeds and fruit (5–50 mg) are harvested by larger ants that nest in the soil and may be cached a few centimetres to metres below the soil surface, or concentrated in large refuse piles by leaf-cutter ants. On Barro Colorado Island (BCI), Panama, large num-bers of *Miconia argentea* (Melastomataceae) fruit (20–100 mg) containing numerous tiny seeds are collected from the canopy and soil surface by leaf-cutter ants (*Atta* spp.) to provision their fungus gardens. Densities of > 1000 viable seeds/g are subsequently deposited in refuse piles on the soil surface by *A. colombica* and in underground cham-bers by *A. cephalotes* (Dalling and Wirth, 1998), where they can remain viable for several months (Farji-Brener and Medina, 2000). In addition to *Miconia* fruits, larger individual seeds (> 10 mg) may also be removed by larger ponerine litter ants such as *Ectatomma ruidum* and carried to nests below the surface (J. Dalling, personal observation). It remains unclear, however, whether these larger seeds are actually predated by ants as their seed coats may be too thick for ants to penetrate (as is the case for some desert annuals, O'Dowd and Hay, 1980).

In addition to ants, seeds defecated by mammals may also be secondarily dispersed by dung beetles. Dung beetles are attracted to clumps of faeces, which they bury at depths of 0.5–12 cm and use as a food source for their larvae (Estrada and Coates-Estrada, 1991; Andresen, 2002). Dung is often detected and buried by beetles within a few hours of defecation and seeds are inci-dentally buried along with it. Rapid burial provides an important escape for seeds from rodents, as predation rates for large seeds remaining in faeces on the soil surface or within the top centimetre of soil are very high (Estrada and Coates-Estrada, 1991; Andresen, 1999; Andresen and Feer, Chapter 20, this volume). The extent to which rodents are attracted to the smaller seeds of light-demanding species remains unclear. Small murid mice have been observed eating *Cecropia* seeds < 1 mg seed mass in Amazonian Ecuador (Paula Barriga, Quito, 2003, personal communication), and rodents are reported as seed predators for *Cecropia* in montane forest in Costa Rica (Murray and Garcia, 2002). More generally, if ants and rodents compete for seed resources, as has been shown in desert ecosystems (Brown and Davidson, 1977; Midgley and Anderson, Chapter 11, this vol-ume), then rodents might be predicted to be more important sources of predation in high elevation sites where ant abundance is low.

Seed Persistence and Seed Dormancy in the Soil

Once in the soil, seeds can remain viable for periods ranging from several weeks to sev-eral decades. Experimental burial of seeds in mesh bags in the soil showed that in NE Australia 22 of 25 species of pioneer and early successional species retained > 10% germinability after 2 years (Hopkins and Graham, 1987), while in Panama 10 of 14 pioneer species retained similar germin-ability after 18 months (Dalling *et al.*, 1997). These results are consistent with those of a study comparing annual inputs of seed rain with standing soil seed bank densities in a Costa Rican montane forest, which pro-vided evidence of super-annual accumula-tion of seeds in the soil for 13 of 23 pioneer species (Murray and Garcia, 2002). Thus in general, more than half of pioneer species establish persistent seed banks (*sensu* Baskin and Baskin, 1998).

None of these studies, however, shows the negative relationship between seed per-sistence and seed mass typically found in temperate grassland communities (Thomp-son and Grime, 1979; Rees, 1993; Thompson *et al.*, 1993; Funes *et al.*, 1999). This suggests that diverse seed predator communities in tropical forests preclude size-dependent predation. At this stage we can only speculate on the morphological and ecological correlates of seed persistence that might exist in tropical forests.

Seed persistence in the soil may be mediated by either physical or chemical defensive traits, or may possibly be constrained by the availability of seed reserves and the rate of metabolism (Garwood and Lighton, 1990). Within a single genus, reported rates of seed persistence can vary substantially. For example, seeds of *Cecropia insignis, C. obtusifolia* and *C. polyphlebia* persist for little more than a year in the soil and rapidly lose viability when air-dried (Alvarez-Buylla and Martínez-Ramos, 1990; Dalling *et al.*, 1998b; Murray and Garcia, 2002). In contrast, seeds of *C. sciadophylla* remain viable for > 4 years (Holthuijzen and Boerboom, 1982), perhaps reflecting differences in the chemical composition and thickness of the pericarp among *Cecropia* species (Lobova *et al.*, 2003).

On BCI, Panama, long persistent seed banks appear to be more common among species with larger seed mass (> 3 mg) and with thick seed coats. Small-seeded species and wind-dispersed species tended to show rapid declines in germinability during the first year of burial (Dalling *et al.*, 1997). Indeed, some species with thick seed coats are capable of extraordinary seed persistence. ^{14}C dating of seeds extracted from natural seed banks, using accelerator mass spectrometry (Moriuchi *et al.*, 2000), has shown that seeds of four pioneer species with seed mass > 3 mg (*Trema micrantha, Hyeronima laxiflora, Zanthoxylum eckmannii* and *Croton bilbergianus*) buried at depths of < 3 cm can be > 30 years old (J. Dalling and T. Brown, 2003, unpublished results). Similarly, Murray and Garcia (2002) argue that the high densities of thick-seed-coated *Phytolacca rivinoides* in the seed bank at Monteverde, Costa Rica, suggest accumulation in the soil over many decades.

The adoption of a hard seed coat may provide a broadly effective defence against predators analogous to quantitative defensive traits such as lignin and tannins that reduce the palatability of leaves and shoots to a broad array of herbivores (Feeny, 1976). Hard seed coats, however, are only likely to be effective above a minimum thickness, thus constraining the effectiveness of this defence to larger-seeded species. Persistent small seeds are therefore likely to be chemically defended. An additional potential constraint imposed by a hard seed coat is that its impermeability and toughness may impose physical dormancy. All four species on BCI with decade-long seed persistence have been shown to have a very low initial germination rate under favourable conditions of high red : far red ratio irradiance and moisture availability (Dalling *et al.*, 1997; Silvera *et al.*, 2003). For three of these species which are insensitive to light, presumed physical dormancy can be broken with high temperature or scarification (Acuña and Garwood, 1987), but fresh *Trema micrantha* seeds cannot be induced to germinate, perhaps reflecting physiological rather than physical dormancy. How long initial dormancy remains for seeds buried in the soil remains unclear, but presumably the permeability of seeds increases as the seed coat degrades in the soil.

If physical dormancy represents an unavoidable cost associated with the defence against predators, then we might expect that some shade-tolerant species with hard seed coats would also show super-annual seed persistence in the soil. On BCI, at least two species, the palm *Attalea butyracea* (= *Scheelea zonensis*), and the canopy tree *Vantanea occidentalis* (Humiriaceae) with exceptionally thick, hard endocarps, germinate slowly over at least 3 years (Harms and Dalling, 1995). However, the dormancy mechanisms used by these species are unclear, and may be morphological or physiological rather than physical (*sensu* Baskin and Baskin, 1998). Long-term studies of the germination ecology of shade-tolerant tree species are seldom performed in moist tropical forests (but see Garwood, 1983), and prolonged dormancy, although rare, may be under-reported among these species.

Biotic Constraints on Seed Survival in the Soil

Whereas seeds on the soil surface and in the litter interact with ants, beetles and rodents, seeds buried in the soil encounter a different, but potentially overlapping suite of potential mortality agents. As yet, however, most studies have only documented survival rates of buried seeds, and little attempt has been made to attribute seed losses to particular taxa. The limited evidence that earthworms and fungi contribute to seed mortality is reviewed here.

Earthworms have been ignored as potential sources of seed dispersal and predation, perhaps because their abundance in tropical forest soils was initially underestimated (Fragoso and Lavelle, 1992). More recent reviews estimate their densities at 10–400 individuals/m², with mean values similar to temperate forests (Fragoso and Lavelle, 1992; González and Zou, 1999). Earthworms can exert important effects on seed bank dynamics as seeds are ingested in the soil mass (primarily in sites with richer soils), or in the litter (primarily in oligotrophic soils; Fragoso and Lavelle, 1992), and are subsequently deposited deep in the soil or in casts on the soil surface (Grant, 1983). Seeds of temperate grassland species vary in their susceptibility to breakdown during gut passage, but seeds of at least a few species are defecated in a viable state (McRill and Sagar, 1973; Willems and Huijsmans, 1994). In temperate grassland, Thompson et al. (1994) demonstrated that earthworms can be quite size-selective in the movement of seeds. Species found in surface worm casts were significantly smaller than species found in the soil as a whole, with one species, Cerastium fontanum, accounting for > 85% of seeds in surface casts.

Studies of earthworm effects on seed viability and distribution have not been reported in tropical forests. However, experiments in tropical savannas indicate that earthworms there have mutualistic–antagonistic effects. Decaëns et al. (2003) found that earthworms transport a subset of seed bank seeds to surface casts, but that viability of seeds from casts was up to 40 times lower than in the surrounding soil. Seeds may therefore be degraded during gut passage or perhaps infected by pathogens dispersed with seeds (Toyota and Kimura, 1994). Standing vegetation in savanna and pastures, however, was more similar in composition to worm casts than to the seed bank as a whole, suggesting that worm casts may provide particularly favourable microsites for seedling recruitment (Decaëns et al., 2003).

Fungal pathogens have frequently been invoked as important sources of mortality for tropical forest plants (Augspurger, 1984; Clark and Clark, 1984; Khan and Tripathi, 1991; Gilbert, 2002), with most work focusing on the infection of seedlings (e.g. Alexander and Mihail, 2000; Gilbert et al., 2001). Fungi have been similarly implicated as sources of mortality for seeds in the soil, particularly for light-demanding species whose small seeds fill the forest seed bank (Alvarez-Buyulla and Martínez-Ramos, 1990; Dalling et al., 1998a,b). Despite their implied importance, however, direct evidence that fungi account for seed losses in the soil is remarkably scarce (Baskin and Baskin, 1998), and information on the taxonomic identity, diversity, host affinity, and ecological importance of seed-infecting fungi represents a 'major lacuna in research' (Gilbert, 2002).

Very few studies have linked seed mortality in the soil to fungi. Crist and Friese (1993) and Masaki et al. (1998) attributed 30% of the mortality of five shrub-steppe species, and 20% of the mortality of the warm-temperate tree Cornus contraversa to pathogenic and decomposer fungi based on evidence of necrosis of seeds extracted from the soil. Application of fungicide to buried seeds has shown modest increases in seed survival. Lonsdale (1993) found that the fungicide Benlate reduced mortality of seeds of the exotic shrub Mimosa pigra in northern Australia by 10–16%, while Blaney and Kotanen (2002) found that the fungicide Captan significantly increased survival of seeds of 39 species of native and alien grasses and forbs in a Canadian oldfield site, but only by 5–10%. Finally, Leishman et al. (2000) showed that a cocktail of fungicides was able to increase survival of *Medicago lupulina* seeds by 30% in a British grassland.

Our own studies suggest that fungi are responsible for most seed losses of common pioneer species in the soil seed bank. At BCI, Panama, we found that mortality of seeds of two common pioneer tree species (*Miconia argentea* and *Cecropia insignis*) buried in mesh bags for 6 months was reduced from 90–95% for untreated seeds to 50% for seeds treated with the broad-spectrum fungicide Captan (Dalling *et al.*, 1998a). Similar studies conducted in the growing house at La Selva, Costa Rica, using Benomyl fungicide are consistent with this pattern (Fig. 3.1).

Given the great scarcity of experimental and taxonomic work on seed-associated fungi, the major questions regarding their ecological importance remain unanswered. If pathogens show strong host preferences and are patchily distributed in the soil, then they may contribute strongly to the maintenance of diversity by locally reducing the density of seed banks of susceptible hosts. In contrast, if fungi are broad host-generalists and are uniformly distributed through space, then they may have a more neutral effect by limiting community-wide recruitment. Preliminary studies focused on one host tree, *Cecropia insignis*, suggest limited opportunity for escape from pathogens. Seeds of this species buried at BCI at below maternal crown sites, below conspecific males, and at a range of distances ≤ 50 m from conspecifics showed broadly similar mortality rates (Dalling *et al.*, 1998b; R. Gallery and J. Dalling, 2003, unpublished results).

Initial attempts to isolate fungi from *Cecropia* seeds incubated in the soil have revealed 32 distinct morphotypes emerging in culture from 250 seeds, with seven fungal morphotypes accounting for 72% of the total isolates (R. Gallery and J. Dalling, 2003, unpublished results). The absolute diversity of seed-infecting fungi, however, is likely to be substantially higher as many seed-infecting fungi may be unculturable (e.g. Bridge and Spooner, 2001). Molecular analyses of the seven most abundant fungal morphotypes collected from *C. insignis* seeds indicate that they comprise representatives of several orders of Ascomycota, including species of *Fusarium*, *Chaetomium*, *Dictyochaeta*, *Glionectria* and *Colletotrichum* (A.E. Arnold, North Carolina, 2003, personal communication). Intriguingly, several of these taxa have high sequence affinity with known endophytic fungi that have been isolated from tropical foliage (Arnold *et al.*, 2000; A.E. Arnold, North Carolina, 2003, personal communication). This suggests that some fungi might be vertically transmitted from the maternal host during seed development rather than infecting seeds in the soil (e.g. Kirkpatrick and Bazzaz, 1979). Further possibilities are that some seed-infecting fungi may act as benevolent mutualists rather than deleterious pathogens, or may switch from endophytes to pathogens according to the ontogenetic stage or identity of the host.

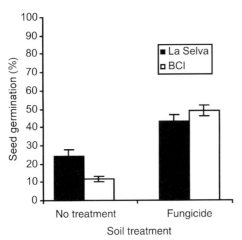

Fig. 3.1. Effect of fungicide treatment on seed germination in *Cecropia insignis*. Shaded bars show results from an experiment conducted at La Selva, Costa Rica, in which seeds were buried for 5 months in soil cores maintained in a shaded growing house. Monthly treatment of cores with Benomyl fungicide had a significant effect on germination, compared with untreated controls ($F = 12.4$, df = 1.33, $P < 0.01$; R. Gallery and J. Dalling, 2004, unpublished data). Open bars show similar results from an experiment conducted at Barro Colorado Island, Panama, for seeds buried for 5 months beneath *Cecropia* tree crowns and treated monthly with Captan fungicide ($F = 29.6$, df = 1.36, $P < 0.01$; Dalling *et al.*, 1998b).

In addition to fungi, many other micro-organisms are seed-borne and are reported as diseases of important agricultural crops, including nematodes, viruses, spiroplasmas and bacteria (reviewed in Agarwal and Sinclair, 1996). Most of these pathogens induce symptoms in developing seedling tissues, with limited effects on seed viability. Bacterial seed rot diseases, however are quite diverse, including *Bacillus*, *Curtobacterium* and *Pseudomonas* (Sinclair, 1978; Agarwal and Sinclair, 1996). More recently developed techniques, including polymerase chain reaction (PCR) technology, are likely to detect many more seed-borne diseases, and shed light on the role of antagonistic relationships between bacteria and fungi that may mediate seed infection in the soil (Kremer, 1987). As yet, however, molecular techniques have had only limited application to studies of the microbial seed ecology of non-crop species (e.g. Jacobson *et al.*, 1998).

Seed Germination and Seedling Emergence

Traits that influence seed burial and seed persistence also influence the microsites in which seeds can germinate and the probability that seedlings successfully emerge. Seed mass is correlated with maximal emergence depth in the soil because limits to seed reserves dictate the resources available for the extension of a hypocotyl through the soil (Bond *et al.*, 1999). For the smallest seeded species (seed mass < 0.1 mg), maximal effective burial depths for emergence are < 5 mm (Dalling *et al.*, 1995; Pearson *et al.*, 2002), providing a major constraint on effective seed bank densities.

Seed mass also affects emergence and establishment success by influencing the susceptibility of seedlings to drought. In large canopy gaps, the soil surface may reach temperatures of > 40°C after only a few hours of exposure to direct irradiance, resulting in rapid drying of the top few millimetres of soil even when moisture levels in the bulk soil are close to field capacity. Dry, clear weather lasting just a few days can be sufficient to cause significant mortality of newly emerging seedlings as soils dry out at a faster rate than roots can grow down through the soil profile (Daws, 2002). This has been shown in large (400 m^2) gaps, where the survival of newly emerged seedlings of five small-seeded species was 20–35% higher in irrigated treatments versus non-irrigated controls 10 days after initiating experimental treatments (Engelbrecht *et al.*, 2001).

Finally, seed mass also influences the range of microsites available for seedling establishment through selection on gap detection mechanisms (Fig. 3.2). Small-seeded light-demanding species (< 2 mg) have been shown to exhibit photoblastic seed germination (Pearson *et al.*, 2002), which is constrained by the limited penetration of light through the surface few millimetres of moist soil (Tester and Morris, 1987). Photoblastic seeds detect small shifts in the ratio of red : far red irradiance (Smith, 2000) associated with light interception by canopy vegetation or leaf litter on the soil surface (Vázquez-Yanes *et al.*, 1990). Continuous variation in red : far red irradiance allows for the discrimination of a range of disturbance sizes including litter-free microsites, and sites at the edge of gaps that are not exposed to direct irradiance.

Conversely, larger seeded light-demanding species (> 2 mg) are capable of emerging from at least 5 cm below the soil surface (Pearson *et al.*, 2002). These species are typically not photoblastic, so must use alternative cues to detect gaps. Four of eight larger seeded species studied on BCI have been shown to respond to temperature fluctuations associated with soil warming during the day in gaps. These diel fluctuations in temperature can be measured at least 10 cm below the soil surface in large gaps and therefore constitute an effective germination cue for species with significant seed reserves (Pearson *et al.*, 2002). A disadvantage, however, of a temperature-based germination cue is that it fails to detect small gaps and gap edge microsites that do not receive sufficient direct irradiance to warm the soil, and therefore potentially limits recruitment

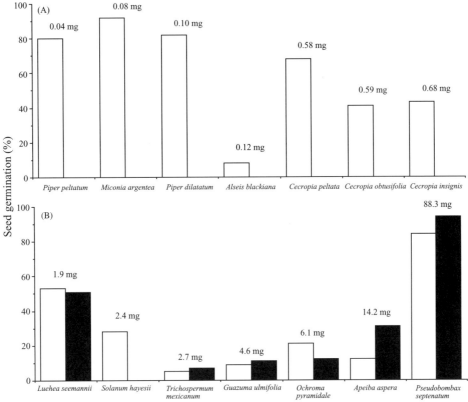

Fig. 3.2. Effect of light conditions on seed germination of pioneer species from Barro Colorado Island, Panama, varying in seed mass. Open bars show seed germination in the light and filled bars show germination in the dark under constant temperature conditions (30°C). (A) Seven species with seed mass < 1 mg all have significantly higher germination in the light than in the dark (where no germination occurred; G test, $P < 0.01$). (B) For six species with seed mass > 1 mg germination is unaffected by light conditions. One exception is *Solanum hayesii*, where germination is triggered by high red : far red irradiance or by fluctuating temperature in the dark (Pearson *et al.*, 2003). Figure modified from Pearson *et al.* (2002).

of these species to larger canopy disturbances.

Conclusions and Future Work

This review reveals a number of apparently correlated traits that can be sketched out as potential syndromes directing the fate of small seeds present in seed banks (Table 3.1). As yet, however, studies of seed ecology in tropical forests are very incomplete, and these associations are therefore best viewed as hypotheses to be tested from wider comparative studies. Research to date

has been concentrated in the neotropical lowlands, most notably at sites in Panama, Costa Rica and Mexico, and has also focused primarily on a narrow taxonomic group with many studies of vertebrate-dispersed tiny-seeded species in the genus *Cecropia*, and the family Melastomataceae.

A future research priority should be to expand studies of seed ecology in the Old World tropics, and to incorporate greater variation in habitat conditions beyond the moist lowland forest. At present, we have little idea of how gradients in temperature and moisture availability influence seed persistence in the soil, or how shifts in composition and abundance of guilds of

Table 3.1. Suggested linkages among traits influencing the fate of small-seeded light-demanding species in tropical forests according to seed mass.

	Seed mass	
	< 1 mg	> 1 mg
Dispersal mode	No relationship	
Probability of seed burial	No relationship	
Seed persistence in soil	Mostly short? (< 1 year)	Short–long (up to 30+ years)
Seed dormancy	Rare	Common
Litter invertebrate predation	Susceptible[a]	Less susceptible
Pathogen infection	Susceptible	Resistant?
Germination cue	Photoblastic	Non-photoblastic

[a]Limited evidence suggests that wind-dispersed seeds are less susceptible to predation by litter invertebrates (principally ants).

invertebrate and vertebrate seed predators influence seed fate in the litter and rates of incorporation into the seed bank. Finally, our inferences of how seed morphology and physiology influence fate are potentially biased by a lack of phylogenetic control over interspecific comparisons. Many light-demanding species belong to speciose genera (e.g. *Cecropia*, *Miconia*, *Piper*, *Macaranga*) with wide geographic distributions and potentially provide model taxa for investigating how seed traits have radiated in response to environmental variation.

References

Acuña, P. and Garwood, N.C. (1987) Efecto de la luz y de la escarificación en al germinación de semillas de cinco especies de arboles tropicales secundarios. *Revista de Biología Tropical* 35, 203–207.

Agarwal, V.K. and Sinclair, J.B. (1996) *Principles of Seed Pathology*, 2nd edn. CRC Press, Boca Raton, Florida, 539 pp.

Alexander, H.M. and Mihail, J.D. (2000) Seedling disease in an annual legume: consequences for seedling mortality, plant size, and population seed production. *Oecologia* 122, 346–353.

Alvarez-Buylla, E.R. and Martínez-Ramos, M. (1990) Seed bank versus seed rain in the regeneration of a tropical pioneer tree. *Oecologia* 84, 314–325.

Andresen, E. (1999) Seed dispersal by monkey and the fate of dispersed seeds in a Peruvian rain forest. *Biotropica* 31, 145–158.

Andresen, E. (2002) Dung beetles in a Central Amazonian rainforest and their ecological role as secondary seed dispersers. *Ecological Entomology* 27, 257–270.

Arnold, A.E., Maynard, Z., Gilbert, G.S., Coley, P.D. and Kursar, T.A. (2000) Are tropical fungal endophytes hyperdiverse? *Ecology Letters* 3, 267–274.

Augspurger, C.K. (1984) Seedling survival of tropical tree species: interactions of dispersal distance, light gaps, and pathogens. *Ecology* 65, 1705–1712.

Augspurger, C.K. (1988) Mass allocation, moisture content and dispersal capacity of wind-dispersed tropical diaspores. *New Phytologist* 108, 357–368.

Augspurger, C.K. and Kitajima, K. (1992) Experimental studies of seedling recruitment from contrasting seed distributions. *Ecology* 73, 1270–1284.

Baskin, C.C. and Baskin, J.M. (1998) *Seeds: Ecology, Biogeography, and Evolution of Dormancy and Germination*. Academic Press, San Diego, California, 666 pp.

Bekker, R.M., Bakker, J.P., Grandin, U., Kalamees, R., Milberg, P., Poschlod, P., Thompson, K. and Willems, J.H. (1998) Seed size, shape and vertical distribution in the soil: indicators of seed longevity. *Functional Ecology* 12, 834–842.

Blaney, C.S. and Kotanen, P.M. (2002) Persistence in the seed bank: the effects of fungi and invertebrates on seeds of native and exotic plants. *Ecoscience* 9, 509–517.

Bond, W.J., Honig, M. and Maze, K.E. (1999) Seed size and seedling emergence: an allometric relationship and some ecological implications. *Oecologia* 120, 132–136.

Bridge, P. and Spooner, B. (2001) Soil fungi: diversity and detection. *Plant and Soil* 232, 147–154.

Brown, J.H. and Davidson, D.W. (1977) Competition between seed-eating rodents and ants in desert ecosystems. *Science* 196, 880–882.

Brühl, C.A., Mohamed, M. and Linsenmair, K.E. (1999) Altitudinal distribution of leaf litter ants along a transect in primary forests on Mount Kinabalu, Sabah, Malaysia. *Journal of Tropical Ecology* 15, 265–277.

Burkey, T.V. (1994) Tropical tree species diversity: a test of the Janzen-Connell model. *Oecologia* 97, 533–540.

Byrne, M. and Levey, D.J. (1993) Removal of seeds from frugivore defecations by ants in a Costa Rican rain forest. *Vegetatio* 107/108, 363–374.

Clark, D.A. and Clark, D.B. (1984) Spacing dynamics of a tropical rain-forest tree: evaluation of the Janzen-Connell Model. *American Naturalist* 123, 626–641.

Crist, T.O. and Friese, C.F. (1993) The impact of fungi on soil seeds: implications for plants and granivores in semiarid shrub-steppe. *Ecology* 74, 2231–2239.

Dalling, J.W. and Hubbell, S.P. (2002) Seed size, growth rate and gap microsite conditions as determinants of recruitment success for pioneer species. *Journal of Ecology* 90, 557–568.

Dalling, J.W. and Wirth, R. (1998) Dispersal of *Miconia argentea* seeds by the leaf-cutting ant, *Atta colombica*. *Journal of Tropical Ecology* 14, 705–710.

Dalling, J.W., Swaine, M.D. and Garwood, N.C. (1995) Effect of soil depth on seedling emergence in tropical soil seed-bank investigations. *Functional Ecology* 9, 119–121.

Dalling, J.W., Swaine, M.D. and Garwood, N.C. (1997) Soil seed bank community dynamics in seasonally moist lowland tropical forest, Panama. *Journal of Tropical Ecology* 13, 659–680.

Dalling, J.W., Hubbell, S.P. and Silvera, K. (1998a) Seed dispersal, seedling emergence and gap partitioning in gap-dependent tropical tree species. *Journal of Ecology* 86, 674–689.

Dalling, J.W., Swaine M.D. and Garwood, N.C. (1998b) Dispersal patterns and seed bank dynamics of pioneer trees in moist tropical forest. *Ecology* 79, 564–578.

Dalling, J.W., Muller-Landau, H.C., Wright, S.J. and Hubbell, S.P. (2002) Role of dispersal in the recruitment limitation of neotropical pioneer species. *Journal of Ecology* 90, 714–727.

Daws, M.I. (2002) Mechanisms of plant species coexistence in a semi-deciduous tropical forest in Panamá. PhD thesis, University of Aberdeen, UK.

Decaëns, T., Mariani, L., Betancourt, N. and Jiménez, J.J. (2003) Seed dispersion by surface casting activities of earthworms in Colombian grasslands. *Acta Oecologica* 24, 175–185.

Engelbrecht, B.M.J., Dalling, J.W., Pearson, T.R.H, Wolf, R.L., Galvez, D.A., Koehler, T., Ruiz, M.C. and Kursar, T.A. (2001) Short dry spells in the wet season increase mortality of tropical pioneer seedlings. In: Ganeshaiah, K.N., Shaanker, U. and Bawa, K.S. (eds) *Tropical Ecosystems: Structure, Diversity and Human Welfare. Proceedings of the International Conference on Tropical Ecosystems: Structure, Diversity and Human Welfare*. Oxford and IBH Publishing Company, New Delhi, pp. 665–669.

Estrada, A. and Coates-Estrada, R. (1991) Howler monkeys (*Alouatta palliata*), dung beetles (Scarabaeidae) and seed dispersal: ecological interactions in the tropical rain forest of Los Tuxtlas, Mexico. *Journal of Tropical Ecology* 7, 459–474.

Farji-Brener, A.G. and Medina, C.A. (2000) The importance of where to dump the refuse: seed banks and fine roots in nests of the leaf-cutting ants *Atta cephalotes* and *A. colombica*. *Biotropica* 32, 120–126.

Feeny, P. (1976) Plant apparency and chemical defense. *Recent Advances in Phytochemistry* 10, 1–40.

Foster, R.B. (1982) The seasonal rhythm of fruitfall on Barro Colorado Island. In: Leigh, E.G. Jr, Rand, A.S. and Windsor, D.M. (eds) *The Ecology of a Tropical Forest. Seasonal Rhythms and Long-term Changes*. Smithsonian Institution Press, Washington, DC, pp. 151–172.

Foster, S.A. and Janson, C.H. (1985) The relationship between seed size and establishment conditions in tropical woody plants. *Ecology* 66, 773–780.

Fragoso, C. and Lavelle, P. (1992) Earthworm communities of tropical rain forests. *Soil Biology and Biochemistry* 24, 1397–1408.

Fragoso, J.M.V. (1997) Tapir-generated seed shadows: scale-dependent patchiness in the Amazon rain forest. *Journal of Ecology* 85, 519–529.

Funes, G., Basconcelo, S., Diaz, S. and Cabido, M. (1999) Seed size and shape are good predictors of seed persistence in soil in temperate mountain grasslands of Argentina. *Seed Science Research* 9, 341–345

Garwood, N.C. (1983) Seed germination in a seasonal tropical forest in Panama: a community study. *Ecological Monographs* 53, 159–181.

Garwood, N.C. (1989) Tropical soil seed banks: a review. In: Leck, M., Parker, V. and Simpson, R. (eds) *The Ecology of Soil Seed Banks.* Academic Press, San Diego, California, pp. 149–209.

Garwood, N.C. and Lighton, J.R.B. (1990) Physiological ecology of seed respiration in some tropical species. *New Phytologist* 115, 549–558.

Gilbert, G.S. (2002) Evolutionary ecology of plant diseases in natural ecosystems. *Annual Review of Phytopathology* 40, 13–43.

Gilbert, G.S., Harms, K.E., Hamill, D.N. and Hubbell, S.P. (2001) Effects of seedling size, El Niño drought, seedling density, and distance to nearest conspecific adult on 6-yr survival of *Ocotea whitei* seedlings in Panama. *Oecologia* 127, 509–516.

González, G. and Zou, X. (1999) Plant and litter influences on earthworm abundance and community structure in a tropical wet forest. *Biotropica* 31, 486–493.

Grant, D.J. (1983) The activities of earthworms and the fate of seeds. In: Satchell, J.E. (ed.) *Earthworm Ecology.* Chapman & Hall, London, pp. 107–122.

Hall, J.B. and Swaine, M.D. (1980) Seed stocks in Ghanaian forest soils. *Biotropica* 12, 256–263.

Hammond, D.S. and Brown, V.K. (1995) Seed size of woody plants in relation to disturbance, dispersal, and soil type in wet neotropical forests. *Ecology* 76, 2544–2561.

Harms, K.E. and Dalling, J.W. (1995) Seasonal consistency in germination timing in the palm *Scheelea zonensis. Principes* 39, 104–106.

Holthuijzen, A.M.A. and Boerboom, J.H.A. (1982) The *Cecropia* seedbank in the Surinam lowland rain forest. *Biotropica* 14, 62–68.

Hopkins, M.S. and Graham, A.W. (1987) The viability of seeds of rainforest species after experimental soil burials under tropical wet lowland forest in north-eastern Australia. *Australian Journal of Ecology* 12, 97–108.

Horvitz, C.C. and Schemske, D.W. (1986) Seed removal and environmental heterogeneity in a neotropical myrmecochore: variation in removal rates and dispersal distance. *Biotropica* 18, 319–323.

Horvitz, C.C. and Schemske, D.W. (1994) Effects of dispersers, gaps, and predators on dormancy and seedling emergence in a tropical herb. *Ecology* 75, 1949–1958.

Jacobson, D.J., LeFebvre, S.H., Ojerio, R.S., Berwald, N. and Heikkinen, E. (1998) Persistent,

systemic, asymptomatic infections of *Albugo candida,* an oomycete parasite, detected in three wild crucifer species. *Canadian Journal of Botany* 76, 739–750.

Janzen, D.H. (1982) Seeds in tapir dung in Santa Rosa National Park, Costa Rica. *Brenesia* 19/20, 129–135.

Julliot, C. (1997) Impact of seed dispersal of red howler monkeys *Alouatta seniculus* in the understory of tropical rain forest. *Journal of Ecology* 85, 431–440.

Kaspari, M. (1993) Removal of seeds from neotropical frugivore droppings: ant responses to seed number. *Oecologia* 95, 81–88.

Kaspari, M. (1996) Worker size and seed size selection by harvester ants in a neotropical forest. *Oecologia* 105, 397–404.

Khan, M.L. and Tripathi, R.S. (1991) Seedling survival and growth of early and late successional tree species as affected by insect herbivory and pathogen attack in subtropical humid forest stands of north-east India. *Acta Oecologica* 12, 569–579.

Kirkpatrick, B.L. and Bazzaz, F.A. (1979) Influence of certain fungi on seed germination and seedling survival of four colonizing annuals. *Journal of Applied Ecology* 16, 515–527.

Kremer, R.J. (1987) Identity and properties of bacteria inhabiting seeds of selected broadleaf weed species. *Microbial Ecology* 14, 29–37.

Lawton, R.O. and Putz, F.E. (1988) Natural disturbance and gap phase regeneration in a wind-exposed tropical cloud forest. *Ecology* 69, 764–777.

Leishman, M.R., Masters, G.J., Clarke, I.P. and Brown, V.K. (2000) Seed bank dynamics: the role of fungal pathogens and climate change. *Functional Ecology* 14, 293–299.

Levey, D.J. and Byrne, M.M. (1993) Complex ant–plant interactions: rain forest ants as secondary dispersers and post-dispersal seed predators. *Ecology* 74, 1802–1812.

Lobova, T.A., Mori, S.A., Blanchard, F., Peckham, H. and Charles-Dominique, P. (2003) *Cecropia* as a food resource for bats in French Guiana and the significance of fruit structure in seed dispersal and longevity. *American Journal of Botany* 90, 388–403.

Lonsdale, W.M. (1993) Losses from the seed bank of *Mimosa pigra*: soil-microorganisms vs temperature-fluctuations. *Journal of Applied Ecology* 30, 654–660.

Marshall, T.J., Holmes, J.W. and Rose, C.W. (1996) *Soil Physics,* 3rd edn. Cambridge University Press, Cambridge, 472 pp.

Masaki, T., Tanaka, H., Shibata, M. and Nakashizuka, T. (1998) The seed bank dynamics of *Cornus controversa* and their role in regeneration. *Seed Science Research* 8, 53–63.

McCoy, E.D. (1990) The distribution of insects along elevational gradients. *Oikos* 58, 313–322.

McRill, M. and Sagar, G.R. (1973) Earthworms and seeds. *Nature* 243, 482.

Metcalfe, D.M. (1996) Germination of small-seeded tropical rain forest plants exposed to different spectral compositions. *Canadian Journal of Botany* 74, 516–520.

Metcalfe, D.M. and Grubb, P.J. (1997) The responses to shade of seedlings of very small-seeded tree and shrub species from tropical rain forest in Singapore. *Functional Ecology* 11, 215–221.

Milton, K. (1984) The role of food-processing factors in primate food choice. In: Rodman, P.S. and Cant, J.G.H. (eds) *Adaptations of Foraging in Nonhuman Primates*. Columbia University Press, New York, pp. 249–279.

Moles, A.T., Hudson, D.W. and Webb, C.J. (2000) Seed size and shape and persistence in the soil in the New Zealand flora. *Oikos* 89, 541–545.

Molino, J.-F. and Sabatier, D. (2001) Tree diversity in tropical rain forests: a validation of the Intermediate Disturbance Hypothesis. *Science* 294, 1702–1704.

Moriuchi, K.S., Venable, D.L., Pake, C.E. and Lange,T. (2000) Direct measurement of the seed bank age structure of a Sonoran desert annual plant. *Ecology* 81, 1133–1138.

Murray, K.G. and Garcia, M. (2002) Contributions of seed dispersal and demography to recruitment limitation in a Costa Rican cloud forest. In: Levey, D.J., Silva, W.R. and Galetti, M. (eds) *Seed Dispersal and Frugivory: Ecology, Evolution and Conservation*. CAB International, Wallingford, UK, pp. 323–338.

Ng, F.S.P. (1980) Germination ecology of Malaysian woody plants. *Malaysian Forester* 43, 406–438.

O'Dowd, D.J. and Hay, M.E. (1980) Mutualism between harvester ants and a desert ephemeral: seed escape from rodents. *Ecology* 61, 531–540.

Olson, D.M. (1994) The distribution of leaf litter invertebrates along a Neotropical altitudinal gradient. *Journal of Tropical Ecology* 10, 129–150.

Passos, L. and Ferreira, S.O. (1996) Ant dispersal of *Croton priscus* (Euphorbiaceae) seeds in a tropical semideciduous forest in southeastern Brazil. *Biotropica* 28, 697–700.

Pearson, T.R.H., Burslem, D.F.R.P., Mullins, C.E. and Dalling, J.W. (2002) Germination ecology of neotropical pioneers: interacting effects of environmental conditions and seed size. *Ecology* 83, 2798–2807.

Pearson, T.R.H., Burslem, D.F.R.P., Mullins, C.E. and Dalling, J.W. (2003) Functional significance of photoblastic germination in neotropical pioneer trees: a seed's eye view. *Functional Ecology* 17, 394–402.

Peart, M.H. (1984) The effects of morphology, orientation and position of grass diaspores on seedling survival. *Journal of Ecology* 72, 437–453.

Putz, F.E. and Appanah, B. (1987) Buried seeds, newly dispersed seeds, and the dynamics of a lowland forest in Malaysia. *Biotropica* 19, 326–339.

Rees, M. (1993) Trade-offs among dispersal strategies in the British flora. *Nature* 366, 150–152.

Samson, D.A., Rickart, E.A. and Gonzales, P.C. (1997) Ant diversity and abundance along an elevational gradient in the Philippines. *Biotropica* 29, 349–363.

Silvera, K., Skillman, J.B. and Dalling, J.W. (2003) Seed germination, seedling growth and habitat partitioning in two morphotypes of the tropical pioneer tree *Trema micrantha* in a seasonal forest in Panama. *Journal of Tropical Ecology* 19, 27–34.

Sinclair, J.B. (1978) The seed-borne nature of some soybean pathogens, the effect of *Phomopsis* spp., and *Bacillus subtilis* on germination and their occurrence in Illinois. *Seed Science and Technology* 6, 957.

Smith, H. (2000) Phytochromes and light signal perception by plants – an emerging synthesis. *Nature* 407, 585–591.

Tester, M. and Morris, C. (1987) The penetration of light through soil. *Plant, Cell and Environment* 10, 281–286.

Thompson, K. (1987) Seeds and seed banks. In: Rorison, I.H., Grime, J.P., Hunt, R., Hendry, G.A.F. and Lewis, D.H. (eds) *Frontiers of Comparative Plant Ecology*. *New Phytologist* 106 (Suppl.), 23–34.

Thompson, K. and Grime, J.P. (1979) Seasonal variation in the seed banks of herbaceous species in ten contrasting habitats. *Journal of Ecology*, 67, 893–921.

Thompson, K., Band, S.R. and Hodgson, J.G. (1993) Seed size and shape predict persistence in soil. *Functional Ecology* 7, 236–241.

Thompson, K., Green, A. and Jewels, A.M. (1994) Seeds in soils and worm casts from a neutral grassland. *Functional Ecology* 8, 29–35.

Toyota, K. and Kimura, M. (1994) Earthworms disseminate a soil-borne pathogen, *Fusarium oxysporum* f. sp. *raphani*. *Biology and Fertility of Soils* 18, 32–36.

Vázquez-Yanes, C.A., Orozco-Segovia, A., Rincon, E., Sanchez-Coronado, M.E., Huante, R., Toledo, J.R. and Barradas, U.L. (1990) Light beneath the litter in a tropical forest: effect on seed germination. *Ecology* 71, 1952–1958.

Wehncke, E.V., Hubbell, S.P., Foster, R.B. and Dalling, J.W. (2003) Seed dispersal patterns produced by White-faced Monkeys: implications for the dispersal limitation of neotropical tree species. *Journal of Ecology* 91, 677–685.

Wehncke, E.V., Numa Valdez, C. and Domínguez, C.A. (2004) Seed dispersal and defecation patterns of *Cebus capucinus* and *Alouatta palliata*: consequences for seed dispersal effectiveness. *Journal of Tropical Ecology* 20, 535–543.

Willems, J.H. and Huijsmans, K.G.A. (1994) Vertical seed dispersal by earthworms – a quantitative approach. *Ecography* 17, 124–130.

Zhang, S.Y. and Wang, L.X. (1995) Fruit consumption and seed dispersal of *Ziziphus cinnamomum* (Rhamnaceae) by two sympatric primates (*Cebus apella* and *Ateles paniscus*) in French Guiana. *Biotropica* 27, 397–401.

4 Frugivore-mediated Interactions Among Bruchid Beetles and Palm Fruits at Barro Colorado Island, Panama: Implications for Seed Fate

Kirsten M. Silvius

Environmental Center, Krauss Annex #19, 2500 Dole St., University of Hawaii, Honolulu, HI 96822-2303, USA

Introduction

The microhabitat and distance to which seeds are carried from parent trees are important determinants of seed fate and have attracted considerable attention from researchers (Janzen, 1970; Connell, 1971; Howe and Smallwood, 1982; Forget, 1990, 1992, 1993, 1996; Fleming and Estrada, 1993; Terborgh et al., 1993; Cintra, 1997; Peres et al., 1997; Wenny and Levey, 1998; Andresen, 1999). Less research has been directed at the complex interactions occurring among insects, microbes, fruits and vertebrates after primary seed dispersal and prior to seed germination or death. Nevertheless, these interactions can significantly affect a seed's probability of mortality (Herrera, 1982, 1984, 1986, 1989; Janzen, 1982; Bucholz and Levey, 1990; Traveset, 1992, 1993; García et al., 1999; also see reviews in Sallabanks and Courtney (1992) and Silvius and Fragoso (2002)).

Several researchers have shown that husk and pulp removal from palm fruits by vertebrates affects subsequent insect predation of the seeds (Janzen, 1971; Wright, 1983, 1990; Delgado et al., 1997; Silvius and Fragoso, 2002). For the palm *Attalea*

maripa in northern Brazil (Maracá Island Ecological Reserve, henceforth 'Maracá'), Silvius and Fragoso (2002) showed that different species of vertebrate frugivores removed different amounts of pulp from the palm endocarp, and that the pattern of pulp removal affected subsequent oviposition behaviour by bruchid beetles, the primary agent of mortality for the seeds. Intact fruits that fell to the ground without first being handled by primates, birds or agoutis escaped bruchid predation, because bruchids did not oviposit on the fruit husk. These fruits also had a high probability of removal by tapirs (*Tapirus terrestris*), a long-distance seed disperser that feeds on intact fruits from the ground (Silvius and Fragoso, 2002; Fragoso et al., 2003). Fruits whose husk had been entirely removed, and their pulp partially removed, by primates, agoutis (*Dasyprocta leporina*) and pacas (*Agouti paca*) received the largest numbers of eggs, whereas fruits whose pulp was either intact or completely removed received low and intermediate numbers of eggs, respectively (Silvius and Fragoso, 2002). These results indicate that both intact fruit pulp and intact husks deter insect seed predators, and that by removing or partially removing

©CAB International 2005. *Seed Fate*
(eds P.-M. Forget, J.E. Lambert, P.E. Hulme and S.B. Vander Wall)

these structures vertebrates increase the susceptibility of the fruit to insect predation. They also suggest that the timing and rate of fruit husk removal over the course of the fruiting season will affect the timing of bruchid oviposition and emergence, which could in turn affect the length of time to which fruits are exposed to predation by bruchids. The composition of the frugivore community at a given site could thus affect the pattern of emergence of beetles by affecting the ratio of intact to vertebrate-handled fruits.

Attalea palms occur throughout South and Central America. To test for the pervasiveness of these vertebrate–bruchid–fruit interactions, I repeated the fruit manipulation experiments originally carried out at Maracá (Silvius and Fragoso, 2002) on Barro Colorado Island, Panama, where a different Attalea species occurs, the relative abundance of two bruchid species is reversed, and the relative abundance of terrestrial frugivores is radically different (tapirs are rare, white-lipped peccaries are absent, and agouti and squirrel populations are probably at higher densities (Smythe et al., 1982; Giacalone-Madden et al., 1990; but see Wright et al., 1994)). My objectives were to determine:

1.   Whether the same processes occur in related species at a geographically distinct location.
2.   Whether delayed oviposition by beetles on intact fruits results in a staggered emergence of adult beetles from the general palm seed population during the course of the palm's reproductive season.

Study System

Seeds of palms in the genus *Attalea* (includes *Maximiliana*, *Orbygnia* and *Scheelea*; Henderson, 1995) are attacked by several species of bruchid beetles throughout South and Central America, but most commonly by *Pachymerus cardo* and *Speciomerus giganteus* (Nilsson and Johnson, 1993; Delobel *et al.*, 1995; Johnson *et al.*, 1995; Quiroga-Castro and Roldán,

2001). These beetles are often the primary source of mortality for the seeds. On Maracá, shortly after fruit fall between 70 and 90% of all *Attalea maripa* seeds remaining within 30 metres of a parent plant are infested by *P. cardo*, while infestation rates reach 100% inside experimental vertebrate exclosures (Fragoso, 1997; Silvius, 2002; Silvius and Fragoso, 2002; Fragoso *et al.*, 2003). In Panama and Costa Rica, mortality rates for *Attalea butyracea* fruits range from 35% to 77% under different experimental and natural settings (Janzen, 1971, 1972; Wilson and Janzen, 1972; Bradford and Smith, 1977; Wright, 1983, 1990; Forget *et al.*, 1994; Wright *et al.*, 2000; Wright and Huber, 2001). When confined to seeds directly beneath the parent tree, mortality is greater than 80% on Barro Colorado Island (BCI) (Wright, 1983). *P. cardo* and *S. giganteus* both occur on BCI and Maracá, but *S. giganteus* is more common than *P. cardo* at BCI (Wright, 1983, 1990; Wright *et al.*, 2000; Wright and Huber, 2001), while the opposite is true at Maracá (Silvius and Fragoso, 2002).

Attalea spp. palm fruits drop intact to the ground when they are ripe, but many are knocked down prior to abscission by primates, carnivores and birds, often with the pulp only partly removed. Many more fall intact once they are fully ripe. On the ground, rodents, ungulates, and a variety of insects feed on the pulp (Wright, 1983, 1990; Fragoso, 1997; Silvius, 2002; Silvius and Fragoso, 2002). Wright (1983, 1990) and others (e.g. Janzen, 1971, 1972) indicated that in Central America bruchids oviposit only on fallen *Attalea* fruits whose husk has been removed by animals or rot. Approximately 5–7 days after oviposition, the first instar larvae burrow through the endocarp, leaving a visible entry hole. Although several larvae usually enter a seed, only one survives what appears to be competition within the seed (note that this applies only to *Pachymerus* and *Speciomerus* on *Attalea*; other bruchid species on other palms have different reproductive strategies; K.M. Silvius, personal observation). The surviving larva feeds on the endosperm and embryo, pupates, and emerges as an adult through a hole in the endocarp carved by the last instar larva.

Study Site

Leigh *et al.* (1982) have described in detail the physical and ecological characteristics of BCI. The 1500 ha isolated mountaintop in the Panama Canal zone supports moist tropical forest, receives about 2600 mm of rain per year, with a marked dry season from December to April. Old growth forest (200+ years old; Foster and Brokaw, 1982) with no *Attalea* palms, and old forest with low *Attalea* palm abundances occur on the island, as well as secondary forest (about 100 years old), usually with mid- to high *Attalea* palm abundances (Foster and Brokaw, 1982; Wright and Huber, 2001; K.M. Silvius, personal observation). Experimental exclosures in this study were all located in secondary forest.

Individual *A. butyracea* on BCI produce from one to three bunches with 100–600 fruits from late April to early December, with a peak in fruit production from July to August (De Steven *et al.*, 1987; Wright, 1990). Fruits are 4.5–8.5 cm long and 3–4.5 cm in diameter. Seeds are surrounded by a thick endocarp with fibres, thick oily yellow pulp and a resistant fibrous husk. Fruits contain from one to three seeds, though one seed is most common. Fourteen per cent of endocarps at BCI are two- to three-seeded, but 64% of trees produce at least some multiple-seeded endocarps ($n = 67$ trees, or 1214 endocarps, with a minimum of 15 endocarps opened per tree; K.M. Silvius, unpublished data). I refer to the endocarp with enclosed seed(s) as 'endocarp' throughout this chapter.

Methods

Pulp manipulation exclosures

In June 2000, wire-mesh exclosures containing ten palm fruits each were set at the edge of the fruit shadows of each of seven fruiting *A. butyracea* trees. Two of the trees were within 30 m of each other; all others were at least 500 m apart. The edge of the fruit shadow was defined as the point where only solitary fruits were found on the forest floor, rather than groups of fruits; the distance was usually 2–5 m from the parent tree. Exclosures were meant to prevent vertebrates from reaching the fruits, while allowing access to insects. Each exclosure contained a different fruit manipulation type (as per Silvius and Fragoso, 2002): Intact Pulp fruits (IP; husk removed but pulp untouched); Gashed fruits (G; husk removed, pulp gashed in four strips to expose the endocarp, imitating treatment by primates); Bare Endocarps (BE; husk and pulp completely removed, exposing the entire endocarp); Intact Husk fruits (IH; nothing removed) and Rodent fruits (R; pulp gnawed by agoutis and pacas, leaving fibre with some pulp attached to the endocarp). IH exclosures suffered high fruit removal rates by vertebrates, reducing sample size to only three incomplete replicates (two IP and one G exclosure were also breached by vertebrates, but only one or two fruits were removed in those cases). R exclosures were set at only three trees, due to the difficulty of obtaining agouti-gnawed fruits on BCI; during the peak fruiting season at Maracá agoutis leave gnawed *Attalea* fruits lying near the parent tree, whereas at BCI agoutis carried these fruits away and scatterhoarded them. Fruits were exposed to oviposition for 6 nights, after which bruchid eggs were counted on each fruit, the fruits were replaced in their original position, and the exclosure (except IH fruits) was covered with mosquito netting to prevent further oviposition by beetles. This is necessary because after a week of exposure, pulp degrades due to rotting and removal by insects, altering the initial pulp treatment. After 3–6 weeks (when larvae were large enough to be easily visible inside the endocarp), endocarps were collected and the number of bruchid larva entry holes counted. Entry holes were assessed visually, not tested with a needle as per Forget *et al.* (1994). These counts represent an estimate of the number of eggs that survive to give rise to larvae that attempt to enter the endocarp, and are thus a measure of the relative hostility of the fruit surface to egg survival. The number of larvae that actually enter is not important in this case because only one larva survives per seed. Endocarps were cracked open to determine the number of seeds and their infestation status.

Numbers of eggs and entry holes on different fruit types and on different trees were compared with one-factor ANOVAs on log or log + 1 transformed data. A one-factor ANOVA was used to compare the number of infested endocarps per fruit type for the three fruit types that had ten fruits in all exclosures (IP, G, BE). Interaction between tree and fruit type was not analysed statistically due to the unbalanced design that resulted from loss of endocarps to vertebrates; this interaction is therefore explored qualitatively. Fisher's PLSD *post hoc* tests at the 0.01 significance level (as per Silvius and Fragoso, 2002) were used to identify significant differences among pairs of fruit types or trees.

Long-term emergence exclosures

In July 2000, 29 wire-mesh exclosures containing a variable number of *A. butyracea* fruits (range 25–100, depending on the number of fruits available by the tree) were set on the ground at the edge of the natural fruit fall shadow of 20 fruiting trees. Fruit condition varied among the exclosures: seven exclosures contained Intact Husk fruits (IH), while 18 contained Partial Pulp fruits (PP), i.e. rotting fruits from which the husk and part of the pulp were manually removed in the laboratory, or ripe fruits with pulp removed by animals in the field. Each IH exclosure was paired with a PP exclosure, separated by no more than 2 cm. By December 2000, vertebrates had removed all endocarps from several exclosures, so that only four of the original seven IH exclosures remained, and 12 of the 18 PP exclosures. All exclosures that survived through December were located in 'young forest' within 2 km of the station

buildings. At this time a subset of six endocarps was removed from remaining exclosures that contained > 30 endocarps, and all endocarps from an exclosure in which only six endocarps remained. The sub-sampled endocarps were cracked open to determine infestation rates and the development stage of the larvae; these endocarps were each cut several times to expose as much of the endosperm as possible so that even very small larvae could be detected. After sub-sampling, the 15 remaining exclosures and their endocarps were left undisturbed until late April 2001, when at least ten endocarps were opened from each exclosure and examined. The proportions of each fruit type that contained adult *Pachymerus*, adult *Speciomerus*, larvae, pupa, or other material were compared for each sample period using chi-squared tests.

Results

Pulp manipulation exclosures

Vertebrates and ants removed or covered up fruits in several exclosures, and sample sizes (number of trees and number of endocarps per tree) therefore vary with fruit type (Table 4.1). Vertebrates removed nearly half of the fruits in IH exclosures. Only one of the surviving fruits received one egg after 6 nights' exposure, indicating that oviposition is delayed on IH fruits. Twenty-two of the original 50 IH endocarps survived for 3 weeks; of these, nine had low numbers of entry holes, and only one endocarp was infested (Table 4.1).

There were significant differences in the numbers of eggs ($F = 360.06$, df = 4.197, $P < 0.0001$) and entry holes ($F = 87.99$, df = 4.242, $P < 0.0001$) among the five fruit types. Egg number differed significantly between all pairs of fruit types ($P < 0.0001$; Fisher's PLSD) except between G-R and BE-R ($P = 0.025$ and 0.0131, respectively). At all trees G fruits received higher numbers of eggs than BE fruits, and BE fruits higher numbers of eggs than IP fruits (Table 4.1). The same sequence was true for entry holes, with the exception of two trees, one of which had slightly more eggs on BE than G fruits, with IP receiving the lowest number of eggs, the other of which had slightly more eggs on IP than BE fruits, with G receiving the highest number of eggs. All contrasts among fruit types in entry hole number were significantly different (Fisher's PLSD $P < 0.0001$), except for G-R ($P = 0.501$).

There was a significant difference in the mean number of eggs per fruit, averaged across all fruit types, received at the different experimental trees ($F = 4.18$, df = 5.196; $P = 0.0012$) and in the mean number of entry holes ($F = 24.79$, df = 6.240, $P < 0.0001$). *Post hoc* tests (Fisher's PLSD, at the 0.01% significance level) showed that both differences are due only to tree 69 (Fig. 4.1A,B), which had only BE and G exclosures, the

two fruit types that receive the highest number of eggs. Tree × fruit type interactions are therefore unlikely to be important in this study.

There was a significant difference in the mean number of infested endocarps per fruit type for IP, G and BE fruits (number of endocarps with at least one seed killed by a larva; $F = 4.558$, df = 2.17, $P = 0.026$). Pairs of fruit types do not differ statistically at the 0.01 significance level, but the trend is for G fruits to have a higher number of infested endocarps than IP fruits ($P = 0.0078$), with little difference between BE and G fruits ($P = 0.1961$) and BE and IP fruits ($P = 0.1029$). The effect of fruit type on larval infestation rates is more marked if only multiple-seeded endocarps are considered: 11 out of 16 multiple-seeded IP fruits had at least one

Table 4.1. Differences in oviposition, entry holes and infestation rates among the five fruit types (see text for fruit type definition).

Fruit type	Mean number of eggs ± SD (n = endocarps)	Mean number of entry holes ± SD (n = endocarps)	Mean number of endocarps infested[a] ± SD (n = trees)
IP	12.95 ± 8.5 (n = 40) range 6.3–29.9	5.23 ± 4.62 (n = 60) range 2.5–9.1	8.3 ± 1.37 (n = 6 trees) range 6–10
G	42.9 ± 19.2 (n = 50) range 34.5–72.3	19.16 ± 11.17 (n = 70) range 11.4–41	10 (n = 6 trees) range 10
BE	28.0 ± 13.3 (n = 60) range 16.8–48.7	11.0 ± 5.6 (n = 69) range 1.9–16.6	9.3 ± 1.2 (n = 7 trees) range 7–10
IH	0.038 ± 0.196 (n = 26) range 0–1	1.32 ± 1.91 (n = 22) range 0–6	10 (n = 4 trees) range 0–1.7
R	35.54 ± 11.09 (n = 26) range 15–66	19.73 ± 7.57 (n = 26) range 10–33	0.41 ± 0.85 (n = 3 trees) range 0–1.7

[a]At least one seed infested for multiple-seeded endocarps. When fewer than ten endocarps remained in exclosure, number out of ten was calculated by using the equivalency 1/actual surviving = x/10.

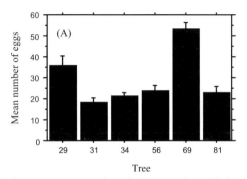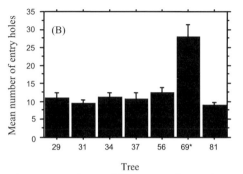

Fig. 4.1. Mean number of eggs (A) and entry holes (B) per fruit, averaged over all fruit type exclosures for each tree. The only significant difference shown by *post hoc* tests on log + 1 transformed data is between tree 69 and tree 31.

seed surviving, while 0 out of 10 and 2 out of 12 multiple seeded endocarps in G fruits and BE respectively had at least one seed surviving ($\chi^2 = 16.516$, df = 2, $P = 0.0003$).

Long-term emergence exclosures

Significantly more PP fruits were infested in December than IH fruits ($\chi^2 = 55.6$, df = 2, $P < 0.0001$). Only one IH endocarp was infested ($n = 6$ endocarps from each of four exclosures), while all but one PP endocarps sampled were either infested or had rotted; no seeds survived. Mean infestation rate per exclosure for PP fruits was 90.7%. At this time, all larvae were fully grown and had consumed all the endosperm within their endocarps. None showed the thickening and yellowing of the skin, reduction in size and reduction in mobility that characterizes the transition to the pupal stage.

In April 2001, only six exclosures (one IH, five PP) contained endocarps with exit holes, but all exclosures had infested endocarps. Only three of the 163 endocarps opened were intact (all in IH fruits); 23 had unidentifiable contents (degraded/rotten matter), and the remainder contained bruchid beetle larvae, pupae and mature beetles. Mean infestation rate for the three surviving IH exclosures was 65%, while mean infestation rate for the 12 surviving PP exclosures was 90% ($\chi^2 = 15.093$, df = 2, $P < 0.0005$).

Beetles were identified using voucher specimens previously identified by John Kingsolver. Of 76 identifiable mature and near mature beetles, 69 were *Speciomerus* and 14 were *Pachymerus*. The distribution of *Pachymerus*, *Speciomerus*, pupae and larvae among IH and PP fruits was significantly different ($\chi^2 = 70.02$, df = 3, $P < 0.0001$). All but one *Pachymerus* occurred in IH fruits, while all *Speciomerus* occurred in PP fruits. In the three IH exclosures, three pupae and four larvae remained unidentified. None of the pupae were of a size that matched the large size of pupae found in PP exclosures containing

only *Speciomerus* adults. The ratio of adults to larvae in exclosures containing *Pachymerus* was much lower (14/13) than those containing only *Speciomerus* (69/10), perhaps due to the much later oviposition by *Pachymerus* on IH fruits (post-December).

Discussion

The pulp manipulation experiments conducted at BCI with *A. butyracea* and two beetle species confirmed the findings obtained at Maracá for *A. maripa* (Silvius and Fragoso, 2002). At BCI, husks completely deterred oviposition as long as they remained intact; the presence of complete pulp reduced oviposition; and partial pulp removal resulted in high oviposition. For multiple-seeded fruits, the protection conferred by pulp alone was sufficient to ensure the survival of at least one seed in 70% of the cases. At Maracá, this effect was not noted because fruits were not protected with mosquito netting. This supports the argument by Bradford and Smith (1977) and Harms and Dalling (2000) that bruchid beetles, acting alone, exert direct selective pressure favouring multiple-seeded fruits. Note, however, that in the presence of white-lipped peccaries (*Tayassu pecari*), which open *Attalea* endocarps to feed on bruchid beetle larvae rather than seeds (Silvius, 2002), bruchid beetles enclosed in single-seeded, thick endocarps have a better chance of surviving predation by white-lipped peccaries than thinner, multiple-seeded endocarps (Fragoso, 1994). Therefore, in the presence of white-lipped peccaries, bruchid beetles should prefer single-seeded endocarps, with the indirect effect of white-lipped peccaries reversing the expected direction of selection by bruchids.

The pulp manipulation experiment discussed above shows that *Speciomerus* attacking *A. butyracea* fruits in central Panama behaves very similarly to *Pachymerus* attacking *A. maripa* fruits in northern Brazil; both avoid husks and pulp, and favour endocarps with some exposed surface but some pulp attached. The results of the long-term exclosures, however, indicate that *Pachymerus* on BCI behaves very differently from *Pachymerus* at Maracá, occurring in Intact Husk fruits but not in fruits with the husk and some pulp removed.

The small sample sizes, due to loss of more than half the IH exclosures, and the location of all IH exclosures within 500 m of each other make it difficult to interpret the results with regard to species separation. Nevertheless, the near complete separation of *Pachymerus* and *Speciomerus* by fruit type suggests that the two beetle species partition the *Attalea* seed resource by fruit condition. *Pachymerus* but not *Speciomerus* appears able to use IH fruits. *Speciomerus* may not be able to use Intact fruits because: (i) its eggs will not stick on the husk; (ii) the first instar larvae cannot carve through the husk; (iii) the last instar larvae cannot carve through the husk when making their preliminary exit hole; or (iv) ovipositing adults do not recognize IH fruits as appropriate laying substrates (i.e. a pulp-associated cue is absent in IH fruits).

On the other hand, it is unclear whether *Pachymerus* ignores PP fruits on BCI altogether, is out-competed by the larger *Speciomerus* at these fruits, or oviposits later in the year when the only uninfested fruits available are IH fruits. The fact that *Pachymerus* emerges at approximately the same time as *Speciomerus* in early fruiting season indicates that adults of both species are present in the forest during the early to peak fruit fall period. Competition for fruits between the two species is very likely, as both species are restricted to *Attalea* palms at BCI (Wright, 1990), and many more eggs are laid on each endocarp during the peak laying season (up to 70 eggs per endocarp; Table 4.1) than will survive to adulthood in the endocarp. Endocarps continue to receive eggs during the first 2 weeks of exposure, indicating that different females continue to lay on endocarps that already have many eggs on them (Silvius, 1999), although they selectively lay on endocarps with fewer eggs (Wright, 1983). In addition to size, behavioural differences may account for the inequality in competitive ability between

the two species. When endocarps were cracked open, any near-adult or adult *Speciomerus* responded very actively to the disturbance, crawling out from the fragments, perching on the tips of high fragments or other objects, and struggling when held. Adult or near-adult *Pachymerus*, on the other hand, either lay still or crawled to protected, covered corners of the container, often hiding under intact endocarps.

Timing of beetle emergence

Contrary to Wright's (1990) finding that peak beetle oviposition occurs in the early to peak fruiting period, and no oviposition occurs late in the season, the present study suggests that if the right kinds of fruits are available (i.e. intact fruits) then these may receive bruchid (*Pachymerus*) eggs outside the fruiting season. *Attalea* trees at BCI show consistency at the time of fruiting, and trees that fruit early or late in the season do so year after year (Wright, 1990). Given a fixed development time (longer for *Speciomerus* than *Pachymerus*), at least *Speciomerus* beetles can 'expect' to emerge at the right time to oviposit at their individual host tree, as long as it fruits that year. For example, beetles emerged from one early fruiting tree in March 2000, and these newly emerged beetles were already ovipositing on the same tree's new fruit crop in April 2000. Beetles developing in fruits dropped later in the year will have to wait to complete their development time before they can emerge. Thus, there will be continuous emergence of beetles up until the peak of the fruiting season.

Mammal community structure and beetle populations

Niche differentiation in bruchids as a function of the condition of the fruit is not unexpected, and has been documented by Johnson (1981) for legume bruchids. In the case of *Attalea* on BCI, however, differentiation is linked to the manipulation of fruits by animals, rather than to the unassisted release of seeds from fruits, as in legumes. With changes in the animal community, the availability of intact versus manipulated fruits should also change, potentially influencing the abundance of *Pachymerus* relative to *Speciomerus*. Given that resource use and competitor avoidance can evolve rapidly (Thompson, 1998; Kinnison and Hendry, 2001; Reznick and Ghalambor, 2001), the possibility exists for the two beetle species to show differing cycles of abundance, perhaps influenced by changes in the frugivore community.

At the scale of one or a few trees, the abundance of IH fruits can be determined by the relative frequency of fruiting trees. When several neighbouring trees drop fruits simultaneously in high *Attalea* density habitat, not all fruits will be removed by mammals and some remain as intact fruits on the ground. Trees that fruit at a time when food is abundant, or are located near preferred food sources, may also retain the husk for long periods of time. These factors will result in spatial patchiness in Intact Husk availability regardless of the composition of the frugivore community (Silvius, 2002).

At a larger scale, e.g. all of BCI or one of its peninsulas, the abundance of IH fruits can be determined by the overall abundance of key animals that strip the fruits of their husk, i.e. squirrels, agoutis, primates, pacas, coatis, deer (*Mazama* spp.) or collared peccaries (*Pecari tajacu*). Differences in fruiting at the population scale could have the same effect – palm populations that have more extended fruiting seasons, so that more fruits are available at times of year when mammals are not using them, could also give rise to higher availability of intact fruits in the environment. This scale is more likely to affect the relative abundance of the two beetle species, mediate competition and lead to niche differentiation over evolutionary time.

Wright *et al.* (2000) and Wright and Huber (2001) found decreased seed dispersal for *Attalea* palms at small isolated islands and defaunated mainland sites near BCI when compared to BCI. This difference is correlated with lower densities of dispersers and predators at mainland sites experiencing hunting and at habitat fragments with depauperate frugivore faunas. Changes in animal communities at this scale, due to anthropogenic disturbance, could imitate natural patchiness in fruit type availability at a large scale, and lead to patchiness in the relative abundance of the two beetle species in sites with different degrees of defaunation. For example, islands lacking all mammals but spiny rats and opossums (*Didelphis* spp.) should have a high availability of IH fruits relative to mammal-manipulated fruits, and may be favourable to *Pachymerus*, allowing it to out-compete *Speciomerus*.

References

Andresen, E. (1999) Seed dispersal by monkeys and the fate of dispersed seeds in a Peruvian rain forest. *Biotropica* 31, 145–158.

Bradford, D.F. and Smith, C.C. (1977) Seed predation and seed number in *Scheelea* palm fruits. *Ecology* 58, 667–673.

Bucholz, R. and Levey, D.J. (1990) The evolutionary triad of microbes, fruits, and seed dispersers: an experiment in fruit choice by cedar waxwings, *Bombycilla cedrorum*. *Oikos* 59, 200–204.

Cintra, R. (1997) A test of the Janzen–Connell model with two common tree species in Amazonian forest. *Journal of Tropical Ecology* 13, 641–658.

Connell, J.H. (1971) On the role of natural enemies in preventing competitive exclusion in some marine animals and in rain forests. In: den Boer, P.J. and Gradwell, G.R. (eds) *Dynamics of Populations*. Centre for Agricultural Publishing and Documentation, Wageningen, The Netherlands, pp. 298–312.

Delgado, C., Couturier, G. and Delobel, A. (1997) Oviposition of seed-beetle *Caryoborus serripes* (Sturm) (Coleoptera: Bruchidae) on palm (*Astrocaryum chambira*) fruits under natural conditions in Peru. *Annals of the Society of Entomology, France (N.S.)* 33, 405–409.

Delobel, A., Couturier, G., Khan, F. and Nilsson, J.A. (1995) Trophic relationships between palms and bruchids (Coleoptera: Bruchidae: Pachymerini) in Peruvian Amazonia. *Amazoniana* 8, 209–219.

De Steven, D., Windsor, D.M., Putz, F.E. and de León, B. (1987) Vegetative and reproductive phonologies of a palm assemblage in Panama. *Biotropica* 19, 342–356.

Fleming, T.H. and Estrada, A. (1993) *Frugivory and Seed Dispersal: Ecological and Evolutionary Aspects*, Vol. 107/108. Kluwer Academic Publishers, Dordrecht, The Netherlands, 392 pp.

Forget, P.-M. (1990) Seed dispersal of *Vouacapoua americana* (Caesalpiniaceae) by caviomorph rodents in French Guiana. *Journal of Tropical Ecology* 6, 459–468.

Forget, P.-M. (1992) Seed removal and seed fate in *Gustavia superba* (Lecythidaceae). *Biotropica* 24, 408–414.

Forget, P. (1993) Post-dispersal predation and scatterhoarding of *Dipteryx panamensis* (Papillonaceae) seeds by rodents in Panama. *Oecologia* 94, 255–261.

Forget, P.-M. (1996) Removal of seeds of *Carapa procera* (Meliaceae) by rodents and their fate in rainforest in French Guiana. *Journal of Tropical Ecology* 12, 751–761.

Forget, P.-M., Muñoz, E. and Leigh, E.G. Jr (1994) Predation by rodents and bruchid beetles on seeds of *Scheelea* palms on Barro Colorado Island, Panama. *Biotropica* 26, 420–426.

Foster, R.B. and Brokaw, N.V.L. (1982) Structure and history of the vegetation of Barro Colorado Island. In: Leigh, E.G. Jr, Rand, A.S. and Windsor, D.M. (eds) *Ecology of a Tropical Forest: Seasonal Rhythms and Long-Term Changes*. Smithsonian Institution Press, Washington, DC, pp. 67–82.

Fragoso, J.M.V. (1994) Large mammals and the community dynamics of an Amazonian rain forest. PhD thesis, University of Florida, Gainesville, Florida.

Fragoso, J.M.V. (1997) Tapir-generated seed shadows, scale-dependent patchiness in the Amazon rain forest. *Journal of Ecology* 85, 519–529.

Fragoso, J.M.V., Silvius, K.M. and Correa, J.A. (2003) Long distance seed dispersal by tapirs increases seed survival and aggregates tropical trees. *Ecology* 84, 1998–2006.

García, D., Zamora, R., Gómez, J.M. and Hódar, J.A. (1999) Bird rejection of unhealthy fruits reinforces the mutualism between juniper and its avian dispersers. *Oikos* 85, 536–544.

Giacalone-Madden, J., Glanz, W.E. and Leigh, E.G. Jr (1990) Fluctuaciones poblacionales a largo plazo de *Sciurus granatensis* en relación con la disponibilidad de frutos. In: Leigh, E.G. Jr, Rand, A.S. and Windsor, D.M. (eds) *Ecología de un Bosque Tropical*.

Smithsonian Institution Press, Washington, DC, pp. 331–336.

Harms, K.E. and Dalling, J.W. (2000) A bruchid beetle and a viable seedling from a single diaspore of *Attalea butyracea*. *Journal of Tropical Ecology* 16, 319–325.

Henderson, A. (1995) *The Palms of the Amazon*. Oxford University Press, New York, 388 pp.

Herrera, C.M. (1982) Defense of ripe fruit from pests: its significance in relation to plant–disperser interactions. *American Naturalist* 120, 218–247.

Herrera, C.M. (1984) Avian interference of insect frugivory: an exploration into the plant–bird–fruit pest evolutionary triad. *Oikos* 42, 203–210.

Herrera, C.M. (1986) Vertebrate-dispersed plants: why they don't behave the way they should. In: Estrada, A. and Fleming, T.H. (eds) *Frugivores and Seed Dispersal*. Dr W. Junk Publishers, Dordrecht, The Netherlands, pp. 5–18.

Herrera, C.M. (1989) Vertebrate frugivores and their interaction with invertebrate fruit predators: supporting evidence from a Costa Rican dry forest. *Oikos* 54, 185–188.

Howe, H.F. and Smallwood, J. (1982) Ecology of seed dispersal. *Annual Review of Ecology and Systematics* 13, 201–228.

Janzen, D.H. (1970) Herbivores and the number of tree species in tropical forests. *American Naturalist* 104, 501–528.

Janzen, D.H. (1971) The fate of *Scheelea rostrata* fruits beneath the parent tree: predispersal attack by bruchids. *Principes* 15, 89–101.

Janzen, D.H. (1972) Predation on *Scheelea* palm seeds by bruchid beetles: seed density and distance from the parent palm. *Ecology* 53, 954–959.

Janzen, D.H. (1982) Simulation of *Andira* fruit pulp removal by bats reduces seed predation by *Cleogonus* weevils. *Brenesia* 20, 165–170.

Johnson, C.D. (1981) Interactions between bruchid (Coleoptera) feeding guilds and behavioural patterns of pods of the Leguminosae. *Environmental Entomology* 10, 249–253.

Johnson, C.D., Zona, S. and Nilsson, J.A. (1995) Bruchid beetles and palm seeds: recorded relationships. *Principes* 39, 25–35.

Kinnison, M.T. and Hendry, A.P. (2001) The pace of modern life II: from rates of contemporary microevolution to pattern and process. *Genetica* 112/113, 145–164.

Leigh, E.G. Jr, Rand, A.S. and Windsor, D.M. (1982) *Ecology of a Tropical Forest: Seasonal Rhythms and Long-Term Changes* (2nd edn).

Smithsonian Institution Press, Washington, DC, 480 pp.

Nilsson, J.A. and Johnson, C.D. (1993) A taxonomic revision of the palm bruchids (Pachymerini) and a description of the world genera of Pachymerinae (Coleoptera: Bruchidae). *Memoirs of the American Entomological Society* 41, 1–104.

Peres, C., Schiesari, L.C. and Dias-Leme, C.L. (1997) Vertebrate predation of Brazil-nuts (*Bertholletia excelsa*, Lecythidaceae), an agouti-dispersed Amazonian seed crop: a test of the escape hypothesis. *Journal of Tropical Ecology* 13, 69–79.

Quiroga-Castro, V.D. and Roldán, A.I. (2001) The fate of *Attalea phalerata* (Palmae) seeds dispersed to a tapir latrine. *Biotropica* 33, 472–477.

Reznick, D.N. and Ghalambor, C.K. (2001) The population ecology of contemporary adaptations: what empirical studies reveal about the conditions that promote adaptive evolution. *Genetica* 112/113, 183–198.

Sallabanks, R. and Courtney, S.P. (1992) Frugivory, seed predation, and insect–vertebrate interactions. *Annual Review of Entomology* 37, 377–400.

Silvius, K.M. (1999) Interactions among *Attalea* palms, bruchid beetles, and Neotropical terrestrial fruit-eating mammals: implications for the evolution of frugivory. PhD thesis, University of Florida, Gainesville, Florida.

Silvius, K.M. (2002) Spatio-temporal patterns of palm endocarp use by three Amazonian forest mammals: granivory or 'grubivory'? *Journal of Tropical Ecology* 90, 1024–1032.

Silvius, K.M. and Fragoso, J.M.V. (2002) Pulp handling by vertebrate seed dispersers increases palm seed predation by bruchid beetles in the northern Amazon. *Journal of Ecology* 90, 1024–1032.

Smythe, N., Glanz, W.E. and Leigh, E.G. Jr (1982) Population regulation in some terrestrial frugivores. In: Leigh, E.G. Jr, Rand, A.S. and Windsor, D.M. (eds) *Ecology of a Tropical Forest: Seasonal Rhythms and Long-Term Changes.* Smithsonian Institution Press, Washington, DC, pp. 227–238.

Terborgh, J., Losos, E., Riley, M.P. and Bolaños Riley, M. (1993) Predation by vertebrates and invertebrates on the seeds of five canopy tree species of an Amazonian forest. *Vegetatio* 107/108, 375–386.

Thompson, J.N. (1998) Rapid evolution as an ecological process. *Trends in Ecology and Evolution* 13, 329–332.

Traveset, A. (1992) Effect of vertebrate frugivores on bruchid beetles that prey on *Acacia farnesiana* seeds. *Oikos* 63, 200–206.

Traveset, A. (1993) Weak interactions between avian and insect frugivores: the case of *Pistacia terebinthus* L. (Anacardiaceae). *Vegetatio* 107/108, 191–203.

Wenny, D.G. and Levey, D.J. (1998) Directed seed dispersal by bellbirds in a tropical cloud forest. *Proceedings of the National Academy of Sciences USA* 95, 6204–6207.

Wilson, D.E. and Janzen, D.H. (1972) Predation on *Scheelea* palm seeds by bruchid beetles: seed density and distance from the parent palm. *Ecology* 53, 954–959.

Wright, S.J. (1983) The dispersion of eggs by a bruchid beetle among *Scheelea* palm seeds and the effect of distance to the parent palm. *Ecology* 64, 1116–1021.

Wright, S.J. (1990) Cumulative satiation of seed predator over the fruiting season of its host. *Oikos* 58, 272–276.

Wright, S.J., Gompper, M.E. and DeLeon, B. (1994) Are large predators keystone species in Neotropical forests? The evidence from Barro Colorado Island. *Oikos* 71, 279–294.

Wright, S.J. and Huber, H.C. (2001) Poachers and forest fragmentation alter seed dispersal, seed survival and seedling recruitment in the palm *Attalea butyracea*, with implications for tropical tree diversity. *Biotropica* 33, 583–595.

Wright, S.J, Zeballos, H., Dominguez, I., Gallardo, M.M., Moreno, M.C. and Ibañez, R. (2000) Poachers alter mammal abundance, seed dispersal and seed predation in a Neotropical forest. *Conservation Biology* 14, 227–239.

5 Patterns of Seed Predation by Vertebrate versus Invertebrate Seed Predators among Different Plant Species, Seasons and Spatial Distributions

Evan M. Notman[1] and Ana C. Villegas[2]

[1]*Organization for Tropical Studies, Apartado 676-2050, San Pedro, Costa Rica;*
[2]*US State Department, Bureau of Oceans, Environmental, and Scientific Affairs, Washington, DC, USA*

Introduction

Seed predation has been shown to be a major source of seed mortality in a wide range of habitats (see Hulme and Kollmann, Chapter 2, this volume). Animals that consume seeds are, however, a diverse group, and differences among predators may have significantly divergent impacts on patterns of seed survival. Changes in the seed predator community either spatially or temporally can therefore potentially result in changes in the plant community (Brown and Heske, 1990; Davidson, 1993; Leigh et al., 1993; Terborgh and Wright, 1994; Asquith et al., 1997; Wright et al., 2000; Silman et al., 2003).

Although considerable attention has been given to assessing the impact of seed predators on plant recruitment by examining the overall levels of seed predation for single or multiple species across different conditions such as habitat (Notman et al., 1996; Hulme and Kollmann, Chapter 2, this volume), temporal (Wright, 1983; Forget et al., 1999) or spatial distribution pattern of seeds (Augspurger and Kitajima, 1992; Burkey, 1994; Akashi, 1997), fewer studies

have attempted to separate the relative importance of multiple predators under these different conditions (see Tamura et al., Chapter 14, this volume).

In this chapter we examine how the relative importance of two broad groups of seed predators, insects and mammals, changes depending on seed species, habitat, and spatial and temporal distribution of seeds. Despite the great diversity of insect and mammalian seed predators, general morphological, physiological and behavioural traits separate these two classes of predators. For instance, vertebrates and invertebrates differ in their range of sizes; how they search for, select and treat seeds; and how they are affected by disturbance (Hulme and Kollmann, Chapter 2, this volume). Although differential seed predation by insects and mammals across the full range of seed sizes in the community may have important consequences for plant communities (e.g. Brown and Heske, 1990), our comparison will focus only on seeds large enough to be attacked by both of these groups. We present the results of studies comparing seed predation by insects and mammals across a range of different conditions aiming to answer the following questions:

1. Does seed predation by insects and mammals change within and between forests depending on the timing of seed dispersal with respect to seasonal differences in climate and fruit production?

2. How does the distance of seeds from parent trees influence insect and mammal seed predators?

3. How is insect and mammal seed predation influenced by forest type and level of disturbances?

4. Are within-forest differences in insect and mammal seed predation consistent among different plant species?

We first compared patterns of insect and mammal seed predation on *Socratea exorrhiza* (hereafter *Socratea*) a single large-seeded palm species, for seeds dispersed at different times of the year with seasonal differences in climate and fruit production.

Differences in the generation times of mammals and insects may influence how they respond to temporal changes in the availability of resources. For instance, insects typically have shorter generation times than mammals, and may be able to adjust their population size more rapidly to respond to temporal changes in the abundance of resources. Mammals and insects also differ in the degree to which they specialize on particular resources. Vertebrate herbivores and seed predators are, with a few exceptions, generalists consuming a wide range of seeds from a taxonomically diverse range of plants, as well as other plant or animal resources (Janzen, 1971; Freeland and Janzen, 1974; Dearing *et al.*, 2000; Hulme and Kollmann, Chapter 2, this volume). In contrast, specialization on a single or a few closely related species is probably quite common among invertebrates (Janzen, 1969, 1970; Bernays and Chapman, 1994).

We expected to see greater seed predation by mammals for seeds produced at the beginning of the peak in community, and *S. exorrhiza*, fruit production when mammals take advantage of relatively abundant resources after a period of resource

scarcity and broad dietary generalists such as peccaries switch from other resources to fruits and seeds (Forget *et al.*, 2002; see Beck, Chapter 6, this volume). We also expected a decrease in seed predation, as well as a possible increase in seed caching to occur during the mid-peak of fruit production due to satiation, followed by an increase in seed predation at the end of the fruiting peak when resource availability declines (Forget *et al.*, 2002). In contrast we predicted that insect infestation would be lowest at the beginning of the fruiting peak when populations of specialized insect predators are still low, and predation would increase throughout the peak of fruit production as insect populations build up.

Second, we compared seed predation by insects and mammals on *Socratea* seeds in relation to distance from parent trees in two contrasting forests with distinct patterns of seasonality (Costa Rica and Panama). We expected to find higher levels of seed predation by insects closer to parent trees, as insects are likely to use parent trees as cues to find resources (Janzen, 1972; Hammond and Brown, 1998). We also expected predation, or alternatively seed caching, by mammals to increase near parent trees because of the higher density of seeds around the base of parents (Janzen, 1972). We also predicted that in Panama, where peak fruit production is more synchronized and occurs within a more narrow time period, mammals would cache a larger proportion of seeds and also differ more in the relative proportion of seeds cached between the mid- and late fruiting peaks than in La Selva.

Finally we compared seed predation on multiple species by insects and mammals between mature and disturbed forests. Mammals and insects may differ in their susceptibility to human disturbance for several reasons. Faster generation times may allow many insects populations to recover quickly after disturbances, while many vertebrates, particularly larger mammals may take years to recover from a large reduction in population. Vertebrates are also more likely to suffer from human disturbances due to larger minimum foraging areas and increased risk from hunting. Local changes in environmental conditions due to human disturbance (e.g. agricultural clearings and selective logging) may also affect the abundance and behaviour of insects and mammal seed predators differentially. For instance small mammal seed predators have been shown to avoid open areas without understorey where they are more visible to predators (Hulme and Kollmann, Chapter 2, this volume) and insects may be more susceptible to changes in humidity and temperature (Didham, 1997).

Based on previous studies of seed predation in a variety of habitats (e.g. De Steven, 1991; Terborgh *et al.*, 1993; Hammond, 1995; Asquith *et al.*, 1997; Holl and Lulow, 1997; Hulme, 1997; Blate *et al.*, 1998; Guariguata *et al.*, 2000; Notman and Gorchov, 2001; Westerman *et al.*, 2003), we predicted to find generally greater seed predation by mammals than insects. However, we expected to find greater differences in predation between mature and disturbed forests for insects than for mammals because insects are more likely to be specialists, and more likely to attack seeds in habitats where adult host plants occur. In contrast, we expected mammal predation to be more similar between forest types since they are mostly generalist seed predators.

Methods

Does seed predation by insects and mammals change within forest sites with fruiting season and distances from parents? (Experiment 1)

Study species

Socratea exorrhiza is a sub-canopy to canopy palm widely distributed from Nicaragua to Brazil from 0 to 1000 m in elevation (Henderson *et al.*, 1995). Fruits are ellipsoid and about 2–4 cm in diameter containing a single ellipsoid seed surrounded by dense starchy white mesocarp. We have observed the fruits being eaten from trees in one or both of the study areas by white-throated capuchin monkeys (*Cebus capucinus*), Central American spider monkeys (*Ateles

geoffroyi*), red-tailed squirrel (*Sciurus granatensis*), keel-billed toucan (*Ramphastos sulfuratus*), chestnut-mandibled toucan (*Ramphastos swainsonii*) and yellow-eared toucanet (*Selenidera spectabilis*). We have also observed fallen fruits and/or seeds being eaten by collared peccary (*Pecari tajacu*), and Central American agouti (*Dasyprocta punctata*). Teeth marks on seeds also indicate that the seeds are eaten by several smaller rodents, including the spiny rat (*Proechimys semispinosus*). Seeds are generally quickly removed or eaten from under trees although occasionally large numbers of seeds can be found at the base of individual trees (Notman and Villegas, personal observation).

Beetles in the genus *Coccotrypes* (Scolytidae) infest seeds of *Socratea* and other tropical palms. *Coccotrypes* females are small beetles (approximately 1 mm in length) that disperse from their natal seed to a new seed where they lay multiple eggs and form a new colony (Kirkendall *et al.*, 1997). Over 100 offspring may be produced from a single infested seed. We have observed seed infestation by *Coccotrypes* only after seeds reach the ground although it is possible that very low levels of pre-dispersal infestation also occur. The likelihood of survival for infested seeds largely depends on the time and location of the point of infestation. Infestation that occurs near to the embryo, or through the point of germination, often leads to rapid mortality while those infested through the endosperm or the side of seeds can potentially remain alive for months and may occasionally be able to germinate and establish (Notman and Villegas, unpublished data).

Study sites

This study was conducted at La Selva Biological Station (hereafter 'La Selva') owned by the Organization for Tropical Studies in the Caribbean lowlands of Costa Rica (10°26′N, 84°00′W, elevation 37–150 m). La Selva has an annual rainfall of 4010 mm with two rainfall peaks of more than 400 mm in June–July and November–December. The dry season is relatively short and mild (i.e. rainfall rarely less than 100 mm per month) lasting from January to April or May (Sanford et al., 1994). Rainfall during 2001 was fairly typical with respect to this pattern. For a more detailed site description see Hartshorn (1983).

A community-wide increase in fruit abundance occurs from mid- to late rainy season with canopy species fruiting in October and November (Frankie et al., 1974) and understorey plants between August and January (Levey, 1988). The number of species fruiting also varies during the year with the fewest understorey species fruiting in June at the end of the dry season (Levey, 1988). Many species, however, produce fruits over an extended period (two or more consecutive seasons) and thus even in times of low fruit production there is a minimal baseline source of food for frugivores at La Selva (Frankie et al., 1974).

Arborescent palms have been found to account for 25.5% of all stems greater than 10 cm DBH and 8% of the basal area in permanent plots in mature forest at La Selva (Lieberman et al., 1985). Density of adult *Socratea* ranges from 5 to 23 stems per hectare depending on soil type at La Selva (Clark et al., 1995).

Experimental design

FRUIT PRODUCTION   At La Selva fruit production was monitored on 92 adult *Socratea* trees within four plots (95 m × 60 m) located in mature forest, similar residual volcanic soils and slope 1–10 degrees. Plots ranged from 600 m to 2000 m away from one another. The presence of racemes with mature fruits was documented for 2 years almost every month from February 2001 to May 2003 (Fig. 5.1). Fruit production

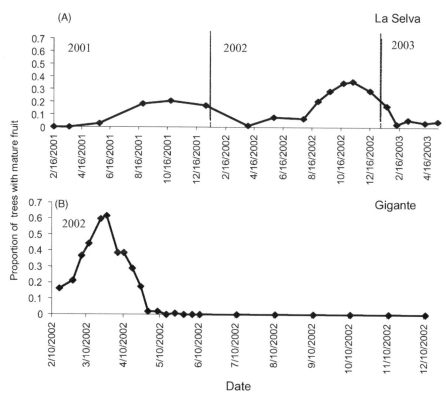

Fig. 5.1.   Fruiting of *Socratea exorrhiza*. The percentage of *S. exorrhiza* adult trees with racemes with mature seeds was recorded monthly for (A) 92 palms at La Selva Biological Station, Costa Rica for 2 years and (B) 53 palms at Gigante Peninsula, Panama for 1 year.

estimates were based on the number of palm trees with mature racemes and not on the number of mature fruits/racemes per tree because abundance estimates for mature fruits changed too rapidly due to fruits falling or being eaten.

SEED FATE   In order to evaluate the interaction of time of seed dispersal and predator type on seed mortality we placed *Socratea* seeds in the field at three different times of the year at La Selva (hereafter placement dates). We chose placement dates that correspond with periods of contrasting fruit abundance and rainfall to explore the role of resource availability on the relative role of seed predator type. We experimentally dispersed recently produced seeds on 6 February 2001 (low community fruit production and the start of the dry season), 7 September 2001 (high community fruit production and a short period of lower rainfall during the wet season, 'veranillo'); and 7 December 2001 (mid-to-late peak of community fruit production and high rainfall towards the end of the wet season). The time of *Socratea* fruit production coincides with the community fruiting pattern.

At each placement date 160 seeds were collected from between 13 and 28 trees. Only newly fallen seeds (i.e. seeds with fresh fruit pulp or with a shiny colour and slippery feel) were used in this experiment. Seeds that had holes (i.e. sign of insect infestation) or that floated were rejected. Seeds were placed in the forest within 10 days of collection. During the first placement date, however, the majority of seeds used had a single small hole (evidence of bark beetle infestation), probably because these seeds were mixed together during storage with older infested seeds that may have acted as a source of infestation. For the rest of the experiments seeds were stored in separate plastic bags to prevent infestation before placement in the forest.

Seeds were placed in the forest within the same four plots used to monitor fruit production. Within each plot seeds were placed along three parallel 95 m transects separated by 10 m with a different transect being used for each placement date. Pairs of seeds were placed 1 m from each other, perpendicular to the transect and 5 m from the next pair along each transect. For each pair of seeds, one seed was placed without protection (open treatment), and one seed was protected from vertebrate predators (exclosure treatment). The exclosure treatment consisted of a 20 cm radius cylinder cage made of hardware cloth (1 cm mesh) placed over single seeds. The base of each cage was buried 2–3 cm in the soil and staked down with galvanized wire stakes.

In order to follow the fate of control seeds and measure secondary dispersal each seed was glued with instant bonding glue ('Superbonder'™) to a thin, white nylon thread spooled on a plastic bobbin suspended on a wire inside a plastic film container. Glue was also placed on seeds in the exclosure treatment to control for any effect of the glue. Seeds in both treatments were covered with a nearby fallen leaf to ensure that they were exposed to similar conditions.

Seeds were checked every 3 days for the first 3 weeks, then weekly for 5–7 weeks, then bi-weekly for at least 6 weeks and monthly for the remaining study time period. At each census we checked if seeds had been eaten, moved, infested by beetles or had germinated. All seeds that were removed from their threads were assumed eaten. Moved seeds that were not eaten were left in place and the new location marked. Seeds that appeared to be dead (i.e. signs of rot, heavy beetle infestation, blackened or withered germination tip) were taken out from the field.

Statistical analysis

Final seed mortality could not be analysed statistically because very few seeds survived after a year. Rate of seed mortality was compared among placement dates (February, September and December 2001) and protection treatment (open vs. exclosure) using a Cox proportional hazard model. A proportionate hazard model was used because in this experiment treatments can change the chance of mortality but not the time of highest mortality risks. Protection treatment and date of placement were included in the model as independent categorical variables together with all inter-action terms. Non-significant interaction terms were dropped for the final analysis.

The model was right-censored to account for any seeds still surviving after 1 year (Fox, 1999). Six seeds in the exclosure treatment were eaten after their protective cages were tipped over; these seeds were also censored at the date on which they were eaten. The Cox model calculates the survival function, which represents the proportion of the initial seed population still present at each successive census. The parameters of the distribution and the regression coefficients are determined using maximum likelihood methods (Fox, 1999). This procedure combines repeated censuses into a single test, and correctly incorporates right-censored data.

Separate pairwise comparisons of each placement date/treatment combination were done using the Mantel–Haenzel test (log-rank test) to compare survival functions (i.e. equality of survival curves) in order to determine which placement dates were significantly different from each other (Fox, 1999). All statistical analysis was done using JMP 3.2 statistical software (SAS Institute, 1989).

Does seed predation by insects and mammals change between forest sites with fruiting season and distances from parents? (Experiment 2)

Study sites

This study was carried out at La Selva Biological Station and Gigante Peninsula (hereafter 'Gigante'), a field site run by the Smithsonian Institute within the Barro Colorado Nature Monument (BCNM) in the Pacific lowlands of Panama. Gigante Penin-sula is within 1 km of Barro Colorado Island (BCI), which receives an average annual rainfall of 2700 mm in a well-marked bimodal distribution, with a strong dry sea-son usually occurring from late December until April, during which average monthly

rainfall is less than 100 mm (Croat, 1978; Leigh et al., 1982; Gentry, 1990). BCI exhib-its strong periodicity in flowering and fruit-ing coincident with seasonal alterations of dry and rainy seasons (Foster, 1982; Garwood, 1983; Augspurger, 1985; Leigh, 1999). The plant community has two fruit-ing peaks, the first peak in April–May (both wind- and animal-dispersed species) and the second one during September–October (animal-dispersed species only) (Foster, 1982). A period of food scarcity for frugi-vorous animals extends from November, late in the rainy season, to February, the middle of the dry season (Smythe, 1970; Foster, 1982).

Socratea is also common at Gigante and appears to have a somewhat more clumped distribution than at La Selva (personal observation). Although hunting is prohib-ited at Gigante, hunting does occur outside the borders of the park and possibly also occasionally within (Wright et al., 2000). Frequency of one important seed predator, the collared peccary, appears to be much lower at Gigante than La Selva based on frequency of observation (personal observation).

Experimental design

FRUIT PRODUCTION   Data from Experiment 1 (see above) were used to estimate fruit production at La Selva. At Gigante fruit production was monitored on 53 Socratea adult trees along the two main trails in the Gigante Peninsula using the same method as at La Selva. Fruit production was measured at least once every month from February 2001 to December 2002 (Fig. 5.2).

SEED FATE   Seeds were collected for this experiment following the same procedures described in Experiment 1. Ten fruiting Socratea trees were chosen as focal trees. To minimize the effects of human distur-bances and fruiting conspecifics, all focal trees were located farther than 10 m from the trails and other Socratea fruiting palms. At La Selva all trees were selected in areas of primary forest with volcanic soil and flat topography. In Gigante all selected

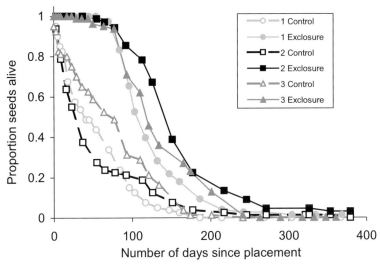

Fig. 5.2. Experiment 1. Proportion of *Socratea exorrhiza* seeds alive. Seeds were placed in the field at La Selva at three different times in 2001 (*n* = 160 each time): (1) end of fruiting season and beginning of dry season (February); (2) peak of fruiting season and 'veranillo', a moderate dry time in the middle of the wet season (September); (3) declining fruiting season and end of the rainy season (December). Each time half the seeds were placed either under a protection from mammals in an 'exclosure' treatment or exposed to all predators.

trees were located in mature forest and approximately 3 km from each other but not less than 20 m apart.

In order to test for the effect of forest site, at La Selva, seeds were placed in the field on 10 October 2001 (peak of *Socratea* fruit production) and 8 February 2002 (end of *Socratea* fruit production). In Gigante seeds were placed on 20 March 2002 (start of *Socratea* fruit production) and 2 May 2002 (middle of *Socratea* fruit production). Tree selection followed similar procedures in both sites, except that on 2 May only nine focal trees were used at Gigante. In addition, in order to test for the effect of distance to parent trees, at La Selva in February 2001 and in Gigante in March and May 2002 seeds were placed at each of the focal trees, 20 seeds were placed in a circle of approximately 1.25 m radius at both a near (1 m) and far (10 m) distance treatments. Seeds in the near treatment were placed with the focal tree as the centre point of the circle while seeds in the far treatment were placed in a circle 10 m from the focal tree or any other conspecific adult tree. At each distance, seeds were alternately placed in the open

and exclosure treatment, ten seeds in each treatment.

All seeds placed in the field were checked at least every other week for 6 weeks and then monthly until 1 year from time of placement. Seeds placed at La Selva in October 2002 were only placed in the near distance treatment and only the seeds in the control treatment were included in the analysis. In October fewer stakes were used to secure the exclosure treatment cages and over 70% of the cages were knocked over and the seed was removed. Seeds placed at Gigante were checked every week for 13 weeks and monthly after that for a year.

Statistical analysis

At Gigante, levels of seed mortality after 264 days were compared among placement dates, protection treatment and distance from focal tree using logistic regression. Seed survival was modelled as a binomial response (alive or dead). Tree was included in the analysis as a categorical variable to examine the effect of local spatial variation. Placement date, protection treatment, distance and tree were all included as categorical predictor variables. All two- and three-way interactions were included in the initial model and removed if not significant. At La Selva survival after 1 year was near zero in almost all treatments, thus the use of any statistical procedures to compare final levels of seed mortality was not possible.

We analysed the effect of site (La Selva vs. Gigante), date of placement (October 2001 vs. February 2002 in La Selva and March 2002 vs. May 2002 in Gigante), distance from parent tree (1 m vs. 10 m), and protection treatment (open vs. exclosure) on the rate of seed mortality using a Cox proportional hazard model.

We ran a total of five separate analyses; in La Selva we compared the effect of distance and treatment for those seeds placed in February 2002, and we compared placement dates only for seeds in the open treatment near the parent tree because the effect of distance and vertebrate exclusion were not included in October 2001.

In Panama we analysed the effect of date of placement, distance from parent tree and protection treatment on seed mortality. Finally, we compared the rate of seed mortality between sites, protection treatment, date of placement, distance, and protection treatment in two separate comparisons. We compared mortality rate in Costa Rica in February 2002 to both placement dates in Panama (March 2002 and May 2002) because survival in Panama between these two dates was significantly different. We did not include seeds placed in October 2001 in Costa Rica in the analysis because neither distance nor vertebrate exclusion was tested during this period.

How are insect and mammal seed predators influenced by forest disturbance and plant species? (Experiment 3)

Details of the study site and methods of this study have been published elsewhere (Notman and Gorchov, 2001) and will therefore be only briefly summarized here.

In this study we compared seed predation by insects and mammals for 26 species of trees and lianas (see Table 5.4) in mature upland forest and areas of young forest regenerating after swidden agriculture (fallowed for 2.5–3 years). The study was conducted in and around the area of the Centro de Investigaciones Jenaro Herrera (CIJH), approximately 140 km south of Iquitos, Peru.

Seeds were placed along eight transects in four plots per forest type in two separate treatments, unprotected and protected from mammals. Unprotected seeds were placed on the ground and covered by a thin layer of litter. Seeds in the protected treatment were treated similarly but were covered by a cage constructed of 1.27 cm mesh hardware cloth. Four replicate treatment pairs of each species were placed in each plot. For 25 of the 26 species, five seeds were placed in each treatment replicate, for the largest-seeded species (*Licania* sp. 1) only three seeds were used per treatment.

In order to ensure that seeds removed from the open treatment were in fact eaten and not cached, a 50 cm nylon thread with flagging at one end was glued to seeds. Seed survival was checked approximately every 2 weeks. Missing seeds, and those removed from their flagging, were assumed to be eaten by vertebrate predators.

Results

Experiment 1

Fruit production

Mature *Socratea* fruits were present most of the year in La Selva, but were significantly lower between February and June and highest between November and January at the end of the wet season when rainfall peaks. Patterns of *Socratea* fruit production were relatively similar for 2001–2002 and 2002–2003 although total production was higher in 2002–2003 (Fig. 5.1A). Peak *Socratea* fruit production coincided with the peak in community-wide fruit production at La Selva.

Predator identity

Although we were generally unable to determine the identity of vertebrate predators with certainty, by following the trail of moved seeds underneath low logs or through dense underbrush we were able to determine that seeds were frequently moved by small rodents. We also frequently observed evidence of collared peccaries, such as footprints near the area of consumed seeds. Collared peccaries appeared to move seeds only short distances (< 2 m) before eating them.

Nearly all (94%) of the seeds protected from vertebrates that died showed evidence of infestation by bark beetles. The majority of beetles collected were *C. carpophagus*, but individuals of *C. palmarum* were also collected from seeds (identification by Larry Kirkendall).

Table 5.1. Experiment 1. Results of Cox proportional hazard tests to compare the rate of seed mortality between two protection treatments (open or protected from vertebrates) and three placement dates (1 February, September and December). The effect of replicate plots was also included in the model.

Source	df	Chi-squared	P
Placement date	2	7.17	< 0.05
Protection treatment	1	97.87	< 0.001
Plot	3	7.39	0.06
Placement × protection	2	9.42	< 0.01
Pairwise comparisons			
Protected (insects)			
Dec. vs. Sept.	1	7.59	< 0.01
Dec. vs. Feb.	1	3.8	0.05
Feb. vs. Sept.	1	1.39	0.23
Unprotected (mammals)			
Dec. vs. Sept.	1	7.0	< 0.01
Dec. vs. Feb.	1	1.8	0.18
Feb. vs. Sept.	1	15.8	< 0.001

Final survival

Survival of *Socratea* seeds after 1 year at La Selva was extremely low during all three placement periods for both unprotected and protected seeds (Fig. 5.2), and ranged from 0 to 0.025% in any of the treatment/ placement period combinations. This very low survival made statistical comparison inappropriate, but suggests that insects and mammals had a similar impact on long-term seed survival regardless of placement period.

Predation rate

The rate of seed mortality due to mammals was significantly faster than mortality due to insects, with a median survival time of 64 days for seeds unprotected from mammals and 132 days for protected seeds (Table 5.1; Fig. 5.2). Placement date also had a significant effect on survival time, but this effect was not consistent between mammal and insect predators (Table 5.1; Fig. 5.2). The rate of seed predation by mammals was significantly lower for seeds placed in December 2001 than for both September and February (Table 5.1). Rates of mammal seed predation for February 2001 and September 2001 were not significantly different (Table 5.1; Fig. 5.2). Rate of insect predation for September was significantly lower than for February and December, but rates of predation for these 2 months were not different from each other (Table 5.1; Fig. 5.2).

Seed movement

The proportion of seeds moved from the unprotected treatment was 16.9%, 28.8% and 23.8% for seeds placed in February, September and December 2001, respectively. The majority of moved seeds were moved only a short distance and then consumed (mean distance = 228 ± 26.2 cm, median = 130 cm), and only three seeds were moved further than 10 m the furthest seed being moved 17 m. Few seeds were cached in any placement period (8, 12 and 13, respectively), although it is possible that seeds were cached for short periods and thus not detected in the time period between censuses.

Experiment 2

Fruit production in Panama

Socratea fruit production in Panama was highly seasonal, and peaked between February and June during the dry and early wet season. No mature fruits were present from May to December during the wet season at Gigante (Fig. 5.1B). Fruiting of *Socratea* was more synchronized at Gigante than at La Selva with both a shorter peak fruiting period and a much higher percentage of trees producing fruits during this period (Fig. 5.1). Although peak fruiting of *Socratea* coincided with peak fruiting in the community at both sites, timing of peak fruiting was opposite with respect to periods of seasonal differences in rainfall at the two sites. In Gigante peak fruiting occurred during the dry season while in La Selva peak fruiting occurred during the rainy season.

Seed predation in La Selva

All seeds were eaten or killed at the end of 1 year regardless of predator type, distance from focal tree, or placement date, but rates of seed mortality did differ between treatments.

As in Experiment 1, the mortality rate of seeds in La Selva exposed to mammals in February 2002 was significantly faster than for seeds exposed only to insects ($\chi^2 = 69.86$, df = 1, $P < 0.001$; Fig 5.3B). Also similarly to Experiment 1, seed mortality due to insects increased rapidly once seeds began to germinate (Fig. 5.3B). Although seed predation by mammals in La Selva was faster for seeds set out in October 2001 than in February 2002 this difference was not significant ($\chi^2 = 13.46$, df = 9, $P = 0.14$; Fig. 5.3A, B).

The rate of seed mortality was significantly different between seeds 1 and 10 m from focal trees, but there was a significant interaction between the effect of distance

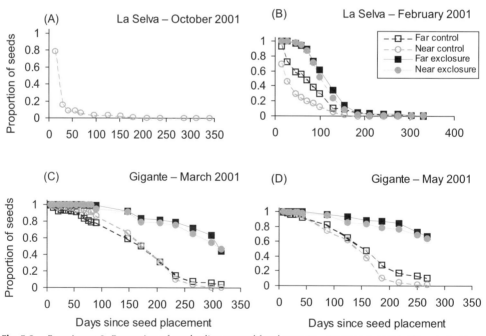

Fig. 5.3. Experiment 2. Proportion of seeds alive around focal trees (*n* = 10 trees except in Gigante in May with only nine trees). Seeds were placed at two distances from conspecific trees (near and far) and under two protection treatments (control and exclosure) at La Selva, Costa Rica and in Gigante, Panama. First placement date was at the peak of *Socratea exorrhiza* fruit production (October at La Selva and March at Gigante) and second at the end of the fruiting peak (February at La Selva and May in Panama). Seeds placed at La Selva in October were only placed in the near treatment.

and predator type, indicating that insects and mammals did not respond to the distance of seeds from the parent tree in the same way (Fig. 5.3B; Table 5.2, La Selva). The rate of seed predation by mammals was significantly faster near the parent tree than 10 m away ($\chi^2 = 26.65$, df = 1, $P < 0.000$), while the rate of seed predation by insects was not different between seeds near and far from the parent tree ($\chi^2 = 1.67$, df = 1, $P < 0.20$).

Seed predation in Gigante

Final seed predation after 264 days by mammals was significantly higher than insect predation ($\chi^2 = 187.34$, df = 1, $P < 0.001$), but there were no significant differences in seed predation by either insects or mammals between placement dates ($\chi^2 = 0.67$, df = 1, $P = 0.413$; Fig. 5.3C, D). Seed survival was also not significantly different between seeds placed 1 or 10 m from parent trees ($\chi^2 = 3.66$, df = 1, $P = 0.06$), although predation by mammals was slightly lower for seeds placed 10 m from trees (Fig. 5.3C, D).

The rate of seed mortality due to mammals was also significantly faster than mortality due to insects (Fig. 5.3C, D; Table 5.2, Gigante). The effect of placement date on the rate of predation differed between these two predator groups; pairwise comparison of mammal and insect predation rates found that the rate of mammal seed predation was significantly faster in May at the end of the fruiting peak than it was in March during the middle of the fruiting peak ($\chi^2 = 6.47$, df = 1,

$P < 0.01$), but there was no difference between placement dates for insect predation ($P > 0.50$).

Seed movement

Seed movement from around focal trees in La Selva was higher than for dispersed seeds in Experiment 1, particularly for seeds placed out in October 2001, but the percentage of seeds cached remained low (Fig. 5.4A) and seeds were generally moved only short distances (mean distance = 168 ± 14.7 cm, median = 107 cm).

On Gigante seed movement as well as the percentage of seeds cached was higher in March 2001 than May 2001 (Fig. 5.4C, D, respectively). Although levels of seed movement were similar between Gigante and La Selva the percentage of seeds cached was higher on Gigante (Fig. 5.4).

La Selva versus Gigante

Seed predation by mammals was very high at both La Selva and Gigante with very few seeds remaining by the end of the experiment in either site. Seed predation by insects was, however, much lower in Panama regardless of the date of placement (Fig. 5.3). Evidence of infestation by bark beetles was observed much less frequently at Gigante (64% and 51% of protected seeds in March 2001 and May 2001, respectively) than at La Selva (100% of protected seeds) and the bark beetle observed attacking seeds in Gigante appears to be a distinct

Table 5.2. Experiment 2. Results of Cox proportional hazard tests to compare rate of seed mortality between protection treatment (open or protected from vertebrates) and distance from conspecific adult (0 vs. 10 m) at La Selva (Costa Rica). In Gigante (Panama) we also compared seed predation at two placement dates (March and May). Only significant interactions were left in the model. Seeds were placed around ten replicate focal trees at La Selva and Gigante in May and nine trees in March.

Comparison	Source	df	Chi-squared	P
La Selva February 2002	Distance	1	13.70	< 0.001
	Protection treatment	1	69.86	< 0.001
	Focal trees	9	13.46	0.14
	Distance × protection	1	7.25	< 0.01
Gigante March and May 2002	Placement date	1	4.23	< 0.05
	Distance	1	3.00	0.08
	Protection treatment	1	277.01	< 0.001

species of *Cocotrypes* (L. Kirkendall, Costa Rica, 2002, personal communication). Rates of both insect and mammal seed predation were also faster in La Selva than for seeds placed out in either March 2002 or May 2002 in Gigante (Fig. 5.3; Table 5.3).

The effect of distance was only partly consistent between La Selva and Gigante

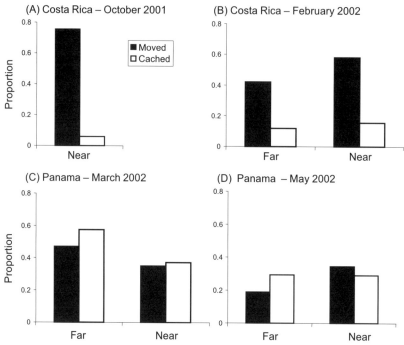

Fig. 5.4. Experiment 2. Proportion of all *Socratea exorrhiza* seeds moved and the proportion of moved seeds that were cached in La Selva, Costa Rica and Gigante, Panama during separate dates corresponding to the peak fruiting season (October – Costa Rica, March – Panama) and late fruiting season (February – Costa Rica, May – Panama).

Table 5.3. Experiment 2. Results of Cox proportional hazard tests to compare the rate of seed mortality between sites (Gigante in Panama vs. La Selva in Costa Rica), location with respect to parent tree (1 m vs. 10 m), protection treatment (open vs. vertebrate exclosure) and interactions. Only significant interactions were left in the model. La Selva seeds placed in February were compared independently to seeds placed in Gigante in March and seeds placed in May.

Comparison	Source	df	Chi-squared	P
La Selva February vs. Gigante March	Site	1	456.89	< 0.001
	Distance	1	8.14	< 0.005
	Protection treatment	1	179.62	< 0.001
	Distance × protection	1	5.67	< 0.05
	Site × distance	1	11.40	< 0.001
	Site × protection	1	11.98	< 0.001
La Selva February vs. Gigante May	Site	1	401.70	< 0.001
	Distance	1	12.56	< 0.001
	Protection treatment	1	174.91	< 0.001
	Distance × protection	1	8.75	< 0.005
	Site × distance	1	3.58	0.05
	Site × protection	1	16.54	< 0.001

because there was not a consistent effect of distance between the two placement dates in Gigante. In March 2002 at Gigante, neither insect nor mammal predation rates differed with distance from the parent tree, but during May 2002 in Gigante and February 2002 at La Selva, vertebrate predators attacked seeds significantly faster near the parent than 10 m away while rates of insect predation were unaffected by distance (Fig. 5.3; Table 5.3).

Experiment 3

Predator identity

Based on evidence of teeth marks found on partially eaten shells, as well as by trapping (Notman, 2000), mammal seed predation appeared to be mostly due to several species of spiny rats (*Proechimys* spp.) as well as smaller numbers of rice rats (*Oryzomys* spp.) and spiny mice (*Neacomys* sp.). Determining the identity of insect seed predators was more difficult because we were typically unable to identify any characteristic marks on seed remains. The insect most frequently observed actually eating seeds was an unidentified species of cricket that buried the seeds under the soil before consuming it. Several unidentified species of small ants were also observed eating seeds, but only after they were first damaged by rodents. Seeds of a number of species were also killed by unobserved insects that cut, but did not eat, the developing hypocotyl and/or epicotyl.

Seed predation by mammals was extremely high in both forest types (average 92.0%), and was significantly higher than insect predation (average 16.2%). There was also, however, significant variation in levels of both insect and mammal predation among species (Table 5.4). Predation by mammals in both forest types ranged from a low of 77.5% for *Leonia glycycarpa* to a high of 100% for *Bactris gasipes*, while average total predation by insects ranged from a low of 0% for *Licania urceatoris* to a high of 51.6% for *Pouteria* sp. (Table 5.4).

Species-specific differences in predation were not consistent between insects and mammals. Although predation by mammals was significantly greater than insect predation for almost all species, the difference in mammal predation relative to insects varied considerably. Species with the highest levels of seed predation by mammals did not have the highest levels of insect seed predation (Table 5.4).

Forest disturbance

Mammal and insect seed predators were not equally affected by forest disturbance (Table 5.4). Average predation by mammals across all species was not significantly different between young and mature forest, but average predation by insects was significantly higher in mature forest than young forest (see Table 5.4).

Discussion

We found seed predation by mammals to be consistently greater than predation by insects for all of the species examined in Peru, as well as for *Socratea* in Panama. These results are similar to other studies of seed predation in tropical forests, particularly those that have followed the fate of seeds over longer time periods (e.g. Holl and Lulow, 1997; Blate *et al.*, 1998; Guariguata *et al.*, 2000).

But very high levels of insect seed predation by bark beetles in La Selva also show that insects can be important post-dispersal predators even for dispersed seeds. High levels of palm seed infestation by bruchid beetles have also been found for other palm species (Smythe, 1989; Wright, 1990; Forget *et al.*, 1994; Silvius, Chapter 4, this volume), although these studies examined seeds in high densities near the base of parent trees. Studies of insect predation by bark beetles (family Scolytidae) on other seed species have also found very high levels of seed mortality (Janzen, 1972; Hammond *et al.*, 1999), suggesting that seed specialists in this

Table 5.4. Experiment 3. Twenty-one seed species were placed in both mature and young forests in Peru. Seeds were placed under two protection treatments to separate the effect of insects versus mammals on seed survivorship. Columns from the left are as follows: species name, family name, habitat where species occurs (Habitat), species habit, seed species weight (mean ± standard deviation), and proportions of dead seeds (mean ± standard deviation) under each habitat where seeds were placed and given protection treatment.

Species	Family	Habitat[a]	Habit[b]	Wet wt	Mature forest		Young forest	
					Insects	Mammals	Insects	Mammals
Licania urceatoris	Chrysoblanaceae	M	T	33.4 ± 1.0	0.00 ± 0.00	0.85 ± 0.07	0.00 ± 0.00	0.92 ± 0.06
Inga sp.	Mimosaceae	M	T	8.2 ± 0.02	0.54 ± 0.08	0.94 ± 0.04	0.15 ± 0.08	0.88 ± 0.08
Garcinia sp.	Clusiaceae	M	T	11.1 ± 0.1	0.00 ± 0.00	1.00 ± 0.00	0.04 ± 0.03	0.88 ± 0.06
Minquartia guianensis	Olacaceae	M	T	11.6 ± 0.1	0.10 ± 0.05	0.89 ± 0.07	0.01 ± 0.01	0.89 ± 0.05
Ormosia sp.	Fabaceae	M	T	8.8 ± 0.02	0.16 ± 0.05	0.96 ± 0.02	0.10 ± 0.04	0.89 ± 0.05
Telitoxicum sp.	Menispermaceae	M	L	13.4 ± 0.1	0.08 ± 0.04	0.99 ± 0.01	0.00 ± 0.00	0.89 ± 0.06
Mendoncia glabra	Acanthaceae	Y	L	6.6 ± 0.01	0.41 ± 0.08	0.89 ± 0.07	0.23 ± 0.08	0.95 ± 0.05
Enterolobium cyclocarpum	Mimosaceae	Y/M	T	10.3 ± 0.03	0.16 ± 0.05	0.89 ± 0.06	0.05 ± 0.03	0.96 ± 0.03
Sacoglottis sp.	Humiriaceae	M	T	14.7 ± 0.01	0.06 ± 0.04	0.61 ± 0.11	0.01 ± 0.01	0.71 ± 0.10
Pouteria sp.	Sapotaceae	M	T	7.1 ± 0.02	0.69 ± 0.07	0.99 ± 0.01	0.34 ± 0.07	0.98 ± 0.03
Virola elongata	Myristicaceae	M	T	12.3 ± 0.1	0.34 ± 0.09	0.96 ± 0.03	0.16 ± 0.07	1.00 ± 0.00
Bactris gasipes	Areacaceae	Y/M	T	16.4 ± 0.1	0.21 ± 0.07	1.00 ± 0.00	0.03 ± 0.02	1.00 ± 0.00
Hevea sp.	Euphorbiaceae	M	T	17.3 ± 0.04	0.21 ± 0.08	0.86 ± 0.08	0.05 ± 0.03	0.93 ± 0.06
Aniba sp.	Lauraceae	M	T	14.1 ± 0.01	0.49 ± 0.08	0.91 ± 0.04	0.14 ± 0.04	1.00 ± 0.00
Ocotea sp.	Lauraceae	M	T	7.9 ± 0.03	0.16 ± 0.06	0.99 ± 0.01	0.08 ± 0.04	0.99 ± 0.01
Virola sp.	Myristicaceae	M	T	12.7 ± 1.1	0.38 ± 0.10	0.95 ± 0.04	0.09 ± 0.04	0.99 ± 0.01
Sclerolobium sp.	Caesalpiniaceae	M	T	28.2 ± 0.01	0.15 ± 0.08	0.80 ± 0.06	0.06 ± 0.04	0.86 ± 0.06
Leonia glycycarpa	Violaceae	M/Y	T	8.8 ± 0.02	0.70 ± 0.09	0.78 ± 0.07	0.21 ± 0.04	0.78 ± 0.06
Heisteria duckei	Olacaceae	M	T	10.4 ± 0.2	0.08 ± 0.04	0.91 ± 0.05	0.03 ± 0.03	1.00 ± 0.00
Licania cf.	Chrysobalanacae	M	T	18.0 ± 0.01	0.11 ± 0.04	0.91 ± 0.05	0.05 ± 0.03	1.00 ± 0.00
Protium sp.	Burseraceae	M	T	11.0 ± 0.01	0.18 ± 0.07	0.86 ± 0.04	0.18 ± 0.06	0.91 ± 0.04
Salacia sp.	Hippocrateaceae	M	L	8.4 ± 0.1	0.39 ± 0.09	1.00 ± 0.00	0.11 ± 0.04	0.93 ± 0.05
Wettinia augusta	Arecaceae	M	T	13.3 ± 0.1	0.01 ± 0.01	0.99 ± 0.01	0.00 ± 0.00	1.00 ± 0.00
Tapirira sp.	Anacardiaceae	M	T	10.7 ± 0.01	0.04 ± 0.04	0.84 ± 0.07	0.00 ± 0.00	0.94 ± 0.06
Rhigospira quadrangularis	Apocynaceae	M	T	6.7 ± 0.01	0.10 ± 0.04	0.94 ± 0.05	0.03 ± 0.02	1.00 ± 0.00
Rourea sp.	Connaraceae	M	L	7.2 ± 0.01	0.06 ± 0.02	0.93 ± 0.06	0.05 ± 0.03	1.00 ± 0.00

[a]Y, young forest; M, mature forest.
[b]T, tree; L, liana.

family have the potential to be major seed predators for some plant species.

We also found that mammal and insect seed predators did not respond to differences in spatial and temporal patterns of seed availability in the same manner.

Effect of the fruiting season on seed fate

Seed predation by insects and mammals was not affected by date of placement in a similar manner at either La Selva or Gigante. At La Selva mammal predation of dispersed seeds was significantly lower during the peak of fruit production than at either the start or end of the peak. In contrast, insect predation was lowest during the beginning of the fruiting season and increased towards the end of the fruiting peak. Mammal seed predation from around trees was also lower during the fruiting peak at Gigante, but insect predation did not differ between placement dates.

Lower mammal seed predation during the peak of the fruiting season agrees with our expectation that mammals may respond to increases in Socratea seed production by temporarily concentrating foraging efforts on this resource at the start of the fruiting peak, but may have been temporarily satiated either by Socratea production or the production of fruits in the community as a whole during December. These results agree with findings by Forget et al. (2002), who found that predation by caviomorph rodents declined during the peak in community fruiting and seed caching increased. We did not, however, find any increase in the percentage of seeds cached during the fruiting peak at either site, suggesting that mammal seed predators may have been shifting to other resources rather than storing excess Socratea fruits during this peak production period.

Seed predation by mammals did not differ between placement dates at La Selva for seeds at the base of trees, although fruit production was significantly higher in

October than in February. The lack of consistency between mammal seed predation in the two experiments may be due to differences in the species of mammals that preyed on dispersed seeds and seeds at the base of trees. Based on evidence of tracks and the movement patterns of seeds it appeared that collared peccaries foraged more commonly around the base of trees than at dispersed seeds (see also Kiltie, 1981), and may respond differently to changes in resources from the small to medium-sized rodents that appeared to be more common predators of dispersed seeds.

Higher levels of insect seed predation at the middle and end of the Socratea fruiting peak agreed with our predictions. If Socratea is a key host species for C. carpophagus, lower rates of predation may have been due to fewer adult females emerging during September because of reduced availability of suitable host seeds in the previous months when Socratea seed production was low. Although we have collected C. carpophagus from the seeds of at least six other palm species, Socratea is infested significantly faster than the two other species we have studied (Welfia regia and Iriartea deltoidea), which make up the most widespread and abundant large-seeded palm species at La Selva (Lieberman et al., 1985).

The development time of related Cocotrypes species has been reported to be 30–45 days (Kirkendall et al., 1997). If development time is similar in C. carpophagus populations, C. carpophagus could have increased for several generations during a period of high Socratea seed production in September and October before we placed the next set of seeds in December. Wilson and Janzen (1972) found a similar increase in rate of seed infestation by C. carpophagos on seeds of Euterpe globosa in Puerto Rico from the start of the fruiting season towards the peak, and suggested that this was due to a drop in the populations of C. carpophagos during the dry season when host seed availability was very limited. In contrast, Wright (1990) found that infestation by bruchid beetles of the palm seeds of

Scheelea zonensis decreased towards the end of the fruiting season and suggested that this is because bruchid development is longer than the fruiting season, and thus most infestation occurs at the start of the fruiting season by females that emerged from the previous year's crop.

Because seeds placed in February 2001 were mostly already infested prior to placement, interpretation of these data must take into consideration that mortality rate due to insects is likely to be an overestimation. It should be noted, however, that infestation by beetles during the other placement dates was often very rapid (< 2 weeks), and mortality rate for February was not significantly different from seeds placed in December, none of which were infested at the time of placement. This suggests that the time till initial infestation may not be as important as timing and intensity of subsequent infestation.

Effect of distance from parent tree

Distance from the parent tree did not influence predation by insects at either La Selva or Gigante. In contrast, predation by mammals was greater near the parent tree both at La Selva and at Gigante in May. This pattern contradicts the results of many previous studies that have generally found that insects respond to distance while mammals and other vertebrates do not (see review by Hammond and Brown, 1998).

The contrasting patterns of predation found in this study may be due to the particular species of predators studied or, alternatively, may reflect particular characteristics of *Socratea*. Previous studies of beetles in the genus *Cocotrypes* have failed to find evidence of strong distance-dependent infestation (Wilson and Janzen, 1972; Silva and Tabarelli, 2001), but Hammond *et al.* (1999) found that the time to infestation of *Chlorocardium rodiei* (Lauraceae) by bark beetles in the genus *Sternobothrus* decreased with increasing distance, while the time to vertebrate predation was not related to distance. Because

vertebrate predation of seeds is also likely to kill any developing larva inside, it is possible that selection has acted against beetles using parent trees as cues for species of plants in which vertebrate predation is distance dependent.

The spatial scale at which distance influences infestation of *Socratea* seeds by *Cocotrypes* may also be greater than the 10 m examined in this experiment. Wright (1983) found no effect of distance on palm seed infestation by bruchid beetles from 0 to 16 m, but did find reduced infestation levels at 100 m, and Hammond *et al.* (1999) examined infestation rate over a continuous range of distances from 0 to 80 m. We chose to compare predation at 1 m and 10 m because the high density and somewhat clumped distribution of *Socratea* made it difficult to find sites further than 10 m from another adult *Socratea*. In addition, because *Cocotrypes* appear capable of infesting seeds of several very common palm species at La Selva, it may be particularly difficult to find areas in which fruiting host trees are separated by distances greater than 10 m.

Although distance-dependent predation by vertebrates appears to be generally less common than by invertebrates, several other studies have also found distance-dependent predation by vertebrates on palm seeds including *Welfia exhoriza* seeds in La Selva (Schupp and Frost, 1989) and *Scheelea zonensis* palms on BCI, Panama (Forget *et al.*, 1994). Patterns of vertebrate predation may depend greatly on the species of vertebrates. In La Selva greater predation by collared peccaries at the base of palm trees may have been responsible for the faster rate of predation we found near trees. Seasonal variation in the effect of distance on vertebrate predation in Gigante suggests that the principal predators of *Socratea* seeds may shift during the year in response to changes in fruit production of *Socratea* or the availability of other resources. The slightly higher proportion of cached seeds near the base of trees in Gigante may also have been due to differences in the species of mammals most likely to find seeds near and far from trees.

La Selva versus Gigante

Seed survival was in general higher at Gigante than at La Selva, and the relative importance of insect and mammal seed predators was very different between these two sites. It is likely that the much higher levels of insect predation at La Selva than Gigante are due to differences in the abundance of *Cocotrypes* at the two sites. Populations of bark beetles at Gigante may be lower than at La Selva due to the lack of a constant source of host seeds throughout the year at Gigante. Low levels of *Socratea* fruits were available at La Selva for at least 9 months of the year in contrast to approximately 4 months at Gigante. In addition, we have also observed *Cocotrypes* infesting seeds of other palm species at La Selva, including *Iriartea deltoidea*, which fruits out of synchrony with *Socratea*. Although *Cocotrypes* infests *I. deltoidea*, seeds in lower frequencies (Notman and Villegas, unpublished data), this alternative host species may maintain populations of *Cocotrypes* during low *Socratea* fruiting.

Although we did not have sufficient information about the relative abundances of mammal seed predators at these two sites to make predictions about which site might have higher levels of predation, we did expect to find higher rates of seed caching at Gigante because of the greater seasonality in fruit production resulting in a longer period of low resource availability through which animals must survive. Our results agreed with this prediction but not with our prediction that higher proportions of caching would occur at the end of the fruiting peak.

It is interesting to note that *Socratea* fruit production matches community fruiting peaks at both of these sites, despite the fact that at La Selva peak fruit production occurs during the rainy season and at Gigante peak production occurs during the dry season. This suggests that *Socratea* may be responding to different cues to trigger flower and fruit production across its range, and it is possible that timing of fruit production has been selected to occur during the peak in community production to escape higher levels of mammal seed predation outside this peak.

Effect of plant species identity

As has been found by many other studies (Terborgh *et al.*, 1993; Hammond, 1995; Asquith *et al.*, 1997; Holl and Lulow, 1997; Blate *et al.*, 1998; Jones *et al.*, 2003) levels of seed predation varied considerably among different plant species. Species-specific differences in predation were not consistent between insects and mammals and species with the highest levels of seed predation by mammals did not have the highest levels of insect seed predation. Nor was there any clear negative correlation between levels of insect predation and mammal predation. Such a negative correlation might have been expected if the seeds selected had included smaller seeds below the size at which mammals could efficiently harvest them. Seed size was, however, not related to predation by either insects or mammals (Notman, 2000).

Effect of forest disturbance

Although predation by mammals was not different between young and mature forest, insect predation was significantly greater overall in mature forest. This matches our prediction that insect predation would be greater in mature forests due to the presence of adult host trees, but greater insect predation in mature forest may also have been due to abiotic conditions, such as temperature, that affected insect community composition and foraging (e.g. Didham, 1997). Although the relative importance of insects did not increase in mature forest for all species there was no relationship between difference in insect predation between the two sites and the forest type that adult plants occurred in (Table 5.4).

Despite the general trend of higher insect predation in disturbed forest, there was still considerable variation among species in the relative importance of mammal and insect predation in young and mature forest. These species-specific differences suggest that changes in predator abundance associated with habitat disturbance are likely to favour regeneration of distinct suites of species, although other factors, including dispersal and abiotic conditions, are also likely to play an equal or greater role in determining plant recruitment to disturbed areas (Schupp et al., 1989).

Conclusions

We found that insect and mammal seed predators do not respond to factors such as resource abundances, the spatial distribution of seeds, and forest age in the same way. This suggests that these two predator groups are likely to exert very different selective forces on the species of seeds that they attack, and a greater understanding of the differences between insect and mammal seed predators is important for understanding the forces shaping the evolution of seeds and seed dispersal systems.

Differences in seed predators may also affect patterns of distribution across the range of a plant species. Levels of insect predation on *Socratea* differed greatly between La Selva and Gigante and in both of these sites mammals appeared to attack seeds in a distance-dependent manner while insects did not. The very high levels of insect predation at La Selva regardless of distance from parent trees is likely to reduce any effect that mammals have on the spatial distribution of surviving seeds. Although we do not have data to compare the distribution of adults at these two sites it is possible that consistently lower levels of insect predation at Gigante could favour the survival of dispersed seeds where mammals are less likely to find them.

Understanding differences between insect and mammal predation is also important for assessing the possible impact of human disturbance. A number of studies have shown that changes in the abundance of vertebrates due to hunting or other human disturbance can affect levels of seed

predation and subsequently the community composition of plants (Terborgh and Wright, 1994; Brewer et al., 1997; Guariguata et al., 2000; Wright et al., 2000; Silman et al., 2003). Reduction in mammal seed predators by hunting is not only likely to favour regeneration of species less susceptible to insect predation but may also have more complex ecological effects such as to alter the effect of timing of fruit and seed production on survival.

Acknowledgements

We would like to thank Juanita Zeledon, Julio Contreras and Lisa Patrick for help with fieldwork at La Selva. Antonio Trabuco also provided help with GIS at La Selva. Natalia Montes, Arturo Morris and Evelyn Sanchez did the majority of fieldwork on Gigante. This study received funds from the Andrew Mellon Foundation, and the Organization for Tropical Studies. Pierre-Michel Forget, Philip Hulme and two anonymous reviewers provided helpful comments on an early draft of the manuscript.

References

Akashi, N. (1997) Dispersion pattern and mortality of seeds and seedlings of *Fagus crenata* Blume in a cool temperate forest in Western Japan. *Ecological Research* 12, 159–165.

Asquith, M.N., Wright, J. and Clauss, J.M. (1997) Does mammal community composition control recruitment in neotropical forests? Evidence from Panama. *Ecology* 78, 941–946.

Augspurger, C.K. (1985) A cue for synchronous flowering. In: Leigh, E.G. Jr, Rand, A.S. and Windsor, D.M. (eds) *The Ecology of a Tropical Forest: Seasonal Rhythms and Long-term Changes.* Smithsonian Institution Press, Washington, DC, pp. 133–150.

Augspurger, C.K. and Kitajima, K. (1992) Experimental studies of seedling recruitment from contrasting seed distributions. *Ecology* 73, 1270–1284.

Bernays, E.A. and Chapman, R.F. (1994) *Host-plant Selection by Phytophagous Insects.* Chapman & Hall, New York.

Blate, G.M., Peart, D.R. and Leighton, M. (1998) Post-dispersal predation on isolated seeds: a comparative study of 40 tree species in a Southeast Asian rainforest. *Oikos* 82, 522–538.

Brewer, S.W., Rejmanek, M., Johnstone, E.E. and Caro, T.M. (1997) Top-down control in tropical forests. *Biotropica* 29, 364–367.

Brown, J.H. and Heske, E.J. (1990) Control of a desert–grassland transition by a keystone rodent guild. *Science* 250, 1705–1707.

Burkey, T.V. (1994) Tropical tree species-diversity – a test of the Janzen-Connell model. *Oecologia* 97, 533–540.

Clark, D.A., Clark, D.B., Sandoval, R. and Castro, M.V. (1995) Edaphic and human effects on landscape-scale distributions of tropical rain-forest palms. *Ecology* 76, 2581–2594.

Croat, T.B. (1978) *Flora of Barro Colorado Island*. Stanford University Press, Stanford, California, 943 pp.

Davidson, D.W. (1993) The effects of herbivory and granivory on terrestrial plant succession. *Oikos* 68, 23–35.

Dearing, M.D., Mangione, A.M. and Karasov, W.H. (2000) Diet breadth of mammalian herbivores: nutrient versus detoxification constraints. *Oecologia* 123, 397–405.

De Steven, D. (1991) Experiments on mechanisms of tree establishment in old-field succession – seedling emergence. *Ecology* 72, 1066–1075.

Didham, R.K. (1997) The influence of edge effects and forest fragmentation on leaf litter invertebrates in Central Amazonia. In: Laurance, W.F. and Bierregaard, R.O. (eds) *Tropical Forest Remnants: Ecology, Management, and Conservation of Fragmented Communities*. University of Chicago Press, Chicago, Illinois, pp. 55–70.

Forget, P.-M., Munoz, E. and Leigh, E.G. Jr (1994) Predation by rodents and bruchid beetles on seeds of *Scheelea* palms on Barro-Colorado Island, Panama. *Biotropica* 26, 420–426.

Forget, P.-M., Kitajima, K. and Foster, R.B. (1999) Pre- and post-dispersal seed predation in *Tachigali versicolor* (Caesalpiniaceae): effects of timing of fruiting and variation among trees. *Journal of Tropical Ecology* 15, 61–81.

Forget, P.-M., Hammond, D.S., Milleron, T. and Thomas, R. (2002) Seasonality of fruiting and food hoarding by rodents in Neotropical forests: consequences for seed dispersal and seedling recruitment. In: Levey, D.J., Silva, W.R. and Galetti, M. (eds) *Seed Dispersal and Frugivory: Ecology, Evolution*

and Conservation. CAB International, Wallingford, UK, pp. 241–253.

Foster, R. (1982) The seasonal rhythm of fruitfall on Barro Colorado Island. In: Leigh, E.G. Jr, Rand, A.S. and Windsor, D.M. (eds) *The Ecology of a Tropical Forest: Seasonal Rhythms and Long-term Changes*. Smithsonian Institution Press, Washington, DC, pp. 151–172.

Fox, A.G. (1999) Failure-time analysis: emergence, flowering, survivorship, and other waiting times. In: Scheiner, S.M. and Gurevitch, J. (eds) *Design and Analysis of Ecological Experiments*. Chapman & Hall, New York, pp. 253–289.

Frankie, G.W., Baker, H.G. and Opler, P.A. (1974) Comparative phenological studies of tree in tropical wet and dry forests in the lowlands of Costa Rica. *Journal of Ecology* 62, 81–91.

Freeland, W.J. and Janzen, D.H. (1974) Strategies in herbivory by mammals: the role of plant secondary compounds. *The American Naturalist* 108, 269–289.

Garwood, N.C. (1983) Seed-germination in a seasonal tropical forest in Panama – a community study. *Ecological Monographs* 53, 159–181.

Gentry, A.H. (1990) *Four Neotropical Rainforests*. Yale University Press, New Haven, Connecticut.

Guariguata, M.R., Adame, J.J.R. and Finegan, B. (2000) Seed removal and fate in two selectively logged lowland forests with constrasting protection levels. *Conservation Biology* 14, 1046–1054.

Hammond, D.S. (1995) Post-dispersal seed and seedling mortality of tropical dry forest trees after shifting agriculture, Chiapas, Mexico. *Journal of Tropical Ecology* 11, 295–313.

Hammond, D.S. and Brown, V.K. (1998) Disturbance, phenology and life-history characteristics: factors influencing distance/density-dependent attack on tropical seeds and seedlings. In: Newbery, D.M., Prins, H.H.T. and Brown, N.D. (eds) *Dynamics of Tropical Communities*, Blackwell Science, Oxford, pp. 51–78.

Hammond, D.S., Brown, V.K. and Zagt, R. (1999) Spatial and temporal patterns of seed attack and germination in a large-seeded neotropical tree species. *Oecologia* 119, 208–218.

Hartshorn, G.S. (1983) Plants. In: Janzen, D.H. (ed.) *Costa Rican Natural History*. University of Chicago Press, Chicago, Illinois, p. 118.

Henderson, A., Galeano, G. and Bernal, R. (1995) *Field Guide to the Palms of the Americas*. Princeton University Press, Princeton, New Jersey, 363 pp.

Holl, K.D. and Lulow, M.E. (1997) Effects of species, habitat, and distance from edge on post-dispersal seed predation in a tropical rainforest. *Biotropica* 29, 459–468.

Hulme, P.E. (1997) Post-dispersal seed predation and the establishment of vertebrate dispersed plants in Mediterranean scrublands. *Oecologia* 111, 91–98.

Janzen, D.H. (1969) Seed-eaters versus seed size, number, dispersal, and toxicity. *Evolution* 23, 1–27.

Janzen, D.H. (1970) Herbivores and the number of tree species in tropical forests. *American Naturalist* 104, 501–528.

Janzen, D.H. (1971) Seed predation by animals. *Annual Review of Ecology and Systematics* 2, 465–492.

Janzen, D.H. (1972) Association of a rainforest palm and seed eating beetles in Puerto Rico. *Ecology* 53, 258–261.

Jones, F.A., Peterson, C.J. and Haines, B.L. (2003) Seed predation in Neotropical pre-montane pastures: site, distance, and species effects. *Biotropica* 35, 219–225.

Kiltie, R.A. (1981) Distribution of palm fruits on a rain-forest floor – why white-lipped peccaries forage near objects. *Biotropica* 13, 141–145.

Kirkendall, L.R., Kent, D.S. and Raffa, K.F. (1997) Interactions among males, females and offspring in bark and ambrosia beetles: the significance of living in tunnels for the evolution of social behavior. In: Choe, J.C. and Crespi, B.J. (eds) *The Evolution of Social Behavior in Insects and Arachnids.* Cambridge University Press, Cambridge, pp. 181–215.

Leigh, E.G. Jr (1999) *Tropical Forest Ecology.* Oxford University Press, New York, 245 pp.

Leigh, E.G. Jr, Rand, A.S. and Windsor, D.M. (1982) *The Ecology of a Tropical Forest. Seasonal Rhythms and Long-term Changes.* Smithsonian Institution Press, Washington, DC.

Leigh, E.G. Jr, Wright, S.J., Herre, E.A. and Putz, F.E. (1993) The decline of tree diversity on newly isolated tropical islands – a test of a null hypothesis and some implications. *Evolutionary Ecology* 7, 76–102.

Levey, J.D. (1988) Spatial and temporal variation in Costa Rican fruit and fruit-eating bird abundance. *Ecological Monographs* 58, 251–269.

Lieberman, D., Lieberman, M., Peralta, R. and Hartshorn, G.S. (1985) Mortality patterns and stand turnover rates in a wet tropical forest in Costa Rica. *Journal of Ecology* 73, 915–924.

Notman, E. (2000) The influence of seed characters on post-dispersal seed predation by vertebrates and invertebrates in mature and regenerating Peruvian lowland tropical forest. MSc thesis, Miami University, Oxford, Ohio.

Notman, E. and Gorchov, D.L. (2001) Variation in post-dispersal seed predation in mature Peruvian lowland tropical forest and fallow agricultural sites. *Biotropica* 33, 621–636.

Notman, E., Gorchov, D.L. and Cornejo, F. (1996) Effect of distance, aggregation, and habitat on levels of seed predation for two mammal-dispersed neotropical rain forest tree species. *Oecologia* 106, 221–227.

Sanford, J.R.L., Luvall, J., Paaby, P. and Phillips, E. (1994) The La Selva ecosystem: climate, geomorphology, and aquatic systems. In: McDade, L., Bawa, K., Hartshorn, G. and Hespenheide, H. (eds) *La Selva: the Ecology and Natural History of a Neotropical Rainforest.* University of Chicago Press, Chicago, Illinois, pp. 19–33.

SAS Institute (1989) SAS/STAT User's Guide, Version 6, 4th edn, Vol. 1. SAS Institute Inc., Cary, North Carolina, 943 pp.

Schupp, E.W. and Frost, E.J. (1989) Differential predation of *Welfia georgii* seeds in treefall gaps and the forest understory. *Biotropica* 21, 200–203.

Schupp, E.W., Howe, H.F., Augspurger, C.K. and Levey, D.J. (1989) Arrival and survival in tropical treefall gaps. *Ecology* 70, 562–563.

Silman, M.R., Terborgh, J.W. and Kiltie, R.A. (2003) Population regulation of a dominant-rain forest tree by a major seed-predator. *Ecology* 84, 431–438.

Silva, G.M. and Tabarelli, M. (2001) Seed dispersal, plant recruitment and spatial distribution of *Bactris acanthocarpa* Martius (Arecaceae) in a remnant of Atlantic forest in northeast Brazil. *Acta Oecologica* 22, 259–268.

Smythe, N. (1970) Relationship between fruiting seasons and seed dispersal methods in a neotropical forest. *The American Naturalist* 104, 25–35.

Smythe, N. (1989) Seed survival in the palm *Astrocaryum standleyanum*: evidence for dependence upon its seed dispersers. *Biotropica* 21, 50–56.

Terborgh, J. and Wright, S.J. (1994) Effects of mammalian herbivores on plant recruitment in 2 neotropical forests. *Ecology* 75, 1829–1833.

Terborgh, J., Losos, E., Riley, M.P. and Riley, M.B. (1993) Predation by vertebrates and invertebrates on the seeds of 5 canopy tree species in an Amazonian forest. *Vegetatio* 108, 375–386.

Westerman, P.R., Hofman, A., Vet, L.E.M. and van der Werf, W. (2003) Relative importance of vertebrates and invertebrates in epigamic weed seed predation in organic cereal fields. *Agriculture, Ecosystems and Environment* 95, 417–425.

Wilson, D.E. and Janzen, D.E. (1972) Predation on *Scheelea* palm seeds by bruchid beetles: seed density and distance from the parent palm. *Ecology* 53, 1954–1959.

Wright, S.J. (1983) The dispersion of eggs by a bruchid beetle among *Scheelea* palm seeds and the effect of distance to the parent palm. *Ecology* 64, 1016–1021.

Wright, S.J. (1990) Cumulative satiation of a seed predator over the fruiting season of its host. *Oikos* 58, 272–276.

Wright, S.J., Zeballos, H., Dominguez, I., Gallardo, M.M., Moreno, M.C. and Ibanez, R. (2000) Poachers alter mammal abundance, seed dispersal, and seed predation in a Neotropical forest. *Conservation Biology* 14, 227–239.

6 Seed Predation and Dispersal by Peccaries throughout the Neotropics and its Consequences: a Review and Synthesis

Harald Beck

Duke University, Center for Tropical Conservation, Nicholas School of the Environment and Earth Sciences, PO Box 90381, Durham, NC 27708-0381, USA

Introduction

Studies on plant–frugivore interactions are crucial for the understanding of evolutionary and ecological processes at species, population, and community levels. These interactions are more complex in the tropics than in other ecosystems because of their higher species richness of both frugivores and plants. A comprehensive understanding of these processes, plus the development of new hypotheses and future research can only be achieved if ecologists consider existing results and examine interactions among many different taxa. However, studies on frugivorous mammals have been biased towards small-bodied vertebrate species, primarily rodents and bats, whereas the megafauna such as deer and peccaries have been largely ignored (Ulbrich, 1928; Ridley, 1930; Müller, 1955; Pijl, 1969; Janzen, 1971, 1983c, 1984; Howe and Smallwood, 1982; Gautier-Hion et al., 1985; Estrada and Fleming, 1986; Howe, 1986; Murray, 1986; Roth, 1987; Fleming and Estrada, 1993; Bongers et al., 2001; Levey et al., 2002). This bias and the relative lack of attention to the existing literature on large frugivorous vertebrates may preclude a comprehensive understanding of the processes and consequences of plant–frugivore interactions. For example, although early studies demonstrated that peccaries disperse seeds of numerous species (e.g. Enders, 1930, 1935; Chapman, 1931, 1936; Janzen, 1975), the perception that peccaries are seed predators persists in the literature (Janzen, 1983c; Sowls, 1983; Hallwachs, 1986; Howe, 1989; March, 1990; Schmidt, 1990; Cullen et al., 2001; Forget and Hammond, 2005). Thus, if we consider that peccaries have a food retention time of up to 3 days (Strey et al., 1985; Comizzoli et al., 1997) and move large distances, sometimes over 10 km per day (i.e. Kiltie and Terborgh, 1983), seeds could be dispersed over a wider range of habitats and farther away from parent trees compared to the action of most other dispersal vectors.

In this chapter, I first review the literature on seed predation and dispersal by collared peccaries (*Pecari tajacu*) and white-lipped peccaries (*Tayassu pecari*) (nomenclature after Voss et al., 2001) throughout the Neotropics, and synthesize the results of 143 studies published between 1836 and 2003. I then discuss the ecological implications of seed predation and dispersal by both peccary species, taking into consideration their interactions with 212 plant species. Subsequently, I summarize different mechanisms of seed dispersal and examine

©CAB International 2005. *Seed Fate*
(eds P.-M. Forget, J.E. Lambert, P.E. Hulme and S.B. Vander Wall)

whether peccaries play a key role among frugivores by creating a bimodal rather than a typical leptokurtic seed shadow. Then, I review the empirical evidence and ask if peccaries compete with other frugivores for food resources. And finally, I synthesize the findings in a conceptual model and deduce some hypotheses of peccary–plant and peccary–frugivore interactions and relate the results to conservation efforts.

Autecology of Peccaries: a Brief Overview

Three species of peccaries (Artiodactyla: Tayassuidae) occur in the Neotropics. The Chacoan peccary (*Catagonus wagneri*), which occurs mainly in xeric habitat within Bolivia and Paraguay (Mayer and Wetzel, 1986; Hutterer, 1998), will not be further considered here. The collared peccary has the widest geographic distribution, ranging from Arizona to Argentina (Mayer and Brandt, 1982; Sowls, 1997; Eisenberg and Redford, 1999). The collared peccary is a habitat generalist and can be found in tropical rainforests, swamps and dry forests, as well as xeric thorn forests, and open areas such as arid deserts, dry savannas and grasslands (Mayer and Brandt, 1982; Mayer and Wetzel, 1987; Sowls, 1997; Eisenberg and Redford, 1999). Collared peccaries form groups of five to more than 50 individuals (Castellanos, 1983; Judas and Henry, 1999), and have a density of up to 35 individuals/km[2] (Mena *et al.*, 2000). Individuals may weigh as much as 30 kg (Sowls, 1997), and their biomass ranges from 125 kg/km[2] (Terborgh, 1983) to 373 kg/km[2] (Eisenberg, 1980). Collared peccaries have high site fidelity, and their home ranges vary from 38 ha (Castellanos, 1985) to 685 ha (Taber *et al.*, 1994; Carrillo *et al.*, 2002). In 1987, the Convention for International Trade in Endangered Species listed the collared peccary in Appendix II (CITES, 2003).

White-lipped peccaries range from southern Mexico to northern Argentina (Mayer and Brandt, 1982; Mayer and Wetzel, 1987; Eisenberg and Redford, 1999), and inhabit primarily tropical rainforests, riparian and swamp forests, and grasslands (Mayer and Brandt, 1982; Sowls, 1997; Eisenberg and Redford, 1999). They live in groups of 10–400 individuals (Kappler, 1887; Miller, 1930; Enders, 1935; Husson, 1978; Kiltie and Terborgh, 1983; Sowls, 1997; Altrichter and Almeida, 2002). However, groups of over 1000 individuals were reported historically (Jardine, 1836; Perry, 1970; Mayer and Wetzel, 1987; Sowls, 1997). The maximum density recorded was 16 individuals/km[2] in Venezuela (Eisenberg, 1980). The body mass of white-lipped peccaries can reach up to 50 kg (J. Terborgh, North Carolina, 2002, personal communication) and, depending on the geographic region, their biomass varies from 105 kg/km[2] (Terborgh, 1983) to 219 kg/km[2] (Cullen *et al.*, 2001). White-lipped peccaries are known to travel between 5 and 13 km per day and some authors have suggested they are a nomadic or migratory species (Rengger, 1830; Hershkovitz, 1969; Kiltie, 1980; Kiltie and Terborgh, 1983; Bodmer, 1990; Sowls, 1997; Fragoso, 1998b; Altrichter and Almeida, 2002). White-lipped peccaries require over 1000 km[2] of mosaic forests (Terborgh, 1992; Peres, 2001), and their home ranges encompass from 23 km[2] (Fragoso, 1998a) to over 200 km[2] (Kiltie and Terborgh, 1983). In 1987, the Convention for International Trade in Endangered Species listed white-lipped peccaries in Appendix II (CITES, 2003).

In an intact forest, white-lipped peccaries represent the largest terrestrial biomass, up to 230 kg/km[2] (Terborgh, 1983; Peres, 1996; Silman *et al.*, 2003), and thereby form the largest foraging groups among New World mammals (Eisenberg and Redford, 1999). The complex socio-ecological organization of peccaries may enable individuals to forage more efficiently and minimize predation by jaguars and pumas (Kiltie and Terborgh, 1983). A large amount of food consisting mainly of fruits and seeds is required to sustain peccaries. Furthermore, peccaries consume a wide range of other resources, including leaves, tubers, roots, rhizomes, terrestrial invertebrates, turtle and bird eggs, frogs, fishes, snakes and small mammals (Kappler, 1887; Brehm, 1891; Gaumer, 1917;

Roosevelt, 1920; Enders, 1935; Cutright, 1940; Carr, 1967; Duke, 1967; Frädrich, 1968; Leopold, 1972; Schaller, 1983; Mayer and Wetzel, 1987; Sowls, 1997; Fragoso, 1998a, 1999). Analyses of stomach contents and faeces of peccaries indicated that fruits and seeds were the primary food resources followed by leaves and roots (Table 6.1). Therefore, several authors consider peccaries to be primarily frugivores or omnivores (Kiltie, 1981a; Bodmer, 1989, 1991a,b; Altrichter et al., 2000, 2001; Keuroghlian, 2003).

Data Acquisition

I surveyed for publications reporting plant consumption by both peccary species in various databases including Science Citation Index, Web of Science, and Biological Abstracts. Furthermore, I used other avenues to expand my search: in 2003, at the annual meeting of the Association for Tropical Biology and Conservation (ATBC), I presented a poster summarizing the existing findings and requested additional information about peccary–plant interactions. Similarly, I had announcements in the journal *Palms* (in 2003) and in *Tropinet* (ATBC supplementary publication, in 2004).

From each publication and personal communication, I recorded the following data: plant family and species; life form (i.e. tree, liana); part used (fruit pulp and/or seed eaten); seed fate (alive or killed); the country where the study was conducted; the method employed (i.e. field observation, stomach analysis); and the habitat (i.e. tropical rainforest, dry forest). If some of these data were

Table 6.1. Percentage of food resources eaten by collared peccaries (*Pecari tajacu*) and white-lipped peccaries (*Tayassu pecari*) classified into five categories. Data which provide a range of values indicate a seasonal variation in % of used resource (when provided by the original articles). The categories 'Leaves' and 'Roots' are combined if authors classified resources categories differently, i.e. as 'plant material, vegetative plant parts or fibre'. Unidentified resources are excluded. For more details see footnotes.

Leaves (%)	Roots (%)	Fruits (%)	Seeds (%)	Animals (%)	Site	Method	Habitat	Reference
Collared peccary								
37–45	41–52	8–12	ND	0	M	FA	D	Martínez-Romero and Mandujano (1995)
0.8	31	29[a]		2	V	SC	D	Barreto et al. (1997)
50–90[b]		63–70	ND	ND	CR	FA	D	McCoy et al. (1985)
22–28[b,c]		18–20[c]	40–50[c]	10–12[c]	FG	SC	R	Henry (1997)
21–90[c]		3–59[c]	ND	7–20[c]	P	SC	R	Bodmer (1991a)
71		29	ND	Trace	P	SC	R	Kiltie (1981a)
28[b]		17	45	10	FG	SC	R	Feer et al. (2001)
22[c]		60[c]	ND	18[c]	P	SC	R	Bodmer (1989, 1991b)
22[d]		78	ND	ND	B	FA	R	Keuroghlian (2003)
White-lipped peccary								
2–17[d]		80–83	ND	ND	B	FA	R	Keuroghlian (2003)
38		62	0.4		CR	FA	H	Altrichter et al. (2000, 2001)
39		61	Trace		P	SC	R	Kiltie (1981a)
22[c]		66[c]	12[c]		P	SC	R	Bodmer (1991b)

[a]Authors combined fruits and seeds.
[b]Authors combined leaves and roots.
[c]Values estimated from graphs.
[d]I combined multiple classes (fibre, leaf, and grass) into one category (leaves); ND, not determined.
Site: B, Brazil; CR, Costa Rica; FG, French Guiana; M, Mexico; P, Peru; V, Venezuela.
Methods: FA, faecal analysis; SC, stomach contents.
Habitat: H, humid tropical forest; D, tropical dry forest; R, tropical rainforest.

not provided, whenever possible I contacted the author to complete the data set. I did not include plant species that authors 'suspected' or speculated were utilized by peccaries. I also compiled a separate database when authors did not identify the species of peccary that consumed a given plant species. I excluded these data from most statistical analyses, except to compare seed predation versus seed dispersal across peccary species. To avoid pseudoreplication, I excluded plants that were not identified to species level from any statistical analyses.

Data Analyses

I used two-way chi-squared contingency analyses to test if the frequency distributions of consumed plant families differed between peccary species across the Neotropics, and within South and Central America separately. I used this approach because of differences in the floral composition between these geographic regions (e.g. Gentry, 1990; Gentry and Terborgh, 1990; Hartshorn and Hammel, 1994).

The different feeding habits of peccaries determine whether they act as seed predators or dispersers. Peccaries can consume a whole fruit and either destroy the seeds by mastication and/or chemical breakdown, ingest the seeds without destroying them, or spit them out. To test for differences in seed predation and dispersal between peccaries, I analysed the data across the Neotropics, and South and Central America by first comparing the frequency distributions of plant species whose seeds survived and then considering only the distributions of species whose seeds were destroyed. For all of these analyses I used two-way contingency tables and pooled the tail categories of each distribution to ensure an expected value of greater than five for each plant family. In addition, I performed goodness-of-fit statistics to test for differences in seed predation between both peccary species for the Neotropics, and South and Central America separately. I performed all statistical analyses with SPSS version 11.5 (SPSS, 2002).

I calculated Pianka's Measure, Percentage Overlap and the Morisita's Measure (Krebs, 1999) to compare the dietary niche overlap between peccary species by geographic regions using *Ecological Methodology* version 5.2 (Kenney and Krebs, 2000).

Potential biases

I located 143 studies published between 1836 and 2003 that mentioned seed predation and/or dispersal by peccary species. Despite this large number of papers the data found might be biased. For example, authors might have focused on large fruit or seed species of trees, ignoring liana and vine species. Another bias did 'hit the deck' because many studies did not specify which peccary species consumed a given plant species, or whether seeds were consumed, or did not identify the consumed plant species. Those papers are only partially useful in any review.

Consumption of Fruits and Seeds by Peccaries

Overall, both peccary species consumed fruits and seeds from 207 plant species belonging to 55 families. Collared peccaries consumed 128 plant species belonging to 40 families (Appendix 6.1), while white-lipped peccaries fed on 144 plant species from 38 families (Appendix 6.2). Papers where authors did not specify which species of peccary consumed fruits and/or seeds of a given plant species are summarized in Appendix 6.3. This appendix revealed the usage of an additional 18 new plant species belonging to 15 families. The large numbers of used plant species demonstrate that both peccary species are unequivocally opportunistic animals able to utilize fruits and seeds from a wide morphological and taxonomical range. Comparisons of the frequency distributions of plant families used between peccary species revealed no differences across the Neotropics ($\chi_5^2 = 0.714$, $P = 0.714$, Fig. 6.1), within

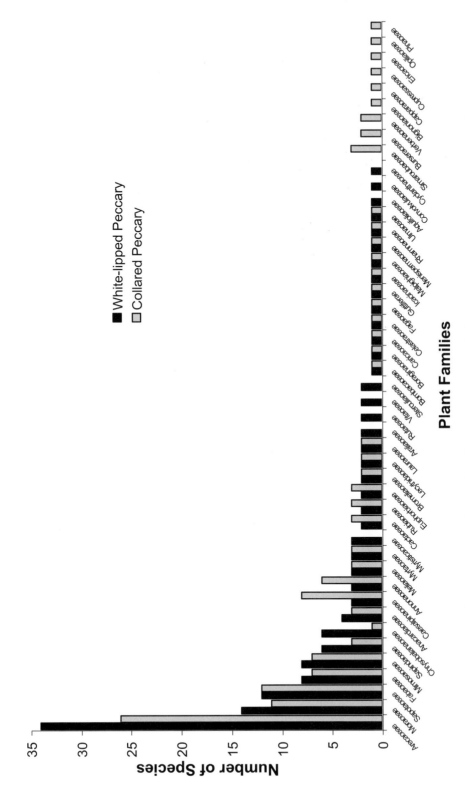

Fig. 6.1. Frequency distribution of plant species by family eaten by collared (*Pecari tajacu*) and white-lipped (*Tayassu pecari*) peccaries across the Neotropics.

South America ($\chi^2_3 = 0.001$, $P = 1.000$) and Central America ($\chi^2_4 = 0.000$, $P = 1.000$).

In the Neotropics the rank order of consumed fruits and seeds of the four most important plant families differed between peccaries (Fig. 6.1). For collared peccaries the order was Arecaceae (20%), Sapotaceae (9%), Moraceae (9%) and Fabaceae (5%). The rank order for white-lipped peccaries was Arecaceae (24%), Moraceae (10%), Sapotaceae (8%) and Fabaceae (6%). In fact, the palm family, Arecaceae, was most frequently used across all geographic regions. Why do palms constitute such a large fraction of the peccary's diet? At the species level, palms have higher densities than trees (Gentry, 1988; Kahn, 1991; Hartshorn and Hammel, 1994; Lieberman and Lieberman, 1994; Peres, 1994; Henderson, 1995, 2002; Terborgh *et al.*, 1996; Pitman *et al.*, 2001, 2003), and their highly nutritious fruits are available year-round, including the dry season when fruits are generally scarce. Although most palms have a hard-shelled endocarp to protect their seeds, peccaries possess a powerful mastication apparatus, i.e. interlocking teeth, thick enamel layer, specialized muscles and a jaw morphology that creates a bite force of over 300 kg (Herring, 1972; Kiltie, 1981c, 1982), enabling them to crush the majority of these seeds. Thus, it is not surprising that palm fruits represent a key resource for peccaries (Terborgh, 1986a,b; Sist, 1989b; Zona and Henderson, 1989; Kahn, 1991; Kahn and de Granville, 1992; Peres, 1994; Bodmer *et al.*, 1997; Painter, 1998; Altrichter *et al.*, 2001; Cullen *et al.*, 2001; Beck and Terborgh, 2002; Silman *et al.*, 2003).

The Effects of Seed Predation and its Consequences

Plants employ numerous defence strategies to decrease their vulnerability to seed predation while simultaneously maximizing their attractiveness for seed dispersal (e.g. Janzen, 1969, 1972, 1977; Harper *et al.*, 1970; Bodmer, 1991a). Seeds can contain distasteful (e.g. *Calatola* sp., Freeland and Janzen, 1974; Kiltie, 1982) or toxic components (e.g. Janzen, 1977; Bell, 1984; Waterman, 1984), possess mechanical structures such as protective fibrous lignin (e.g. *Spondias* spp., Kiltie, 1982; Janzen, 1985; *Jessenia bataua*, Bodmer, 1991a) or hard coats (e.g. *Phytelephas* spp., Spruce, 1871; Kiltie, 1982), or be too small to be masticated (e.g. *Ficus* spp., Janzen, 1982; Bodmer, 1991a). Moreover, seeds may escape predation when mast fruiting events cause predator satiation (e.g. Janzen, 1974; Beck and Terborgh, 2002). However, seed predators have co-evolved strategies to overcome plant defence strategies (e.g. Janzen, 1982; Janzen and Martin, 1982; Price *et al.*, 1991; Traveset, 1998; Herrera and Pellmyr, 2002). For example, peccaries have a strong mastication apparatus (e.g. Kiltie, 1982) and pregastric fermentation (Langer, 1978) that enable them to destroy seeds from over 73% of all consumed plant species. In the Neotropics, from a total of 107 plant species consumed, collared peccaries destroyed the seeds of 79 species (73%, $\chi^2_1 = 30.868$, $P < 0.0001$, Table 6.2). Similarly, white-lipped peccaries killed the seeds of 97 species from 124 plant species consumed (78%, $\chi^2_1 = 42.982$, $P < 0.0001$, Table 6.2). Comparisons between both peccary species indicated no difference in the frequency of destroyed seeds per plant species across the Neotropics ($\chi^2_4 = 0.000$, $P = 0.995$), in South America ($\chi^2_2 = 0.009$, $P = 1.000$), or Central America ($\chi^2_2 = 0.000$, $P = 1.000$).

Can seed predation by peccaries affect the recruitment, spatial distribution and population dynamic of plants? Silman *et al.* (2003) compared the abundance and spatial distribution of *Astrocaryum murumuru* seedlings in Cocha Cashu, Peru, before and during local extinction, and after re-colonization of white-lipped peccaries. During peccary extinction, *Astrocaryum* seedling density increased 1.7-fold and had a more uniform spatial distribution. After 12 years of re-colonization, the density and spatial distribution of *Astrocaryum* seedlings were comparable to those prior to peccary extinction. Wyatt (2002) reported parallel results for *Astrocaryum* and *Iriartea*

Table 6.2. Comparison of numbers of plant families, species and their percentage used by collared peccaries (*Pecari tajacu*) and white-lipped peccaries (*Tayassu pecari*) across Neotropics, South America and Central America. Data were estimated from the total number of plant species used (given in brackets). Data are categorized by utilization patterns: first, all plant species where peccaries used the fruit pulp and/or the seeds; second, plant species whose seeds survived (i.e. seeds were spat out); third, plant species whose seeds were killed.

Utilization patterns	Collared peccaries		White-lipped peccaries	
	Number of unique families (number of spp.)	Percentage of spp. (total number of spp.)	Number of unique families (number of spp.)	Percentage of spp. (total number of spp.)
Neotropics				
All plants used	9 (12)	9.2 (127)	8 (18)	13.0 (140)
Seeds alive	5 (6)	17.2 (29)	1 (1)	3.4 (29)
Seeds killed	5 (5)	6.8 (74)	8 (17)	17.4 (92)
South America				
All plants used	3 (3)	4.3 (70)	6 (8)	8.5 (82)
Seeds alive	0 (0)	0.0 (29)	1 (1)	3.1 (32)
Seeds killed	2 (2)	5.0 (40)	4 (7)	13.3 (45)
Central America				
All plants used	16 (18)	31.5 (58)	9 (19)	35.2 (54)
Seeds alive	3 (13)	25.0 (12)	1 (1)	11.1 (9)
Seeds killed	5 (5)	16.7 (33)	9 (17)	40.0 (42)

deltoidea seedlings. On the other hand, seedling densities can also be affected by the trampling of foraging peccaries. Roldán and Simonetti (2001) demonstrated that *Astrocaryum* seedling mortality was significantly higher in the presence of peccaries. Numerous studies have reported similar effects of peccaries on plant species (Clark and Clark, 1989; Asquith *et al.*, 1997; Fragoso, 1997; Wright *et al.*, 2000; Wright and Duber, 2001; Ticktin, 2003). Because peccaries destroy a large number of seeds and seedlings of many plant species, they may play a fundamental role in regulating recruitment, demography, and the spatial distribution of plants, thereby reducing competitive exclusion among plants and promoting plant diversity (Fig. 6.2).

Another not yet investigated effect of peccaries is their potential impact on insect seed predators. Because peccaries prefer seeds infested with nutritional insect larvae such as bruchid beetles (e.g. Kiltie, 1981b; Fragoso, 1994; Silvius, 2002; Jansen, 2003), they may control insect populations in a top-down fashion, and indirectly enhance seed survival (Fig. 6.2).

The Effects of Seed Dispersal and its Consequences

Although most of the emphasis in the literature has focused on peccaries as predators, they can be effective seed dispersers via three mechanisms: endozoochory of small seeds, expectoration of large seeds, and epizoochory.

Endozoochory of small seeds

Many small-seeded species (0.1–1 cm in length) escape mastication and are resistant to the chemical breakdown of peccaries' digestive systems (Appendices 6.1, 6.2, 6.3). Seeds that successfully germinated from peccary faecal samples include: *Enterolobium cyclocarpum* (Janzen and Higgins, 1979), *Ficus continifolia* (Janzen in Jordano, 1983), *F. maxima*, *F. zarzalensis*, *Psidium guajaba* (Altrichter *et al.*, 1999), *Enterolobium contortisiliquum*, *Ficus cfr. monkii*, *Gleditsia amorphoides*, *Panicum laxum*, *Phytolacca dioica*, *Solanum*

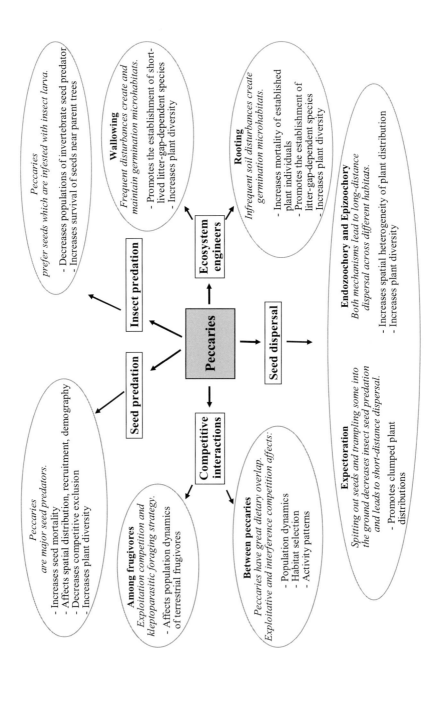

Fig. 6.2. Conceptual model hypothesizing peccary–plant and peccary–frugivore interactions and their ecological consequences.

claviceps, *Sonchus* sp. and *Syagrus ramanzoffianum* (Varela and Brown, 1995). Several other intact seeds were found in faecal matter (but no germination tests were performed), such as *Brosimum* spp. (Bodmer, 1991a), *Copernica tectum* (Barreto *et al.*, 1997), *Dipteryx panamensis* (Chapman, 1931; Enders, 1935) and *Ficus* spp. (Sowls, 1983). Numerous large seeds, such as *Acacia macracantha*, *Genipa americana*, *Guazuma ulmifolia*, *Pithecellobium saman*, *P. carabobense* (Barreto *et al.*, 1997), *Iriartea deltoidea*, *Socratea* sp. (Kiltie, 1981a) and *Prosopis* sp. (Sowls, 1984), were found undamaged in stomach samples.

Because peccaries can travel over 10 km per day (e.g. Kiltie and Terborgh, 1983), and have a long food retention of up to 52 h (Strey *et al.*, 1985; Comizzoli *et al.*, 1997), seeds can be dispersed across the landscape and into different habitats. Barreto *et al.* (1997) found seeds of *Copernica tectum* in faecal samples of white-lipped peccaries; however, this plant did not occur within their study site. Thus, peccaries can disperse seeds of numerous plant species over long distances and generate a large seed shadow, which may influence the spatial distribution of plants (Fig. 6.2). Moreover, peccaries could promote gene flow among plant populations, and subsequently increase genetic diversity.

S. purpurea and *S. radlkoferi* (Janzen, 1985; Barreto *et al.*, 1997; Altrichter *et al.*, 1999). Seeds from distasteful or poisonous species, including *Calatola* sp. (Kiltie, 1982) and *Ormosia arborea* (Galetti, 2002), are also expectorated as animals move from one feeding tree to another.

Foraging peccaries and other large mammals can bury some of these expectorated seeds by accidentally trampling them into the ground. Several studies demonstrated that trampled seeds have decreased seed predation, particularly from insects (Fragoso, 1997, 1998b; Silvius, 1999; Mitja and Ferraz, 2001; Silvius and Fragoso, 2002). Therefore, peccaries can enhance germination success in close proximity to parent trees, which could be one mechanism leading to clumped species distributions (e.g. Hubbell, 1979; Howe, 1989; Fragoso, 1997, 1998b). In contrast, scatterhoarding rodents including *Proechimys* spp. (Forget, 1990, 1991), *Myoprocta* spp. (Morris, 1962; Forget, 1991), and *Dasyprocta* spp. (Smythe, 1970a,b, 1978, 1986; Silvius and Fragoso, 2003) disperse most seeds within a 100 m radius of the parent tree (Jansen and Forget, 2001; Forget *et al.*, 2002), thus creating a leptokurtic seed shadow (Howe and Smallwood, 1982). Peccaries generate a plant species-specific bimodal seed shadow; endozoochory leads to long-distance dispersal and expectoration leads to short-distance dispersal (Fig. 6.2).

Expectoration of large seeds

Peccaries only chew off the fruit pulp of seeds that are too hard or large to be masticated or swallowed, and frequently expectorate (spit) them near the parent trees. Examples include *Crescentia alata* (Janzen, 1982), *Enterolobium cyclocarpum* (Janzen and Higgins, 1979), *Guazuma ulmifolia*, *Hymenaea courbaril* (Hallwachs, 1986), *Jessenia bataua* (Bodmer, 1991a), *Mauritia flexuosa* (Bodmer, 1991a; Fragoso, 1997), *Maximiliana maripa* (Fragoso, 1997, 1998b), *Orbignya martiana* (Anderson, 1983), *Scheelea rostrata* (Janzen and Martin, 1982), *Spondias mombin* (Kiltie, 1982),

Epizoochory

Epizoochory – a largely ignored mechanism – is the dispersal of seeds by adhesion to the feathers or fur of animals (Hildebrand, 1873; Ulbrich, 1928; Ridley, 1930; Pijl, 1969; Agnew and Flux, 1970; Sorensen, 1986; Castillo-Flores and Calvo-Irabién, 2003). I located only one published study describing epizoochory of peccaries. This is surprising considering that many investigators killed peccaries (Roosevelt, 1920), had access to dead animals from local hunters (Mayer and Brandt, 1982; Henry, 1997) or sedated them in the field (McCoy *et al.*,

1985; Fragoso, 1998a, 1999). In Colombia, Izawa and Kobayashi (1997) found that 38 animal species, including collared peccaries, dispersed attached seeds from *Pharus virescens*, a herbaceous bambusoid grass. In Peru and Bolivia, seeds of the genera *Bidens, Desmodium, Pavonia, Petiveria* and *Pharus* that had attached to peccaries were recovered around their wallows (R.B. Foster, Florida, 2003, personal communication). Therefore, peccaries disperse and may facilitate the establishment of epizoochorous species (Fig. 6.2). Investigators planning on sedating peccaries should consider carrying a hairbrush in their toolbox and quantifying the number of seeds and species attached to their fur.

Peccaries as ecosystem engineers

Peccaries can be considered ecosystem engineers (*sensu* Jones et al., 1994, 1997), because their rooting and wallowing behaviour leads to the removal of soil and leaf-litter. Soil and leaf-litter can act as physical or chemical barriers to the establishment of small-seeded and litter-gap-dependent species (e.g. Putz, 1983; Vázquez-Yanes et al., 1990; Metcalfe, 1996; Metcalfe and Turner, 1998; Farji-Brener and Illes, 2000). Thereby peccaries create new habitats which may permit the establishment of many small-seeded and litter-gap-dependent species. Peccary wallows are also critical breeding habitats for several amphibian species (Zimmerman and Bierregaard, 1986; Gascon, 1991; Zimmerman and Simberloff, 1996), which go locally extinct shortly after peccaries are extirpated (Simberloff, 1992).

Some evolutionary considerations

The earliest fossil records of peccaries in the New World date from the late Pliocene and Pleistocene (Woodburne, 1968, 1969; Simpson, 1984). All three present peccary species evolved in, rather than immigrated to, the Neotropics (Woodburne, 1968; Wetzel, 1977). Therefore, plant species in

this region should have had sufficient time to evolve mechanisms that minimize seed mortality and maximize dispersal by peccaries. For example, seeds of the genus *Couroupita* (Lecythidaceae) have evolved exotestal hairs that protect the embryo while passing through the digestive system of animals (Tsou and Mori, 2002). Furthermore, Prance and Mori (1978) and Tsou and Mori (2002) hypothesized that *C. guianensis* developed this unique seed coat as an adaptation to endozoochory by peccaries. Clearly, future studies need to consider evolutionary adaptations between peccaries and plant species.

Competitive Interactions

Between peccary species

Analyses of the diet of both peccary species indicated no significant differences between the frequency distributions of plant species consumed, either by geographic region or by utilization pattern. Furthermore, the dietary niche overlap between both peccaries was over 78% in the Neotropics (Table 6.3, see also Kiltie, 1981a; Bodmer, 1991b; Barreto et al., 1997; Painter, 1998). This level of resource overlap suggests that exploitative competition is potentially high between both peccary species (e.g. Schaller, 1983; Taber et al., 1993; Barreto et al., 1997; Fig. 6.2).

However, when comparing the species whose seeds survived with species whose seeds were destroyed, all niche measures indicate a slightly reduced overlap (Table 6.3). These findings are also consistent with those of the uniquely consumed plant families and species (Table 6.2). This decline might indicate some resource partitioning based on the ability to prey upon harder seed species. Evidence also comes from the observation that throughout the Neotropics white-lipped peccaries destroyed mostly hard-seeded species such as *Mauritia flexuosa, Attalea* spp. and *Socratea* spp., which was rarely the case for collared peccaries (Table 6.2; Appendices 6.1, 6.2). This

Table 6.3. Comparison of dietary niche overlap between collared peccaries (*Pecari tajacu*) and white-lipped peccaries (*Tayassu pecari*) calculated using Pianka's Measure, Percentage Overlap and Morisita's Measure for Neotropics, South America and Central America. Niche overlaps were calculated for three different utilization patterns. First, all plant species whose fruits were consumed by both peccaries. Second, plant species whose seeds survived (i.e. seeds were spat out). Third, plant species whose seeds were killed.

Utilization patterns	Pianka's measure	Percentage overlap	Morisita's measure
Neotropics			
All plants used	0.954	78.6	1.035
Seeds alive	0.964	79.3	1.123
Seeds killed	0.947	78.7	1.057
South America			
All plants used	0.968	80.3	1.087
Seeds alive	0.945	83.0	1.100
Seeds killed	0.946	80.0	1.114
Central America			
All plants used	0.838	58.7	1.038
Seeds alive	0.827	66.7	1.586
Seeds killed	0.819	59.5	1.025

may be explained by the 1.3 times stronger bite force of white-lipped peccaries as compared to collared peccaries (Kiltie, 1982). Therefore, Kiltie (1981c, 1982) and Kiltie and Terborgh (1983) hypothesized that palm seed hardness is crucial for determining diet niche divergence for sympatric peccary species. Other studies provide empirical evidence for resource partitioning based on seed hardness (Bodmer, 1989; Barreto *et al.*, 1997; Fragoso, 1998b, 1999; Painter, 1998).

Competition with other frugivorous species

This review illustrates that across the Neotropics both peccary species feed on fruits and seeds from over 207 species belonging to 55 families (Appendices 6.1, 6.2, 6.3). Therefore, I hypothesize that peccaries compete with other terrestrial frugivores for food resources in two ways (Fig. 6.2). First, via interference and/or exploitative competition, and second, via a kleptoparasitic foraging strategy. Empirical evidence for the

competition hypothesis comes from feeding observations and experiments on other frugivores such as rice rats (*Oryzomys* spp.), spiny rats (*Proechimys* spp.), paca (*Agouti paca*), agouti (*Dasyprocta* spp.), brocket deer (*Mazama* spp.) and tapir (*Tapirus* spp.) (Mori, 1970; Smythe, 1970a,b, 1978; Boucher, 1979, 1981a; Janzen, 1979, 1982; Guillotin, 1982; Russell, 1985; Smythe *et al.*, 1985; Hallwachs, 1986; Zona and Henderson, 1989; Bodmer, 1991a; Varela and Brown, 1995; Adler and Kestell, 1998; Forget *et al.*, 1998; Painter, 1998; Beck-King *et al.*, 1999; Silvius, 1999, 2002; Galetti *et al.*, 2001; Beck, 2002; Beck and Terborgh, 2002; Silvius and Fragoso, 2003). Shared plant species include *Astrocaryum* spp., *Attalea* spp., *Bertholletia excelsa*, *Brosimum* spp., *Dipteryx* spp., *Ficus* spp., *Quercus oleoides* and *Spondias* spp.

Empirical evidence supports the hypothesis that peccaries have a kleptoparasitic foraging strategy, by actively searching for seeds hidden by rodents (Kiltie, 1981b; Kiltie and Terborgh, 1983; Hallwachs, 1986; Janzen, 1986; Smythe, 1986, 1989; Ortiz, 1995; Fragoso, 1999; Jansen, 2003; Silman *et al.*, 2003). Numerous rodent genera including *Proechimys* spp., *Myoprocta* spp. (e.g. Morris, 1962; Smythe, 1978; Forget, 1990, 1991; Adler and Kestell, 1998) and *Dasyprocta* spp. (Smythe, 1970a,b, 1978; Hallwachs, 1986; Jansen, 2003) cache individual or multiple seeds, frequently near objects such as fallen logs, which may act as landmarks and thereby facilitate future recovery (e.g. Kiltie, 1981b; Kiltie and Terborgh, 1983; Forget, 1990; Jansen, 2003). Rodent-cached and peccary-recovered seeds include *Astrocaryum* spp. (Kiltie, 1981b; Smythe, 1989; Forget, 1991), *Bertholletia excelsa* (Ortiz, 1995) and *Hymenaea courbaril* (Hallwachs, 1986).

Studies that focus on the ecological consequences of species extinction might provide additional support for the hypothesis of interspecific competition. Local extinction of large-bodied species can lead to competitive release and result in density compensation (*sensu* MacArthur *et al.*, 1972) of smaller species (Emmons, 1984; Gliwicz, 1984; Brewer *et al.*, 1996; Asquith *et al.*,

1997; Bodmer *et al.*, 1997; Malcolm, 1997; Beck-King *et al.*, 1999; Peres and Dolman, 2000; Wright *et al.*, 2000; Roldán and Simonetti, 2001). Terborgh (1988, 1992) and Terborgh *et al.* (1999) hypothesized that top predators such as jaguars control populations of large- and medium-sized terrestrial mammals in a top-down fashion. They compared the densities of mammals from Cocha Cashu, an intact forest, with Barro Colorado Island (BCI), a fragmented forest, where top predators have been extinct, and found that the densities of medium-sized species were eight to 20 times higher on BCI (see also Enders, 1939; Glanz, 1982, 1991; Emmons, 1984). However, Terborgh and colleagues did not take into consideration that not only top predators, but also white-lipped peccaries have been extinct on BCI since the early 1930s (Glanz, 1982). Enders (1939) had already reported that on BCI large mammals were diminishing in numbers while small-bodied mammals increased in densities. Because white-lipped peccaries can have the highest biomass among terrestrial mammals (e.g. Silman *et al.*, 2003), and feed on many plant species that other frugivores use, bottom-up regulation might be an equally parsimonious mechanism to top-down regulation. Future research is needed to elucidate whether top-down and/or bottom-up regulation are occurring and quantify their effects on interspecific competition.

Epilogue

Both peccary species interact with at least 212 plant species (includes epizoochory) by killing or dispersing their seeds over short and/or long distances. Burying seeds, trampling of seedlings, ecosystem engineering, intra- and interspecific competition further increase the number and magnitude of peccary–plant and peccary–animal interactions. Because of these numerous (direct and indirect) positive and negative effects of peccaries on other organisms, their extinction will affect the spatiotemporal distribution and the demography of animals and plants; as well as the community composition and diversity of tropical ecosystems (see also De Steven and Putz, 1984; Sork, 1987; Terborgh, 1988, 1992; Dirzo and Miranda, 1990, 1991; Asquith *et al.*, 1997; Terborgh *et al.*, 1997, 1999, 2001; Wright *et al.*, 2000; Wright and Duber, 2001; Silvius and Fragoso, 2002; Silman *et al.*, 2003; Ticktin, 2003).

Peccary populations are continuously declining, and they are one of the most endangered species throughout the Neotropics (e.g. Ceballos and Navarro, 1991; Taber, 1991; Sowls, 1997). Depressingly, too many studies report local extinction of peccaries because of overhunting and/or habitat fragmentation (Enders, 1935; Leopold, 1972; IUCN, 1978; Coe and Diehl, 1980; Mayer and Brandt, 1982; Anderson, 1983; Broad, 1984; Emmons, 1984; Ceballos, 1985; Vaughan, 1985; Redford *et al.*, 1990; Anderson *et al.*, 1991; Ceballos and Navarro, 1991; Dirzo and Miranda, 1991; Glanz, 1991; Oliver and Santos, 1991; Robinson and Redford, 1991; Taber, 1991, 1993; Redford, 1992; March, 1993; Estrada *et al.*, 1994; Taber *et al.*, 1994; Timm, 1994; Ergueta and Morales, 1996; Peres, 1996, 2001; Bodmer *et al.*, 1997; Terborgh *et al.*, 1997; Chiarello, 1999; Wong *et al.*, 1999; Cullen *et al.*, 2000, 2001; Robinson and Bennett, 2000; Timm and LaVal, 2000; Wright *et al.*, 2000; Roldán and Simonetti, 2001).

I hope this chapter has not only reviewed and synthesized the literature on interactions of peccaries with other species, but also provides some focus for future research and conservation strategies, thus enabling future generations to enjoy what was once thought to be '. . . public enemy No. 1 among the mammals of tropical America . . .' (Chapman, 1936, p. 409).

Acknowledgements

I thank Mariana Altricher, Douglas Boucher, Steven Brewer, Francois Feer, Pierre-Michel Forget, Robin Foster, Mauro Galetti, Alexine Keuroghlian, Scott Mori, Steven Oberbauer, Fabio Olmos, Enrique Ortiz, Carlos Peres, Miles Silman, John

Vandermeer and Scott Zona for providing unpublished field observations or pointing out publications. Andrew Henderson helped with palm taxonomy. Special thanks to Dave Janos, Joe Wright and John Terborgh for stimulating discussions. Ted Fleming, Pierre-Michel Forget, Mike Gaines, Karim Ledesma, John Terborgh, Larry Wilson and three anonymous reviewers provided valuable comments on previous drafts. And finally a special thank-you to the employees of the interlibrary loan at the University of Miami for their enduring service in retrieving hundreds of resources.

References

Adler, G.H. and Kestell, D.W. (1998) Fates of Neotropical tree seeds influenced by spiny rats (*Proechimys semispinosus*). *Biotropica* 30, 677–681.

Agnew, A.D.Q. and Flux, J.E.C. (1970) Plant dispersal by hares (*Lepus capensis* L.) in Kenya. *Ecology* 51, 735–737.

Altrichter, M. and Almeida, R. (2002) Exploitation of white-lipped peccaries *Tayassu pecari* (Artiodactyla: Tayassuidae) on the Osa Peninsula, Costa Rica. *Oryx* 36, 126–132.

Altrichter, M., Sáenz, J. and Carrillo, E. (1999) Chanchos cariblancos (*Tayassu pecari*) como depredadores y dispersores de semillas en el Parque Nacional Corcovado, Costa Rica. *Brenesia* 52, 53–59.

Altrichter, M., Sáenz, J.C., Carrillo, E. and Fuller, T.K. (2000) Dieta estacional del *Tayassu peccari* (Artiodactyla: Tayassuidae) en el Parque Nacional Corcovado, Costa Rica. *Revista de Biología Tropical* 48, 689–702.

Altrichter, M., Carrillo, E., Sáenz, J. and Fuller, T.K. (2001) White-lipped peccary (*Tayassu pecari*, Artiodactyla: Tayassuidea) diet and fruit availability in a Costa Rican rain forest. *Revista de Biología Tropical* 49, 1183–1192.

Anderson, A.B. (1983) The biology of *Orbignya martiana* (Palmae), a tropical dry forest dominant in Brazil. PhD thesis, University of Florida, Gainesville, Florida.

Anderson, A.B., May, P.H. and Balick, M.J. (1991) *The Subsidy from Nature. Palm Forests, Peasantry, and Development on an Amazonian Frontier.* Columbia University Press, New York, 233 pp.

Asquith, N.M., Wright, S.J. and Clauss, M.J. (1997) Does mammal community composition control recruitment in Neotropical forests? Evidence from Panama. *Ecology* 78, 941–946.

Baker, R.H. (1983) *Sigmodon hispidus* (rata algodonera hispida), hispid cotton rat. In: Janzen, D.H. (ed.) *Costa Rica Natural History.* University of Chicago Press, Chicago, Illinois, pp. 490–492.

Barreto, G.R., Hernandez, O.E. and Ojasti, J. (1997) Diet of peccaries (*Tayassu tajacu* and *T. pecari*) in a dry forest of Venezuela. *Journal of Zoology* 241, 79–284.

Beck, H. (2002) Population dynamics and biodiversity of small mammals in treefall gaps within an Amazonian rainforest. PhD thesis, University of Miami, Miami, Florida.

Beck, H. and Terborgh, J. (2002) Groves versus isolates: how spatial aggregation of *Astrocaryum murumuru* palms affects seed removal. *Journal of Tropical Ecology* 18, 275–288.

Beck-King, H., Helversen, O.V. and Beck-King, R. (1999) Home range, population density, and food resources of *Agouti paca* (Rodentia: Agoutidae) in Costa Rica: a study using alternative methods. *Biotropica* 31, 675–685.

Belbenoit, P., Poncy, O., Sabatier, D., Prévost, M.-F., Riéra, B., Blanc, P., Larpin, D. and Sarthou, C. (2001) Floristic checklist of the Nouragues area. In: Bongers, F., Charles-Dominique, P., Forget, P.-M. and Théry, M. (eds) *Nouragues. Dynamics and Plant–Animal Interactions in a Neotropical Rainforest.* Kluwer Academic Publishers, Dordrecht, The Netherlands, pp. 301–341.

Bell, E.A. (1984) Toxic compounds in seeds. In: Murray, D.R. (ed.) *Seed Physiology*, Volume 1. Academic Press, Sydney, pp. 245–262.

Bodmer, R.E. (1989) Frugivory in Amazonian artiodactyla: evidence for the evolution of ruminant stomach. *Journal of Zoology* 219, 457–467.

Bodmer, R.E. (1990) Responses of ungulates to seasonal inundations in the Amazon floodplain. *Journal of Tropical Ecology* 6, 191–201.

Bodmer, R.E. (1991a) Strategies of seed dispersal and seed predation in Amazonian ungulates. *Biotropica* 23, 255–261.

Bodmer, R.E. (1991b) Influence of digestive morphology on resource partitioning in Amazonian ungulates. *Oecologia* 85, 361–365.

Bodmer, R.E., Eisenberg, J.F. and Redford, K.H. (1997) Hunting and the likelihood of extinction of Amazonian mammals. *Conservation Biology* 11, 460–466.

Bonaccorso, F.J., Glanz, W.E. and Sandford, C.M. (1980) Feeding assemblies of mammals at fruiting *Dipteryx panamensis*

(Papilionaceae) trees in Panama: seed predation, dispersal, and parasitism. *Revista de
Biología Tropical* 28, 61–72.

Bongers, F., Charles-Dominique, P., Forget, P.-M.
and Théry, M. (2001) *Nouragues. Dynamics
and Plant–Animal Interactions in Neotropical Rainforest.* Kluwer Academic Publishers,
Dordrecht, The Netherlands, 421 pp.

Borgtoft Petersen, H. and Balslev, H. (1990)
Ecuadorean Palms for Agroforestry. AAU
Reports 23, Botanical Institute Aarhus
University, Denmark, 122 pp.

Boucher, D.H. (1979) Seed predation and dispersal
by animals in a tropical dry forest. PhD
thesis, University of Michigan, Ann Arbor,
Michigan.

Boucher, D.H. (1981a) The 'real' disperser of
Swartzia cubensis. Biotropica 13, 77–78.

Boucher, D.H. (1981b) Seed predation by
mammals and forest dominance by *Quercus
oleoides*, a tropical lowland Oak. *Oecologia*
49, 409–414.

Brehm, A.E. (1891) *Die Säugetiere.* Dritter Band.
Bibliographisches Institut, Leibzig and
Vienna, 744 pp.

Brewer, S.W. (2001) Predation and dispersal of
large and small seeds of a tropical palm.
Oikos 92, 245–255.

Brewer, S.W., Rejmanek, M., Johnstone, E.E.
and Caro, T.M. (1996) Top-down control in
tropical forests. *Biotropica* 29, 364–367.

Broad, S. (1984) The peccary skin trade. *Traffic
Bulletin* 6, 27–28.

Carr, A.F. (1953) *High Jungles and Low.* University
of Florida Press, Gainesville, Florida, 226 pp.

Carr, A. (1967) *So Excellent a Fishe. A Natural
History of Sea Turtles.* The American
Museum of Natural History Press, Garden
City, New York, 248 pp.

Carrillo, E., Saenz, J.C. and Fuller, T.K. (2002)
Movements and activities of white-lipped
peccaries in Corcovado National Park, Costa
Rica. *Biological Conservation* 108, 317–324.

Castellanos, H.G. (1983) Aspectos de la
organización social del baquiro de collar,
Tayassu tajacu L., en el estado Guarico–
Venezuela. *Acta Biológica Venezuelica* 11,
127–143.

Castellanos, H.G. (1985) Home range size and habitat selection of the collared peccary in the
state of Guarico, Venezuela. In: Ockenfels,
R.A., Day, I.G. and Supplee, V.C. (eds)
Proceedings of the Peccary Workshop. University of Arizona, Tucson, Arizona, p. 50.

Castillo-Flores, A.A. and Calvo-Irabién, L.M.
(2003) Animal dispersal of two secondary-
vegetation herbs into the evergreen rain forest
of south-eastern Mexico. *Journal of Tropical
Ecology* 19, 271–278.

Ceballos, G. (1985) The importance of riparian
habitats for the conservation of endangered
mammals in Mexico. In: Johnson, R.R.,
Ziebell, C.D., Patton, D.R., Ffolliott, P.F. and
Hamre, R.H. (eds) *Riparian Ecosystems and
their Management: Reconciling Conflicting
used.* First North American Riparian
Conference, Tucson, Arizona, pp. 96–100.

Ceballos, G. and Navarro, D.L. (1991) Diversity
and conservation of Mexican mammals.
In: Mares, M.A. and Schmidly, D.J. (eds)
*Latin American Mammalogy. History, Bio-
diversity, and Conservation.* University
of Oklahoma Press, Norman, Oklahoma,
pp. 167–198.

Chapman, F.M. (1931) Seen from a tropical air
castle. *Natural History* 31, 347–358.

Chapman, F.M. (1936) White-lipped peccary.
Natural History 38, 408–413.

Chiarello, A.G. (1999) Effects of fragmentation
of Atlantic forest mammal communities in
south-eastern Brazil. *Biological Conservation*
89, 71–82.

Cintra, R. (1997) Leaf litter effects on seed and
seedling predation of the palm *Astrocaryum
murumuru* and the legume tree *Dipteryx
micrantha* in Amazonian forest. *Journal of
Tropical Ecology* 13, 709–725.

Cintra, R. (1998) Sobrevivência pós-dispersão
de sementes e plântulas de três espécies
de palmeiras em relação a presença de
componentes da complexidade estrutural
da floresta Amazônica. In: Gascon, C. and
Moutinho, P. (eds) *Floresta Amazônica:
Dinâmica, Regeneração e Manejo.* Ministério
da Ciência e Technologia, Manaus, pp. 83–98.

Cintra, R. and Horna, V. (1997) Seed and seedling
survival of the palm *Astrocaryum murumuru*
and the legume tree *Dipteryx micrantha* in
gaps in Amazonian forest. *Journal of Tropical
Ecology* 13, 257–277.

CITES (2003) Convention on International
Trade in Endangered Species of Wild
Fauna and Flora. Available at: http://www.
cites.org

Clark, D.B. and Clark, D.A. (1989) The role of
physical damage in the seedling mortality
regime of a Neotropical rain forest. *Oikos* 55,
225–230.

Coe, M.D. and Diehl, R.A. (1980) *In the Land of the
Olmec. The People of the River.* Volume 2.
University of Texas Press, Austin, Texas,
198 pp.

Comizzoli, P., Peiniau, J., Dutertre, C., Planquette,
P. and Aumaitre, A. (1997) Digestive

utilization of concentrated and fibrous diets by two peccary species (*Tayassu pecari, Tayassu tajacu*) raised in French Guyana. *Animal Feed Science Technology* 64, 215–226.

Cullen, L. Jr, Bodmer, R.E. and Valladares-Padua, C. (2000) Effects of hunting in habitat fragments of the Atlantic forests, Brazil. *Biological Conservation* 95, 49–56.

Cullen, L. Jr, Bodmer, R.E. and Valladares-Padua, C. (2001) Ecological consequences of hunting in Atlantic forest patches, São Paulo, Brazil. *Oryx* 35, 137–144.

Cutright, P.R. (1940) *The Great Naturalists Explore South America*. The Macmillan Company, New York, 340 pp.

De Steven, D. and Putz, F.E. (1984) Impact of mammals on early recruitments of a tropical canopy tree, *Dipteryx panamensis*, in Panama. *Oikos* 43, 207–216.

Dirzo, R. and Miranda, A. (1990) Contemporary neotropical defaunation and forest structure, function, and diversity – a sequel to John Terborgh. *Conservation Biology* 4, 444–447.

Dirzo, R. and Miranda, A. (1991) Altered patterns of herbivory and diversity in the forest understory: a case study of the possible consequences of contemporary defaunation. In: Price, P.W., Lewinsohn, T.M., Fernandes, G.W. and Benson, W.W. (eds) *Plant–Animal Interaction: Evolutionary Ecology in Tropical and Temperate Regions*. John Wiley & Sons, New York, pp. 273–287.

Donkin, R.A. (1985) The peccary – with observations on the introduction of pigs to the New World. *Transactions of the American Philosophical Society*, Vol. 75.

Duke, J.A. (1967) *Mammal Dietary*. Battelle Memorial Institute, Columbus, Ohio, 33 pp.

Eisenberg, J.F. (1980) The density and biomass of tropical mammals. In: Soulé, M.E. and Wilcox, B.A. (eds) *Conservation Biology an Evolutionary-Ecological Perspective*. Sinauer Associates, Inc., Sunderland, Massachusetts, pp. 35–55.

Eisenberg, J.F. and Redford, K.H. (1999) *Mammals of the Neotropics. The Central Neotropics Ecuador, Peru, Bolivia, Brazil*, Vol. 3. The University of Chicago Press, Chicago, Illinois, 609 pp.

Emmons, L.H. (1984) Geographic variation in densities and diversities of non-flying mammals in Amazonia. *Biotropica* 16, 210–222.

Enders, R.K. (1930) Notes on some mammals from Barro Colorado Island, Canal Zone. *Journal of Mammalogy* 11, 280–292.

Enders, R.K. (1935) Mammalian life histories from Barro Colorado Island, Panama. *Bulletin of the Museum of Comparative Zoology* 78, 385–502.

Enders, R.K. (1939) Changes observed in the mammal fauna of Barro Colorado Island, 1929–1937. *Ecology* 20, 104–106.

Ergueta, P.S. and Morales, C. de (1996) *Libro Rojo de los Vertrebrados de Bolivia*. Centro de Datos para la Conservación, La Paz, Bolivia, 347 pp.

Estrada, A. and Fleming, T.H. (1986) *Frugivores and Seed Dispersal*. Dr W. Junk Publishers, Dordrecht, The Netherlands, 392 pp.

Estrada, A., Coates-Estrada, R. and Meritt, D. Jr (1994) Non flying mammals and landscape changes in the tropical rain forest of Los Tuxtlas, Mexico. *Ecography* 17, 229–242.

Farji-Brener, A.G. and Illes, A.E. (2000) Do leaf-cutting ant nets make 'bottom-up' gaps in neotropical rain forest? A critical review of the evidence. *Ecology Letters* 3, 219–227.

Feer, F. and Forget, P.-M. (2002) Spatio-temporal variations in post-dispersal seed fate. *Biotropica* 34, 555–566.

Feer, F., Henry, O., Forget, P.-M. and Gayot, M. (2001) Frugivory and seed dispersal by terrestrial mammals. In: Bongers, F.P., Charles-Dominique, P., Forget, P.-M. and Théry, M. (eds) *Nouragues. Dynamics and Plant–Animal Interactions in a Neotropical Rainforest*. Kluwer Academic Publishers, Dordrecht, The Netherlands, pp. 227–232.

Fleming, T.H. and Estrada, A. (1993) *Frugivory and Seed Dispersal: Ecological and Evolutionary Aspects*. Kluwer Academic Publishers, Dordrecht, The Netherlands, 392 pp.

Forget, P.-M. (1990) Seed-dispersal of *Vouacapoua americana* (Caesalpiniaceae) by caviomorph rodents in French Guiana. *Journal of Tropical Ecology* 6, 459–468.

Forget, P.-M. (1991) Scatterhoarding of *Astrocaryum paramaca* by Proechimys in French Guiana: comparison with *Myoprocta exilis*. *Tropical Ecology* 32, 155–167.

Forget, P.-M. (1996) Removal of seeds of *Carapa procera* (Meliaceae) by rodents and their fate in rainforest in French Guiana. *Journal of Tropical Ecology* 12, 751–761.

Forget, P.-M. and Hammond, D.S. (2005) Vertebrates and food plant diversity in Guianan rainforest. In: Hammond, D.S. (ed.) *Tropical Rainforests of the Guiana Shield*. CAB International, Wallingford, UK (in press).

Forget, P.-M., Milleron, T. and Feer, F. (1998) Patterns in post-dispersal seed removal by neotropical rodents and seed fate in relation to seed size. In: Newbery, D.M., Prins, H.H.T. and Brown, N.D. (eds) *Dynamics of Tropical Communities.* Blackwell Science, Oxford, pp. 25–49.

Forget, P.-M., Randin-de Merona, J.M. and Julliot, C. (2001) The effects of forest type, harvesting and stand refinement on early seedling recruitment in a tropical rain forest. *Journal of Tropical Ecology* 17, 593–609.

Forget, P.-M., Hammond, D.S., Milleron, T. and Thomas, R. (2002) Seasonality of fruiting and food hoarding by rodents in Neotropical Forests: consequences for seed dispersal and seedling recruitments. In: Levey, D.J., Silva, W.R. and Galetti, M. (eds) *Seed Dispersal and Frugivory: Ecology, Evolution and Conservation.* CAB International, Wallingford, UK, pp. 241–256.

Frädrich, H. (1968) Swine and peccaries. In: Grzimek, B. (ed.) *Grzimek's Animal Life Encyclopedia*, Volume 13. Van Nostrand Reinhold Company, New York, pp. 76–108.

Fragoso, J.M.V. (1994) Large mammals and the community dynamics of an Amazonian rain forest. PhD thesis, University of Florida, Gainesville, Florida.

Fragoso, J.M.V. (1997) Tapir-generated seed shadows: scale-dependent patchiness in the Amazon rain forest. *Journal of Ecology* 85, 519–529.

Fragoso, J.M.V. (1998a) Home range and movement patterns of white-lipped peccary (*Tayassu pecari*) herds in the northern Brazilian Amazon. *Biotropica* 30, 458–469.

Fragoso, J.M.V. (1998b) White-lipped peccaries and palms on the Ilha de Maracá. In: Milliken, W. and Ratter, J.A. (eds) *Maracá: The Biodiversity and Environment of an Amazonian Rainforest.* John Wiley & Sons Ltd, Chichester, UK, pp. 151–162.

Fragoso, J.M.V. (1999) Perception of scale and resource partitioning by peccaries: behavioral causes and ecological implications. *Journal of Mammalogy* 80, 993–1003.

Freeland, W.J. and Janzen, D.H. (1974) Strategies in herbivory by mammals: the role of plant secondary compounds. *The American Naturalist* 108, 269–289.

Galetti, M. (2002) Seed dispersal of mimetic fruits: parasitism, mutualism, aposematism or exaptation. In: Levey, D.J., Silva, W.R. and Galetti, M. (eds) *Seed Dispersal and Frugivory: Ecology, Evolution and Conservation.* CAB International, Wallingford, UK, pp. 177–191.

Galetti, M., Zipparro, V.B. and Morellato, P.C. (1999) Fruiting phenology and frugivory on the palm *Euterpe edulis* in a lowland Atlantic forest of Brazil. *Ecotropica* 5, 115–122.

Galetti, M., Keuroghlian, A., Hanada, L. and Morato, M.I. (2001) Frugivory and seed dispersal by the lowland tapir (*Tapirus terrestris*) in Southeast Brazil. *Biotropica* 33, 723–726.

Gascon, C. (1991) Population- and community-level analyses of species occurrences of Central Amazonian rainforest tadpoles. *Ecology* 72, 1731–1746.

Gaumer, G.F. (1917) *Monografía de los Mamíferos de Yucatán.* Departamento de Talleres Gráficos de la Secretaría de Fomento, México, 331 pp.

Gautier-Hion, A., Duplantier, J.-M., Quris, R., Feer, F., Sourd, C., Decoux, J.-P., Dubost, G., Emmons, L., Erard, C., Hecketsweiler, P., Moungazi, A., Roussilhon, C. and Thiollay, J.-M. (1985) Fruit characters as a basis of fruit choice and seed dispersal in a tropical vertebrate community. *Oecologia* 65, 324–337.

Gentry, A.H. (1988) Changes in plant community diversity and floristic composition on geographical and environmental gradients. *Annals of the Missouri Botanical Garden* 75, 1–34.

Gentry, A.H. (1990) Floristic similarities and differences between southern central America and upper and central Amazonia. In: Gentry, A.H. (ed.) *Four Neotropical Rainforests.* Yale University Press, New Haven, Connecticut, pp. 141–157.

Gentry, A.H. (1993) *A Field Guide to the Families and Genera of Woody Plants of Northwest South America.* Conservation International, Washington, DC, 894 pp.

Gentry, A.H. and Terborgh, J. (1990) Composition and dynamics of the Cocha Cashu 'mature' floodplain forest. In: Gentry, A.H. (ed.) *Four Neotropical Rainforests.* Yale University Press, New Haven, Connecticut, pp. 524–564.

Glanz, W.E. (1982) The terrestrial mammal fauna of Barro Colorado Island: censuses and long-term changes. In: Leigh, E.G. Jr, Rand, A.S. and Windsor, D.M. (eds) *The Ecology of a Tropical Forest: Seasonal Rhythms and Long-term Changes.* Smithsonian Institution Press, Washington, DC, pp. 455–468.

Glanz, W.E. (1991) Mammalian densities at protected versus hunted sites in central Panama. In: Robinson, J.G. and Redford, K.H. (eds) *Neotropical Wildlife Use and Conservation.* The University of Chicago Press, Chicago, Illinois, pp. 163–173.

Gliwicz, J. (1984) Population dynamics of the spiny rat *Proechimys semispinosus* on

Orchid Island (Panama). *Biotropica* 16, 73–78.

Gottsberger, G. and Silberbauer-Gottsberger, I. (1983) Dispersal and distribution in the Cerrado vegetation of Brazil. *Sonderbände des Naturwissenschaftlichen Vereins* 7, 315–352.

Guillotin, M. (1982) Rythmes d'activité et régimes alimentaires de *Proechimys cuvieri* et d'*Oryzomys capito velutinus* (Rodentia) en forêt Guyanaise. *Revue d'Ecologie (La Terre et la Vie)* 36, 337–371.

Hallwachs, W. (1986) Agoutis (*Dasyprocta punctata*): the inheritors of Guapinol (*Hymenaea courbaril*: Leguminosae). In: Estrada, A. and Fleming, T.H. (eds) *Frugivores and Seed Dispersal*. Dr W. Junk Publishers, Dordrecht, The Netherlands, pp. 285–304

Harper, J.L., Lovell, P.H. and Moore, K.G. (1970) The shapes and size of seeds. *Annual Review of Ecology and Systematics* 1, 327–356.

Hartshorn, G. and Hammel, B.E. (1994) Vegetation types and floristic patterns. In: McDade, L.A., Bawa, K.S., Hespenheide, H.A. and Hartshorn, G.S. (eds) *La Selva. Ecology and Natural History of a Neotropical Rain Forest*. The University of Chicago Press, Chicago, Illinois, pp. 73–89.

Henderson, A. (1990) Arecaceae. Part I. Introduction and the Iriarteinae. *Flora Neotropica Monograph* 53, 1–100.

Henderson, A. (1995) *The Palms of the Amazon*. Oxford University Press, New York, 362 pp.

Henderson, A. (2002) *Evolution and Ecology of Palms*. The New York Botanical Garden Press, New York, 259 pp.

Henderson, A., Galeano, G. and Bernal, R. (1995) *Field Guide to the Palms of the Americas*. Princeton University Press, Princeton, New Jersey, 352 pp.

Henry, O. (1997) The influence of sex and reproductive state on diet preference in four terrestrial mammals of the French Guianan rain forest. *Canadian Journal of Zoology* 75, 929–935.

Hermann, R. (1836) Report of an Expedition into the interior of British Guayana in 1835–6. *The Journal of the Royal Geographic Society of London* 6, 224–284.

Herrera, C.M. and Pellmyr, O. (2002) *Plant–Animal Interactions. An Evolutionary Approach*. Blackwell Science, Oxford, 313 pp.

Herring, S.W. (1972) The role of canine morphology in the evolutionary divergence of pigs and peccaries. *Journal of Mammalogy* 53, 500–512.

Hershkovitz, P. (1969) The evolution of mammals on southern continents. *The Quarterly Review of Biology* 44, 1–70.

Hildebrand, F. (1873) *Die Verbreitungsmittel der Pflanzen*. Verlag von Wilhelm Engelmann, Leipzig, 162 pp.

Hopkins, H.C. and Hopkins, M.J.G. (1983) Fruit and seed biology of the neotropical species of *Parkia*. In: Sutton, S.L., Whitmore, T.C. and Chadwick, A.C. (eds) *Tropical Rain Forest: Ecology and Management*. Blackwell Scientific Publications, Oxford, pp. 197–209.

Howe, H.F. (1986) Seed dispersal by fruit-eating birds and mammals. In: Murray, D.R. (ed.) *Seed Dispersal*. Academic Press, Sydney, pp. 123–189.

Howe, H.F. (1989) Scatter- and clump-dispersal and seedling demography: hypothesis and implications. *Oecologia* 79, 417–426.

Howe, H.F. and Smallwood, J. (1982) Ecology of seed dispersal. *Annual Review of Ecology and Systematics* 13, 201–228.

Hubbell, S.P. (1979) Tree dispersion, abundance and diversity in a tropical dry forest. *Science* 203, 1299–1309.

Husson, A.M. (1978) *The Mammals of Suriname*. Rijksmuseum van Natuurlijke Historie, Leiden, The Netherlands, 569 pp.

Hutterer, R. (1998) Diversity of mammals in Bolivia. In: Barthlott, W. and Winiger, M. (eds) *Biodiversity. A Challenge for Development and Research Policy*. Springer, Berlin, pp. 279–288.

IUCN (1978) The Chacoan peccary. In: *Red Data Book for Mammals*. International Union for the Conservation of Nature and Natural Resources, Morges, Switzerland, pp. 19.121.1a.1.

Izawa, K. and Kobayashi, M. (1997) Seed dispersers of the Monocarpic herbaceous Bamboo *Pharus virescence* (Poaceae: Bambusoideae) found in the Neotropical rain forest of La Macarena and Trinigua National Parks of Colombia. *Tropics* 7, 153–159.

Jansen, P.A. (2003) Scatterhoarding and tree regeneration: ecology of nut dispersal in a neotropical rainforest. PhD thesis, Wageningen University, Wageningen, The Netherlands.

Jansen, P.A. and Forget, P.-M. (2001) Scatter-hoarding rodents and tree regeneration. In: Bongers, F.P., Charles-Dominique, P., Forget, P.-M. and Théry, M. (eds) *Nouragues. Dynamics and Plant–Animal Interactions in a Neotropical Rainforest*. Kluwer Academic

Publishers, Dordrecht, The Netherlands, pp. 275–288.

Jansen, P.A., Bartholomeus, M., Bongers, F., Elzinga, J.A., den Ouden, J. and Wieren, S.E.V. (2002) The role of seed size in dispersal by a scatter-hoarding rodent. In: Levey, D.J., Silva, W.R. and Galetti, M. (eds) *Seed Dispersal and Frugivory: Ecology, Evolution and Conservation*. CAB International, Wallingford, UK, pp. 209–225.

Janzen, D.H. (1969) Seed-eaters versus seed size, number, toxicity and dispersal. *Evolution* 32, 1–27.

Janzen, D.H. (1971) Seed predation by animals. *Annual Review of Ecology and Systematics* 2, 456–492.

Janzen, D.H. (1972) Escape in space by *Sterculia apetala* seeds from the bug *Dysdercus fasciatus* in a Costa Rican deciduous forest. *Ecology* 53, 350–361.

Janzen, D.H. (1974) Tropical blackwater rivers, and mast fruiting by the Dipterocarpaceae. *Biotropica* 6, 69–103.

Janzen, D.H. (1975) Intra- and interhabitat variations in *Guazuma ulmifolia* (Sterculiaceae) seed predation by *Amblycerus cistelinus* (Bruchidae) in Costa Rica. *Ecology* 56, 1009–1013.

Janzen, D.H. (1977) Why fruits rot, seeds mold, and meat spoils. *The American Naturalist* 111, 691–713.

Janzen, D.H. (1979) How to be a fig. *Annual Review of Ecology and Systematics* 10, 13–51.

Janzen, D.H. (1982) Seeds in tapir dung in Santa Rosa National Park, Costa Rica. *Brenesia* 19/20, 129–135.

Janzen, D.H. (1983a) Hymenaea courbaril (Guapinol, Stinking Toe). In: Janzen, D.H. (ed.) *Costa Rican Natural History*. University of Chicago Press, Chicago, Illinois, pp. 253–256.

Janzen, D.H. (1983b) *Pithecellobium saman* (Cenízero, Genízero, Raintree). In: Janzen, D.H. (ed.) *Costa Rican Natural History*. University of Chicago Press, Chicago, Illinois, pp. 305–306.

Janzen, D.H. (1983c) Dispersal of seeds by vertebrate guts. In: Futuyma, D.J. and Slatkin, M. (eds) *Coevolution*. Sinauer Associates, Sunderland, Massachusetts, pp. 231–262.

Janzen, D.H. (1984) Dispersal of small seeds by big herbivores: foliage is the fruit. *The American Naturalist*, 123, 338–353.

Janzen, D.H. (1985) *Spondias mombin* is culturally deprived in megafauna-free forest. *Journal of Tropical Ecology* 1, 131–155.

Janzen, D.H. (1986) Mice, big mammals, and seeds: it matters who defecates what where. In: Estrada, A. and Fleming, T.H. (eds) *Frugivores and Seed Dispersal*. Dr W. Junk Publishers, Dordrecht, The Netherlands, pp. 251–271.

Janzen, D.H. and Higgins, M.L. (1979) How hard are *Enterolobium cyclocarpum* (Leguminosae) seeds? *Brenesia* 16, 61–76.

Janzen, D.H and Martin, P.S. (1982) Neotropical anachronisms: the fruits the Gomphotheres ate. *Science* 215, 19–27.

Janzen, D.H. and Wilson, D.E. (1983) Mammals. In: Janzen, D.H. (ed.) *Costa Rican Natural History*. University of Chicago Press, Chicago, Illinois, pp. 426–440.

Jardine, W.B. (1836) *The Naturalist's Library. Mammalia. Thick-skinned Quadrupeds*, Vol. XXIII. Henry G. Bohn, London, 248 pp.

Jones, C.G., Lawton, J.H. and Shachak, M. (1994) Organisms as ecosystem engineers. *Oikos* 69, 373–386.

Jones, C.G., Lawton, J.H. and Shachak, M. (1997) Positive and negative effects of organisms as physical ecosystem engineers. *Ecology* 78, 1946–1957.

Jordano, P. (1983) Fig-seed predation and dispersal by birds. *Biotropica* 15, 38–41.

Judas, J. and Henry, O. (1999) Seasonal variation of home range of collared peccaries in tropical rain forests of French Guiana. *Journal of Wildlife Management* 63, 546–552.

Kahn, F. (1991) Palms as key swamp forest resources in Amazonia. *Forest Ecology and Management* 38, 133–142.

Kahn, F. and de Granville, J.-J. (1992) *Palms in Forest Ecosystems of Amazonia*. Springer-Verlag, Berlin, 226 pp.

Kappler, A. (1887) *Surinam, sein Land, seine Natur, Bevölkerung und seine Kultur-Verhältnisse mit Bezug auf Kolonisation*. J.G. Cotta'sche Verlagsbuchhandlung, Stuttgart, 383 pp.

Kenney, A.J. and Krebs, C.J. (2000) *Programs for Ecological Methodology*. Version 5.2. University of British Columbia, Vancouver.

Keuroghlian, A. (2003) The response of peccaries to seasonal fluctuations in an isolated patch of tropical forest. PhD thesis, University of Nevada, Reno, Nevada.

Kiltie, R.A. (1980) More on Amazon cultural ecology. *Current Anthropology* 21, 551–546.

Kiltie, R.A. (1981a) Stomach contents of rain forest peccaries (*Tayassu tajacu* and *T. pecari*). *Biotropica* 13, 234–236.

Kiltie, R.A. (1981b) Distribution of palm fruits on a rain forest floor: why white-lipped peccaries forage near objects. *Biotropica* 13, 141–145.

Kiltie, R.A. (1981c) The function of inter-locking canines in rain forest peccaries (Tayassuidae). *Journal of Mammalogy* 62, 459–469.

Kiltie, R.A. (1982) Bite force as a basis for niche differentiation between rain forest peccaries (*Tayassu tajacu* and *T. pecari*). *Biotropica* 14, 188–195.

Kiltie, R.A. and Terborgh, J. (1983) Observations on the behavior of rain forest peccaries in Peru: why do white-lipped peccaries form herds? *Zeitschrift für Tierpsychologie* 62, 214–255.

Krebs, C.J. (1999) *Ecological Methodology*, 2nd edn. Addison Wesley Longman, Menlo Park, California, 626 pp.

Langer, P. (1978) Anatomy of the stomach of collared peccary, *Dicotyles tajacu* (L. 1758) (Artiodactyla: Mammalia). *Zeitschrift für Säugetierkunde* 44, 321–333.

Leopold, A.S. (1972) *Wildlife in Mexico*, 2nd edn. University of California Press, Berkeley, California, 568 pp.

Levey, D.J., Silva, W.R. and Galetti, M. (2002) *Seed Dispersal and Frugivory: Ecology, Evolution and Conservation*. CAB International, Wallingford, UK, 511 pp.

Lieberman, M. and Lieberman, D. (1994) Patterns of density and dispersion of forest trees. In: McDade, L.A., Bawa, K.S., Hespenheide, H.A. and Hartshorn, G.S. (eds) *La Selva. Ecology and Natural History of a Neotropical Rain Forest*. The University of Chicago Press, Chicago, Illinois, pp. 106–119.

Lleellish, M., Amanzo, J., Hooker, Y. and Yale, S. (2003) Evaluacíon poblacional de pecaríes en la región del Alto Purús. In: Pitman, R.L., Pitman, N. and Álvarez, P. (eds) *Alto Purús biodiversidad, conservación y manejo*. Impresso Gráfica S.A., Lima, pp. 137–145.

MacArthur, R.H., Diamond, J.M. and Karr, J.R. (1972) Density compensation in island faunas. *Ecology* 53, 330–342.

Malcolm, J.R. (1997) Biomass and diversity of small mammals in Amazonian forest fragments. In: Laurance, W.F. and Bierregaard, R.O. Jr (eds) *Tropical Forest Remnants. Ecology, Management, and Conservation of Fragmented Communities*. The University of Chicago Press, Chicago, Illinois, pp. 207–221.

Mandujano, S., Gallina, S. and Bullock, S.H. (1994) Frugivory and dispersal of *Spondias purpurea* (Anacardiaceae) in a tropical

deciduous forest in Mexico. *Revista de Biología Tropical* 42, 107–114.

March, M.I.J. (1990) Monographie des Weißbart-pekaris (*Tayassu pecari*). *Bongo, Frädrich-Jubiläumsband* 18, 151–170.

March, M.I.J. (1993) The white-lipped peccary (*Tayassu pecari*). In: Oliver, W.L.R. (ed.) *Pigs, Peccaries, and Hippos: Status Survey and Conservation Action Plan*. IUCN, World Conservation Union, Gland, Switzerland, pp. 7–13.

Martínez-Gallardo, R. and Cordero, V.S. (1997) Historia natural de algunas especies de mamíferos terrestres. In: Soriano, E.G., Dirzo, R. and Vogt, R.C. (eds) *Historia Natural de Los Tuxtlas*. Universidad Nacional Autónoma de México, México, pp. 591–624.

Martínez-Ramos, M. (1997) *Astrocaryum mexicanum*. In: Soriano, E.G., Dirzo, R. and Vogt, R.C. (eds) *Historia Natural de Los Tuxtlas*. Universidad Nacional Autónoma de México, México, pp. 92–95.

Martínez-Romero, L.E. and Mandujano, S. (1995) Habitos alimentarios del pecari de collar (*Pecari tajacu*) en un bosque tropical caducifolio de Jalisco, México. *Acta Zoológica Mexicana* 64, 1–20.

Mayer, J.J. and Brandt, P.N. (1982) Identity, distribution, and natural history of the peccaries, Tayassuidae. In: Mares, M.A. and Genoways, H.H. (eds) *Mammalian Biology in South America*. The Pymatuning Symposia in Ecology, Volume 6. University of Pittsburgh, Pittsburgh, Pennsylvania, pp. 433–455.

Mayer, J.J. and Wetzel, R.M. (1986) *Catagonus wagneri*. *Mammalian Species* 259, 1–5.

Mayer, J.J. and Wetzel, R.M. (1987) *Tayassu pecari*. *Mammalian Species* 293, 1–7.

McCoy, M.B., Vaughan, C.S., Rodriguez, M.A. and Kitchen, D. (1985) Movement, activity and diet of collared peccaries in Costa Rican dry forest. In: Ockenfels, R.A., Day, I.G. and Supplee, V.C. (eds) *Proceedings of the Peccary Workshop*. University of Arizona, Tucson, Arizona, pp. 52–53.

McCoy, M.B., Vaughan, C.S., Rodriguez, M.A. and Kitchen, D. (1990) Seasonal movement, home range, activity and diet of collared peccaries (*Tayassu tajacu*) in Costa Rican dry forest. *Vida Silvestre Neotropical* 2, 6–20.

McHargue, L.A. and Hartshorn, G.S. (1983) Seed and seedling ecology of *Carapa guianensis*. *Turrialba* 33, 399–404.

Mena, P.V., Stallings, J.R., Regalado, J.B. and Cueva, R.L. (2000) The sustainability of current hunting practices by the Huaorani. In: Robinson, J.G. and Bennett, E.L. (eds) *Hunting for Sustainability in Tropical Forests.* Columbia University Press, New York, pp. 57–78.

Metcalfe, D.J. (1996) Germination of small-seeded tropical rain forest plants exposed to different spectral compositions. *Canadian Journal of Botany* 74, 516–520.

Metcalfe, D.J. and Turner, I.M. (1998) Soil seed bank from lowland rain forest in Singapore: canopy-gap and litter-gap demanders. *Journal of Tropical Ecology* 14, 103–108.

Miller, C.J. (1990) Natural history, economic botany, and germplasm conservation of the Brazil nut tree (*Bertholletia excelsa*, Humb & Bonpl.). MSc thesis, University of Florida, Gainesville, Florida.

Miller, F.W. (1930) Notes on some mammals of southern Matto Grosso, Brazil. *Journal of Mammalogy* 11, 10–22.

Mitja, D. and Ferraz, I.D.K. (2001) Establishment of Babassu in pastures in Pará, Brazil. *Palms* 45, 138–147.

Mori, S. (1970) The ecology and uses of the species of *Lecythis* in Central America. *Turrialba* 20, 344–350.

Morris, D. (1962) The behaviour of the green acouchi (*Myoprocta pratti*) with special reference to scatter hoarding. *Proceedings of the Zoological Society of London* 139, 701–733.

Müller, P. (1955) *Verbreitungsbiologie der Blütenpflanzen.* Verlag Hans Huber, Bern, 152 pp.

Murray, D.R. (1986) *Seed Dispersal.* Academic Press, Sydney, 322 pp.

Oliver, W.L.R. and Santos, I.B. (1991) Threatened endemic mammals of the Atlantic forest region of south-east Brazil. *Special Science Report 4.* Jersey Wildlife Preservation Trust, Jersey, UK.

Olmos, F. (1993) Diet of sympatric Brazilian caatinga peccaries (*Tayassu tajacu* and *T. pecari*). *Journal of Tropical Ecology* 9, 255–258.

Olmos, F., Pardini, R., Boulhosa, R.L.P., Bürgi, R. and Morsello, C. (1999) Do tapirs steal food from palm seed predators or give them a lift? *Biotropica* 31, 375–379.

Ortiz, E.G. (1995) Survival in a nutshell. *Américas* September/October, 7–17.

Painter, R.L.E. (1998) Gardeners of the forest: plant–animal interactions in a neotropical forest ungulate community. PhD thesis, University of Liverpool, Liverpool, UK.

Peres, C.A. (1994) Composition, density, and fruiting phenology of arborescent palms in an Amazonian Terra Firme Forest. *Biotropica* 26, 285–294.

Peres, C.A. (1996) Population status of white-lipped *Tayassu pecari* and collared peccaries *T. tajacu* in hunted and unhunted Amazonian forests. *Biological Conservation* 77, 115–123.

Peres, C.A. (2001) Synergistic effects of subsistence hunting and habitat fragmentation on Amazonian forest vertebrates. *Conservation Biology* 15, 1490–1505.

Peres, C.A. and Dolman, P.M. (2000) Density compensation in neotropical primate communities: evidence from 56 hunted and nonhunted Amazonian forests of varying productivity. *Oecologia* 122, 175–189.

Perry, R. (1970) *The World of the Jaguar.* Taplinger Press, New York, 168 pp.

Pijl, L. van der (1969) *Principles of Dispersal in Higher Plants.* Springer-Verlag, Berlin, 153 pp.

Pitman, N.C.A., Terborgh, J., Silman, M.R., Nuòez, P., Neill, V.D.A., Ceron, C.E., Palacios, W.A. and Aulestia, M. (2001) Dominance and diversity of tree species in upper Amazonian terra firme forests. *Ecology* 82, 2101–2117.

Pitman, N.C.A., Terborgh, J., Nuòez, P. and Valenzuela, M. (2003) Los árboles de la cuenca del Río Alto Purús. In: Pitman, R.L., Pitman, N. and Álvarez, P. (eds) *Alto Purús Biodiversidad, Conservación y Manejo.* Impresso Gráfica S.A., Lima, pp. 53–61.

Prance, G.T. and Mori, S.A. (1978) Observations on the fruits and seeds of neotropical Lecythidaceae. *Brittonia* 30, 21–33.

Prance, G.T. and Mori, S.A. (1979) Lecythidaceae – Part I. *Flora Neotropica* 21, 1–270.

Price, P.W., Fernandes, G.W., Lewinsohn, T.M. and Benson, W.W. (1991) *Plant Animal Interactions. Evolutionary Ecology in Tropical and Temperate Regions.* John Wiley & Sons, New York, 639 pp.

Putz, F.E. (1983) Treefall pits and mounds, buried seeds, and the importance of disturbance to pioneer trees on Barro Colorado Island, Panama. *Ecology* 64, 1069–1074.

Quiroga-Castro, V. and Roldán, A.I. (2001) The fate of *Attalea phalerata* (Palmae) seeds dispersed to a tapir latrine. *Biotropica* 33, 472–477.

Redford, K.H. (1992) The empty forest. *Bioscience* 42, 412–423.

Redford, K.H., Taber, A. and Simonetti, J.A. (1990) There is more to biodiversity than the tropical rain forests. *Conservation Biology* 4, 328–330.

Reis, A. (1995) Dispersão de sementes de *Euterpe edulis* Mautius–(Palmae) em uma floresta ombrófila densa montana da encosta Atlântica em Blumenau, SC. PhD thesis, Universidade Estadual de Campinas, São Paulo, Brazil.

Rengger, J.R. (1830) *Naturgeschichte der Saeugethiere von Paraguay*. Schweighauserschen Buchhandlung, Basel, 394 pp.

Ridley, H.N. (1930) *The Dispersal of Plants throughout the World*. L. Reeve & Co., Ashford, UK, 744 pp.

Robinson, J.G. and Bennett, E.L. (2000) *Hunting for Sustainability in Tropical Forests*. Columbia University Press, New York, 582 pp.

Robinson, J.G. and Eisenberg, J.F. (1985) Group size and foraging habitats of the collared peccary *Tayassu tajacu*. *Journal of Mammalogy* 66, 153–155.

Robinson, J.G. and Redford, K.H. (1991) *Neotropical Wildlife Use and Conservation*. The University of Chicago Press, Chicago, Illinois, 520 pp.

Roldán, A.I. and Simonetti, J.A. (2001) Plant–mammal interactions in tropical Bolivian forests with different hunting pressures. *Conservation Biology* 15, 617–623.

Roosevelt, T. (1920) *Through the Brazilian Wilderness*. Charles Scribers's Sons, New York, 410 pp.

Roth, I. (1987) *Stratification of a Tropical Forest as Seen in Dispersal Types*. Dr W. Junk Publishers, Dordrecht, The Netherlands, 324 pp.

Russell, J.K. (1985) Timing and reproduction by coatis (*Nasua narica*) in relation to fluctuation in food resources. In: Leight, E.G. Jr, Rand, A.S. and Windsor, D.M. (eds) *The Ecology of a Tropical Forest: Seasonal Rhythms and Long-term Changes*, 2nd edn. Smithsonian Institution Press, Washington, DC, pp. 413–431.

Sánchez-Cordero, V. and Martinez-Gallardo, R. (1998) Post-dispersal fruit and seed removal by forest-dwelling rodents in a lowland rainforest in Mexico. *Journal of Tropical Ecology* 14, 139–151.

Schaller, G.B. (1983) Mammals and their biomass on a Brazilian ranch. *Arquivos de Zoologia* 31, 1–36.

Schmidt, C.R. (1990) Monographie des Halsbandpekaris (*Tayassu tajacu*). *Bongo, Frädrich-Jubiläumsband* 18, 171–190.

Silman, M.R., Terborgh, J.W. and Kiltie, R.A. (2003) Population regulation of a dominant rain forest tree by a major seed predator. *Ecology* 84, 431–438.

Silvius, K.M. (1999) Interactions among *Attalea* palms, bruchid beetles, and neotropical terrestrial fruit-eating mammals: implications for the evolution of frugivory. PhD thesis, University of Florida, Gainesville, Florida.

Silvius, K.M. (2002) Spatio-temporal patterns of palm endocarp use by three Amazonian forest mammals: granivory or 'grubivory'? *Journal of Tropical Ecology* 18, 707–723.

Silvius, K.M. and Fragoso, J.M.V. (2002) Pulp handling by vertebrate seed dispersers increases palm seed predation by bruchid beetles in the northern Amazon. *Journal of Ecology* 90, 1024–1032.

Silvius, K.M. and Fragoso, J.M.V. (2003) Red-rumped Agouti (*Dasyprocta leporina*) home range use in an Amazonian forest: implications for the aggregated distribution of forest trees. *Biotropica* 35, 74–83.

Simberloff, D. (1992) Do species–area curves predict extinction in fragmented forests? In: Whithmore, T.C. and Sayer, J.A. (eds) *Tropical Deforestation and Species Extinction*. Chapman & Hall, London, pp. 75–89.

Simpson, C.D. (1984) Artiodactyls. In: Anderson, S. and Jones, J.K. Jr (eds) *Recent Mammals of the World. A Synopsis of Families*. Ronald Press, New York, pp. 563–588.

Sist, P. (1989a) Demography of *Astrocaryum sciophilum*, an understory palm of French Guiana. *Principes* 33, 142–151.

Sist, P. (1989b) Peuplement et phénologie des palmiers en forêt guyanaise (Piste de St Elie). *Revue d'Ecologie (La Terre et la Vie)* 44, 113–151.

Smith, N.J.H. (1976) Utilization of game along Brazil's transamazon highway. *Acta Amazonica* 6, 455–466.

Smythe, N. (1970a) Ecology and behavior of the agouti (*Dasyprocta punctata*) and related species on Barro Colorado Island, Panama. PhD thesis, University of Maryland, Baltimore, Maryland.

Smythe, N. (1970b) Relationships between fruiting seasons and seed dispersal methods in a Neotropical forest. *The American Naturalist* 104, 25–35.

Smythe, N. (1978) The natural history of the Central American agouti (*Dasyprocta punctata*). *Smithsonian Contribution Zoology* 257, 1–52.

Smythe, N. (1986) Competition and resource partitioning in the guild of Neotropical terrestrial frugivorous mammals. *Annual Review of Ecology and Systematics* 17, 169–188.

Smythe, N. (1989) Seed survival in the palm *Astrocaryum standleyanum*: evidence for dependence upon its seed dispersers. *Biotropica* 21, 50–56.

Smythe, N., Glanz, W.E. and Leigh, E.G. Jr (1985) Population regulation in some terrestrial frugivores. In: Leigh, E.G. Jr, Rand, A.S. and Windsor, D.M. (eds) *The Ecology of a Tropical Forest: Seasonal Rhythms and Long-term Changes*, 2nd edn. Smithsonian Institution Press, Washington, DC, pp. 227–238.

Sorensen, A.E. (1986) Seed dispersal by adhesion. *Annual Review of Ecology and Systematics* 17, 443–463.

Sork, V.L. (1987) Effects of predation and light on seedling establishment in *Gustavia superba*. *Ecology* 68, 1341–1350.

Sowls, L.K. (1983) *Tayassu tajacu* (Saino, Collared Peccary). In: Janzen, D.H. (ed.) *Costa Rican Natural History*. University of Chicago Press, Chicago, Illinois, pp. 497–498.

Sowls, L.K. (1984) *The Peccaries*. University of Arizona Press, Tucson, Arizona, 251 pp.

Sowls, L.K. (1997) *Javelinas and other Peccaries: Their Biology, Management, and Use*, 2nd edn. Texas A&M University Press, College Station, Texas, 325 pp.

Spruce, R. (1871) On equatorial-American palms. *The Journal of the Linnean Society* 11, 65–183.

SPSS (2002) *SPSS for Windows*. Version 11.5, SPSS, Chicago, Illinois.

Standley, P.C. (1928) *Flora of the Panama Canal Zone. Contributions from the United States National Herbarium*, Vol. 27. Smithsonian Institution, Washington, DC, 416 pp.

Stevens, G. (1983) *Bursera simaruba* (Indio Desnudo, Jinocuave, Gumbo Limbo). In: Janzen, D.H. (ed.) *Costa Rican Natural History*. University of Chicago Press, Chicago, Illinois, pp. 201–202.

Strey, O.F., Carl, G.R., Priebe, J.C. and Brown, R.D. (1985) *In vivo* and *in vitro* digestion trials for collared peccaries in South Texas. In: Ockenfels, R.A., Day, G.I. and Supplee, V.C. (eds) *Peccary Workshop Proceedings*. Arizona Game and Fish Department, Tucson, Arizona, p. 57.

Taber, A.B. (1991) The status and conservation of the Chacoan peccary in Paraguay. *Oryx* 25, 147–155.

Taber, A.B. (1993) The Chacoan peccary (*Catagonus wagneri*). In: Oliver, W.L.R. (ed.) *Status Survey and Conservation Plan. Pigs, Peccaries and Hippos*. IUCN, The World Conservation Union, Gland, Switzerland, pp. 22–29.

Taber, A.B., Doncaster, C.P., Neris, N.N. and Colman, F. (1993) Ranging behavior and population dynamics of the Chacoan peccary, *Catagonus wagneri. Journal of Mammalogy* 74, 443–454.

Taber, A.B., Doncaster, C.P., Neris, N.N. and Colman, F. (1994) Ranging behaviour and activity patterns of two sympatric peccaries, *Catagonus wagneri* and *Tayassu tajacu*, in Paraguayan Chaco. *Mammalia* 1, 61–71.

Terborgh, J. (1983) *Five New World Primates: a Study in Comparative Ecology*. Princeton University Press, Princeton, New Jersey, 260 pp.

Terborgh, J. (1986a) Community aspects of frugivory in tropical forests. In: Estrada, A. and Fleming, T.H. (eds) *Frugivores and Seed Dispersal*. Dr W. Junk Publishers, Dordrecht, The Netherlands, pp. 371–384.

Terborgh, J. (1986b) Keystone plant resources in the tropical forest. In: Soulé, M.E. (ed.) *Conservation Biology*. Sinauer, Sunderland, Massachusetts, pp. 330–344.

Terborgh, J.W. (1988) The big things that run the world – a sequel to E. O. Wilson. *Conservation Biology* 2, 402–403.

Terborgh, J. (1992) Maintenance of diversity in tropical forests. *Biotropica* 24, 283–292.

Terborgh, J. and Kiltie, R.A. (1976) *Ecology and Behavior of Rain Forest Peccaries in Southeastern Peru*. Research Report, National Geographic Society, Vol. 17, Washington, DC.

Terborgh, J., Foster, R.B. and Nuñez, P.V. (1996) Tropical tree communities: a test of the nonequilibrium hypothesis. *Ecology* 77, 561–567.

Terborgh, J., Lopez, L., Tello, J., Yu, D. and Bruni, A.R. (1997) Transitory states in relaxing ecosystems of land bridge islands. In: Laurance, W. and Bierregaard, R.O. Jr (eds) *Tropical Forest Remnants. Ecology, Management, and Conservation of Fragmented Communities*. The University of Chicago Press, Chicago, Illinois, pp. 256–274.

Terborgh, J., Estes, J.A., Paquet, P., Ralls, K., Boyd-Heger, D., Miller, B.J. and Noss, R.F. (1999) The role of top carnivores in regulation of terrestrial ecosystems. In: Soulé, M.E. and Terborgh, J. (eds) *Continental Conservation. Scientific Foundation of Regional Reserve Networks*. Island Press, Washington, DC, 227 pp.

Terborgh, J., Lopez, L., Nuñez, P.V., Rao, M., Shahabuddin, G., Orihuela, G., Riveros, M., Ascanio, R., Adler, G.H., Lambert, T.D. and Balbas, L. (2001) Ecological meltdown in predator-free forest fragments. *Science* 294, 1923–1925.

Ticktin, T. (2003) Relationship between El Niño Southern Oscillation and demographic patterns in a substitute food for collared peccaries in Panama. *Biotropica* 35, 189–197.

Timm, R.M. (1994) The mammal fauna. In: Dade, L.A., Bawa, K.S., Hespenheide, H.A. and Harshorn, G.S. (eds) *La Selva: Ecology and Natural History of a Neotropical Rain Forest.* The University of Chicago Press, Chicago, Illinois, pp. 229–237.

Timm, R.M. and LaVal, R.K. (2000) Mammals. In: Nadkarni, N.M. and Wheelwright, N.T. (eds) *Monteverde: Ecology and Conservation of a Tropical Cloud Forest.* Oxford University Press, New York, pp. 223–244.

Traveset, A. (1998) Effect of seed passage through vertebrate frugivores' guts on germination: a review. *Perspective in Plant Ecology, Evolution and Systematics* 1/2, 151–190.

Tsou, C.-H. and Mori, S.A. (2002) Seed coat anatomy and its relationship to seed dispersal in subfamily Lecythidoideae of the Lecythidaceae (the Brazil nut family). *Botanical Bulletin of Academia Sinica* 43, 37–56.

Ulbrich, E. (1928) *Biologie der Früchte und Samen (Karpobiologie).* Verlag von Julius Springer, Berlin, 230 pp.

Urrego, G.L.E. (1987) Estudio preliminar de la fenología de la Cananguchi (*Mauritia flexuosa* L. f.). MSc thesis, Universidad Nacional del Colombia, Facultad de Agronomía, Medellín.

Vandermeer, J.H., Stout, J. and Risch, S. (1979) Seed dispersal of a common Costa Rican forest palm. *Tropical Ecology* 20, 17–26.

Varela, R.O. and Brown, A.D. (1995) Tapires y pecaríes como dispersores de plantas de los bosques húmedos subtropicales de Argentina. In: Brown, A.D. and Grau, H.R. (eds) *Investigación, Conservación y Desarrollo en Selvas Subtropicales de Montana.* Universidad Nacional de Tucuman, Horco Molle, Argentina, pp. 129–140.

Vaughan, C. (1985) Central America. In: Ockenfels, R.A., Day, I.G. and Supplee, V.C. (eds) *Proceedings of the Peccary Workshop.* University of Arizona, Tucson, Arizona, pp. 28–35.

Vázquez-Yanes, C., Orozco-Segovia, A., Rincón, E., Sánchez-Coronado, M.E., Huante, P., Toledo, J.R. and Barrados, V.L. (1990) Light beneath the litter in a tropical forest: effects on seed germination. *Ecology* 71, 1952–1958.

Voss, R.S., Lunde, D.P. and Simmons, N.B. (2001) The mammals of Paracou, French Guiana: a Neotropical lowland rainforest fauna. Part 2. Nonvolant species. *Bulletin of the American Museum of Natural History* 263, 1–236.

Waterman, P.G. (1984) Food acquisition and processing as a function of plant chemistry. In: Chivers, D.J., Wood, B.A. and Bilsborough, A. (eds) *Food Acquisition and Processing in Primates.* Plenum Press, New York, pp. 177–211.

Wenny, D.G. (1999) Two-stage dispersal of *Guarea glabra* and *G. kunthiana* (Meliaceae) in Monteverde, Costa Rica. *Journal of Tropical Ecology* 15, 481–496.

West, R.C. (1957) *The Pacific Lowlands of Colombia. A Negroid Area of American Tropics.* Louisiana State University Press, Baton Rouge, Louisiana, 278 pp.

Wetzel, R.M. (1977) The Chacoan peccary, *Catagonus wagneri* (Rusconi). *Bulletin of the Carnegie Museum of Natural History* 3, 1–36.

Wong, G., Saenz, J.C. and Carrillo, E. (1999) *Mammals of Corcovado National Park, Costa Rica.* Instituto Nacional de Biodiversidad (INBio), Heredia, Costa Rica, 54 pp.

Woodburne, M.O. (1968) The cranial mycology and osteology of *Dicotyles tajacu*, the collared peccary, and its bearing on classification. *Memoirs of the Southern California Academy of Sciences*, Vol. 7.

Woodburne, M.O. (1969) A late Pleistocene occurrence of the collared peccary, *Dicotyles tajacu*, in Guatemala. *Journal of Mammalogy* 50, 121–125.

Wright, S.J. and Duber, H.C. (2001) Poachers and forest fragmentation alter seed dispersal, seed survival, and seedling recruitment in the palm *Attalea butyraceae*, with implications for tropical tree diversity. *Biotropica* 33, 583–595.

Wright, S.J., Zeballos, H., Domínguez, I., Gallardo, M.M., Moreno, M.C. and Ibáñez, R. (2000) Poachers alter mammal abundance, seed dispersal and seed predation in a Neotropical forest. *Conservation Biology* 14, 227–239.

Wyatt, J. (2002) Density and distance-dependence in three Neotropical palms. MSc thesis, Wake Forest University, Winston-Salem, North Carolina.

Yumoto, T. (1999) Seed dispersal by Salvin's Curassow, *Mitu salvini* (Cracidae), in a tropical forest of Colombia: direct measures of dispersal distance. *Biotropica* 31, 654–660.

Zimmerman, B.L. and Bierregaard, R.O. (1986) Relevance of the equilibrium theory of island biography and species–area relations to conservation with a case from Amazonia. *Journal of Biogeography* 13, 133–143.

Zimmerman, B.L. and Simberloff, D. (1996) An historical interpretation of habitat use by frogs in a Central Amazonian forest. *Journal of Biogeography* 23, 27–46.

Zona, S. and Henderson, A. (1989) A review of animal-mediated seed dispersal of palms. *Selbyana* 11, 6–21.

Appendix 6.1. Summary of studies on fruit consumption, seed predation and seed dispersal by collared peccary (*Pecari tajacu*) throughout the Neotropics. For more details see footnotes to Appendix 6.3.

Family and species	Life form	Part used	Seed fate	Site	Method	Habitat	Reference
Anacardiaceae							
Spondias mombin	T	F	A	V, P, CR	O, FE	D	Barreto *et al.* (1997), Kiltie (1982), Janzen (1985)
Spondias purpurea	T	F, S	A, K	M	O, FA	D	Martínez-Romero and Mandujano (1995), Mandujano *et al.* (1994)
Spondias tuberosa	T	F	ND	B	O	X	Olmos (1993)
Annonaceae							
Annona cacans	T	F, S	A, K	B	O, FA	R	Keuroghlian (2003)
Annona jahnii	T	F	ND	V	O	G	Robinson and Eisenberg (1985)
Annona purpurea	T	F	A	V	O	D	Barreto *et al.* (1997)
Cymbopetalum baillonii	T	F, S	K	M	O, FE	R	Martínez-Gallardo and Cordero (1997), Sánchez-Cordero and Martínez-Gallardo (1998)
Duguetia lanceolata	T	F, S	A, K	B	O, FA	R	Keuroghlian (2003)
Rollinia sylvatica	T	F, S	A, K	B	O, FA	R	Keuroghlian (2003)
Arecaceae[a]							
Acrocomia aculeata[b]	P	F, S	ND	Pa	O, FA	R	Enders (1930, 1935)
Acrocomia vinifera	P	F, S	K	CR	O	R	Baker (1983)
Astrocaryum aculeatum	P	F	A	B	O	R	Silvius (1999)
Astrocaryum chambira	P	F	ND	P	O	R	Lleellish *et al.* (2003)
Astrocaryum jauari	P	F, S	K	E	O	R	Borgtoft Petersen and Balslev (1990)
Astrocaryum mexicanum	P	F, S	K	Be, M	O, FE	R	Brewer (2001, pers. com.), Sánchez-Cordero and Martínez-Gallardo (1998), Martínez-Ramos (1997)
Astrocaryum murumuru	P	F, S	K	P, Bo	O, SC	R	Kiltie (1981a,b,c, 1982), Kiltie and Terborgh (1983), Roldán and Simonetti (2001), Cintra (1997)
Astrocaryum paramaca	P	F, S	K	FG	FE	–	Feer (pers. com.)
Astrocaryum sp.	P	F, S	K, ND	P, B	O, FE	R	Terborgh and Kiltie (1976), Schaller (1983)
Astrocaryum standleyanum	P	F, S	K	Pa	O	R	Smythe (1970a,b, 1978, 1989)
Attalea butyracea[c]	P	F, S	A, K, ND	Pa, M, CR	O	R, D	Enders (1930, 1935), Coe and Diehl (1980), Janzen and Martin (1982)
Attalea maripa[d]	P	F	A	B	O	R	Fragoso (1997, 1999), Silvius (1999)
Attalea phalerata	P	F, S	K	Bo	O	R	Quiroga-Castro and Roldán (2001)
Attalea speciosa[e]	P	F	A	B	O	D	Anderson (1983)
Attalea sp.[f]	P	F, S	A, ND	B, P, H	O, FE, SC	R	Duke (1967), Kiltie (1981a,b, 1982), Schaller (1983)

continued

Appendix 6.1. *Continued.*

Family and species	Life form	Part used	Seed fate	Site	Method	Habitat	Reference
Bactris maraja	P	S	K	B	O	R	Fragoso (1999)
Copernicia tectorum	P	F	ND	V	O	G	Robinson and Eisenberg (1985)
Euterpe edulis	P	F, S	A, K	B, P	O	R	Reis (1995), Galetti *et al.* (1999), Keuroghlian (2003), Lleellish *et al.* (2003)
Euterpe sp.	P	F	ND	P	SC	R	Bodmer (1990)
Iriartea deltoidea[g]	P	F, S	A, K	P	O, SC	R	Kiltie (1981a,b, 1982), Cintra (1998)
Jessenia bataua	P	F	A	P	O, FE	R	Bodmer (1991a)
Jessenia sp.	P	F, S	K	P	O, FE, SC	R	Terborgh and Kiltie (1976), Kiltie (1981a,b,c, 1982), Kiltie and Terborgh (1983)
Mauritia flexuosa	P	F, S	A, K	E, B, FG, P	O, FE	R	Kiltie (1981a, 1982), Urrego (1987), Bodmer (1991a), Peres (1994, pers. com.), Fragoso (1999), Lleellish *et al.* (2003), Feer (pers. com.)
Oenocarpus bataua	P	F	ND	P	O	R	Lleellish *et al.* (2003)
Oenocarpus sp.	P	F	ND	H	O	R	Duke (1967)
Phytelephas microcarpa	P	F	A	P	O, FE	R	Kiltie (1982)
Raphia taedigera	P	F, S	K	CR	O	R	Janzen and Wilson (1983, pers. com.)
Socratea exorrhiza	P	F	ND	P	O	R	Lleellish *et al.* (2003)
Syagrus oleracea	P	F	A	B	O, FA	R	Keuroghlian (2003)
Syagrus romanzoffiana	P	F, S	A, K	B	O, FA, FE	R	Olmos *et al.* (1999, pers. com.), Keuroghlian (2003)
Welfia georgii	P	F, S	K	CR	O	R	Vandermeer *et al.* (1979, pers. com.)
Bignoniaceae							
Crescentia alata	T	F, S	A	CR	O	D	Janzen (1982)
Bombacaceae							
Quararibea sp.	T	F	ND	CR	O	R	Duke (1967)
Boraginaceae							
Cordia sp.	T	F, S	K	B	O	R	Keuroghlian (2003)
Bromeliaceae							
Ananas sp.	B	F, S	ND	B	O	R	Schaller (1983)
Bromelia chrysanta	S	F, S	A	V	FA	D	Barreto *et al.* (1997)
Bromelia sp.	S	F, S	ND	B	O	R	Schaller (1983)
Burseraceae							
Bursera simaruba	T	F, S	A	CR	FA	D	Stevens (1983)

Species							Reference
Cactaceae							
Opuntia excelsa	C	F, S	ND	M	FA	D	Martínez-Romero and Mandujano (1995)
Opuntia sp.	C	F, S	ND	M	O	D	Leopold (1972)
Pereskia aculeate	S	F, S	A	B	FA	R	Keuroghlian (2003)
Pereskia guamacho	S	F, S	A, K	V	SC	D	Barreto et al. (1997)
Caesalpiniaceae							
Caesalpinia coriari	T	F, S	K	CR, V	O, SC	D	Janzen (1982), Barreto et al. (1997)
Cassia biglandularis	T	F, S	K	V	SC	D	Barreto et al. (1997)
Copaifera langsdorfii	T	F, S	K	B	O	R, X	Olmos (1993), Keuroghlian (2003)
Holocalix balansae	T	F, S	K	B	O	R	Keuroghlian (2003)
Hymenaea courbaril	T	F, S	K	B, CR	O, FA	D, R	Boucher (1979, pers. com.), Sowls (1983), Hallwachs (1986), Janzen (1983a, 1986), Keuroghlian (2003)
Prioria sp.	T	F	ND	H	O	R	Duke (1967)
Swartzia cubensis	T	F, S	K	CR	O	D	Boucher (1979, 1981b, pers. com.)
Swartzia sp.	T	S	K	B	O	R	Fragoso (1999)
Vouacapoua americana	T	F, S	K	FG	O	R	Forget (1990), Jansen (2003)
Capparaceae							
Capparis odoratissima	T	F, S	K	V	SC, FA	D	Barreto et al. (1997)
Caricaceae							
Jacaratia spinosa	T	F, S	K	B	O, FA	R	Keuroghlian (2003)
Celastraceae							
Schaefferia frutescens	T	F, S	K	V	O, FA	D	Barreto et al. (1997)
Chrysobalanaceae							
Licania alba	T	F, S	K	FG	O	R	Jansen (2003)
Clusiaceae							
Calophyllum brasiliense	T	F, S	K	B	O	R	Keuroghlian (2003)
Cupressaceae							
Juniperus sp.	T	F, S	ND	M	O	D	Leopold (1972)
Ericaceae							
Arctostaphylos sp.	S	F, S	ND	M	O	D	Leopold (1972)
Euphorbiaceae							
Manihot caerulescens	T	S	K	B	O	X	Olmos (1993)
Margaritaria nobilis	T	F, S	K	B	O, FA	R	Keuroghlian (2003)
Omphalea oleifera	L	F, S	K	M	FE	–	Dirzo and Miranda (1990)
Leguminosae	?	F	ND	P	SC	R	Bodmer (1990)

continued

Appendix 6.1. *Continued.*

Family and species	Life form	Part used	Seed fate	Site	Method	Habitat	Reference
Fabaceae							
Dipteryx micrantha	T	F	A	P	O, FE	R	Kiltie (1982)
Dipteryx panamensis	T	F, S	A, K	Pa	O	R	Chapman (1931, 1936), Enders (1935), Bonaccorso *et al.* (1980), De Steven and Putz (1984)
Dussia mexicana	T	F, S	K	M	FE	–	Dirzo and Miranda (1990)
Lonchocarpus costaricensis	T	S	K	CR	O	D	Janzen (1986)
Phaseolus lunatus	V	S	K	CR	O	D	Janzen (1986)
Phaseolus vulgaris	V	S	K	CR	O	D	Janzen (1986)
Prosopis sp.	T	F, S	ND	H, M	O	D	Leopold (1972)
Fagaceae							
Quercus oleoides	T	F, S	K	CR	O	D, R	Boucher (1979, 1981a), Sowls (1983)
Quercus sp.	T	S	ND	H, M	O	D, R	Duke (1967), Leopold (1972)
Icacinaceae							
Calatola sp.	T	F	A	P	FE	–	Kiltie (1982)
Lauraceae							
Nectandra ambigens	T	F, S	K	M	O, FE	R	Dirzo and Miranda (1990), Martínez-Gallardo and Cordero (1997), Sánchez-Cordero and Martínez-Gallardo (1998)
Nectandra spp.	T	F, S	K	B, CR	O	R	Keuroghlian (2003), Oberbauer (pers. com.)
Ocotea sp.	T	F, S	K	B	O	R	Keuroghlian (2003)
Lecythidaceae							
Bertholletia excelsa	T	S	K	B, P	O	R	Miller (1990), Ortiz (1995, pers. com.)
Gustavia superba	T	F, S	ND	Pa	O	R	Enders (1935)
Gustavia sp.	T	F	ND	H	O	R	Duke (1967)
Malpighiaceae							
Dicella bracteosa	L	F, S	K	B	O	R	Keuroghlian (2003)

Meliaceae

Species							Reference
Carapa guianensis	T	F, S	K	CR	O	R	McHargue and Hartshorn (1983)
Carapa procera	T	F, S	K, ND	FG	O	R	Forget et al. (2001), Jansen et al. (2002), Jansen (2003)
Guarea kunthiana	T	F, S	K	CR	O	Cl	Wenny (1999)
Menispermaceae	?	F	ND	P	SC	R	Bodmer (1990)
Mimosaceae							
Acacia sp.	T	F, S	ND	M	O	D	Leopold (1972)
Enterolobium contortisiliquum	T	F, S	K	B	O	R	Keuroghlian (2003)
Ormosia arborea	T	F, S	K	B	O	R	Galetti (2002), Keuroghlian (2003)
Pithecellobium carabobense	T	F, S	A	V	FA	D	Barreto et al. (1997)
Pithecellobium mangense	T	F, S	K	CR	O	D	Boucher (1979, pers. com.)
Pithecellobium saman	T	F, S	A, K	CR, V	O, FA, SC	D	Boucher (1979, pers. com.), Janzen (1982), McCoy et al. (1985, 1990), Barreto et al. (1997)
Pithecellobium sp.	T	F	ND	H	O	R	Duke (1967)
Enterolobium cyclocarpum	T	F, S	A, K	B, CR, P	O, FA, FE	D, R	Terborgh and Kiltie (1976), Boucher (1979, pers. com.), Janzen (1982, 1986), Janzen and Higgins (1979), Kiltie (1981c), Sowls (1983), McCoy et al. (1985, 1990), Fragoso (1999)
Moraceae							
Brosimum alicastrum	T	F, S	K, ND	Be, CR,M	O, FE	D, R	Sowls (1983), Dirzo and Miranda (1990), Martínez-Romero and Mandujano (1995), Martínez-Gallardo and Cordero (1997), Sánchez-Cordero and Martínez-Gallardo (1998), Brewer (pers. com.)
Brosimum sp.	T	F	A	P	SC	R	Bodmer (1991a)
Clorophora tinctoria	T	F	ND	CR	FA	D	McCoy et al. (1985, 1990)
Ficus colubrinae	T	F, S	A	CR	O	R	Oberbauer (pers. com.)
Ficus continifolia	T	F, S	A	CR	O, FA	R	Jordano (1983)
Ficus insipida	T	F, S	A	M	O	R	Martínez-Gallardo and Cordero (1997)
Ficus spp.	T	F, S	A, ND	B, CR, M, Pa	O, FA	D, R	Enders (1935), Schaller (1983), Sowls (1983), Martínez-Romero and Mandujano (1995), Keuroghlian (2003)
Ficus sphenophylla	T	F, S	A	C	O	R	Yumoto (1999)

continued

Appendix 6.1. *Continued.*

Family and species	Life form	Part used	Seed fate	Site	Method	Habitat	Reference
Ficus trigona	T	F, S	ND	V	O	G	Robinson and Eisenberg (1985)
Ficus yoponensis	T	F, S	K	M	FE	–	Sánchez-Cordero and Martínez-Gallardo (1998)
Poulsenia armata	T	F, S	K	M	O	R	Martínez-Gallardo and Cordero (1997)
Pseudolmedia oxyphyllaria	T	F, S	K	M	O	R	Martínez-Gallardo and Cordero (1997)
Myrtaceae							
Eugenia sp.	?	F, S	A, K	B	O, FA	R	Keuroghlian (2003)
Myrcia sp.	?	F, S	A, K	B	O, FA	R	Keuroghlian (2003)
Psidium sp.	?	F, S	ND	B	O	R	Fragoso (1999)
Opiliaceae							
Agonandra microcarpa	S	F	A	CR	FA	D	McCoy *et al.* (1985, 1990)
Pinaceae							
Pinus sp.	T	S	ND	M	O	D	Leopold (1972)
Rhamnaceae							
Zizyphus guatemalensis	T	F	ND	CR	FA	D	McCoy *et al.* (1985, 1990)
Rosaceae							
Rubus sp.	S	F, S	ND	H	O	R	Carr (1953)
Rubiaceae							
Chomelia spinosa	T	F, S	K	CR	FA	D	McCoy *et al.* (1985, 1990)
Genipa americana	T	F, S	A, ND	V	O, FA	G, D	Robinson and Eisenberg (1985), Barreto *et al.* (1997)
Randia armata	S	F, S	K	CR	O	D	Boucher (1979, pers. com.)
Sapindaceae							
Paullinia granatensis	L	F, S	K	CR	O	R	Oberbauer (pers. com.)
Paullinia sp.	L	F	ND	P	SC	R	Bodmer (1990)
Sapindus saponaria	T	F, S	K	V	O, SC, FA	D	Barreto *et al.* (1997)
Talisia esculenta	T	F, S	ND	B	O	R	Schaller (1983)

Sapotaceae							
Chrysophyllum gonocarpum	T	F, S	K	B	O	R	Keuroghlian (2003)
Chrysophyllum lucentifolium	T	F, S	K	FG	O	R	Feer and Forget (2002, pers. com.), Jansen *et al.* (2002)
Chrysophyllum venezuelanense	T	F, S	K	Be	O	R	Brewer (pers. com.)
Manilkara bidentata	T	F, S	K	FG	O	R	Brewer (pers. com.)
Manilkara chicle	T	F, S	K	Be	O	R	Brewer (pers. com.)
Manilkara huberi	T	F, S	K	FG	O	R	Forget (pers. com.), Brewer (pers. com.)
Manilkara staminodella	T	F, S	K	Be	O	R	Brewer (pers. com.)
Manilkara zapota	T	F, S	K	CR, M	O	D, R	Janzen (1982), Sowls (1983), Martínez-Gallardo and Cordero (1997)
Pouteria surumensis	T	F	A	B	O	R	Silvius (1999)
Pouteria venosa	T	F, S	K	B	O	R	Fragoso (1999)
Pradosia surinamensis	T	F, S	K	B	O	R	Fragoso (1999)
Sideroxylon capiri	T	F, S	ND	M	O	D	Martínez-Romero and Mandujano (1995)
Simaroubaceae							
Apeiba aspera	T	S	ND	Pa	O	R	Enders (1935)
Guazuma ulmifolia	T	F, S	A, K	CR, V	O, FA, SC	D	Boucher (1979, pers. com.), Janzen (1982), Sowls (1983), McCoy *et al.* (1985, 1990), Barreto *et al.* (1997)
Simaruba glauca	T	F, S	K	CR	O	D	Boucher (1979, pers. com.)
Ulmaceae							
Ampelocera hottlei	T	S	K	Be	O	R	Brewer (pers. com.)
Verbenaceae							
Vitex mollis	T	F, S	ND	M	O	D	Martínez-Romero and Mandujano (1995)
Vitex orinocensis	T	F	ND	V	O	G	Robinson and Eisenberg (1985)

Appendix 6.2. Summary of studies on fruit consumption, seed predation and seed dispersal by white-lipped peccary (*Tayassu pecari*) throughout the Neotropics. For more details see footnotes to Appendix 6.3.

Family and species	Life form	Part used	Seed fate	Site	Method	Habitat	Reference
Anacardiaceae							
Anacardium excelsum	T	F, S	K	CR	O, FA	H	Altrichter *et al.* (1999, 2000)
Spondias mombin	T	F	A	CR, P, V	O, FA, FE	H, D	Kiltie (1982), Barreto *et al.* (1997), Altrichter *et al.* (1999, 2000)
Spondias purpurea	T	F	A	CR	O, FA	H	Altrichter *et al.* (1999)
Spondias radlkoferi	T	F	A	CR	O, FA	H	Altrichter *et al.* (1999, 2000)
Spondias spp.	T	F	A	CR	O, FA	H	Altrichter *et al.* (2000)
Annonaceae							
Annona cacans	T	F, S	K	B	O, FA	R	Keuroghlian (2003)
Duguetia lanceolata	T	F, S	A, K	B	O, FA	R	Keuroghlian (2003)
Rollinia sylvatica	T	F, S	K	B	O, FA	R	Keuroghlian (2003)
Aquifoliaceae							
Ilex jenmani	S	F	ND	B	O	R	Fragoso (1999)
Araliaceae							
Dendropanax spp.[a]	T	F, S	K	CR	O, FA	H	Altrichter *et al.* (1999, 2000)
Arecaceae							
Astrocaryum aculeatum	P	F, S	K	Bo, B	O	R	Painter (1998), Silvius (1999)
Astrocaryum chambira	P	F	ND	P	O	R	Lleellish *et al.* (2003)
Astrocaryum jauari	P	F, S	K	E	O	R	Borgtoft Petersen and Balslev (1990)
Astrocaryum mexicanum	P	F, S	K	Be	O	R	Brewer (2001, pers. com.)
Astrocaryum murumuru	P	F, S	K	P	O, FE, SC	R	Kiltie (1981a,b,c, 1982), Kiltie and Terborgh (1983), Cintra (1997, 1998), Cintra and Horna (1997), Wyatt (2002)
Astrocaryum paramaca	P	F, S	K	FG	FE	–	Feer (pers. com.)
Astrocaryum sciophilum	P	F, S	K	FG	O	R	Sist (1989a)
Astrocaryum spp.	P	F, S	K, ND	B, CR, H, P	O, FA, FE	H, R	Duke (1967),Terborgh and Kiltie (1976), Moskovits *et al.* in Kiltie (1982), Schaller (1983), Altrichter *et al.* (1999, 2000)
Astrocaryum standleyanum	P	F, S	K	Pa	O	R	Smythe (1970a,b, 1989)
Astrocaryum tucuma	P	F, S	ND	B	O	R	Peres (1996)
Astrocaryum vulgare	P	F, S	ND	S	O	R	Husson (1978), Penard *et al.* in Husson (1978)
Attalea butyracea[c]	P	F, S	K	CR, P	O, FA	H, R	Altrichter *et al.* (2000), Wyatt (2002)

Species							References
Attalea maripa[d]	P	F, S	A, K, ND	B, S	O	R	Husson (1978), Penard *et al.* in Husson (1978), Fragoso (1998a,b, 1999), Silvius (1999, 2002)
Attalea phalerata	P	F, S	K	Bo	O	R	Quiroga-Castro and Roldán (2001)
Attalea speciosa[e]	P	F	A	B	O	D	Anderson (1983)
Attalea sp.[f]	P	F, S	A, ND	B, P	O, FE	R	Kiltie (1981b, 1982), Schaller (1983)
Bactris sp.	P	S	K	CR	O, FA	R	Altrichter *et al.* (1999, 2000)
Copernicia tectorum	P	F	A	V	FA	H	Barreto *et al.* (1997)
Crysophila guagara	P	S	K	CR	O, FA	D	Altrichter *et al.* (1999, 2000)
Dictyocaryum lamarckianum	P	F, S	K	Pa, P	O	H	Mori in Henderson (1990), Silman (pers. com)
Elaeis guineensis	P	F	A	FG	FE	–	Feer (pers. com.)
Euterpe edulis	P	F, S	A, K	B, P	O, FA	R	Reis (1995), Galetti *et al.* (1999), Keuroghlian (2003), Lleellish *et al.* (2003)
Euterpe oleroceca	P	F, S	K	B	O	R	Smith (1976)
Euterpe precatoria	P	F, S	K	B	O	R	Silvius (1999)
Iriartea deltoidea[g]	P	F, S	K	CR, P	O, FA, FE, SC	H, R	Kiltie (1981a,b,c, 1982), Kiltie and Terborgh (1983), Cintra (1998), Altrichter *et al.* (1999, 2000), Wyatt (2002)
Iriartea sp.	P	F, S	K, ND	B, P	O, FE, SC	R	Smith (1976), Terborgh and Kiltie (1976), Bodmer (1990)
Jessenia bataua	P	F	A	P	O, FE	R	Bodmer (1991a)
Jessenia sp.	P	F, S	K	P	O, SC, FE	R	Terborgh and Kiltie (1976), Kiltie (1981a,c, 1982), Kiltie and Terborgh (1983)
Mauritia flexuosa	P	F, S	A, K	B, E, FG, P	O, FE	R	Kiltie (1981a,c, 1982), Kiltie and Terborgh (1983), Urrego (1987), Bodmer (1991a), Fragoso (1998a,b, 1999), Silvius (1999), Lleellish *et al.* (2003), Feer (pers. com.)
Oenocarpus bataua	P	F	ND	H, P	O	R	Duke (1967), Lleellish *et al.* (2003)
Orbignya phalerata	P	F, S	ND	B	O	D	Peres (1996)
Phytelephas microcarpa	P	F	A	P	O, FE	R	Kiltie (1982)
Raphia taedigera	P	F, S	K	CR	O	R	Janzen and Wilson (1983, pers. com.)
Socratea durissima	P	F, S	K	P	O	R	Kiltie (1981c), Kiltie and Terborgh (1983)
Socratea exorrhiza	P	F, S	A, K	B, Bo, FG, P	O, FE, SC	R	Kiltie (1981a,c, 1982), Painter (1998), Silvius (1999), Lleellish *et al.* (2003), Feer (pers. com.)
Socratea sp.	P	F, S	K	P	FE	–	Terborgh and Kiltie (1976)
Syagrus oleracea	P	F	A	B	O, FA	R	Keuroghlian (2003)
Syagrus romanzoffiana	P	F, S	A, K, ND	B	O, FA, FE	R	Cullen *et al.* (2001), Keuroghlian (2003), Olmos (pers. com.)
Syagrus sancona	P	F, S	K	Bo	O	R	Painter (1998)

continued

Appendix 6.2. *Continued.*

Family and species	Life form	Part used	Seed fate	Site	Method	Habitat	Reference
Bombacaceae							
Quararibea asterolepis	T	S	K	CR	O, FA	H	Altrichter et al. (1999, 2000)
Boraginaceae							
Cordia sp.	T	F, S	K	B	O	R	Keuroghlian (2003)
Bromeliaceae							
Ananas sp.	S	F, S	ND	B	O	R	Schaller (1983)
Bromelia chrysanta	S	F, S	A	V	O	D	Barreto et al. (1997)
Bromelia sp.	S	F, S	ND	B	O	R	Schaller (1983)
Cactaceae							
Pereskia aculeate	T	F, S	A, K	B	O, FA	R	Keuroghlian (2003)
Pereskia guamacho	T	F, S	K	V	SC	D	Barreto et al. (1997)
Caesalpiniaceae							
Holocalix balansae	T	F, S	K	B	O	R	Keuroghlian (2003)
Hymenaea courbaril	T	F, S	K	B, Bo, CR	O	D, R	Hallwachs (1986), Painter (1998), Keuroghlian (2003)
Hymenaea sp.	T	S	ND	B	O	R	Silvius (1999)
Vouacapoua americana	T	F, S	K	FG	O	R	Forget (1990), Jansen (2003)
Caricaceae							
Jacaratia spinosa	T	F, S	A, K	B	O, FA	R	Keuroghlian (2003)
Celastraceae							
Schaefferia frutescens	T	F, S	K	V	O, SC	D	Barreto et al. (1997)
Chrysobalanaceae							
Licania alba	T	F, S	K	FG	O	R	Jansen (2003)
Licania operculipetala	T	F, S	K	CR	O, FA	H	Altrichter et al. (1999, 2000, 2001)
Licania platypus	T	F, S	K	CR	O, FA	H	Altrichter et al. (1999, 2000)
Licania sp.	T	F, S	K	CR	O, FA	H	Altrichter et al. (1999, 2001)
Maranthes panamensis	T	F, S	K	CR	O, FA	H	Altrichter et al. (1999, 2000)
Parinari parvifolia	T	F, S	K	CR	O, FA	H	Altrichter et al. (1999, 2000)
Convolvulaceae							
Maripa nicaraguensis	L	F, S	K	CR	O, FA	H	Altrichter et al. (1999, 2000)
Cyclanthaceae							
Carludovica palmate	H	F	ND	H	O	R	Duke (1967)

Taxon							Reference
Euphorbiaceae							
Manihot caerulescens	T	S	K	B	O	X	Olmos (1993)
Omphalea oleifera	L	F, S	K	M	FE	—	Dirzo and Miranda (1990)
Fabaceae							
Copaifera langsdorffii	T	F, S	K	B	O	R	Keuroghlian (2003)
Mucuna sp.	L	F, S	K	B	O, FA	R	Keuroghlian (2003)
Dipteryx micrantha	T	F	A	P	O, FE	R	Kiltie (1982)
Dipteryx odorata	T	F, S	K	H	O	R	Carr (1953)
Dipteryx panamensis	T	F, S	A, K	Pa	O	R	Chapman (1931, 1936)
Dussia mexicana	T	F, S	K	M	FE	—	Dirzo and Miranda (1990)
Ormosia arborea	T	F, S	K	B	O	R	Galetti (2002), Keuroghlian (2003)
Fagaceae							
Quercus oleoides	T	S	K	CR	O	D	Vaughan (1985)
Guttiferae							
Calophyllum brasiliense	T	F, S	K	B, P	O	R	Keuroghlian (2003), Beck (unpublished data)
Icacinaceae							
Calatola sp.	T	F	A	P	FE	—	Kiltie (1982)
Lauraceae							
Nectandra ambigens	T	F, S	K	M	FE	—	Dirzo and Miranda (1990)
Ocotea sp.	T	F, S	K	B	O, FA	R	Keuroghlian (2003)
Lecythidaceae							
Bertholletia excelsa	T	S	K	B	O	R	Miller (1990)
Lecythis costaricensis	T	F, S	K	CR	O	R	Mori (1970)
Leguminosae	?	F	ND	P	SC	R	Bodmer (1990)
Malpighiaceae							
Dicella bracteosa	L	F, S	K	B	O	R	Keuroghlian (2003)
Meliaceae							
Carapa guianensis	T	F, S	K	CR	O, FA	H, R	McHargue and Hartshorn (1983), Altrichter *et al.* (1999)
Carapa procera	T	F, S	K	FG	O	R	Forget (1996), Jansen (2003)
Trichilia sp.	T	F, S	ND	B	O	R	Keuroghlian (2003)
Menispermaceae	?	F	ND	P	SC	R	Bodmer (1990)

continued

Appendix 6.2. *Continued.*

Family and species	Life form	Part used	Seed fate	Site	Method	Habitat	Reference
Mimosaceae							
Acacia aroma	S	F, S	K	Par	SC	X	Mayer and Brandt (1982)
Acacia macracantha	T	F, S	A	V	SC	D	Barreto et al. (1997)
Enterolobium contortisiliquum	T	F, S	A, K	B	O, FA	R	Keuroghlian (2003)
Enterolobium cyclocarpum	T	S	K	P	O, FE	R	Terborgh and Kiltie (1976), Kiltie (1982)
Enterolobium sp.	T	NA	NA	B	O	R	Silvius (1999)
Inga spp.	T	S	K	CR	O, FA	H	Altrichter et al. (1999, 2000)
Parkia multijuga	T	F, S	K	B	O	R	Ayers in Hopkins and Hopkins (1983)
Pithecellobium carabobense	T	F, S	A	V	SC	D	Barreto et al. (1997)
Pithecellobium saman	T	F, S	A, K	CR, V	O, SC	D	Janzen (1983b), Barreto et al. (1997)
Prosopis sp.	T	F, S	A, ND	Par	O, SC	X	Sowls (1984, 1997), Taber et al. (1994)
Moraceae							
Brosimum alicastrum	T	F, S	K	Be, CR, M	O, FA, FE	H, R	Dirzo and Miranda (1990), Altrichter et al. (1999, 2000), Brewer (pers. com.)
Brosimum costaricarum	T	S	K	CR	O, FA	H	Altrichter et al. (1999, 2000)
Brosimum spp.	T	F, S	A, K	CR, P	O, FA, SC	H, R	Bodmer (1991a), Altrichter et al. (1999, 2000, 2001)
Brosimum utile	T	S	K	CR	O, FA	H	Altrichter et al. (1999, 2000, 2001)
Clarisia biflora	T	F, S	K	Bo, CR	O, FA	H, R	Altrichter et al. (1999, 2000), Roldán and Simonetti (2001)
Ficus continifolia	T	F, S	A	CR	O, FA	R	Janzen in Jordano (1983)
Ficus eximia	T	F, S	ND	Bo	O	R	Painter (1998)
Ficus maxima	T	F, S	A	CR	O, FA	H	Altrichter et al. (1999, 2000)
Ficus trigona	T	F, S	ND	Bo	O	R	Painter (1998)
Ficus spp.	T	F, S	A, ND	Bo, B, CR	O, FA	H, R	Schaller (1983), Painter (1998), Altrichter et al. (1999, 2000, 2001), Keuroghlian (2003)
Ficus zarzalensis	T	F, S	A	CR	O, FA	H	Altrichter et al. (1999, 2000)
Helicostylis tomentosa	T	F, S	ND	Bo	O	R	Painter (1998)
Perebea mollis	T	F, S	ND	Bo	O	R	Painter (1998)
Pourouma guianensis	T	F, S	ND	Bo	O	R	Painter (1998)
Pseudomedia laevis	T	F, S	ND	Bo	O	R	Painter (1998)
Myristicaceae							
Otoba novagrantensis	T	F, S	K	CR	O, FA	H	Altrichter et al. (2000)
Otoba sp.	T	S	K	CR	O, FA	H	Altrichter et al. (1999)
Virola spp.	T	F, S	K, ND	CR, P	O, FA, SC	H, R	Bodmer (1990), Altrichter et al. (1999, 2000)

Species							Reference
Myrtaceae							
Compsoneura sprucei	T	S	K	CR	O, FA	H	Altrichter et al. (1999, 2000)
Eugenia sp.	?	F, S	A, K	B	O, FA	R	Keuroghlian (2003)
Psidium guajava	T	F, S	A	CR	O, FA	H	Altrichter et al. (1999, 2000)
Rhamnaceae							
Zizyphus mistol	T	F	ND	Par	O	X	Taber et al. (1994)
Rubiaceae							
Genipa americana	T	F, S	A	V	SC	D	Barreto et al. (1997)
Guettarda parviflora	T	F, S	A	V	FA	D	Barreto et al. (1997)
Rutaceae							
Balfourodendron sp.	T	F, S	K	B	O	R	Keuroghlian (2003)
Esenbeckia leiocarpa	T	F, S	K	B	O	R	Keuroghlian (2003)
Sapindaceae							
Paullinia spp.	L	F, S	K, ND	CR, P	O, FA, SC	H, R	Bodmer (1990), Altrichter et al. (1999, 2000)
Sapindus saponaria	T	F, S	K	V	O, SC	D	Barreto et al. (1997)
Serjania sp.	L	F, S	K	CR	O, FA	H	Altrichter et al. (1999, 2000)
Talisia esculenta	T	F, S	ND	B	O	R	Schaller (1983)
Sapotaceae							
Chrysophyllum gonocarpum	T	F, S	K	B	O, FA	R	Keuroghlian (2003)
Chrysophyllum lucentifolium	T	F, S	K	FG	O	R	Feer and Forget (2002, pers. com.), Jansen et al. (2002)
Chrysophyllum venezuelanense	T	F, S	K	Be	O	R	Brewer (pers. com.)
Manilkara bidentata	T	F, S	K	FG	O	R	Brewer (pers. com.)
Manilkara chicle	T	F, S	K	Be	O	R	Brewer (pers. com.)
Manilkara huberi	T	F, S	K	FG	O	R	Brewer (pers. com.), Forget (pers. com.)
Manilkara staminodella	T	F, S	K	Be	O	R	Brewer (pers. com.)
Pouteria spp.	?	S	K	CR	O, FA	H	Altrichter et al. (1999, 2000)
Pouteria surumensis	T	F, S	K	B	O	R	Fragoso (1999), Silvius (1999)
Pouteria venosa	T	F, S	K	B	O	R	Fragoso (1999)
Pradosia surinamensis	T	F, S	K	B	O	R	Fragoso (1999)
Sterculiaceae							
Guazuma ulmifolia	T	F, S	A	V	SC	D	Barreto et al. (1997)
Theobroma speciosum	T	F, S	ND	Bo	O	R	Painter (1998)
Ulmaceae							
Ampelocera ruizii	T	F, S	ND	Bo	O	R	Wallace in Painter (1998)
Vitaceae							
Cissus spp.	L	F, S	K	CR	O, FA	H	Altrichter et al. (1999, 2000)

Appendix 6.3. Summary of studies on fruit consumption, seed predation and seed dispersal by both peccary species (*Pecari tajacu* and *Tayassu pecari*) throughout the Neotropics. Here the authors did not specify which peccary species was involved. This table includes plant species not mentioned in Appendix 6.1 or 6.2. For more details see footnotes.

Family and species	Life form	Part used	Seed fate	Site	Method	Habitat	Reference
Arecaceae[a]							
Acrocomia mexicana	P	F, S	K	M	O	ND	Gaumer (1917)
Butia leiospatha	P	F, S	A	B	O	C	Gottsberger and Silberbauer-Gottsberger (1983)
Jessenia polycarpa	P	F, S	A	C	O	R	West (1957)
Euphorbiaceae							
Hevea braziliensis	L	F, S	K	ND	O	R	Ridley (1930)
Hippomane mancinella	T	F, S	A, K	CR	O	D	Janzen and Martin (1982)
Fabaceae							
Geoffroya spinoza	T	F, S	ND	ND	O	ND	Silverman-Cope in Donkin (1985)
Caesalpiniaceae							
Prioria copaifera	T	F, S	ND	Pa	O	R	Standley (1928)
Gramineae							
Panicum laxum	H	F, S	A	A	FA	M	Varela and Brown (1995)
Juglandaceae							
Juglans australis	T	F, S	K	A	FA	M	Varela and Brown (1995)
Lactuceae							
Sonchus sp.	?	F, S	A	A	FA	M	Varela and Brown (1995)
Lecythidaceae							
Couroupita guianensis	T	F, S	A, ND	B, P	O	R	Prance and Mori (1978), Koepcke in Prance and Mori (1979), Tsou and Mori (2002)
Moraceae							
Ficus cf. *monkii*	T	F, S	A	A	FA	M	Varela and Brown (1995)

Nyctaginaceae						
Neea nov. sp.	T	S		P	O	R	Beck (unpublished data)
Passifloraceae						
Passiflora sp.	?	F	ND	Pa	O	R	Duke (1967)
Phytolaccaceae						
Phytolacca dioica	H	F, S	A	A	FA	M	Varela and Brown (1995)
Sapotaceae						
Gardenia sp.	?	F	ND	ND	O	ND	Pijl (1969)
Solanaceae						
Solanum calaviceps	T	F, S	A	A	FA	M	Varela and Brown (1995)
Sterculiaceae						
Theobroma cacao	T	F	A, ND	G, ND	O	R	Hermann (1836), Schombutgk in Donkin (1985)

Life form: B, bromeliad; C, cactus; H, herb; L, liana; P, palm; S, shrub; T, tree; V, vine. Data were obtained from original papers, or from Gentry (1993) and Belbenoit *et al.* (2001).

Part used: F, fruit pulp; S, seed; ND, not determined.

Seed fate: A, alive and potentially dispersed; K, killed; ND, not determined.

Site: A, Argentina; B, Brazil; Be, Belize; Bo, Bolivia; C, Colombia; CR, Costa Rica; E, Ecuador; FG, French Guiana; G, Guyana; H, Honduras; M, Mexico; Pa, Panama; Par, Paraguay; P, Peru; S, Suriname; V, Venezuela.

Methods: FA, faecal analysis; FE, feeding experiment (here no habitat is indicated because experiments were performed on captive animals); O, observation; SC, stomach contents.

Habitat: C, cerrado; Cl, cloud forest; G, gallery forest; H, humid tropical forest; M, subtropical moist forest; D, tropical dry forest; R, tropical rainforest; X, thorny xeric forest.

[a]I replaced previously used palm species names with their new nomenclature following Henderson *et al.* (1995).
[b]*Acrocomia aculeata = Acrocomia sclerocarpa;* [c]*Attalea butyracea = Attalea gymphococca, Scheelea liebermanii, S. rostrata, S. zonensis or Orbignya martiana;* [d]*Attalea maripa = Attalea regia or Maximiliana maripa;* [e]*Attalea speciosa = Orbignya martiana;* [f]*Attalea* sp. = *Scheelea* sp.; [g]*Iriartea deltoidea = Iriartea ventricosa.*

7 Seed Predation, Seed Dispersal and Habitat Fragmentation: Does Context Make a Difference in Tropical Australia?

Andrew J. Dennis,[1] Geoffrey J. Lipsett-Moore,[2] Graham N. Harrington,[1] Eleanor N. Collins[1,2] and David A. Westcott[1]

[1]CSIRO Sustainable Ecosystems and the Rainforest Cooperative Research Centre, PO Box 780, Atherton, Qld 4883, Australia; [2]Department of Tropical Biology, James Cook University of North Queensland, Townsville, Qld 4810, Australia

Introduction

Seed predation by vertebrates is an important process affecting plant population dynamics. Rates of seed predation can be highly variable for different species and for the same species at different times and in different locations. Some of this variation is due to fluctuations in seed production by plants in time and space, the effect of mass fruiting events on predator populations and the community structure and behaviour of predators and dispersers (Willson, 1988; Willson and Whelan, 1990; Crawley, 1992; Terborgh, 1992; Terborgh *et al.*, 1993; Osunkoya, 1994; Notman *et al.*, 1996; Asquith *et al.*, 1997; Green *et al.*, 1997; Forget *et al.*, 1998; Hammond *et al.*, 1999; Notman and Gorchov, 2001). However, not all plants are equally affected by predation of their seeds. In some plants, recruitment rates seem to respond little to high levels of predation (e.g. Andersen, 1989; Zuidema and Boot, 2002). This complexity makes it difficult to quantify the impacts of seed predation on plant recruitment.

Crawley (1992) suggests that the impact of seed predators on plant population dynamics is dependent on whether these populations are limited by seed density or by the availability of microsites for germination. More recent literature on recruitment limitation focuses on the interactions between these factors and recognizes that plants lie along a continuum of responses for both factors (Hurtt and Pacala, 1995; Muller-Landau *et al.*, 2002; Schupp *et al.*, 2002). Plants that are limited by the number of seeds should show a strong response to a reduction in predation rates if the availability of microsites for germination are not limiting. In contrast, those that have abundant seeds but are restricted by microsite should show little response to a reduction in predation rate. In tropical rainforests, there are likely to be plants from all parts of the spectrum, including differences between species or within a species in different parts of the landscape. However, the available evidence suggests that a majority would be strongly seed limited due to many species being rare, clumped or having their seeds dispersed in a restricted or contagious manner so that seeds are unable to saturate suitable germination sites (Clark *et al.*, 1998; Muller-Landau *et al.*, 2002; Terborgh *et al.*,

2002). Thus, seed predators are expected to have a significant impact for most plants in tropical rainforests.

Many vertebrate seed predators also act as seed dispersers (Price and Jenkins, 1986; Vander Wall, 2001; Hulme, 2002). Dispersal by animals that also act as seed predators has been demonstrated to be an important mode of dispersal for some plants, often those with large, hard seeds (Smith and Reichman, 1984; Price and Jenkins, 1986; Forget, 1990, 1991; Vander Wall, 1990; Forget *et al.*, 1998, 2002; Jansen *et al.*, 2002). Therefore, vertebrate seed predators have a dual impact on plant populations through reduction in the number of propagules available for recruitment and through the dispersal of some seeds to appropriate microsites for germination (e.g. Smith and Reichman, 1984; Feer *et al.*, 2001; see also Lambert and Chapman, Chapter 8, Vander Wall and Longland, Chapter 18, Wenny, Chapter 21, and Zhang *et al.*, Chapter 16, this volume). Thus, species that are dispersed primarily by seed predators should not normally be limited by availability of seeds. However, what is the net impact on the plant community of seed predators that also act as seed dispersers? In this chapter we ask: what are the impacts of seed predators and how do these impacts vary across the landscape?

Habitat fragmentation has significant impacts on many vertebrate populations and on vertebrate community structure (e.g. Glanz, 1982; Pahl *et al.*, 1988; Laurance, 1990, 1994, 1997; Warburton, 1997). Several studies have demonstrated a link between changes in mammal community composition and the survival of seeds and/or recruitment of seedlings (De Steven and Putz, 1984; Sork, 1987; Terborgh and Wright, 1994; Asquith *et al.*, 1997; Corlett, 2002). For example, Asquith *et al.* (1997) demonstrated much higher rates of seed consumption and seedling herbivory on small islands in Gatun Lake in Panama, than on mainland sites or on the large Barro Colorado Island (BCI). Rates of seed predation were also higher on BCI than on the mainland sites, whereas seedling herbivory was lower. These changes in recruitment potential for forest trees were linked to differing levels of

mammal defaunation on islands of different size as compared to the mainland sites. Changes in the plant community structure have also been measured on these islands (Leigh *et al.*, 1993). A driving force behind these changes is thought to be the changes in mammal abundance and diversity and the resultant effects on plant recruitment.

We ask: what are the likely effects on plant community structure in the different parts of the landscape (i.e. fragments of different size versus continuous forest) of changes in seed predator and disperser populations? To do this we present data from one previously published and six new studies to examine the broad implications of vertebrate seed dispersal and predation in Australia's tropical rainforests. More specifically we examine:

1.   The physical characteristics of seeds preferred by a suite of mammalian predators and dispersers.
2.   The effects of location at landscape and micro-site scales on predation, removal and caching rates for a suite of plants.
3.   The number of other dispersers feeding on the same plant taxa as the seed predators.
4.   The distribution of seed predators and dispersers in a fragmented landscape.

These components are then combined to predict the likely effects of changes in animal distributions on the survival and predation of seeds and recruitment of seedlings in fragmented and continuous forests. We then use these data to outline three groups of plants that are likely to respond in different ways to the effects of fragmentation on animal populations. We believe that understanding species interactions and how these influence processes that maintain or drive communities is important to the long-term conservation and management of habitats.

Study Sites and Methods

Environment and study sites

The wet tropics of northeastern Queensland contain the largest continuous rainforest in Australia, covering 6300 km² (Tracey, 1982). The region consists of narrow coastal plains flanked by rugged mountains to 1622 m with extensive upland areas gradually sloping to the west. Approximately one-third of the rainforest present prior to European settlement has been cleared, mostly on the coastal lowlands and tablelands (Winter *et al.*, 1987). Rainfall on areas covered by rainforest varies from 1500 mm to more than 9000 mm annually with over 70% falling in the period December to March. Our study sites received 3000–4000 mm of rainfall annually.

Our study sites include one site in continuous forest at Paluma (19°00′S, 146°10′E) and 14 sites on the Atherton Tableland, seven in habitat fragments and seven in continuous forest (all within 9 km of 17°22′S, 145°40′E). The seven habitat fragments are located in farming land between the townships of Malanda, Jagaan and Butchers Creek. The continuous forest sites are all located in the Wooroonooran National Park on the eastern tablelands near Mount Bartle Frere and not far from Butchers Creek. Dominant woody plant families in north Queensland rainforests are Myrtaceae, Lauraceae, Proteaceae, Elaeocarpaceae and Rutaceae (Hyland *et al.*, 1999). The forests at our study sites consist either of complex mesophyll vine forest on basalt-derived soils (Atherton Tableland sites) or simple notophyll vine forests on granitic soils (Paluma site; Tracey, 1982). All sites were 600–800 m elevation and exhibited high plant species richness, tall stature (24–33 m) and closed evergreen canopies. All habitat fragments were surrounded by a matrix of pastureland, and thus suffered from an altered microclimate, edge effects, impacts from domestic animals and (in some cases) periodic timber harvesting (Laurance, 1991; Turton and Freiburger, 1997).

Fauna

Here we deal primarily with data on the predation and dispersal of seeds by murid rodents and relate their behaviour to a wide range of other dispersers including marsupials, terrestrial and volant birds and fruit bats. Most of the rodents are granivorous but also carry and cache fruits and seeds. They include giant white-tailed rat, *Uromys caudimaculatus* (660 g), masked white-tailed rat, *Uromys hadrourus* (190 g), fawn-footed melomys, *Melomys cervinipes* (90 g), bush rat, *Rattus fuscipes* (150 g), and Cape York rat, *Rattus leucopus* (132 g). Another terrestrial mammal is the musky rat-kangaroo, *Hypsiprymnodon moschatus* (520 g), a frugivorous kangaroo that scatterhoards whole fruit but also consumes some seeds from those fruits (~11% of seeds; Dennis, 2002). Avian dispersers that forage on the forest floor include the southern cassowary, *Casuarius casuarius*, a large flightless ratite, and to a lesser extent bowerbirds (Ptilinorhynchidae) and three pigeons and doves (Columbidae). The latter three are gristmill predators (Crome, 1975), whereas cassowaries and musky rat-kangaroos are two of the few species able to disperse large fruits and seeds. The bowerbirds (all Australian tropical rainforest species) also occasionally carry and cache some large fruits. Species in which the seeds are too large to be swallowed are still dispersed by this group by being carried away from the source, having their flesh consumed and seed dropped elsewhere. In addition, the fruits of strangler figs, particularly those with large fruits, are regularly cached in trees, ideal germination sites for those species (Dennis and Westcott, personal observation). The rest of the volant community are primary dispersers removing fruits directly from trees, vines and shrubs. Although the mammals we consider are partly scansorial, foraging from the forest floor to the upper canopy (*U. caudimaculatuss* and *M. cervinipes*) or coming partway up into the understorey (the two *Rattus* species and *H. moschatus*), here we consider their impacts as dispersers and predators of fallen fruit and therefore treat them as terrestrial.

Feeding behaviour

To determine which fruit structural types are subject to predation by mammals we collected data on feeding behaviour. We did this by examining feeding marks left by animals in fruits and seeds on the forest floor along 7 m × 300 m long belt transects and by offering fruits to captive animals. Transects were sampled monthly from October 1990 to December 1992 at a site on the Ghourka Road on the Atherton Tableland (see Dennis and Marsh, 1997). Each fruit along a transect was examined and its condition recorded. Foragers were identified by comparing feeding marks to a reference collection of feeding marks by known species, and the part eaten (flesh, seed or both) was recorded. The two *Rattus* spp. (*R. fuscipes* and *R. leucopus*) could not be distinguished and were lumped as a single category. Feeding marks that were indistinct were recorded as 'unknown'. Fruits and seeds without any evidence of feeding were also recorded to estimate total availability.

Only rodents were used in captive feeding trials. Animals were live-trapped and housed in individual enclosures for a maximum of 10 days. They were fed commercial rodent food and mixed grain or the fruits of a single species of plant on alternate nights. Quantities of fruits offered depended on the supply available and the size of the fruit but sufficient fruit were always offered to provide the animals' nightly requirement. The morning following a fruit offering, all remaining fruits, seeds and their fragments were retrieved from the tanks. The proportion of flesh or seed eaten was recorded for each fruit based on mass offered and eaten. In this way the behaviour of 60 individual rats (13 *U. caudimaculatus*, 11 *M. cervinipes* and 36 *Rattus* spp.) handling 33 species of fruit was recorded.

We calculated a ranking of preference using data from fruits collected on the forest floor and from fruits used in the captive feeding trials following the method of Johnson (1980). Ranks were calculated separately for each species due to body size differences and these are plotted as separate points on graphs combining all species. In this method, the rank of consumption is subtracted from the food's rank of availability to provide a simple and robust ranking for the quantity of fruit consumed in relation to its availability. Preferences were calculated separately for flesh, seeds and flesh and seeds combined to analyse patterns of preference based on fruit physical characteristics.

To examine the influence of fruit physical characteristics on preferences, we regressed entire fruit mass, flesh mass, seed mass, fruit volume, flesh volume, seed volume, the ratio of flesh mass to seed mass and moisture content on preference ranks. We conducted these analyses separately for individual species and for the terrestrial frugivore/granivore community as a whole. Data collected along transects were treated separately from data collected during captive feeding trials. Separate treatment of these two data sets was necessary on the basis of differences in the type of data collected (see above) and differences in the inherent biases associated with each method. Only seven species were common to both data sets and both provide useful insight to species preferences so we include both sets here. We classified the hardness of the seed coat for each species using the categories of Blate *et al.* (1998): soft, able to be punctured by a fingernail; hard, able to be cut with a knife; and very hard, not able to be cut with a knife. Chi-squared analyses were used to test if the quantity eaten was different from that expected based on the amount available (number of fruits on transects and weight of fruits in captive feeding trials).

Seed removal, predation and caching

We conducted a series of six seed placement trials using the seeds of 16 plant species (Table 7.1) to determine rates of seed removal, predation or caching. Trials were conducted in a range of locations to test small and large-scale effects of seed placement. Trials 1 to 4 examined whether predation or removal rates were affected by

Table 7.1. Plant species used in trials and characteristics of their fruits and seeds.

Species	Family	Fruit traits				Seed traits		
		Length	Width	Mass (g)	Colour	Mass (g)	Hardness	Trial
Corynocarpus cribbianus	Corynocarpaceae	41	47	52	Red	15	Soft	5
Elaeocarpus largiflorens	Elaeocarpaceae	18	12	1.5	Blue	0.7	V. hard	4
Beilschmiedia bancrofti	Lauraceae	50	43	55	Brown	41	V. hard	5
Beilschmiedia tooram	Lauraceae	38	22	14	Black	9	Soft	5
Cryptocarya densiflora	Lauraceae	13	16	~3	Black	~2.5	Soft	1
Cryptocarya pleurosperma	Lauraceae	44	44	53	Red	24	V. hard	6
Endiandra dielsiana	Lauraceae	40	25		Black		Soft	4
Endiandra insignis	Lauraceae	60	57	107	Red/yellow	28	V. hard	5
Endiandra sankeyana	Lauraceae	34	28	16	Black	9	Soft	5
Litsea connorsii	Lauraceae	17	11		Black		Soft	1,2,3
Aglaia sapindina	Meliaceae	23	19	4	Orange	1	Soft	6
Syzygium johnsonii	Myrtaceae	20	15		Purple		Soft	1,2
Athertonia diversifolia	Proteaceae	41	22	19	Blue	7	V. hard	5
Prunus turneriana	Rosaceae	27	30	11	Black	5	Hard	5
Acronychia acronychiodes	Rutaceae	14	14		Yellow		V. hard	1
Pouteria castanosperma	Sapotaceae	50	35	63	Black	16	V. hard	5

small-scale location conditions and were conducted at a single site near Paluma. These trials examined whether rates of predation or removal varied between seed clumps (clumped together in a pile contacting one another) and scattered arrays (30 seeds/0.25 m², mimicking fruit fall). This was done in a range of microsites, including clumps at bowerbird courts (see below), beneath parent trees, beneath allospecific fruiting trees, randomly beneath tree canopies (i.e. non-fruiting trees), near logs and in canopy gaps. We placed seeds in patterns to mimic post-primary dispersal by bowerbirds or cassowaries (as clumps at or away from traditional bowerbird courts), to mimic post-primary dispersal patterns by other frugivores (as scattered arrays at locations away from fruiting trees) and as scattered arrays at fruiting trees to mimic fruit fall. Combined predation or removal rates were recorded each morning for up to 3 days. We were interested in determining if rates of encounter were affected by location so we recorded *in situ* predation and removal either to caches or an alternative site for consumption of the seed without any attempt to differentiate the final fate. Some seed clusters were placed at the display courts of tooth-billed bowerbirds,

Scenopoetes dentirostris. These courts represent a regular seasonal resource where clusters of seeds naturally resulted from the activity of attendant male birds (Frith and Frith, 1995). Clusters placed away from these courts were taken to represent the less predictable dispersal clumps created by *C. casuarius*.

Trial 1 involved placing mixed species clumps at five replicate bowerbird courts to determine the effects of seed mixing on predation and removal rates. Each clump ($n = 10$) consisted of 30 seeds each from three species (90 seeds per clump) and was conducted on 2 nights. Trial 2 examined predation and removal rates for two seed species at five microsites: at bowerbird courts, at conspecific fruit falls, at allospecific fruit falls, at random beneath the canopy (non-fruiting trees) and in canopy gaps. Each microsite contained 30 seeds of each species as scattered allospecific pairs, mimicking dispersed seeds. This pattern was replicated at five sites for a total of 4500 seeds repeated over 3 nights. In Trial 3, we examined predation and removal on seed clumps of *Litsea connorsii* at four microsites: bowerbird courts, at conspecific fruit falls, at random beneath the canopy and in canopy gaps. The latter three placements mimicked dropping piles created by cassowaries at different locations in the forest. Each placement involved 30 seeds, was replicated five times and repeated over 3 days using a total of 1800 seeds. We also repeated this trial in the following year for a total of 3600 seeds. In Trial 4, we continued our examination of post-dispersal seed clumps with another two species. Thirty seeds of each species were placed in clumps at four replicate bowerbird courts, sites near logs beneath the canopy and in canopy gaps, and at sites away from logs beneath the canopy and in canopy gaps. We used a total of 1200 seeds in this experiment. Predation and removal rates are expressed as mean ± 1 SE/day for Trials 1–4 in the results.

Trials 5 and 6 were designed to examine broad-scale effects of degree of forest fragmentation on seed predation and removal rates. All sites were in the same forest structural type on the Atherton Tableland. Trial 5 included four sites in continuous forest in Wooroonooran National Park and four forest fragments (3, 8, 20 and 97 ha in area). The seeds of three species (*Athertonia diversifolia*, *Beilschmiedia bancrofti* and *Pouteria castanosperma*) were individually marked using cotton bobbins allowing us to follow threads to caches or sites where seeds were consumed. The remaining species were monitored for predation or removal only (see Table 7.1 for species). We placed seeds of each species in both scattered arrays (to mimic a fruit fall) and singly (to mimic the result of primary dispersal). Data from this experiment have been presented previously (Harrington *et al.*, 1997) and are re-analysed and presented here using Blate *et al.*'s (1998) removal rate index. Trial 6 included three sites in continuous forest, three in small fragments (6, 11 and 13 ha) and three in large fragments (65, 80 and 80 ha). Because some seeds in this experiment were too small to accommodate cotton bobbins, they were marked using 30 cm of nylon thread with a short flagging tape marker attached at the end and individually numbered. Seed fate was then tracked by searching out the seeds by finding the flagging tape marker. Seeds were placed in scattered arrays of mixed species to mimic deposition following primary

dispersal. We placed three replicates at each site for both Trials 5 and 6. Each replicate in Trial 5 contained a mean of 30 seeds, with some variation in actual numbers based on seed availability, for a total of 4990 seeds. In Trial 6 each replicate contained ten seeds of a species (30 in an array including three species) for a total of 810 seeds.

Other dispersers: diet overlap and landscape sensitivity

Frugivores other than those studied here are effective in dispersing seeds throughout the forest, some seeds arriving at safe sites. In doing so, it is likely that seed survival to germination is enhanced (e.g. Wenny, 2001). In this case, rodents can be predators of the dispersed seeds, although they may also disperse some seeds into caches (e.g. Vander Wall, 1990). To examine which frugivores ate fruits of a given species of plant, we have used a large database of frugivore–fruit interactions (Dennis and Westcott, unpublished). The data were collected in several ways including: published accounts in peer-reviewed journals and texts, 2140 h of direct observation at fruiting trees, and by recording incidental observations of feeding behaviour. The latter two include our own observations and those of other experienced biologists and naturalists. Plants were identified using Hyland *et al.* (2003) or verified by botanists from the Australian National Herbarium's Queensland Research Station. Nomenclature follows Hyland *et al.* (2003). Species of uncertain identification have been excluded from the data and voucher specimens have been collected for undescribed species. The database currently contains 5451 records, 2296 of which are unique interactions between a plant species and a frugivore.

To categorize the sensitivity of avian dispersers to fragmentation we have used Warburton's (1997) classification. Briefly, by counting the occurrence of 60 bird species across an array of 32 forest fragments and continuous forest sites at all seasons for

4.5 years, Warburton (1997) was able to rank the species into a series of sensitivity categories on the basis of their incidence functions (Diamond, 1975). Sensitivity definitions are: (i) species showing little or no consistent response to size of forest fragment; (ii) species occurring in most sites but less frequently in very small forest fragments; (iii) species occurring in continuous forest and large fragments, less frequently in medium-sized fragments and not at all in small fragments; (iv) species occurring in continuous forest but which are missing from some large fragments and occur less frequently as size of fragment decreases; and (v) species that occur only in continuous forest and large or very large forest fragments. Fragment sizes fit into six categories as follows: > 665 ha, 150–665 ha, 33–149 ha, 4.5–32 ha, 1–4.4 ha and isolated paddock trees. For species not included in Warburton's (1997) study, we estimated the landscape sensitivity on the basis of the occurrence of species in 2140 h of focal tree observations in fragments fitting the size range outlined above or by using data from literature that examined species occurrence in fragments (Laurance, 1994, 1997; Harrington et al., 1997, 2001).

Results

Feeding preferences

All terrestrial mammals in our study consumed seeds and, for the species for which we have data (U. caudimaculatus, M. cervinipes, H. moschatus and the two Rattus species), fruit flesh (Fig. 7.1). In captive feeding trials, M. cervinipes and Uromys consumed proportionally more seed than flesh, while the Rattus consumed proportionally more flesh (ANOVA $F_{1,3832}$ = 1196, $P < 0.001$; rodent species–fruit part interaction $F_{2,3832}$ = 139.7, $P < 0.001$). Based on fruit collected along transects, H. moschatus consumed the flesh of 89.6% of fruits and the seeds of only 10.4% ($n = 646$ fruits of 35 species). There was a high degree of overlap in the species consumed along transects and in captive feeding trials

(94%). All rodents consumed most species offered, although they did not necessarily eat the same fruit parts or the same quantities (see Fig. 7.1 for examples). Hypsiprymnodon moschatus ate 58% of the same species as U. caudimaculatus and 71% of the species eaten by the smaller rats.

Data collected from fruits along transects showed a weak positive relationship between fruit size (mass) and the preferences of the terrestrial mammal community ($F_{1,42}$ = 8.18, $P = 0.006$, $r^2 = 0.2$). However, data from captive feeding trials with the rodents did not show a similar relationship, either for the rodent community or its component species. After consideration of the inherent biases in the types of data collection for captive feeding trials and for fruits along transects, we chose to use the captive feeding trial data to characterize the fruits consumed by the rodent community.

Both as a community and as individual species, terrestrial mammals ate fruits and seeds representing almost the full suite of structural characteristics measured and tested. The exceptions to this were that Rattus spp. and H. moschatus were unable to penetrate very hard seeds. Although the flesh and/or seeds of some species were totally avoided, no structural or size restrictions were evident, suggesting that species were probably avoided on the basis of their chemical composition. We found that Rattus spp. showed a preference for smaller fruits and seeds and for soft seeds (fruit size regression $F_{1,15}$ = 7.45, $P = 0.015$, $r^2 = 0.33$; seed hardness $\chi^2 = 140.5$, df = 2, $P < 0.0001$). Uromys caudimaculatus showed a preference for hard and very hard seeds ($\chi^2 = 10.2$, df = 2, $P = 0.006$), and M. cervinipes showed a preference for hard seeds ($\chi^2 = 127.6$, df = 2, $P = 0.0001$). H. moschatus showed preferences for large fruits and soft seeds (Dennis, 2002).

Predation or removal rates

The overall results from the six seed placement experiments are broadly summarized in Table 7.2 using the total proportion of

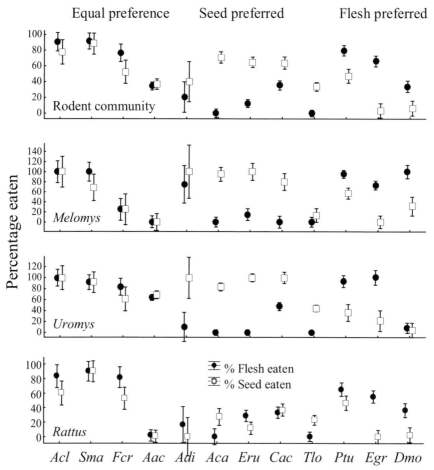

Fig. 7.1. Examples showing the percentages of fruit flesh and seed eaten by rodents during captive feeding trials, displayed for the four rodent species combined and for each genus of rodent separately. Data are means ± SE. Plant species include: *Acl, Acmenosperma clavisporum*; *Sma, Synima macrophylla*; *Fcr, Ficus crassipes*; *Aac, Acronychia acidula*; *Adi, Athertonia diversifolia*; *Aca, Alpinea caerulea*; *Eru, Elaeocarpus ruminatus*; *Cac, Carpentaria acuminata*; *Tlo, Tetrasynandra longipes*; *Ptu, Prunus turneriana*; *Egr, Elaeocarpus grandis*; *Dmo, Dysoxylum mollisimum*.

seeds removed at the end of each trial. The levels of predation and removal for ten of the species were monitored for over 100 days (Trials 5 and 6). For these species, removal ranged between 0 and 100%. Two species were not taken at all. We recorded complete removal for five species, high levels for two species and variable levels for one species. For the species monitored for only 1 night, removal rates ranged from 2 to 88% per day. Four of these species had removal rates < 10% per day and the others 50 and 88% per day.

Landscape context had little effect on the rates of seed removal for the 16 plant species used in these experiments. Instead, predation and removal rates were most consistently affected by the species of plant from which the seeds were derived. Seeds of most species were taken at similar rates regardless of dispersion, microsite or landscape context. The exceptions were few but include a large soft-seeded species, *Corynocarpus cribbianus*, that was taken at slow but consistent rates in large fragments and continuous forest and infrequently in

Table 7.2. Summary of mean seed removal for trials in different sized forest fragments and different patterns of seed dispersion. Statistical analysis for species tested only in continuous forest incorporate microsite effects, none of which was significant. Data are final percentage of seeds removed at the end of each trial, due to different periods of observation and intervals for recording in different trials. Species marked with an asterisk were monitored for 1 day in each trial, the unmarked species were monitored for > 100 days. SF, small fragments (3, 6, 11, 13 and 20 ha); MF, medium fragments (65, 80, 80 and 97 ha); C, continuous forest.

Species	SF	MF	C	Scattered	Clumped	P
Beilschmiedia tooram	0	0	0	0	0	NS
Endiandra sankeyana	0	0	0	0	0	NS
Corynocarpus cribbianus	16	100	99			< 0.001
Aglaia sapindina	92	60	69			NS
Litsea connorsii*			88	26	60	< 0.001
Syzygium johnsonii*			8.5	7	15	NS
Cryptocarya densiflora*			2			NS
Endiandra dielsiana*			4		4	NS
Prunus turneriana	100	100	100			NS
Athertonia diversifolia	100	100	100	99	100	NS
Endiandra insignis	100	100	100	100	100	NS
Pouteria castanosperma	100	100	100	98	97	NS
Beilschmiedia bancroftii	100	100	100	100	99	NS
Cryptocarya pleurosperma	90	86	84			0.005[a]
Acronychia acronychioides*			50		50	NS
Elaeocarpus largiflorens*			5		5	NS

[a]Caching only.

smaller fragments (Trial 5); a large, hard-seeded species, *Cryptocarya pleurosperma*, which was cached at higher rates in small fragments than large fragments or continuous forest (Trial 6); and a small, soft-seeded species, *L. connorsii*, which was eaten at higher levels in clusters than in scattered arrays (comparison of Trials 2 and 3).

The results of seed removal trials are outlined in the following paragraphs. The first four trials examined effects of local scale location and post-primary dispersal at a site near Paluma. In these trials, most seeds were eaten *in situ* with only a few being removed. The results are presented as the combination of eaten and removed seeds summed. The fates of the removed seeds are unknown. Trials 5 and 6 examined the larger scale effects of fragmentation on pre-dispersal and post-dispersal placements.

Trial 1 demonstrated that seeds placed in natural and predictable deposition sites (bowerbird courts) showed consistent predation or removal rates over both nights ($\chi^2 = 2.471$, df = 4, $P = 0.65$). The most distinctive pattern was a clear preference for

seeds of particular species. Animals took *L. connorsii* and *Acronychia acronychioides* seeds most consistently ($88 \pm 1.3\%$ and $50 \pm 14.2\%$, respectively), while *Syzygium johnsonii* and *Cryptocarya densiflora* seeds were seldom eaten ($15\% \pm 3.9$ and $1\% \pm 0.65$, respectively).

Trial 2 also demonstrated that preference for a species was more important than microsite. It was conducted over 3 nights and examined two species in scattered arrays at five microsites (bowerbird courts, beneath parent *L. connorsii*, beneath parent *S. johnsonii*, randomly beneath tree canopies and in canopy gaps). *L. connorsii* seeds were preferred over *S. johnsonii* seeds (ANOVA plant species effect $F_{1,40} = 39.56$, $P < 0.001$). This was the only statistically significant effect in a mixed model analysis of variance that incorporated microsite, species, night and their interactions.

Results from the first year of Trial 3 showed that predation or removal of *L. connorsii* was similar regardless of microsite (i.e. at bowerbird courts, beneath conspecific adults, at random beneath the canopy, and in canopy gaps) (a mixed model ANOVA showed no significant effects). Predation or removal rates were higher in single species clumps ($60 \pm 4\%$) than in mixed clumps or scattered arrays ($25.4 \pm 7.2\%$) used in Trial 2. When repeated in the second year, predation and removal rates were much lower (4.7%) but again lacked any microsite effects and failed to demonstrate any difference between predation rates on different nights.

Comparison of Trial 2 and the first year of Trial 3 for *L. connorsii*, which were conducted 1 week apart, demonstrated significantly higher rates of predation or removal for clumped seeds than for scattered arrays of seeds (ANOVA, clumped vs. scattered effect $F_{1,32} = 29.21$, $P < 0.001$). This difference may have resulted from the presence of a less favoured species in the scattered arrays but being placed in a scattered array should have reduced any effects of seed mixing.

Trial 4 tested the effect of plant species and five microsites on the predation or removal of clumps of seeds. The microsites included bowerbird courts, beneath canopy, canopy gap, adjacent to log and 3 m from a log. Predation or removal rates for both *Endiandra deilsiana* and *Elaeocarpus largiflorens* were low in all microsites (< 8% in total; mixed model ANOVA showed no significant effects). *Ad hoc* observations indicated that *E. largiflorens* seeds, which remained dormant, were heavily predated later in the year during a period of low fruit availability, while *E. deilsiana* seeds suffered high losses to fungal pathogens.

In Trial 5, as in the previously reported trials, seeds of different species were removed at significantly different rates. Four species were removed at high rates, two species were not removed or consumed at all and one species was removed at a low rate in small fragments and at an intermediate rate in fragments > 20 ha and in continuous forest (general linear model: species effect, $F_{5,102} = 112.5$, $P < 0.001$; species × location interaction, $F_{23,102} = 3.8$, $P < 0.001$). Neither seed dispersion nor fragment size affected rates of removal, except in the case mentioned above. For the three species that were

followed using spool and line tracking, 48% of seeds were cached and caching rates were consistent across locations. All cached seeds were eventually located and eaten, usually within a few days but on one occasion after 125 days.

In Trial 6, removal and predation rates were consistently high. This trial examined the final fate of seeds of three species (*Aglaia sapindina*, *Prunus turneriana* and *C. pleurosperma*) in three fragments in each of two size classes (< 15 ha or between 60 and 80 ha) as well as three continuous forest sites. While removal and predation rates were consistently high, *P. turneriana* was removed at a significantly higher rate than the other species (ANOVA: species effect, $F_{2,72} = 7.7$, $P = 0.001$). *Agliaia sapindina*, a species with small, soft seeds (see Table 7.1), was eaten and rarely cached, whereas *P. turneriana*, a species with hard, medium-sized seeds was eaten less and cached more, while the largest, hardest seeds, *C. pleurosperma*, were preferentially cached with only a few being eaten. The only effect of patch size was that *C. pleurosperma* was cached more frequently in small fragments than large fragments or continuous forest (ANOVA: fragment size, $F_{2,18} = 4.8$, $P = 0.018$).

Other dispersers: diet overlap and landscape sensitivity

By combining data derived from captive feeding trials and observations along transects, we derived a list of 59 plant species known to be consumed by the rodent community. Six of the species' fruits were eaten only by rodents. The rest overlapped with the diets of between one and 18 other frugivores. The species with overlapping diets included *H. moschatus*, *C. casuarius*, and a range of volant species including flying foxes and birds, most of which treat the seeds gently and disperse them away from their source. Species that were least preferred by rodents (preferences for captive and transect data combined) were more likely to have a larger number of other frugivores feeding upon them (Fig. 7.2; $r = 0.34$, $P = 0.008$).

The occurrence of non-rodent dispersers in habitat fragments and intact forest across the landscape was variable. The mean sensitivity score of frugivores consuming fruits that were also in the diets of rodents was higher when only a few species overlapped (Fig. 7.3; ANOVA: $F_{11,39} = 2.7$, $P = 0.01$). For 15 plant species with fruits eaten by only one other frugivore, the mean sensitivity score of those additional frugivores

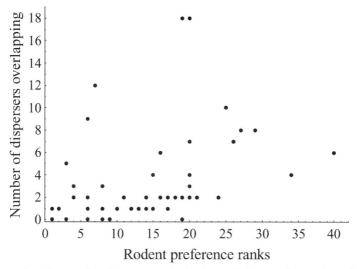

Fig. 7.2. The number of non-rodent disperser species feeding on fruits eaten by members of the granivorous rodent community plotted as a function of the fruits' rodent preference ranks. A score of 1–15 indicates more preferred rodent food while scores of 16–40 are less preferred. Note: because we used a relative ranking system, in some cases multiple plant species have the same preference rank and in other cases rank scores have no plant species.

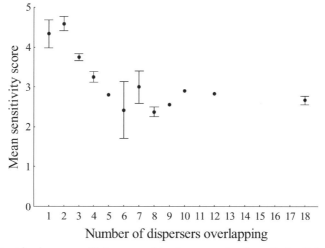

Fig. 7.3. Mean (± SE) landscape sensitivity scores of frugivores overlapping with the diets of terrestrial predators. A high sensitivity score indicates a tendency for species to occupy only very large fragments and continuous forest. Sensitivity scores between two and three indicate species tending to use some of the habitat fragment matrix but avoiding small fragments.

was > 4, meaning a general lack of occurrence in all but the largest fragments for the overlapping frugivores. These additional species were most often *H. moschatus* and *C. casuarius*, which respond poorly to forest fragmentation. Where there were two frugivores with diets overlapping that of the rodents, the mean sensitivity score of the additional frugivores to fragmentation was closer to five, i.e. the species were even less widely distributed in a fragmented landscape. Beyond two overlapping species, the mean landscape sensitivity scores declined to between 2.5 and 3.0. This indicates that, on average, the full complement of dispersers does not occur in all forest fragments whereas the rodents do, to a large degree, occur in all fragments as suggested by the seed placement trials and studies of rodent distribution (Laurance, 1994; Harrington *et al.*, 1997, 2001; see Discussion).

Discussion

Feeding preferences

Our data indicate that the component of the terrestrial mammal community that feeds on fruits in Australia's tropical rainforests represents a continuum of seed interactions from highly predatory species to primarily frugivorous seed dispersers in the order of *U. caudimaculatus* > *M. cervinipes* > *Rattus* spp. > *H. moschatus*. No species is entirely granivorous or frugivorous and all species play roles as seed predators and as seed dispersers. In the past, a dichotomy between seed predators on the one hand and seed dispersers on the other has been described (see Hulme, 2002). Van der Pijl (1972) pointed out that this dual role needs to be incorporated into our attempts to measure and understand the overall effect of an animal on the plant populations with which it interacts. We suggest that the dominant impact of an animal, be it seed dispersal or seed predation, is dependent on plant species, the length of time to germination and the simultaneous impacts of other dispersers in the community.

Both as a community and as individual species, terrestrial mammals consumed fruits and seeds of all sizes and structural characteristics offered, except that *Rattus* spp. and *H. moschatus* were unable to penetrate very hard seeds. This is in some regards surprising given the emphasis placed on the use of large seeded plants by most tropical granivorous rodents and the suggestion that large nutlike fruit are co-evolved with terrestrial granivores (Smith and Reichman, 1984; Vander Wall, 2001; Forget *et al.*, 2002; Jansen *et al.*, 2002). This pattern of fruit and seed consumption may be common throughout the tropics, but an impression that large, nutritious seeds are dominant in the diets of scatterhoarding animals may be biased by the fact that more attention has been paid to the large species that are frequently cached.

The selective pressures that terrestrial scatterhoarders bring to bear on plants are therefore likely to differ with fruit characteristics and the range of other dispersers handling the fruits. For example, according to our data, species with large, hard seeds (e.g. *C. pleurosperma*; Trial 6) are more likely to be cached by granivorous rodents than smaller, soft-seeded species. The latter are more likely to be consumed *in situ* (e.g. *A. sapindina*; Trial 6). *C. pleurosperma* seeds have characteristics that promote caching, such as a hard seed and large size, as well as characteristics that promote dispersal by cassowaries and musky rat-kangaroos, such as thick, juicy flesh. These characteristics would suggest that the latter two benign dispersers and the granivorous rodents are all important in recruitment for this species. However, the small, soft-seeded *A. sapindina* is unlikely to derive any benefits from granivorous rodents and therefore selection should favour individuals less palatable to predators but with flesh that is more attractive to dispersers. The ultimate impact of these potential selective pressures on fruit characteristics will depend on the raft of other potential selective pressures applied by other seed predators, dispersers and the germination requirements of the species.

The effects of landscape context

In our study, location within the land-scape (either microsite or patch size) had little effect on the probability that an animal would encounter and remove a seed. Palatable species were rapidly eaten or cached regardless of microsite and regardless of whether they were in forest fragments or in continuous forest. This is a common result of studies of seed predation and removal rates. However, high levels of variation in patterns have been a hallmark of these studies. In some instances, distinctive patterns of fragment size or microsite effects have been uncovered, while other studies show no consistent patterns (Willson, 1988; Willson and Whelan, 1990; Osunkoya, 1994; Blate et al., 1998; Notman and Gorchov, 2001; Peña-Claros and Deboo, 2002). In our study, the fact that predation or removal rates did not vary significantly with the size of fragments reflects the distribution and abundance of the granivorous rodents in the region (Laurance, 1994; Harrington et al., 2001). While these studies noted statistically lower densities or abundances of U. caudimaculatus in very small frag-ments (Harrington et al., 2001), the differ-ences were minimal, on the order of 1–2 individuals/ha, and fell within the range of variation between sampling times or studies. M. cervinipes, on the other hand, was more abundant in fragments than continuous forest in one study (Laurance, 1994) and equally abundant in the other (Harrington et al., 2001). The two Rattus spp. showed reciprocal responses. Individ-uals identified as R. leucopus were more abundant in fragments than continuous sites, while individuals identified as R. fuscipes showed the reverse trend. However, both species occurred through-out, with R. fuscipes in much higher densities than its congener. Caution is needed in interpreting these patterns due to the difficulty of accurate identification of the species. H. moschatus were absent from forest fragments in these and other studies (Laurance, 1994; Harrington et al., 1997, 2001).

The effects of other dispersers

Here we consider how landscape context affects all dispersal vectors to develop an understanding of how animals influence seed dispersal, seed survival and seedling recruitment. We have demonstrated that the impacts of granivorous rodents are similar across all landscape categories and micro-sites, both through our experimental evidence and from other studies of the dis-tribution of the animals. We have also dem-onstrated that, across plant species, levels of seed predation can be variable and, impor-tantly for some species, can be extremely high. In addition, it seems clear that seeds of some species are cached at high rates while others are not. The activities of seed predators act to reduce the probability of new recruits for some plant species (e.g. Crawley, 1992; Hulme, 2002) but they can also disperse seeds (Theimer, 2001; Dennis, 2003). Although we have demonstrated that dispersed seeds are as likely to be consumed as those at fruiting trees, we maintain that seed dispersal increases the probability of survival for seeds because of the variable effects of mortality caused by agents other than terrestrial vertebrates (e.g. Connell, 1970; Janzen, 1970; Terborgh et al., 2002; Silvius, Chapter 4, this volume). We now explore whether species of plants experience different levels of seed dispersal service in different parts of the landscape. After examining the patterns of granivorous rodent distribution, seed disperser distribu-tion and the overlap in diet, we hypothesize that plant species fall into three distinct categories based on how they are treated by the disperser community.

Group 1 species are those that are con-sumed or cached by predators and are rarely or never dispersed by frugivores. These include the classic 'nut' style fruit that are dispersed primarily by seed-caching animals that also act as seed predators and thus gain enhanced probability of recruit-ment (e.g. Price and Jenkins, 1986; Jansen et al., 2002), and potentially include some wind-dispersed species. Group 2 species are eaten by seed predators, possibly occasion-ally cached, but are dispersed by several species of frugivores (Fig. 7.2). Group 3 species are only rarely eaten by vertebrate seed predators and are generally dispersed only by frugivores (see Table 7.3 for examples for each group). The physical characteristics of fruit and seeds in each group are variable and we define the groups on the basis of rodent granivore preferences and their presence in other disperser diets. However, Group 1 fruits are often woody, have hard to very hard seeds and cover a range of sizes but are commonly large. Group 2 fruits are generally fleshy and thus attractive to frugivores, have soft to hard seeds and also cover a range of sizes. Those not eaten by rats and therefore in Group 3 tend to have soft seeds that are protected by thick latexes and possibly certain chemical defences. These fruits come from a range of size classes but are often small.

The animals involved in the dispersal and predation of Group 1 plant species appear to be little affected by fragmentation and thus the potential for recruitment of plants in this group is likely to be similar in fragmented and continuous forest. This is due to dispersal and predation being carried out by the same suite of animal species (see also Harrington *et al.*, 1997). The dispersal and survival of seeds and recruitment of seedlings in Group 3 species will be affected by fragmentation only to the extent that distribution and abundance of seed-dispersing frugivores are affected by fragmentation. Because many of the birds that disperse these species are not greatly affected by habitat fragmentation, we expect little change in the final dispersal of seeds for many species in this group, particularly those with small fruit. Group 2 plants are likely to be the most affected by fragmentation because the population sizes of animals that disperse them are reduced by fragmentation while their seed predators are relatively unaffected.

There were two patterns of disperser loss in forest fragments for Group 2 plants: complete loss of dispersers (i.e. local extirpation) and partial loss of dispersers (significant population declines or local extirpation of some disperser species but not all). Twenty-one plant species in our data

Table 7.3. Examples of fruit characteristics for tree species falling into three plant groups described in the text. Examples include some species used in experiments and additional species not used in experiments.

| Species | Family | Fruit traits | | | | Seed traits | |
		Length (mm)	Width (mm)	Mass (g)	Colour	Mass (g)	Hardness
Group 1							
Athertonia diversifolia	Proteaceae	41	22	19	Blue	7	V. hard
Beilschmiedia bancrofti	Lauraceae	50	43	55	Green/brown	41	V. hard
Goniothalamus australis	Annonaceae	45	20	–	Green/brown	–	Soft
Hylandia dockrillii	Euphorbiaceae	29	18	9	Green/brown	2	Hard
Aleurites molluccana	Euphorbiaceae	69	50	102	Green/brown	39	Hard
Group 2							
Baileyoxylon lanceolatum	Flacourtiaceae	41	27	16	Orange	8	Hard
Niemeyera prunifera	Sapotaceae	27	27	15	Black	10	Soft
Prunus turneriana	Rosaceae	27	30	11	Black	5	Hard
Acronychia acronychiodes	Rutaceae	14	14	–	Yellow	–	V. hard
Elaeocarpus ruminatus	Elaeocarpaceae	14	12	1.2	Blue	0.6	Hard
Group 3							
Beilschmiedia tooram	Lauraceae	38	22	13	Black	9	Soft
Cryptocarya mackinnoniana	Lauraceae	23	15	2.5	B&W spotted	2	Soft
Polyscias murrayi	Araliaceae	4	6	–	Black/purple	–	Soft
Dysoxylum mollissimum	Meliaceae	17	16	2.5	Red aril	0.9	Soft
Acmena resa	Myrtaceae	12	10	0.8	White	0.6	Soft

set (36%) had only one or two frugivores dispersing them and those frugivores were uncommon or absent in most fragments. Examples of such species include *Fontainea picrosperma* (Euphorbiaceae) and *Austrobaileya scandens* (Austrobaileyaceae), the seeds of which are rapidly consumed by rodent granivores (Dennis, 1997) but are dispersed by only two frugivores, musky rat-kangaroos and cassowaries, which are absent from most forest fragments. We predict that the effect of a partial loss of services for Group 2 plants will fall along a continuum of severity. Less affected will be *Oraniopsis appendiculata* (Arecaceae; preference ranking 4), because they are dispersed by *H. moschatus* and *C. casuarius* in continuous forest but are probably dispersed short distances by spectacled flying-foxes, *Pteropus conspicillatus*, in both continuous and fragmented forest. More affected will be *Neimeyera prunifera* (Sapotaceae; preference ranking 8), which is unlikely to be dispersed by species in forest fragments. Other species, such as *C. pleurosperma* (Lauraceae) and *E. bancroftii* (Elaeocarpaceae), have fleshy fruits and large seeds with very hard seed coats and are also dispersed by musky rat-kangaroos and cassowaries. Because these species are palatable to and cached by *U. caudimaculatus*, and because these animals are less affected by fragmentation, the seeds continue to be dispersed in the absence of the frugivores, although the distances and quality (*sensu* Schupp, 1993) of dispersal are probably reduced. These species are probably examples of plants frequently experiencing two-stage dispersal through endozoochory and then scatterhoarding by rodents (see Vander Wall and Longland, Chapter 18, this volume).

The combination of sustained impact by seed predators and reduced disperser coteries in fragmented forests is likely to cause a change in the trajectory of plant species composition in forest fragments. Our results suggest a scenario in which there would be:

1. A gradual decline of species whose seeds are eaten but rarely dispersed by rodents and normally dispersed by a range of other frugivores that are rare or absent in forest fragments (Group 2 above).

2. A gradual increase in species whose seeds are rarely consumed by rodents but frequently ingested and dispersed by frugivores that show little sensitivity to fragmentation (a proportion of Group 3).

3. Little change for those species consumed and dispersed by rodents (Group 1).

Within the context of these responses, species with seeds that are preferred by predators but generally consumed immediately and not cached (e.g. *F. picrosperma*) might be expected to suffer particularly reduced recruitment in response to forest fragmentation.

Harrington *et al.* (1997) lamented the fact that no response of forest fragmentation on seedling recruitment was measurable in their study of seed predation on large-seeded species. However, the plant species they were studying were primarily dispersed by rodent granivores and therefore were the species that should experience little change in survival to germination (our Group 1 or Class A and B in Harrington *et al.*, 1997). They also included three species seldom eaten by granivorous rodents that would have lost a large component of their disperser coterie through the absence of *C. casuarius* and *H. moschatus*. While these species may show a gradual change in seed dispersion and decline in recruitment, the fact that predators kill few seeds means that many are still available to germinate. What we predict from this study is that medium to small-seeded species that are widely dispersed and of low palatability to rodents will show a relative increase in recruitment, while those that lose a component of their disperser coterie and are palatable to predators will show a relative decline.

Our predictions so far are made in isolation from other factors affecting plant recruitment. The interactions between seed predators and dispersers may be creating a trend favouring Group 3 species, reducing Group 2 species and having little impact on Group 1 species. However, other factors such as levels of disturbance, changes in pollination and fruit set may be having either opposing or synergistic effects, enhancing or reducing the effects of relative changes in predation and dispersal. A recent study by Laurance *et al.* (2003) has demonstrated that, in general, fragmentation does not greatly affect flower and fruit production in plants in a fragmented landscape in the Brazilian Amazon. If a similar pattern occurs in Australia's wet tropics, our predictions should not be complicated by changes in flower and fruit set. The impacts of disturbance and changes in the nature and distribution of recruitment sites associated with fragmentation are likely to be more complex. Most species in tropical rainforests can germinate in shaded conditions and seedlings are responsive to changes in light regimes (Grubb *et al.*, 1998). This suggests that, while pioneer species and other disturbance-adapted plants may undergo changes in abundance and distribution after forest fragmentation (Laurance, 1991), recruitment of the remaining species is likely to be affected primarily by relative changes in seed predation and dispersal.

Conclusions

The potential impacts of granivores on plant population dynamics and the maintenance of diversity in tropical forests have long been recognized (Terborgh *et al.*, 2002). The importance of granivores as dispersers of at least some of the plants with which they interact is also well recognized in both theoretical and empirical studies (Hulme, 2002). Here we have shown that these impacts can be spread across a wide range of species and remain relatively constant in a wide range of rainforest fragments and microsites. The magnitude and direction of the impact of granivores on seeds of a plant species are dependent on the number of other dispersers that interact with those seeds and the rate of caching and recovery of seeds by granivores. Our estimates of a decline in the size of dispersal coteries combined with the evidence of a relatively stable predator regime across the landscape suggest a net loss in recruitment

opportunities for species eaten by rodents and a range of other frugivores. We predict this will lead to long-term changes in the trajectories of plant community structure in fragments, which would vary in intensity depending on the impact of forest fragmentation on dispersers. We expect fragmented tropical rainforests to show a decline in plant species that are fed upon by rodents but rely primarily on animals sensitive to fragmentation for dispersal. Future investigations into the effects of fragmentation on forest tree recruitment may benefit from classifying the impacts of animals on fruits and seeds in a manner similar to that suggested here.

Acknowledgements

We are grateful for support from Matt Bradford and Adam McKeown, Earthwatch Institute volunteers, James Cook University, and Queensland Parks and Wildlife Service. Pete Green, Tad Theimer, Patrick Jansen, Stephen Vander Wall and Pierre-Michel Forget provided helpful comments on an early draft of the manuscript.

References

Andersen, A.N. (1989) How important is seed predation to recruitment in stable populations of long-lived perennials? *Oecologia* 81, 310–315.

Asquith, N.M., Wright, S.J. and Clauss, M.J. (1997) Does mammal community composition control recruitment in neotropical forests? Evidence from Panama. *Ecology* 78, 941–946.

Blate, G.M., Peart, D.R. and Leighton, M. (1998) Post-dispersal predation on isolated seeds: a comparative study of 40 tree species in a Southeast Asian rainforest. *Oikos* 82, 522–538.

Clark, J.S., Macklin, E. and Wood, L. (1998) Stages and spatial scales of recruitment limitation in southern Appalachian forests. *Ecological Monographs* 68, 213–235.

Connell, J.H. (1970) On the role of natural enemies in preventing competitive exclusion in some marine animals and rain forest trees. In: den

Boer, P.J. and Gradwell, G.R. (eds) *Dynamics of Populations.* Centre for Agricultural Publication and Documentation, Wageningen, The Netherlands, pp. 298–312.

Corlett, R.T. (2002) Frugivory and seed dispersal in degraded tropical East Asian landscapes. In: Levey, D.J., Silva, W.R. and Galetti, M. (eds) *Seed Dispersal and Frugivory: Ecology, Evolution and Conservation.* CAB International, Wallingford, UK, pp. 451–465.

Crawley, M.J. (1992) Seed predators and plant population dynamics. In: Fenner, M. (ed.) *Seeds: the Ecology of Regeneration in Plant Communities.* CAB International, Wallingford, UK, pp. 157–191.

Crome, F.H.J. (1975) The ecology of fruit pigeons in tropical northern Queensland. *Australian Wildlife Research* 2, 155–185.

De Steven, D. and Putz, F.E. (1984) Impact of mammals on early recruitment of a tropical canopy tree, *Dipteryx panamensis*, in Panama. *Oikos* 43, 207–216.

Dennis, A.J. (1997) Musky rat-kangaroos, *Hypsiprymnodon moschatus*: cursorial frugivores in Australia's wet tropical rain forests. PhD Thesis, James Cook University of North Queensland, Townsville, Australia.

Dennis, A.J. (2002) The diet of the musky rat-kangaroo, *Hypsiprymnodon moschatus*, a rainforest specialist. *Wildlife Research* 29, 209–219.

Dennis, A.J. (2003) Scatter-hoarding by musky rat-kangaroos, *Hypsiprymnodon moschatus*, a tropical rain forest marsupial from Australia: implications for seed dispersal. *Journal of Tropical Ecology* 19, 619–627.

Dennis, A.J. and Marsh, H.D. (1997) Seasonal reproduction in musky rat-kangaroos, *Hypsiprymnodon moschatus*: a response to changes in resource availability. *Wildlife Research* 24, 561–578.

Diamond, J.M. (1975) The island dilemma: lessons of modern biogeographic studies for the design of natural reserves. *Biological Conservation* 7, 129–146.

Feer, F., Julliot, C., Simmen, B., Forget, P.-M., Bayart, F. and Chauvet, S. (2001) Recruitment, a multi-stage process with unpredictable result: the case of a Sapotaceae in French Guianan forest. *Revue d'Ecologie (la Terre et la Vie)* 56, 119–145.

Forget, P.-M. (1990) Seed-dispersal of *Vouacapoua americana* (Caesalpiniaceae) by caviomorph rodents in French Guiana. *Journal of Tropical Ecology* 6, 459–468.

Forget, P.-M. (1991) Scatterhoarding of *Astrocaryum paramaca* by *Proechimys* in French Guiana: comparison with *Myoprocta exilis*. *Tropical Ecology* 32, 155–167.

Forget, P.-M., Milleron, T. and Feer, F. (1998) Patterns in post-dispersal seed removal by neotropical rodents and seed fate in relation to seed size. In: Newbery, D.M., Prins, H.H.T. and Brown, N.D. (eds) *Dynamics of Tropical Communities.* Blackwell Science, Oxford, pp. 25–49.

Forget, P.-M., Hammond, D.S., Milleron, T. and Thomas, R. (2002) Seasonality of fruiting and food hoarding by rodents in neotropical forests: consequences for seed dispersal and seedling recruitment. In: Levey, D.J., Silva, W.R. and Galetti, M. (eds) *Seed Dispersal and Frugivory: Ecology, Evolution and Conservation.* CAB International, Wallingford, UK, pp. 241–253.

Frith, C.B. and Frith, D.W. (1995) Court site constancy, dispersion, male survival and court ownership in male tooth-billed bowerbirds, *Scenopoetes dentirostris*, (Ptilinorrynchidae). *Emu* 95, 84–98.

Glanz, W.E. (1982) The terrestrial mammal fauna of Barro Colorado island: censuses and long-term changes. In: Leigh, E.G. Jr, Rand, A.S. and Windsor, D.M. (eds) *The Ecology of a Tropical Forest: Seasonal Rhythms and Long-Term Changes.* Smithsonian Institution Press, Washington, DC, pp. 455–468.

Green, P.T., O'Dowd, D.J. and Lake, P.S. (1997) Control of seedling recruitment by land crabs in rain forest on a remote oceanic island. *Ecology* 78, 2474–2486.

Grubb, P.J., Metcalfe, D.J., Grubb, E.A.A. and Jones, G.D. (1998) Nitrogen richness and protection of seeds in Australian tropical rain forest: a test of plant defence theory. *Oikos* 82, 467–482.

Hammond, D.S., Brown, V.K. and Zagt, R. (1999) Spatial and temporal patterns of seed attack and germination in a large-seeded neotropical tree species. *Oecologia* 119, 208–218.

Harrington, G.N., Irvine, A.K., Crome, F.H.J. and Moore, L.A. (1997) Regeneration of large seeded trees in Australian rain forest fragments: a study of higher order interactions. In: Laurance, W.F. and Bierregaard, R.O. (eds) *Tropical Forest Remnants: Ecology, Management and Conservation of Fragmented Communities.* University of Chicago Press, Chicago, Illinois, pp. 292–303.

Harrington, G.N., Freeman, A.N.D. and Crome, F.H.J. (2001) The effects of fragmentation of an Australian tropical rain forest on populations and assemblages of small mammals. *Journal of Tropical Ecology* 17, 225–240.

Hulme, P.E. (2002) Seed-eaters: seed dispersal, destruction and demography. In: Levey, D.J., Silva, W.R. and Galetti, M. (eds) *Seed Dispersal and Frugivory: Ecology, Evolution and Conservation.* CAB International, Wallingford, UK, pp. 257–271.

Hurtt, G.C. and Pacala, S.W. (1995) The consequences of recruitment limitation: reconciling chance, history and competitive differences between plants. *Journal of Theoretical Biology* 176, 1–12.

Hyland, B.P.M., Whiffin, T., Christophel, D.C., Grey, B., Elick, R.W. and Ford, A.J. (1999) *Australian Tropical Rain Forest Trees and Shrubs.* CD ROM and Manual. CSIRO Publishing, Collingwood, Australia.

Hyland, B.P.M., Whiffin, T., Christophel, D.C., Gray, B. and Elick, R.W. (2003) *Australian Tropical Rain Forest Plants: Trees, Shrubs and Vines.* CD ROM and Manual. CSIRO Publishing, Collingwood, Australia.

Jansen, P.A., Bartholomeus, M., Bongers, F., Elzinga, J.A., den Ouden, J. and Van Wieren, S.E. (2002) The role of seed size in dispersal by a scatter-hoarding rodent. In: Levey, D.J., Silva, W.R. and Galetti, M. (eds) *Seed Dispersal and Frugivory: Ecology, Evolution and Conservation.* CAB International, Wallingford, UK, pp. 209–225.

Janzen, D.H. (1970) Herbivores and the number of tree species in tropical forests. *The American Naturalist* 104, 501–528.

Johnson, D.H. (1980) The comparison of usage and availability measurements for evaluating resource preference. *Ecology* 61, 65–71.

Laurance, W.F. (1990) Comparative responses of five arboreal marsupials to tropical forest fragmentation. *Journal of Mammalogy* 71, 641–653.

Laurance, W.F. (1991) Predicting the impacts of edge effects in fragmented habitats. *Biological Conservation* 55, 77–92.

Laurance, W.F. (1994) Rainforest fragmentation and the structure of small mammal communities in tropical Queensland. *Biological Conservation* 69, 23–32.

Laurance, W.F. (1997) Responses of mammals to rainforest fragmentation in tropical Queensland: a review and synthesis. *Wildlife Research* 24, 603–612.

Laurance, W.F., Rankin-de Merona, J.M., Andrade, A., Laurance, S.G., D'Angelo, S., Lovejoy, T.E. and Vasconcelos, H.L. (2003) Rain forest fragmentation and the phenology of Amazonian tree communities. *Journal of Tropical Ecology* 19, 343–347.

Leigh, E.G. Jr, Wright, S.J., Herre, E.A. and Putz, F.E. (1993) The decline of tree diversity on newly isolated tropical islands: a test of null hypothesis and some implications. *Evolutionary Ecology* 17, 76–102.

Muller-Landau, H.C., Wright, S.J., Calderon, O., Hubbell, S.P. and Foster, R.B. (2002) Assessing recruitment limitation: concepts, methods and case studies from a tropical forest. In: Levey, D.J., Silva, W.R. and Galetti, M. (eds) *Seed Dispersal and Frugivory: Ecology, Evolution and Conservation.* CAB International, Wallingford, UK, pp. 35–53.

Notman, E. and Gorchov, D.L. (2001) Variation in post-dispersal seed predation in mature Peruvian lowland tropical forest and fallow agricultural sites. *Biotropica* 33, 621–636.

Notman, E., Gorchov, D.L. and Cornejo, F. (1996) Effect of distance, aggregation, and habitat on levels of seed predation for two mammal-dispersed neotropical rain forest tree species. *Oecologia* 106, 221–227.

Osunkoya, O.O. (1994) Postdispersal survivorship of north Queensland rainforest seeds and fruits: effects of forest, habitat and species. *Australian Journal of Ecology* 19, 52–64.

Pahl, L.I., Winter, J.W. and Heinsohn, G. (1988) Variation in responses of arboreal marsupials to fragmentation of tropical rain forests in north-eastern Australia. *Biological Conservation* 46, 71–82.

Peña-Claros, M. and Deboo, H. (2002) The effect of forest successional stage on seed removal of tropical rain forest tree species. *Journal of Tropical Ecology* 18, 261–274.

Price, M.V. and Jenkins, S.H. (1986) Rodents as seed consumers and dispersers. In: Murray, D.R. (ed.) *Seed Dispersal.* Academic Press, Sydney, pp. 191–235.

Schupp, E.W. (1993) Quantity, quality and effectiveness of seed dispersal by animals. *Vegetatio* 107/108, 15–29.

Schupp, E.W., Milleron, T. and Russo, S.E. (2002) Dissemination limitation and the origin and maintenance of species-rich tropical forests. In: Levey, D.J., Silva, W.R. and Galetti, M. (eds) *Seed Dispersal and Frugivory: Ecology, Evolution and Conservation.* CAB International, Wallingford, UK, pp. 19–33.

Smith, C.C. and Reichman, O.J. (1984) The evolution of food caching by birds and mammals. *Annual Review of Ecology and Systematics* 15, 329–351.

Sork, V.L. (1987) Effects of predation and light on seedling establishment in *Gustavia superba.* *Ecology* 68, 1341–1350.

Terborgh, J.W. (1992) Maintenance of diversity in tropical forests. *Biotropica* 24, 283–292.

Terborgh, J.W. and Wright, S.J. (1994) Effects of mammalian herbivores on plant recruitment in two Neotropical forests. *Ecology* 75, 1829–1833.

Terborgh, J., Losos, E., Riley, M.P. and Riley, M.B. (1993) Predation by vertebrates and invertebrates on the seeds of five canopy tree species of an Amazonian forest. *Vegetatio* 107/108, 375–386.

Terborgh, J., Pitman, N., Silman, M., Schichter, H. and Nunez, V.P. (2002) Maintenance of tree diversity in tropical forests. In: Levey, D.J., Silva, W.R. and Galetti, M. (eds) *Seed Dispersal and Frugivory: Ecology, Evolution and Conservation*. CAB International, Wallingford, UK, pp 1–17.

Theimer, T.C. (2001) Seed scatterhoarding by white-tailed rats: consequences for seedling recruitment by an Australian rain forest tree. *Journal of Tropical Ecology* 17, 177–189.

Tracey, J.G. (1982) *The Vegetation of the Humid Tropical Region of North Queensland*. Commonwealth Scientific and Industrial Research Organisation, Melbourne.

Turton, S.M. and Freiburger, H.J. (1997) Edge and aspect effects on the microclimate of a small tropical forest remnant on the Atherton Tableland, northeastern Australia. In: Laurance, W.F. and Bierregaard, R.O. (eds) *Tropical Forest Remnants: Ecology, Management and Conservation of Fragmented Communities*. University of Chicago Press, Chicago, Illinois, pp. 45–54.

van der Pijl, L. (1972) *Principles of Dispersal in Higher Plants*, 2nd edn. Springer Verlag, Berlin.

Vander Wall, S.B. (1990) *Food Hoarding in Animals*. University of Chicago Press, Chicago, Illinois, 445 pp.

Vander Wall, S.B. (2001) The evolutionary ecology of nut dispersal. *Botanical Review* 67, 74–117.

Warburton, N.H. (1997) Structure and conservation of avifauna in isolated rainforest remnants in tropical Australia. In: Laurance, W.F. and Bierregaard, R.O. (eds) *Tropical Forest Remnants: Ecology, Management and Conservation of Fragmented Communities*. University of Chicago Press, Chicago, Illinois, pp. 190–206.

Wenny, D.G. (2001) Advantages of seed dispersal: a re-evaluation of directed dispersal. *Evolutionary Ecology Research* 3, 51–74.

Willson, M.F. (1988) Spatial heterogeneity of post dispersal survivorship of Queensland rainforest seeds. *Australian Journal of Ecology* 13, 137–145.

Willson, M.F. and Whelan, C.J. (1990) Variation in postdispersal survival of vertebrate-dispersed seeds: effects of density, habitat, location, season, and species. *Oikos* 57, 191–198.

Winter, J.W., Bell, F.C., Pahl, L.I. and Atherton, R.G. (1987) Rainforest clearing in northeastern Queensland. *Proceedings of the Royal Society of Queensland* 98, 41–57.

Zuidema, P.A. and Boot, R.G.A. (2002) Demography of the Brazil nut tree (*Bertholletia excelsa*) in the Bolivian Amazon: impact of seed extraction on recruitment and population dynamics. *Journal of Tropical Ecology* 18, 1–31.

8 The Fate of Primate-dispersed Seeds: Deposition Pattern, Dispersal Distance and Implications for Conservation

Joanna E. Lambert[1] and Colin A. Chapman[2]

[1]Department of Anthropology, University of Wisconsin, Madison, WI 53706, USA;
[2]Department of Zoology, University of Florida, Gainesville, FL 32611, USA, and
Wildlife Conservation Society, 2300 Southern Boulevard, Bronx, NY 10460, USA

Introduction

The role of animals in seed dispersal is well recognized. Indeed, as many as 75% of tropical tree species produce fruits presumably adapted for animal dispersal (Frankie et al., 1974; Howe and Smallwood, 1982) and animals are estimated to move more than 95% of tropical seeds (Terborgh et al., 2002). Some vertebrate groups are argued to be particularly important seed dispersers. Primates, for example, comprise on average between 25 and 40% of the frugivore biomass in tropical forests (Eisenberg and Thorington, 1973; Chapman, 1995), eat large quantities of fruit, and defecate or spit large numbers of viable seeds (Lambert, 1999). Primate frugivory and seed dispersal have been quantified by researchers working in South America (e.g. Garber, 1986; Julliot, 1996; Dew, 2001; Vulinec, 2002), Central America (e.g. Estrada and Coates-Estrada, 1984, 1986; Chapman, 1989), and Africa (e.g. Gautier-Hion, 1984; Gautier-Hion et al., 1985; Wrangham et al., 1994; Chapman and Chapman, 1996; Kaplin and Moermond, 1998; Lambert, 1999; Voysey et al., 1999a,b).

Because primates are relatively large, arboreal mammals requiring the high levels of soluble carbohydrates available in fruit, seed removal from trees by primates can be high. Stevenson (2000), for example, has documented that woolly monkeys (Lagothrix lagothricha) in Colombia disperse 25,000 seeds/km²/day. In northern Costa Rica, spider monkeys (Ateles geoffroyi), howler monkeys (Alouatta palliata) and capuchins (Cebus capucinus) disperse approximately 5600 large seeds/km²/day (Chapman, 1989), and in French Guiana, a single group of howler monkeys (Alouatta seniculus) dispersed more than 1,000,000 seeds per year from approximately 100 plant species (Julliot, 1996). These patterns hold in Africa, too, where in Kibale National Park, Uganda, Lambert (1997, 1999, unpublished) has documented that redtail monkeys (Cercopithecus ascanius), blue monkeys (C. mitis), mangabeys (Lophocebus albigena) and chimpanzees (Pan troglodytes) either spit out or swallow the seeds of approximately 35,000 fruits/km²/day.

We have compelling evidence that primates remove large numbers of fruits from tropical trees. However, data on fruit removal and seed handling must be linked with data on post-dispersal seed fate to fully understand the relationships among primary dispersal and the origin and maintenance of plant species diversity, tropical forest regeneration, and plant demography. Several studies illustrate this point: Balcomb and Chapman (2003), for example, recently quantified patterns of seed removal and seedling recruitment of *Monodora myristica*, a low-fecundity forest tree species in Uganda. The fruit of this species is large, with an average fruit diameter of 18.5 cm, multiple large seeds over 2 cm in length, and a very thick and difficult to penetrate pericarp (Lambert, 1997, 1999; Balcomb, 2001), suggesting dispersal by only the very largest of arboreal frugivores. Balcomb and Chapman (2003) documented fruit and seed removal of *M. myristica* during focal tree watches, quantified the fate of naturally and experimentally dispersed and fallen seeds, and monitored natural seedling, sapling, and pole densities and survivorship of this species at two Kibale sites. As predicted, large-bodied primates played a critical role in primary seed dispersal; *P. troglodytes* and *L. albigena* were the only frugivores that opened the hard-husked fruits and were estimated to disperse over 85% of the mature seeds. However, seed germination and the probability of seedling establishment were highly variable among experimental seed-deposition types within each site, as well as between sites and years, suggesting complexity beyond that which could be explained by primary dispersal of seeds by primates.

Similar complexity has been documented in a study on the long-term consequences of directed seed dispersal by black-and-white-casqued hornbills (*Ceratogymna subcylindricus*) to nests (Paul, 2000). The patterns in which *C. subcylindricus* moves seeds are analogous to what has been described at many primate sleeping sites (Julliot, 1996, 1997). In his study, Paul (2000) assessed the species composition of the advanced plant regeneration below hornbill nests that was identified 12–16 years prior (Kalina, 1988). During nesting, *C. subcylindricus* females and juveniles are sealed into tree cavities and are provisioned by males primarily with ripe fruits. During the 4-month breeding season, numerous seeds are regurgitated or defecated below nests. Many of these seeds establish as seedlings, altering the seedling composition towards hornbill-dispersed species. All woody plants in front of 25 hornbill nests (i.e. sites receiving hornbill dispersed seeds) and behind nests (control sites) were identified. However, there was no difference in total species richness, diversity or density of hornbill-dispersed species in these two microhabitats for any size class larger than the seedling stage. Thus, directed seed dispersal to hornbill nests over a decade did not substantially alter sapling community composition at these sites.

In the Neotropics, both Julliot (1996) and Chapman (1989) have documented the important role that *Alouatta* spp. (howler monkeys) can play in the primary dispersal of seeds. Yet experiments conducted in Santa Rosa National Park, Costa Rica, indicate that within 70 days of dispersal, 97.9% of seeds were removed or destroyed by spiny pocket mice (*Liomys salvini*), peccaries (*Pecari tajacu*), agoutis (*Dasyprocta punctata*) and a diversity of dung beetles (Chapman, 1989). These studies illustrate that processes acting after dispersal can substantially alter dispersion patterns generated by primary dispersers (see also Herrera *et al.*, 1994; Jordano and Herrera, 1995; Schupp and Fuentes, 1995; Kollmann *et al.*, 1998; Rey and Alcantara, 2000; Lambert, 2002) and point to the importance of quantifying seed fate.

Our objective in this review is to consider two aspects of primate primary seed dispersal that have the potential to strongly influence the roles that primates play in influencing plant demography and forest structure.

1.   We consider patterns of seed deposition (i.e. scattered versus clumped; *sensu* Howe, 1989) generated by different primate species and evaluate the probable ecological consequences of these patterns.

2.   We evaluate the distance primates typically move seeds away from the parent tree and the density-dependent mortality associated with parent trees.

Because they allow researchers to link primary dispersal to patterns of seed fate, such data are important for understanding the role of primates in influencing spatial structure of forests, as well as the degree to which they may exert evolutionary selection pressures on plants and seed characteristics. However, given the extreme hunting pressure that primate populations are facing throughout Neo- and Palaeotropical forests, such understanding may be moot. As such, we conclude by contextualizing these data in light of the possible cascading consequences that the removal of primate seed dispersers by hunting will have on the future composition of tree communities.

Pattern of Deposition: Scatter versus Clump Dispersal and Seed Fate

Many primate species exhibit a variety of seed handling behaviours (Corlett and Lucas, 1990; Kaplin and Moermond, 1998; Lambert and Garber, 1998; Lambert, 1999, 2002). For example, seeds themselves can be an important food for primates and, as such, are not dispersed but destroyed. Seed predation occurs in species that consume seeds seasonally, when preferred fruits are not available (e.g. *Cebus* and *Cercopithecus* spp.: Peres, 1991; Gautier-Hion *et al.*, 1993; Kaplin and Moermond, 1998; Lambert and Garber, 1998; Lambert, 1999), as well as in taxa that prefer seeds as food and exhibit adaptations in dentition (e.g. Kinzey and Norconk, 1990; Pitheciines: Kinzey, 1992) and in the gastrointestinal tract (e.g. Colobines: Kay and Davies, 1994; Oates, 1994; Lambert, 1998) for seed predation. It should be noted, however, that even morphologically adapted seed predators such as Pitheciines can be seed dispersers (e.g. Norconk *et al.*, 1998).

Under circumstances when it is fruit pulp that is the sought-out food, there remains the challenge of what to do with the seeds themselves once pulp is removed (Lambert, 1999). Protective seed coats are typically highly refractory to digestion, and seeds can be rife with secondary metabolites

(Janzen, 1971; Fenner, 1992; Chivers, 1994; Waterman and Kool, 1994). Seeds can also account for more than half of the weight of fruits consumed by primates (van Roosmalen, 1984; Garber, 1986). Swallowed seeds may thus represent a significant cost to frugivores in that they not only increase an animal's body mass, but may also displace more readily processed, nutritious digesta from the gut (Corlett and Lucas, 1990; Levey and Grajal, 1991). The seeds of fruits are thereby an unwanted mass, and the adhesive pulp must be removed and seeds discarded in some way.

Despite these constraints, seed swallowing is by far the most common means of primate seed dispersal in the Neotropics (Muskin and Fischgrund, 1981; Estrada and Coates-Estrada, 1984; Garber, 1986; Chapman, 1989; Andresen, 1999; Dew, 2001). In the Palaeotropics many primates also swallow seeds (e.g. *Papio anubis*, Lieberman *et al.*, 1979; *Pan troglodytes*, Wrangham *et al.*, 1994; Lambert, 1999); however, seed spitting is common in African and Asian cheek-pouched monkeys (Cercopithecinae; Gautier-Hion, 1980; Rowell and Mitchell, 1991; Kaplin and Moermond, 1998; Lambert, 1999, 2000).

Spit or dropped seeds are likely to be deposited on the forest floor singly, as fruits are processed one by one. Swallowed seeds, however, can be deposited in either a high-density clump, or just a few seeds, depending on animal size and position in the canopy, defecation size, and the intensity of the feeding bout during which the seeds were swallowed (Kaplin and Moermond, 1998; Andresen, 1999; Lambert, 1999). In Cameroon, for example, Poulsen *et al.* (2001) have found that western lowland gorillas (*Gorilla gorilla*) and chimpanzees average 18 and 41 large (> 2 cm) seeds per faecal sample, respectively, while four frugivorous monkeys had large seeds/faeces averages ranging from 1.0 to 2.1 seeds.

Differences in seed handling strategies result in primary seed deposition patterns that vary depending on fruit species, season, and the primate species handling the seeds (Kaplin and Moermond, 1998; Lambert, 1999, 2002). Because some primates spit or defecate seeds in very low-density seed piles (Andresen, 1999) while other primate species defecate seeds into high-density seed clumps (Julliot, 1996), primates lend themselves to an evaluation of the scatter- versus clump-dispersal hypothesis suggested by Howe (1989) for all fruit-eating animals (Lambert, 2002). Howe (1989) proposed that many tree species are scatter-dispersed by small frugivores that regurgitate, spit or defecate seeds singly. These seeds recruit as isolated individuals and are unlikely to evolve resistance to herbivores, pathogens or other sources of density-dependent seed or seedling mortality. In contrast, other species are dispersed by large frugivores that deposit large numbers of seeds in a single location. Howe (1989) proposed that these seeds germinate in close proximity to one another and thus evolve chemical or morphological defences against seed predators, pathogens and herbivores that act in a density-dependent fashion. These processes are proposed to reflect the spatial distribution of adults, with scatter-dispersed species being widely dispersed and clump-dispersed species being highly aggregated.

One prediction stemming from Howe's (1989) hypothesis is that seeds are likely to be in clumps if frugivores weighing greater than 3 kg disperse them. However, Lambert (1997, 1999, 2004) has demonstrated that a number of variables not related to body size can influence patterns of fruit processing and seed handling in primates in general, and *Pan troglodytes* and *Cercopithecus ascanius* in particular. For example, *C. ascanius* (adult male = 4.2 kg) consistently (77% total fruit-eating events) scatter-disperses the seeds of fruit they consume, depositing them singly within 10 m of the parent tree. Although body size is important, a combination of dental and buccal anatomy with long digestive retention times facilitates this 'scatter' pattern of seed handling and dispersal.

Others argue similarly. In his work on avian seed dispersal, Levey (1986) has suggested that, while disperser body size can be influential, other factors – such as oral and digestive anatomy and the chemical and physical treatment of seeds in the gut – have

important implications for seed treatment and the patterns in which seeds will be dispersed. For example, Levey (1986, 1987; also see Levey and Grajal, 1991) has examined methods of seed processing by nine species of tropical frugivorous birds and found that tanagers and finches masticated fruits and dropped large seeds. Manakins, on the other hand, swallowed fruit intact and dropped no seeds, despite the fact that these birds weigh significantly less than 1 kg. Those seeds that were swallowed by the birds are more likely to be dispersed some distance away from the parent tree and deposited in higher density clumps.

Moreover, Howe's (1989) hypothesis rests on one critical assumption: that the initial seed deposition pattern must persist for density-dependent factors to play a role. Given that processes acting after dispersal have the potential to obscure or even cancel patterns generated by the dispersers (Herrera *et al.*, 1994; Jordano and Herrera, 1995; Schupp and Fuentes, 1995; Kollmann *et al.*, 1998; Rey and Alcantara, 2000; Balcomb and Chapman, 2003), this assumption needs to be substantiated.

As evidenced in chapters throughout this volume, post-dispersal seed predation and secondary dispersal can dramatically affect seed survival and, ultimately, seedling recruitment and the dynamics of plant demography (Feer and Forget, 2002). Numerous studies cite the importance of rodent seed predation on the fate of dispersed seeds and tree regeneration, and this has been known for some time. For example, De Steven and Putz (1984) have documented the influence of mammalian seed predation on the recruitment of a tropical canopy tree (*Dipteryx panamensis*) on Barro Colorado Island, Panama. They documented that predation on unprotected *D. panamensis* seeds and cotyledons exceeded 90%. The authors suggest that predation by the seed-eating mammals can be so extensive that even dispersed seeds have little chance of escape (De Steven and Putz, 1984). Likewise, Forget (1992) examined seed removal and seed fate in *Gustavia superba* on Barro Colorado Island and found that, on average, 86% of the seeds were removed within 28 days.

Estrada and Coates-Estrada (1991) documented that 59% of the seeds dispersed by howler monkeys (*Alouatta palliata*) at Los Tuxtlas, Mexico, disappeared and were presumably killed within 24 h. In Kibale National Park, Uganda, 73% of seeds placed experimentally on the forest floor had disappeared after 6 months (range = 10–100% among species; Chapman and Chapman, 1996).

Rodents demonstrate a preference for dense seed deposits. Price and Jenkins (1986) argue that, in general, absolute numbers and rates of seed removal usually increase with seed density. Other examples support this assertion. For example, Casper (1987) analysed post-dispersal seed predation of the semi-desert annual, *Crytantha falva*, and found that after 3 to 4 days there was a consistent tendency for more seed removal from high-density seed deposits. Kasenene (1980) also found a density effect in Kibale National Park, Uganda, such that a wide variety of rodents appeared to locate and consume larger seed piles than smaller ones. Willson and Whelan (1990) found that high-density seed deposits were more likely to be detected by rodent predators than those containing only a few seeds.

Initial seed deposition patterns by primates are also disrupted by the secondary seed dispersal performed by dung beetles. Andresen (1999) investigated the fate of monkey-dispersed seeds in Peru and found that 27 species of dung beetle visited the dung of *Ateles paniscus* and *Alouatta seniculus* and buried 41% of the seeds they encountered. In Brazil, Vulinec (2002) has argued that dung beetles may be particularly important to regeneration of disturbed forest where there is high primate biomass. In Uganda, Shepherd and Chapman (1998) documented that dung beetles buried 31% of the seeds placed in experimental primate dung.

These studies suggest that clump-dispersed seeds do not remain in a clumped distribution. Other studies provide experimental data in support of this observation. Forget and Milleron (1991) experimentally investigated the fate of dispersed *Virola nobilis* seeds on Barro Colorado Island,

Panama. This aril-bearing fruit is commonly consumed by *Ateles geoffroyi* (Howe, 1981, 1993). These monkeys routinely swallow the seeds whole and defecate them in clumps. Using thread-marked seeds, these authors observed that agoutis (*Dasyprocta punctata*) scatterhoarded *V. nobilis* seeds that they found both singly or in clumps. Seed removal and seed burial rates were strongly affected by features of forest habitats, such as *V. nobilis* tree richness and/or forest age, but not by seed dispersal treatment (scattered versus clumped). Predation (mostly post-dispersal) of unburied seeds by weevils was independent of habitat and dispersal treatment.

Similarly, in the Palaeotropics Lambert (1997, 2001, 2002) determined that most primate-dispersed seeds do not remain at the site of deposition, and that post-dispersal processes dilute the effect of the initial dispersal pattern. For three of seven species that Lambert (2002) studied in Kibale National Park, Uganda, seed density did not have an effect on seed fate. For example, *Aphania senegalensis* had virtually no germinating seeds or fungal attack, and seeds in all treatments (either single seeds, five seeds in clumps or 30 seeds in clumps) were equally vulnerable to predation. Likewise, density of seed deposition did not differentially influence the fate of *Cordia abyssinica* seeds; there was virtually no germination, and seeds were highly vulnerable to fungal attack and predation, independent of treatment. Similarly, *Pseudospondias microcarpa* was highly desirable to rodents, regardless of seed treatment. For another three of the seven species, seed treatment did affect seed fate, though the effects offer little support for Howe's (1989) model. *Monodora myristica* was particularly vulnerable to fungal attack in larger clumps of seeds, suggesting adaptations to scatter dispersal. However, single seeds of this species were also more likely to be killed by rodents than were clumps of seeds. Thus, pathogens and predation in this species offset any selective advantage of the number of seeds in a seed-dispersal event – with more pathogen attack in clumps of seeds and more predation on single seeds. There were also seed treatment effects on the fate of *Linociera johnsonii* and *Celtis durandii*, although, again, the effects did not clearly correspond to those predicted from the scatter- vs. clump-dispersal model. Both species had more seeds damaged when in higher densities. But they also had more seeds removed when in less dense seed stations. Overall, high predation swamped any effect of seed deposition patterns. Indeed, data indicate that rodent seed predation was so intense that seeds never remained in clumps long enough to produce saplings.

Hence, seeds deposited in clumps by primates rarely remain aggregated, suggesting that plant adaptations to particular modes of primary seed deposition are unlikely. None the less, the seed traits proposed by Howe (1989) to be associated with patterns of primary deposition (i.e. scattered or clumped dispersal) appear to exist, although, given the diluting effects of post-dispersal processes, these attributes are probably a result of other factors such as forest microhabitat specialization. For example, the features proposed to be adaptations to scatter dispersal are also traits that favourably dispose seeds to recruiting in forest gaps, while traits proposed to be adaptations to clump dispersal are attributes that facilitate germination and establishment in forest understorey. Species adapted to establishment in large clearings have small seeds to promote wide dispersal, and require high light levels and temperatures for germination and growth (Fetcher *et al.*, 1985; Denslow, 1987; Vázquez-Yanes and Orozco-Segovia, 1993). They also grow rapidly (Kitajima, 1994, 1996), have rapid leaf turnover and, as a result, invest little in secondary metabolites that are presumed to act as deterrents to herbivory (Coley *et al.*, 1985). Conversely, some species do well in forest understorey. These species tend to be larger seeded, can germinate in relatively low light conditions, have a slow rate of growth, invest heavily in seed defences against herbivory, and consequently survive well in the understorey (Coley *et al.*, 1985; Kitajima, 1996; Connell and Green, 2000; Rose and Poorter, 2002).

Seed Dispersal Distance

Additional factors to primary seed deposition, such as the probability of seed survival over time relative to the distance from the parent plant, have been argued to influence plant recruitment and forest structure. Janzen (1970) and Connell (1970) suggested that dispersal away from the parent enhances survival by removing offspring from mortality factors acting in a density- or distance-dependent fashion. The so-called Janzen–Connell model has resulted in speculation regarding the relative fitness advantage to parent trees by primates that spit seeds versus swallow seeds. Indeed, primate seed handling – in combination with patterns of day range and habitat use – directly influences the distance that seeds are removed from parent trees. For example, assuming that fruit pulp is removed soon after a fruit is removed from the tree, spit and dropped seeds are more likely to be deposited in close proximity to the parent tree compared to seeds that are carried in the gut for some time (Rowell and Mitchell, 1991; Chapman, 1995; Lambert, 1999; Kaplin and Lambert, 2002). Data from Uganda (Lambert, 1999) and Kenya (Rowell and Mitchell, 1991) on seed spitting by *Cercopithecus* spp. suggest that seeds are spat out an average distance of 10 m from the parent tree (range 1–100 m). This stands in contrast to Neotropical ceboids – seed swallowers – that defecate seeds much longer distances away from parent trees; average distances range between 151 and 390 m, with a range that – in the case of *Ateles paniscus* (Russo, 2003) – can be up to 1119 m away from parent trees (Estrada and Coates-Estrada, 1984; Garber, 1986; Zhang and Wang, 1995; Julliot, 1996; Stevenson, 2000; Dew, 2001).

Beyond seed handling, habitat use also strongly influences seed shadow and two seed-swallowing primate species in the same habitat will create dissimilar seed shadows as a result of their daily ranging and foraging. In general, primates that travel widely in a day will deposit seeds over a greater area than primates that more intensively exploit a smaller day range. For example, in Kibale National Park, Uganda, redtail monkeys (*C. ascanius*) travel an average of 1178 m and use a total range area of 49 ha very intensively (Lambert, 1999). Over a several-day period they often re-visit trees, and will also use multiple resources from the same tree (e.g. fruit, seeds, leaves, insects). Sympatric *P. troglodytes*, on the other hand, range widely in a day, often moving from one large-fruiting tree to another and feeding along the way on terrestrial herbaceous vegetation. Although determining total day range for chimpanzees is extremely difficult, they can move up to 4–5 km/day, and use a range with an area of up to 2220 ha (Chapman and Wrangham, 1993; Lambert, 1997). These data suggest that, while there is high fruit diet overlap between these two species, they create very different seed shadows as a consequence of overall foraging strategy and habitat use, in addition to differences in seed handling.

Sympatric howler monkeys (*Alouatta* spp.) and spider monkeys (*Ateles* spp.) offer another example. *Alouatta palliata* spends from 28.5 to 50% of its feeding time consuming fruit and is reported to disperse the seeds of several plant species (Estrada and Coates-Estrada, 1984; Chapman, 1988a,b, 1989). However, howlers often spend many hours in the same feeding tree and range from only 10 to 893 m in a day. Chapman (1988b) found that *A. palliata* can stay in close proximity to individual feeding trees for up to 14 days, leaving these trees only to make short excursions. *Ateles geoffroyi* is highly frugivorous (up to 77.7% of feeding time) and is known to disperse many seeds (Chapman, 1989). The ranging patterns of this species suggest that seeds in the dung are unlikely to be defecated below the parent tree and are instead moved throughout the forest in a widely dispersed pattern. Further support of these observations comes from Peru, where Andresen (1999) has found that *Ateles paniscus* spent 22% of its time travelling, while *Alouatta seniculus* spent 9% of its time travelling. *A. seniculus* spent significantly more time in each fruiting tree than the spider monkeys (125 versus 41 min) and as a consequence most defecations were large and contained large aggregations of

seeds over smaller areas than those seed dispersed by *A. paniscus*.

The predictions stemming from the early Janzen (1970) and Connell (1970) articles have been seminal in the history of seed dispersal research and have strongly influenced assessments of primate seed dispersers. For example, some authors have implied that having seeds dispersed by primates further distances away from the parent tree is better in terms of the plant's fitness (Stevenson, 2000; Dew, 2001). However, the degree to which it is necessary for seeds to escape distance- and density-dependent effects near the parent is unclear. A number of species-specific studies examining seedling survival under parent trees have found little or no recruitment under parent trees (Augspurger, 1983, 1984). Howe *et al.* (1985) found that 99.96% of *Virola surinamensis* (also known as *V. nobilis*) fruit that drop under the parent are killed within only 12 weeks. Similarly, Schupp (1988) documented 7% survival of *Faramea occidentalis* seeds under the crown in 30 weeks, in comparison to 24% survival 5 m away from the parent tree. However, other studies reveal relatively small differences in the probability of survival between seeds under parent trees and those dispersed away (De Steven and Putz, 1984; Condit *et al.*, 1992; Chapman and Chapman, 1995, 1996). For example, Chapman and Chapman (1996) investigated primate-dispersed trees in Kibale National Park, Uganda, and found that in *Mimusops bagshawei* seed survivorship increases when dispersed away from adult conspecifics. *Uvariopsis congensis*, on the other hand, experienced 56% more seed predation when dispersed away from parent trees versus seeds directly under a parent tree. Thus, as more demographic and ecological data are garnered on the fate of seeds and seedlings in tropical forests, it is evident that there exists large interspecific variation in the degree to which density-dependent factors influence seed fate.

The Janzen–Connell model has resulted in speculation regarding the relative fitness advantage to parent trees by primates that spit seeds versus swallow seeds and the generalization that seed-swallowing primates are better dispersers (Sussman, 1995; Stevenson, 2000; Dew, 2001). However, given the range in seed survivorship patterns that have been documented for tropical tree species, broad generalizations of 'good' versus 'bad' primate seed dispersal distances are not tenable. For example, many tree species have specific require- ments for germination and growth and tend only to be found in particular habitats (e.g. moist valley bottoms, hillsides) suggesting that long-distance dispersal would result in seeds being carried to unsuitable habitats. Lambert (2001) has found that having seeds spat out near parent trees can be highly favourable for tree species in Kibale National Park, Uganda. For example, redtail monkeys (Cercopithecus ascanius) in Kibale feed extensively on the ripe fruits of Strychnos mitis during some years and some seasons. The monkeys consumed fruits at a rate of approximately two per minute and almost always spit out intact seeds under the crown of the parent tree or within a few metres. S. mitis seeds clearly benefited from monkey processing. Eighty-three per cent of seeds spat out by redtails germinated, while only 12% of the unprocessed fruits survived to germination. Because pulp remained on the seeds, unprocessed fruits were also more likely to be attacked and damaged by seed predators and fungal pathogens. Thus, although seeds were dispersed under or near parent trees, a distance which might be deleterious for some tropical tree species (e.g. Caseraia corymbosa: Howe, 1977), for S. mitis, C. ascanius fruit processing appears to be beneficial – at least at the seed and seedling stage. This indicates the difficulty in generalizing the services of a disperser across tree species.

Conservation Implications

As we have demonstrated, seeds dis- persed by primates rarely remain in their primary deposition sites; primates are thus unlikely to exert selection for seed charac- teristics. We have also indicated that it is

problematic to generalize a primate species' effectiveness as a seed disperser according to whether they disperse seeds far from (via swallowing) or close to (via spitting) parent trees. None the less, primates move literally billions of seeds in forests around the tropics every day and consequently have the capacity to influence numerous post-dispersal processes and initiate innu- merable chains of ecological interactions. Yet, given the precipitous decline in primate populations around the world (Chapman and Peres, 2001), garnering more data to unravel these processes may soon be a luxury. Subsistence and commercial hunting can have profound impacts on forest animal populations, while leaving the physical structure of the original forest largely unaltered (Peres, 1990; Wilkie et al., 1992). Wildlife harvest can be a major source of food for many local communities around the tropics, and primates are often prime targets. For example, a market survey in two cities in Equatorial Guinea, West Africa with a combined human population size of 107,000 recorded 4222 primate carcasses for sale over 424 days (Fa et al., 1995). Peres (1990) documented that a single family of rubber tappers in a remote forest site of western Brazilian Amazonia killed more than 200 woolly monkeys (Lagothrix lagotricha), 100 spider monkeys (Ateles paniscus) and 80 howler monkeys (Alouatta seniculus) over a period of 18 months. Peres (2000) calculated the total game harvest in the Brazilian Amazon by multiplying these values by the size of the zero-income rural population in the entire region. Using the values presented for primates, we estimate that 3.8 million primates are consumed annually in the Brazilian Amazon (range in estimates: 2.2–5.4 million). As dramatic as these figures are, they probably underestimate actual hunting-induced mortality. Harvest estimates from market surveys do not include primates that are consumed in villages (Lahm, 1993). In the Democratic Republic of Congo, 57% of primates are eaten in the villages and do not make it to the market, and, in Liberia, primates were more valuable in rural than urban areas (Colell *et al.*, 1994).

The fact that primates are often both preferred targets in the bushmeat trade and important seed dispersers has led researchers to evaluate the cascading effects of their removal on future forest regeneration and structure (Howe, 1984; Pannell, 1989; Chapman and Chapman, 1995). Chapman and Chapman (1995), for example, estimated the potential loss in plant biodiversity that would result if all large-bodied seed dispersers such as cercopithecines and apes were removed from Kibale National Park, Uganda, resulting in fruit dropping below parent trees rather than being dispersed. On the basis of presence or absence of seedlings and saplings under adults, they concluded that 60% of the 25 tree species they studied would ultimately be lost if primates were removed. Similarly, Chapman and Onderdonk (1998) evaluated intact forest with complete primate communities and fragments around Kibale in which there were no primate seed dispersers. They found fewer seedlings, fewer species, and a higher percentage of small-seeded species in the forest fragments. Other studies corroborate these findings and demonstrate that changes in disperser abundances are associated with changes in seedling densities (Dirzo and Miranda, 1991; Pacheco and Simonetti, 1998; Wright *et al.*, 2000) as well as spatial patterns of seedling recruitment (Pacheco and Simonetti, 1998). However, as there are evident differences in responses to dispersal removal, we clearly have much to learn. For example, in Uganda and Bolivia, reduced numbers of large-bodied primates were correlated with lower seedling densities of large-seeded forest trees species (Chapman and Onderdonk, 1998; Pacheco and Simonetti, 1998) and higher seedling aggregations around parent trees (Pacheco and Simonetti, 1998). In contrast, in Mexico and Panama, seedling densities were higher in areas with depleted mammalian communities (Dirzo and Miranda, 1991; Wright *et al.*, 2000).

In summary, in this review we have documented that seeds are handled by primates in different ways (swallowed or spit), are deposited in different distributions (scattered or clumped), and are subsequently differentially affected by secondary seed dispersers and seed predators. In addition, different primate species disperse seeds in varying distances away from parent trees, and the susceptibility of species to density- or distance-dependent factors is highly variable. There is some evidence that increases in the abundance of some primate species may offset the population declines in other species caused by hunting (Peres and Dolman, 2000). However, if current populations of seed-dispersing primates are hunted to the point of regional decline, it is difficult to argue, a priori, whether other species can fulfil this role in an ecologically equivalent fashion, if at all. Several authors have argued that the role of primates as seed dispersers may be particularly important for large-seeded or hard-husked fruit species, which may be inaccessible to smaller, arboreal taxa (Andresen, 2000; Kaplin and Lambert, 2002). The conservation of primates is thus key to maintaining effective seed dispersal of some species (Andresen, 2000; Entwistle *et al.*, 2000). However, in most situations, variation in post-dispersal seed fate probably makes it impossible to make predictions concerning how a specific tree species will respond to the removal of its primate disperser.

Acknowledgements

Our projects in Kibale have been supported by a Makerere University Grant for Biological Research, Sigma Xi, the American Society of Primatologists, the University of Oregon, the Wildlife Conservation Society, the National Geographic Society, and National Science Foundation Grants (SBR-9617664, SBR-990899). We thank the Government of Uganda, the National Parks Service, and Makerere University for permission to work in Kibale National Park. We thank Lauren Chapman for providing helpful comments on this manuscript.

References

Andresen, E. (1999) Seed dispersal by monkeys and the fate of dispersed seeds in a Peruvian rain forest. *Biotropica* 31, 145–158.

Andresen, E. (2000) Ecological roles of mammals: the case of seed dispersal. In: Entwistle, A. and Dunstone, N. (eds) *Priorities for the Conservation of Mammalian Diversity*. Cambridge University Press, Cambridge, pp. 2–26.

Augspurger, C.K. (1983) Offspring recruitment around tropical trees: changes in cohort distance with time. *Oikos* 40, 189–196.

Augspurger, C.K. (1984) Seedling survival of tropical tree species: interactions of dispersal distance, light-gaps and pathogens. *Ecology* 65, 1705–1712.

Balcomb, S.R. (2001) Patterns of seed dispersal at a variety of scales in a tropical forest system: do post-dispersal processes disrupt patterns established by frugivores? PhD thesis, University of Florida, Gainesville, Florida.

Balcomb, S.R and Chapman, C.A. (2003) Bridging the gap: influence of seed deposition on seedling recruitment in primate–tree interactions. *Ecological Monographs* 73, 625–642.

Casper, B.B. (1987) Spatial patterns of seed dispersal and post dispersal seed predation for *Cryptantha flava* (Boraginaceae). *American Journal of Botany* 74, 1646–1655.

Chapman, C.A. (1988a) Patch use and patch depletion by the spider and howling monkeys of Santa Rosa National Park, Costa Rica. *Behaviour* 105, 99–116.

Chapman, C.A. (1988b) Patterns of foraging and range use by three species of Neotropical primates. *Primates* 29, 177–194.

Chapman, C.A. (1989) Primate seed dispersal: the fate of dispersed seeds. *Biotropica* 21, 148–154.

Chapman, C.A. (1995) Primate seed dispersal: coevolution and conservation implications. *Evolutionary Anthropology* 4, 74–82.

Chapman, C.A. and Chapman, L.J. (1995) Survival without dispersers: seedling recruitment under parents. *Conservation Biology* 9, 675–678.

Chapman, C.A. and Chapman, L.J. (1996) Frugivory and the fate of dispersed and non-dispersed seeds in six African tree species. *Journal of Tropical Ecology* 12, 491–504.

Chapman, C.A. and Onderdonk, D.A. (1998) Forests without primates: primate/plant codependency. *American Journal of Primatology* 45, 127–141.

Chapman, C.A. and Peres, C. (2001) Primate conservation in the new millennium: the role of scientists. *Evolutionary Anthropology* 10, 16–33.

Chapman, C.A. and Wrangham, R.W. (1993) Range use of the forest chimpanzees of Kibale: implications for the evolution of chimpanzee social organization. *American Journal of Primatology* 31, 263–273.

Chivers, D.J. (1994) Functional anatomy of the gastrointestinal tract. In: Davies, G. and Oates, J.F. (eds) *Colobine Monkeys: Their Ecology, Behaviour and Evolution*. Cambridge University Press, Cambridge, pp. 205–228.

Colell, M., Maté, C. and Fa, J.E. (1994) Hunting among Moka Bubis: dynamics of faunal exploitation at the village level. *Biodiversity and Conservation* 3, 939–950.

Coley, P.D., Bryant, J.P. and Chapin, F.S. (1985) Resource availability and plant antiherbivore defense. *Science* 230, 895–899.

Condit, R., Hubbell, S.P. and Foster, R.B. (1992) Recruitment near conspecific adults and the maintenance of tree and shrub diversity in a Neotropical forest. *American Naturalist* 140, 261–286.

Connell, J.H. (1970) On the role of natural enemies in preventing competitive exclusion in some marine animals and rain forest trees. In: den Boer, P.J. and Gradwell, G.R. (eds) *Dynamics of Populations*. Centre for Agricultural Publishing and Documentation, Wageningen, The Netherlands, pp. 298–312.

Connell, J.H. and Green, P.T. (2000) Seedling dynamics over thirty-two years in a tropical rain forest tree. *Ecology* 81, 568–584.

Corlett, R.T. and Lucas, P.W. (1990) Alternative seed-handling strategies in primates: seed spitting by long-tailed macaques (*Macaca fascicularis*). *Oecologica* 82, 166–171.

Denslow, J.S. (1987) Tropical rainforest gaps and tree species diversity. *Annual Review of Ecology and Systematics* 18, 431–451.

De Steven, D. and Putz, F.E. (1984) Impact of mammals on early recruitment of a tropical canopy tree, *Dipteryx panamensis*, in Panama. *Oikos* 43, 207–216.

Dew, J.L. (2001) Synecology and seed dispersal in woolly monkeys (*Lagothrix lagotricha poeppigii*) and spider monkeys (*Ateles belzebuth belzebuth*) in Parque Nacional Uasuni, Equador. PhD thesis, University of California, Davis, California.

Dirzo, R. and Miranda, A. (1991) Altered patterns of herbivory and diversity in the forest understory: a case study of the possible consequences of contemporary defaunation.

In: Price, P., Lewinsohn, T.M., Fernandes, G.W. and Benson, W.W. (eds) *Plant–Animal Interactions: Evolutionary Ecology in Tropical and Temperate Regions.* John Wiley & Sons, Inc., New York, pp. 273–287.

Eisenberg, J.F. and Thorington, R.W. (1973) A preliminary analysis of a neotropical mammal fauna. *Biotropica* 5, 150–161.

Entwistle, A.C., Mickleburgh, S. and Dunstone, N. (2000) Mammal conservation: current contexts and opportunities. In: Entwistle, A. and Dunstone, N. (eds) *Priorities for the Conservation of Mammalian Diversity.* Cambridge University Press, Cambridge, pp. 2–26.

Estrada, A. and Coates-Estrada, R. (1984) Fruit eating and seed dispersal by howling monkeys (*Alouatta palliata*) in the tropical rain forest of Los Tuxtlas, Mexico. *American Journal of Primatology* 6, 77–91.

Estrada, A. and Coates-Estrada, R. (1986). Frugivory in howling monkeys (*Alouatta palliata*) at Los Tuxtlas, Mexico: dispersal and fate of seeds. In: Estrada, A. and Fleming, T.H. (eds) *Frugivores and Seed Dispersers.* Dr W. Junk Publishers, Dordrecht, The Netherlands, pp. 94–104.

Estrada, A. and Coates-Estrada, R. (1991) Howler monkeys (*Alouatta palliata*), dung beetles (Scarabaeidae) and seed dispersal: ecological interactions in the tropical rain forest of Los Tuxtlas, Mexico. *Journal of Tropical Ecology* 7, 459–474.

Fa, J.E., Juste, J., del Val, J.P. and Castroviejo, J. (1995) Impact of market hunting on mammal species in Equatorial Guinea. *Conservation Biology* 9, 1107–1115.

Feer, F. and Forget, P.-M. (2002) Spatio-temporal variation in post-dispersal seed fate. *Biotropica* 34, 555–566.

Fenner, M. (1992) *Seeds: the Ecology of Regeneration in Plant Communities.* CAB International, Wallingford, UK, 373 pp.

Fetcher, N., Oberbauer, S.F. and Strain, B.R. (1985) Vegetation effects on microclimate in lowland tropical forest in Costa Rica. *International Journal of Biometeorology* 29, 145–155.

Forget, P.-M. (1992) Seed removal and seed fate in *Gustavia superba* (Lecythidaceae). *Biotropica* 24, 408–414.

Forget, P.-M. and Milleron, T. (1991) Evidence for secondary seed dispersal by rodents in Panama. *Oecologia* 87, 596–599.

Frankie, G.W., Baker, H.G. and Opler, P.A. (1974) Comparative phenological studies of trees in tropical wet and dry forests in the lowlands of Costa Rica. *Ecology* 62, 881–919.

Garber, P.A. (1986) The ecology of seed dispersal in two species of Callitrichid primates (*Saguinus mystax* and *Saguinus fuscicollis*). *American Journal of Primatology* 10, 155–170.

Gautier-Hion, A. (1980) Seasonal variation of diet related to species and sex in a community of *Cercopithecus* monkeys. *Journal of Animal Ecology* 49, 237–269.

Gautier-Hion, A. (1984) La dissemination des graines par les cercopithecides forestiers Africains. *Terre et Vie* 39, 159–165.

Gautier-Hion, A., Duplantier, J.-M., Quris, R., Feer, F., Sourd, C., Decoux, J.-P., Dubost, G., Emmons, L., Erard, C., Hecketsweiler, P., Moungazi, A., Roussilhon, C. and Thiollay, J.-M. (1985) Fruit characters as a basis of fruit choice and seed dispersal in a tropical forest vertebrate community. *Oecologia* 65, 324–337.

Gautier-Hion, A., Gautier, J.-P. and Maisels, F. (1993) Seed dispersal versus seed predation: an inter-site comparison of two related African monkeys. *Vegetatio* 107/108, 237–244.

Herrera, C.M., Jordano, P., Lopez-Soria, L. and Amat, J.A. (1994) Recruitment of a mast-fruiting, bird-dispersed tree: bridging frugivore activity and seedling establishment. *Ecological Monographs* 64, 315–344.

Howe, H.F. (1977) Bird activity and seed dispersal of a tropical wet forest tree. *Ecology* 58, 539–550.

Howe, H.F. (1981) Removal of wild nutmeg (*Virola surinamensis*) crops by birds. *Ecology* 62, 1093–1106.

Howe, H.F. (1984) Implications of seed dispersal by animals for tropical reserve management. *Biological Conservation* 30, 261–281.

Howe, H.F. (1989) Scatter- and clump-dispersal and seedling demography: hypothesis and implications. *Oecologia* 79, 417–426.

Howe, H.F. (1993) Aspects of variation in a neotropical seed dispersal system. *Vegetatio* 107/108, 149–162.

Howe, H.F. and Smallwood, J. (1982) Ecology of seed dispersal. *Annual Review of Ecology and Systematics* 13, 201–228.

Howe, H.F., Schupp, E.W. and Westley, L.C. (1985) Early consequences of seed dispersal for a neotropical tree (*Virola surinamensis*). *Ecology* 66, 781–791.

Janzen, D.H. (1970) Herbivores and the number of tree species in tropical forests. *American Naturalist* 104, 501–528.

Janzen, D.H. (1971) Seed predation by animals. *Annual Review of Ecology and Systematics* 2, 465–492.

Jordano, P. and Herrera, C.M. (1995) Shuffling the offspring: uncoupling and spatial discordance of multiple stages in vertebrate seed dispersal. *Écoscience* 2, 230–237.

Julliot, C. (1996) Seed dispersal by red howling monkeys (*Alouatta seniculus*) in the tropical rain forest of French Guiana. *International Journal of Primatology* 17, 239–258.

Julliot, C. (1997) Impact of seed dispersal by red howler monkeys *Alouatta seniculus* on the seedling population in the understory of tropical rain forest. *Journal of Ecology* 85, 431–440.

Kalina, J. (1988) Ecology and behavior of the black-and-white casqued hornbill (*Bycanistes subcylindricus subcylindricus*) in Kibale forest, Uganda. PhD thesis, Michigan State University, East Lansing, Michigan.

Kaplin, B.A. and Lambert, J.E. (2002) A review of seed dispersal effectiveness by *Cercopithecus* monkeys: implications for seed input into degraded areas. In: Levey, D.J., Silva, W.R. and Galetti, M. (eds) *Seed Dispersal and Frugivory: Ecology, Evolution and Conservation.* CAB International, Wallingford, UK, pp. 351–364.

Kaplin, B.A. and Moermond, T.C. (1998) Variation in seed handling by two species of forest monkeys in Rwanda. *American Journal of Primatology* 45, 83–101.

Kasenene, J.M. (1980) Plant regeneration and rodent populations in selectively felled and unfelled areas of the Kibale Forest, Uganda. MSc thesis, Makerere University, Kampala, Uganda.

Kay, R.N.B. and Davies, A.G. (1994) Digestive physiology. In: Davies, A.G. and Oates, J.F. (eds) *Colombine Monkeys: their Ecology, Behaviour and Evolution.* Cambridge University Press, Cambridge, p. 229–259.

Kinzey, W.G. (1992) Dietary and dental adaptations in the Pitheciineae. *American Journal of Physical Anthropology* 88, 499–514.

Kinzey, W.G. and Norconk, M.A. (1990) Hardness as a basis of fruit choice in two sympatric primates. *American Journal of Physical Anthropology* 81, 5–15.

Kitajima, K. (1994) Relative importance of photosynthetic traits and allocation patterns as correlates of seedling shade tolerance of 13 tropical trees. *Oecologia* 98, 419–428.

Kitajima, K. (1996) Ecophysiology of tropical tree seedlings. In: Mulkey, S.S., Chazdon, R.L. and Smith, A.P. (eds) *Tropical Forest Plant Ecophysiology.* Chapman & Hall, New York, pp. 559–596.

Kollmann, J., Coomes, D.A. and White, S.M. (1998) Consistencies in post-dispersal seed predation of temperate fleshy-fruited species among seasons, years and sites. *Functional Ecology* 12, 683–690.

Lahm, S.A. (1993) Utilization of forest resources and local variation of wildlife populations in Northeastern Gabon. In: Hladik, C.M., Hladik, A., Linarea, O.F., Pagezy, H., Semple, A. and Hadley, M. (eds) *Tropical Forest, People and Food.* Parthenon Publishing Group, Paris, pp. 213–226.

Lambert, J.E. (1997) Digestive strategies, fruit processing and seed dispersal in the chimpanzees (*Pan troglodytes*) and redtail monkeys (*Cercopithecus ascanius*) of Kibale National Park, Uganda. PhD thesis, University of Illinois, Urbana-Champaign, Illinois.

Lambert, J.E. (1998) Primate digestion: interactions among anatomy, physiology, and feeding ecology. *Evolution Anthropology* 7, 8–20.

Lambert, J.E. (1999) Seed handling in chimpanzees (*Pan troglodytes*) and redtail monkeys (*Cercopithecus ascanius*): implications for understanding hominoid and cercopithecine fruit-processing strategies and seed dispersal. *American Journal of Physical Anthropology* 109, 365–386.

Lambert, J.E. (2000) The fate of seeds dispersed by African apes and cercopithecines. *American Journal of Physical Anthropology* 111 (S30), 204.

Lambert, J.E. (2001) Red-tailed guenons (*Cercopithecus ascanius*) and *Strychnos mitis*: evidence for plant benefits beyond seed dispersal. *International Journal of Primatology* 22, 189–201.

Lambert, J.E. (2002) Exploring the link between animal frugivory and plant strategies: the case of primate fruit-processing and post-dispersal seed fate. In: Levey, D.J., Silva, W.R. and Galetti, M. (eds) *Seed Dispersal and Frugivory: Ecology, Evolution and Conservation.* CAB International, Wallingford, UK, pp. 365–379.

Lambert, J.E. (2004) Competition, predation and the evolution of the cercopithecine cheek pouch: the case of *Cercopithecus* and *Lophocebus. American Journal of Physical Anthropology* (in press).

Lambert, J.E. and Garber, P.A. (1998) Evolutionary and ecological implications of primate seed dispersal. *American Journal of Primatology* 45, 9–28.

Levey, D.J. (1986) Methods of seed processing by birds and seed deposition patterns. In: Estrada, A. and Fleming, T.H. (eds)

Frugivores and Seed Dispersal. Dr W. Junk Publishers, Dordrecht, The Netherlands, pp. 147–158.

Levey, D.J. (1987) Seed size and fruit-handling techniques of avian frugivores. *American Naturalist* 129, 471–485.

Levey, D.J. and Grajal, A. (1991) Evolutionary implication of fruit-processing limitations in cedar waxwings. *American Naturalist* 138, 171–189.

Lieberman, D., Hall, J.B., Swaine, M.D. and Lieberman, M. (1979) Seed dispersal by baboons in the Shai Hills, Ghana. *Ecology* 60, 65–75.

Muskin, A. and Fischgrund, A.J. (1981) Seed dispersal of *Stemmadenia* (Apocynaceae) and sexually dimorphic feeding strategies by *Ateles* in Tikal Guatemala. *Biotropica* 13, 78–80.

Norconk, M.A., Grafton, B.W. and Conklin-Brittain, N.-L. (1998) Seed dispersal by neotropical predators. *American Journal of Primatology* 45, 103–126.

Oates, J.F. (1994) The natural history of African Colombines. In: Davies, A.G. and Oates, J.F. (eds) *Colombine Monkeys: their Ecology, Behaviour and Evolution.* Cambridge University Press, Cambridge, pp. 75–128.

Pacheco, L.F. and Simonetti, J.A. (1998) Consecuencias demograficas para *Inga ingoides* (Mimusoideae) por la perdida de *Ateles paniscus* (Cebidae), uno de sus dispersores de semillas. *Ecologia en Bolivia* 31, 67–90.

Pannell, C.M. (1989) The role of animals in natural regeneration and the management of equatorial rain forests for conservation and timber production. *Commonwealth Forestry Review* 68, 309–313.

Paul, J.R. (2000) Patterns of seed dispersal by animals: influence on sapling composition in a tropical forest. MSc thesis, University of Florida, Gainesville, Florida.

Peres, C.A. (1990) Effects of hunting on western Amazonian primate communities. *Biological Conservation* 54, 47–59.

Peres, C.A. (1991) Seed predation of *Cariniana micrantha* (Lecythidaceae) by brown capuchin monkeys in central Amazonia. *Biotropica* 23, 262–270.

Peres, C.A. (2000) Effects of subsistence hunting on vertebrate community structure in Amazonian forests. *Conservation Biology* 14, 240–253.

Peres, C.A. and Dolman, P.M. (2000) Density compensation in neotropical primate communities: evidence from 56 hunted and nonhunted Amazonian forests of varying productivity. *Oecologia* 122, 175–189.

Poulsen, J.R., Clark, C.J. and Smith, T.B. (2001) Seed dispersal by a diurnal primate community in the Dja Reserve, Cameroon. *Journal of Tropical Ecology* 17, 787–808.

Price, M.V. and Jenkins, B. (1986) Rodents as seed consumers and dispersers. In: Murray, D.R. (ed.) *Seed Dispersal.* Academic Press, Sydney, pp. 191–235.

Rey, P.J. and Alcantara, J.M. (2000) Recruitment dynamics of a fleshy-fruited plant (*Olea europaea*): connecting patterns of seed dispersal to seedling establishment. *Journal of Ecology* 88, 622–633.

Rose, S. and Poorter, L. (2002) The importance of seed mass for early regeneration in tropical forest: a review. In: ter Steege, H. (ed.) *Long Term Changes in Tropical Tree Diversity: Studies from the Guiana Shield, Africa, Borneo and Melanesia*, Tropenbos Series No. 22. Tropenbos International, Wageningen, The Netherlands, pp. 19–35.

Rowell, T.E. and Mitchell, B.J. (1991) Comparison of seed dispersal of guenons in Kenya and capuchins in Panama. *Journal of Tropical Ecology* 7, 269–274.

Russo, S.E. (2003) Linking spatial patterns of seed dispersal and plant recruitment in a neotropical tree, *Virola calophylla* (Myristicaceae). PhD thesis, University of Illinois, Urbana-Champaign, Illinois.

Schupp, E.W. (1988) Seed and early seedling predation in the forest understory and in treefall gaps. *Oikos* 51, 71–78.

Schupp, E.W. and Fuentes, M. (1995) Spatial patterns of seed dispersal and the unification of plant-population ecology. *Écoscience* 2, 267–275.

Shepherd, V.E. and Chapman, C.A. (1998) Dung beetles as secondary seed dispersers: impact on seed predation and germination. *Journal of Tropical Ecology* 14, 199–215.

Stevenson, P.R. (2000) Seed dispersal by woolly monkeys (*Lagothrix lagothricha*) at Tinigua National Park, Colombia: dispersal distance, germination rates and dispersal quantity. *American Journal of Primatology* 50, 275–289.

Sussman, R.W. (1995) How primates invented the rainforest and vice versa. In: Alterman, L., Doyle, G.A. and Izard, M.K. (eds) *Creatures of the Dark: the Nocturnal Prosimians.* Plenum Press, New York, pp. 1–10.

Terborgh, J., Pitman, N., Silman, M., Schichter, H. and Nunez, P.V. (2002) Maintenance of tree diversity in tropical forests. In: Levey, D.J., Silva, W.R. and Galetti, M. (eds) *Seed Dispersal and Frugivory: Ecology, Evolution* and *Conservation.* CAB International, Wallingford, UK, pp. 351–364.

van Roosmalen, M.G.M. (1984) Subcategorizing foods in primates. In: Chivers, D.J., Wood, B.A. and Bilborough, A. (eds) *Food Acquisition and Processing in Primates.* Plenum Press, New York, pp. 167–176.

Vázquez-Yanes, C. and Orozco-Segovia, A. (1993) Patterns of seed longevity and germination in the tropical rainforest. *Annual Review of Ecology and Systematics* 24, 69–84.

Voysey, B.C., McDonald, K.E., Rogers, M.E., Tutin, C.E.G. and Parnell, R.J. (1999a) Gorillas and seed dispersal in the Lope Reserve, Gabon. I: Gorilla acquisition by trees. *Journal of Tropical Ecology* 15, 23–38.

Voysey, B.C., McDonald, K.E., Rogers, M.E., Tutin, C.E.G. and Parnell, R.J. (1999b) Gorillas and seed dispersal in the Lope Reserve, Gabon. II: Survival and growth of seedlings. *Journal of Tropical Ecology* 15, 39–60.

Vulinec, K. (2002) Dung beetle communities and seed dispersal in primary forest and disturbed land in Amazonia. *Biotropica* 34, 297–309.

Waterman, P.G. and Kool, K. (1994) Colobine food selection and plant chemistry. In: Davies, G. and Oates, J.F. (eds) *Colobine Monkeys: Their Ecology, Behaviour and Evolution.* Cambridge University Press, Cambridge, pp. 251–284.

Wilkie, D.S., Sidle, J.G. and Boundzanga, G.C. (1992) Mechanized logging, market hunting and a bank loan in Congo. *Conservation Biology* 6, 570–580.

Willson, M.F. and Whelan, C.J. (1990) Variation in post-dispersal survival of vertebrate-dispersed seeds: effects of density, habitat, location, season and species. *Oikos* 57, 191–198.

Wrangham, R.W., Chapman, C.A. and Chapman, L.J. (1994) Seed dispersal by forest chimpanzees in Uganda. *Journal of Tropical Ecology* 10, 355–368.

Wright, S.J., Zeballos, H., Dominguez, I., Gallardo, M.M., Moreno, M.C. and Ibanez, R. (2000) Poachers alter mammal abundance, seed dispersal and seed predation in a neotropical forest. *Conservation Biology* 14, 227–239.

Zhang, S.Y. and Wang, L.X. (1995) Fruit consumption and seed dispersal of *Ziziphus cinamomum* (Rhamnaceae) by two species of primates (*Cebus apella* and *Ateles paniscus*) in French Guiana. *Biotropica* 27, 397–401.

9 Fallen Fruits and Terrestrial Vertebrate Frugivores: a Case Study in a Lowland Tropical Rainforest in Peninsular Malaysia

Masatoshi Yasuda,[1] Shingo Miura,[1] Nobuo Ishii,[2] Toshinori Okuda[3] and Nor Azman Hussein[4]

[1]Forestry and Forest Products Research Institute, 1 Matsunosato, Tsukuba, Ibaraki 305-8687, Japan; [2]Japan Wildlife Research Center, Shitaya, Taito, Tokyo, Japan; [3]National Institute for Environmental Studies, Onogawa, Tsukuba, Ibaraki, Japan; [4]Forest Research Institute Malaysia, Kepong, Kuala Lumpur, Malaysia

Introduction

Mammals are a dominant component of the species assemblage of frugivores in Malaysian forests. Most Malaysian mammal species depend to varying degrees on wild fruits for food (Harrison, 1954, 1961; McClure, 1966; Lim, 1970; MacKinnon, 1978; Medway, 1978; Payne, 1980; Payne et al., 1985; Corlett, 1998; Kitamura et al., 2002). Although there is a great deal of literature on the food habits of Malaysian frugivorous mammals, only a few studies provide a list of their food items at the plant species level. Such studies allow us to discuss the species–species interactions between plants and animals and the guild structure in the frugivore community.

McClure (1966) studied the animal consumption of fruits from 32 canopy tree species. He found that arboreal mammals, namely giant squirrels (*Ratufa* spp.), gibbons (*Hylobates* spp.) and leaf monkeys (*Presbytis* spp.), commonly consume large fruits in the canopy, whereas various species of frugivorous birds consume small fruits. Payne (1980) compiled observation records

of frugivory in a Malaysian rainforest and presented a list of plants and their fruit consumers; this list included over 300 plant species from 53 families, three primate species, three arboreal squirrel species, and frugivorous bird species from six families. He suggested that primates compete with squirrels and birds for fruits as food.

There are several challenges in synthesizing patterns of plant–frugivore interactions in the tropics. One is that observations are typically of diurnal animals in the canopy, though many researchers mentioned that the consumption of fallen fruits at night can be extensive (McClure, 1966; Payne, 1980). Fruit consumption on the forest floor contributes to important ecological processes, including seed dispersal (Gautier-Hion et al., 1985; Forget, 1990, 1994; Forget et al., 1994). Therefore, 24-h observation of the consumption of fallen fruits is needed.

Another point is that most studies have historically not been quantitative and are often based on the presence or absence of fruit consumers. A quantitative data set and statistical analysis are necessary to reveal the intensity of interactions and the structure of the community. An exception is the work of Gautier-Hion *et al.* (1985), who applied a multivariate analysis to fruit–frugivore interactions in Central Africa. They studied 122 plant species and 39 species of both diurnal and nocturnal frugivorous vertebrates in Gabon, but they used presence/absence data of fruit consumers for each plant species. They suggested that fruit weight and colour are the principal factors in determining the fruit choice of consumers, and that there are two syndromes: bird–monkey and ruminant–rodent–elephant. The fruit characters of the bird–monkey syndrome are bright colour, succulent pulp or arillate seeds and no protective seed cover. Those of the ruminant–rodent–elephant syndrome are large size, dull colour, dry fibrous flesh and well-protected seeds. Kitamura *et al.* (2002) carried out a similar study on fruit–frugivore relationships in a tropical seasonal forest in Thailand, and found that seed size was an important character determining the consumption of a particular fruit by frugivores. Kitamura *et al.* (2002) concluded that the fruits in the Thai forest comprise three dispersal assemblages: birds, birds–mammals, and mammals, which were similar to previously reported 'bird-fruits' and 'mammal-fruits' from tropical forests in the Oriental region (Leighton and Leighton, 1983; Corlett, 1998).

In the present study, we used infrared-triggered remote cameras to carry out a 24-h observation of fruit consumption by vertebrates on the forest floor. Remote cameras have been used widely in wildlife studies (Wemmer *et al.*, 1996), and are ideal for identifying the species inhabiting a particular area, for monitoring relative and absolute abundance of species, and for studying activity patterns (Karanth, 1995; Karanth and Nichols, 1998; McCullough *et al.*, 2000; O'Brien *et al.*, 2003). The method has also been used to answer a variety of ecological and conservation-related questions, for example, frugivory and seed dispersal by vertebrates (Miura *et al.*, 1997; Yasuda *et al.*, 2000; Otani, 2001, 2002). Remote cameras can provide quantitative data on fruit consumption in the number of photos of visiting animals. This allows us to use better

statistical methods to describe the patterns of fruit–frugivore interactions in a forest. On the basis of the results of remote cameras and fruit characters, we attempt here to reveal the niche separation in food resources as it has been suggested to be an important mechanism for the coexistence of frugivores in some tropical forests (Fleming, 1979; Gautier-Hion *et al.*, 1985).

Methods

Study site

The study was carried out in the Pasoh Forest Reserve (2°59′N, 102°19′E, 75–150 m above sea level) in the state of Negeri Sembilan, Peninsular Malaysia. The details of the forest and some ecological studies carried out there are described in Okuda *et al.* (2003). The reserve is an inland forest with a broad expanse of flat land and gently rolling ridges that adjoin the westward side of the Main Range of the Malay Peninsula. The Pasoh forest constitutes about 2500 ha of lowland mixed dipterocarp forests, including areas of primary and regenerating forests. The regenerating forest was selectively logged in the 1950s and has been regenerating naturally since then.

A 50-ha permanent plot is set in the heart of the primary forest area. All the trees in the plot with a diameter at breast height of 1 cm or more were tagged, identified and mapped (Manokaran *et al.*, 1992), and the survival and growth of individual trees are surveyed every half-decade. Published papers on the 50-ha plot cover a wide range of ecological parameters, including distribution patterns of tree species in the plot. Such information is important for seed dispersal studies, since it can provide insight into the dispersal potential of seeds and the recruitment limitation of each plant species.

The vertebrate frugivore fauna in the Pasoh forest has been degraded by human activities, including habitat destruction of the surrounding area, isolation of the forest, and poaching. Sixty-four (41.8%) out of 110 mammal species that were historically potential residents of the forest were now extremely rare or absent (Numata *et al.*, 2005), which include medium- to large-sized frugivores, e.g. the Asian elephant (*Elephas maximus*), Malay tapir (*Tapirus indicus*), sambar deer (*Cervus unicolor*) and barking deer (*Muntiacus muntjak*). The abundance of civets (Viverridae) in the Pasoh forest seems to be lower than in other Malaysian forests with an intact mammalian fauna (M. Yasuda, personal observation).

Plant phenology

Flowering and fruiting phenology were censused monthly from May 1992 to January 1997. One hundred litter traps, each with an open area of 0.5 m², were set in a grid design in a 2-ha study plot in primary forest out of the 50-ha plot. Litter in the traps was collected once a month. The number of litter traps that contained flowers or fruits was recorded.

Frugivory experiments using remote cameras

The remote cameras used for frugivory experiments are described in detail by Miura *et al.* (1997), and consist of a single-lens reflex camera (XR-10M with 35–70 mm zoom lens, Ricoh Co., Ltd, Minato-ku, Tokyo, Japan), an infrared motion sensor (DELCATEC PS-15B, DX Antenna Co., Ltd, Kobe, Hyogo, Japan), a flash (Sunpak Auto25SR, Tocad Energy Co., Ltd, Ohta-ku, Tokyo, Japan) and a battery. The cameras were encased in a plastic box to prevent them from getting wet, with 200 g of silica gel for absorbing moisture. No photographic delay interval was applied to the experiments.

Each experiment was carried out under the parent tree of the fruits tested. Fruit size, weight, colour, shape and ripeness were recorded. The number of fruits used per tree varied from 20 to 100, usually 20 for large-sized fruits (> 100 g), 40–60 for medium-sized fruits (10–100 g) and 50–100 for small-sized fruits (< 10 g). The fruits collected from the ground were placed on the forest floor beneath the parent tree within a circle of 1 m in diameter, and an infrared motion sensor was installed at 1.2 m high above the fruits by hanging on a rope between trees. A camera unit on a tripod aimed at the fruits at a distance of *c.* 2 m and connected to a sensor unit with an electric cable. Cameras and fruits were inspected every day. Each experiment was continued for at least 1 week. When the fruits disappeared within a week, the experiment was continued further during the fruit fall, and was terminated either after two to six rolls of exposed film were obtained or the fruit fall ended.

This study was carried out from July 1993 to May 1995 in a period of normal fruiting years. The study area took in primary and regenerating forests covering 100 ha. In total, 108 plants from 70 species of tree, shrub, liana, epiphyte and rattan in 27 families were studied. The average fresh weight of fruits studied ranged from 0.06 to 464 g. The plant species studied accounted for 8.6% of 814 known species in the 50-ha plot. The families Euphorbiaceae, Myristicaceae, Moraceae, and Fagaceae were studied intensively, because their members bore fruits more frequently during the study period.

Ecological indices and statistical analyses

Of the 108 individuals of 70 plant species studied, 71 individuals of 49 species had ten or more photos of visitors; the data from these individuals were statistically analysed. The number of photos was summed for all replications of the same species for creation of a fruit–frugivore interaction matrix at the species level. No data treatment was applied to series of photos taken within a short period of time since they were probably triggered by the same animal(s); they were simply counted as numbers of photos.

Both cotyledon eaters and pulp eaters were recognized for two plant species. The plants were *Canarium littorale* (Burseraceae) and *Terminalia citrina* (Combretaceae), fruits of which have a hard kernel enveloped in green leathery pulp. The cotyledon eaters were rodents (*Leopoldamys sabanus* and *Lariscus insignis*, Rodentia), and the pulp eaters included *Macaca nemestrina* (Primates) and *Tragulus javanicus* (Artiodactyla). Therefore, the cotyledons and pulp were analysed separately; hereafter we call them by their scientific name plus either cotyledon or pulp. Rattans (*Daemonorops* spp., Arecaceae) were difficult to identify to species, so they were considered a single taxon.

We employed two statistical methods to analyse the fruit–frugivore interaction matrix. The first is a similarity index of species composition, Morisita's similarity index C'_λ (Morisita, 1959, 1971). This can be thought of as a ratio of probabilities: the probability that two individuals chosen from different samples will be of the same species relative to the probability that two individuals chosen from the same sample as each other will be alike. Morisita's similarity index and its modified versions behave appropriately, as its values are nearly independent of sample size (Grassel and Smith, 1976; Wolda, 1981; Magurran, 1988; Krebs, 1998). This index has been used in studies of dietary overlap in frugivores (Gorchov *et al.*, 1995) and insect diversity (Wolda, 1983; Watt *et al.*, 2003).

C'_λ between two communities j and k is derived as

$$C'_\lambda = \frac{1}{\bar{\lambda}} \frac{\sum_i n_{ij} n_{ik}}{N_j N_k}$$

where $\bar{\lambda}$ is given by the equation

$$\bar{\lambda} = \frac{\sum_{i=1} n_{ij}(n_{ij}-1) + \sum_i n_{ik}(n_{ik}-1)}{N_j(N_j-1) + N_k(N_k-1)}$$

where N_j and N_k are the total abundance of the jth and the kth communities, respectively, and n_{ij} and n_{ik} are the abundance of the ith species in the jth and the kth communities, respectively. The generalization for $L \geq 2$ communities is as follows:

$$\bar{\lambda} = \frac{\sum_{i=1}^{L} \sum_i n_{ij}(n_{ij}-1)}{\sum_{j=1}^{L} \sum_i N_j(N_j-1)}$$

and

$$C'_\lambda = \frac{1}{\bar{\lambda}} \frac{\sum_{i=1}^{L-1} \sum_{k>j}^{L} \sum_i n_{ij} n_{ik}}{\sum_{j=1}^{L-1} \sum_{k>j}^{L} \sum_i N_j N_k}$$

respectively. The values of Morisita's similarity index and a dendrogram, which is an application of the above generalization for three or more communities, were computed by the computer program MULVAC'95 (Kobayashi, 1995).

The second statistical method is the two-way indicator species analysis (TWINSPAN), proposed by Hill (1979). This is a classification method that uses a divisive (top-down) strategy, whereas most classification methods, including Morisita's similarity index, use an agglomerative (bottom-up) strategy. TWINSPAN is relatively robust to random errors of samples and leads to very interpretable solutions (Gauch, 1982; Jongman *et al.*, 1995). To make a dichotomy, TWINSPAN uses a set of indicator species that are derived from the partitioning of the first axis of the correspondence analysis. In this study, the cut-off levels were set as 0–5%, 5–10%, 10–20%, 20–50% and more than 50% relative abundance of animal species per fruit species in the photos. TWINSPAN was executed by the TWINSPAN program (Hill, 1979) in Cornell Ecology Programs.

Fruit colour

Fruit colour at time of consumption was divided into two categories, green and coloured (yellow, orange, brown, red, purple, black or white). For example, *Lithocarpus* spp. and *Castanopsis megacarpa* (Fagaceae) were categorized as green, since they are eaten by monkeys and rodents in the tree and on the ground before they turn brown. The colour of dehiscent fruits was the colour of the inner part of the fruits displayed for animals. For example, *Xylopia malayana* (Annonaceae), which is green outside and red inside, was categorized as coloured. Among the 49 fruit species studied, green fruits were most dominant

(18 spp., 36.7%) followed by orange-red fruits (14 spp., 28.6%) and yellow fruits (9 spp., 20.0%). Fruits in brown (3 spp.), black (2 spp.), purple (2 spp.) and white (1) were less common. The fruits categorized as green included: *Alangium ebenaceum*, *Canarium littorale*, *Castanopsis megacarpa*, *Diospyros cauliflora*, *Elaeocarpus stipularis*, *Endospermum malaccense*, *Eugenia* sp., *Lithocarpus* spp., *Parkia speciosa*, *Pithecellobium bubalium*, *Porterandia anisophylla*, *Sapium baccatum*, *Shorea maxima*, *Styrax benzoin* and *Terminalia citrina*. Fisher's exact test was performed to examine whether a given animal species had a preference for green or coloured fruits.

Nutritional analysis of fruits

We analysed the nutrient content of 19 edible parts of 17 selected species belonging to 13 families. Most of these species showed a high rate of fruit disappearance or shed their fruits at high density levels on the forest floor. The nutrients analysed were moisture (W), crude protein (CP), crude fat (EE, after ether extract), crude fibre (CF) and ash (Ash). Nitrogen-free extract (NFE), which represents carbohydrates (starch and sugar), was calculated as the residual fraction of the contents. Gross energy (GE, kJ/g dry matter) was calculated as:

$$GE = (5.61 \times CP + 9.66 \times EE + 4.38 \times NFE + 5.06 \times CF) \times 0.04186$$

where 0.04186 is a constant to transform the unit from calories to joules. The value on a dry matter basis was transformed into that on a fresh matter basis by using the moisture content.

Results

Plant reproductive phenology

The plant reproductive phenology monitored by litter traps is shown in Fig. 9.1. There were two peaks of flowering, in February–April and October. The former peak was dominant, whereas the latter was relatively weak. Flowering condition was normal in 1992–1995 and high in 1996, when a mass flowering occurred. Fruiting showed less seasonality than flowering and fluctuated from year to year. Normal fruiting years continued for the 4 years from 1992 to 1995, followed by a mast fruiting year in 1996.[1] The percentage of traps with fruits was within the range of 2% to 24% during the period of non-mast-fruiting years. The percentage in 1996 increased from May, when immature fruits started to fall, and reached a maximum (76%) in September, when a large number of mature fruits were shed in synchrony.

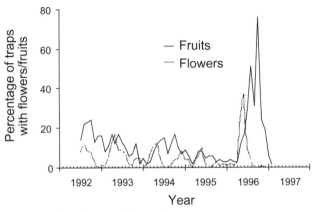

Fig. 9.1. Reproductive phenology of trees in the Pasoh Forest Reserve.

Vertebrate frugivores imaged with remote cameras

Nearly 4000 photos of animals were obtained, and 34 animal species were identified. They belonged to various taxa, including the classes Mammalia, Aves and Reptilia. The most dominant class was Mammalia, including the orders Pholidota (1 species), Insectivora (1), Scandentia (1), Chiroptera (≥ 1), Primates (2), Carnivora (5), Artiodactyla (2) and Rodentia (11). The species list is described in Miura *et al.* (1997).

The 16 most dominant frugivorous species were analysed: 13 species of mammal, two species of bird and one species of reptile (Table 9.1). The pig-tailed macaque (*Macaca nemestrina*) was the most common visitor, using 44 out of the 49 plant species (89.8%, Table 9.1). The second most common animals were small rodents: *Leopoldamys sabanus*, *Maxomys* spp. (Muridae) and

Lariscus insignis (Sciuridae) used 26 species (53.1%). *Maxomys* spp. represents medium-sized rats, either *M. rajah* or *M. surifer*, which were indistinguishable in the photos. Other frequently photographed mammals were the common treeshrew (*Tupaia glis* (Scandentia), 19 spp.), two species of porcupine (*Trichys fasciculata* and *Hystrix brachyura* (Rodentia), 19 and 16 spp., respectively), wild boar (Sus *scrofa* (Artiodactyla), 15 spp.) and lesser mouse-deer (*Tragulus javanicus* (Artiodactyla), 14 spp.). The dusky leafmonkey (*Presbytis femoralis*), a canopy-living primate, sometimes came down to the ground to feed on fallen fruits of four species. Although several species of bird were recorded in this study, only two were relatively common: the crestless fireback (*Lophura erythrophthalma* (Phasianidae), six species) and the emerald pigeon (*Chalcophaps indica* (Columbidae), two species). Monitor lizards, probably including two species of *Varanus* spp.

Table 9.1. Vertebrate frugivores recorded with camera traps baited with fallen fruits on the forest floor. Forty-nine plant species were analysed. Species are sorted in the order of the number of plant species used.

Common name	Species name	Family	Order	Body mass (g)	n	(%) Plant used
Mammalia						
Pig-tailed macaque	*Macaca nemestrina*	Cercopithecidae	Primates	12,000	44	(89.8)
Long-tailed giant rat	*Leopoldamys sabanus*	Muridae	Rodentia	330	26	(53.1)
Three-striped ground squirrel	*Lariscus insignis*	Sciuridae	Rodentia	210	26	(53.1)
Brown/red spiny rats	*Maxomys* spp.	Muridae	Rodentia	150	26	(53.1)
Common treeshrew	*Tupaia glis*	Tupaiidae	Scandentia	150	19	(38.8)
Long-tailed porcupine	*Trichys fasciculata*	Hystricidae	Rodentia	2,000	19	(38.8)
Common porcupine	*Hystrix brachyura*	Hystricidae	Rodentia	8,000	16	(32.7)
Wild boar	*Sus scrofa*	Suidae	Artiodactyla	60,000	15	(30.6)
Lesser mouse-deer	*Tragulus javanicus*	Tragulidae	Artiodactyla	2,000	14	(28.6)
Shrew-faced ground squirrel	*Rhinosciurus laticaudatus*	Sciuridae	Rodentia	240	11	(22.4)
Whitehead's rat	*Maxomys whiteheadi*	Muridae	Rodentia	45	7	(14.3)
Dusky leafmonkey	*Presbytis femoralis*	Cercopithecidae	Primates	7,000	4	(8.2)
Malaysian field rat	*Rattus tiomanicus*	Muridae	Rodentia	80	2	(4.1)
Aves						
Crestless fireback	*Lophura erythrophthalma*	Phasianidae	Galliformes	800	6	(12.2)
Emerald dove	*Chalcophaps indica*	Columbidae	Columbiformes	150	2	(4.1)
Reptilia						
Monitor lizard	*Varanus* spp.	Varinidae	Squamata	5,000	15	(30.6)

(Varanidae), were recorded consuming 15 fruit species.

The number of frugivorous species photographed varied among the 49 plant species, from 1 to 12 per plant species (Fig. 9.2; mean 4.7 ± SE 0.32, median 4, mode 4). The most favoured species was *Sarcotheca monophylla* (*Oxalidaceae*), which bears small red juicy fruit; the fruit was consumed by 12 of the 16 most dominant animal species. *Pithecellobium bubalium* (Mimosaceae) and rattans (Arecaceae) were the second most favoured species and were used by nine animal species. At the opposite extreme were, for example, *Irvingia malayana* (Irvingiaceae), *Neobalanocarpus heimii* (Dipterocarpaceae) and *Quercus argentata* (Fagaceae), whose fruits were only seldomly consumed by vertebrate frugivores on the forest floor, in spite of the high abundance of fruits and the long observation period.

Classification of plant species using Morisita's similarity index

Four clusters were clearly recognized in the classification using Morisita's similarity index (Fig. 9.3 and Table 9.2), though there were small inconsistencies in the dendrogram. The first to the third most dominant species of each cluster on average were *Macaca–Lariscus–Tupaia*, *Leopoldamys–Lariscus–Maxomys* spp., *Presbytis–Hystrix–Maxomys* spp., and *Trichys–Maxomys*

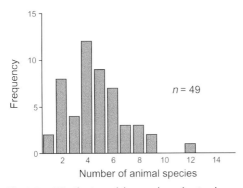

Fig. 9.2. Distribution of the number of animal species per plant species as imaged by remote camera.

spp.–*Macaca*, respectively. Clusters were named after the most dominant animal species in the clusters: namely *Macaca–Lariscus* (consisting of 32 species of plants), *Leopoldamys* (5), *Presbytis–Hystrix* (3) and *Trichys* (11), respectively. Note that the plant species classified into the same cluster do not always share the most dominant species of animal, because the overall similarity of animal occurrence among fruits is considered in Morisita's similarity index.

As shown in Fig. 9.3, the cluster *Leopoldamys* connected with the cluster *Presbytis–Hystrix* ($C'_\lambda = 0.122$), then with the largest cluster *Macaca–Lariscus* at the next step ($C'_\lambda = 0.103$). The cluster *Trichys* was most dissimilar among the four clusters and was connected with the others at the last step ($C'_\lambda = 0.039$). The cotyledons of *Canarium littorale* and *Terminalia citrina* were classified into the cluster *Leopoldamys*, whereas their pulps fell into the cluster *Macaca–Lariscus*.

Classification of plant and animal species using TWINSPAN

Plant species were divided into four groups by TWINSPAN. Group A + B and group C + D were divided at the first step of the dichotomy by five indicator species: *Lariscus*, *Leopoldamys* and *Tupaia* for group A + B, and *Trichys* and *Hystrix* for group C + D. At the next step, the principal indicator species between groups A and B were *Lariscus* (abundance level ≥ 2) and *Leopoldamys* (≥ 4), and between groups C and D was *Presbytis* (≥ 1). Furthermore, group C was divided into two subgroups, C1 and C2, according to the presence of *Macaca* (≥ 4) for C1 and *Trichys* (≥ 4) for C2, respectively. Hereafter, we refer to the groups *Lariscus*, *Leopoldamys* and *Presbytis*, and the subgroups *Macaca* and *Trichys*, according to the principal indicator species.

The plant species classified into the groups and subgroups are shown in Table 9.3. These principal indicator species were

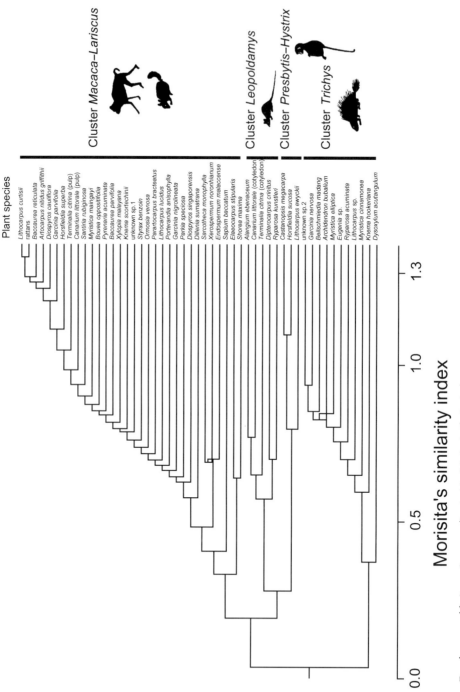

Fig. 9.3. Dendrogram of fruit species according to Morisita's similarity index.

Table 9.2. Clusters of plant species recognized by Morisita's similarity index. Numbers represent the percentage of photos showing the animal species. Species are sorted in the order of similarity in the cluster (see Fig. 9.3).

Code	Plant species	Family	Mammals													Birds		Reptiles
			Tupaia	Presbytis	Macaca	Lariscus	Rhinosciurus	Rattus	Maxomys spp.	M. whiteheadi	Leopoldamys	Trichys	Hystrix	Sus	Tragulus	Chalcophaps	Lophura	Varanus
Cluster Macaca–Lariscus																		
27	Lithocarpus curtisii	Fagaceae			100.0													
39	rattans	Arecaceae			52.7	4.5	0.9		3.6		20.0	1.8		2.7	10.9			2.7
5	Baccaurea reticulata	Euphorbiaceae			96.0	4.0												
3	Artocarpus nitidus griffithii	Moraceae			100.0													
13	Diospyros cauliflora	Ebenaceae			91.3							8.7						
22	Garcinia parvifolia	Clusiaceae	14.3		71.4	7.1					7.1							
24	Horsfieldia superba	Myristicaceae			82.2	2.2			6.7		4.4			6.5	8.7			4.4
48	Terminalia citrina (pulp)	Combretaceae			71.7										19.6			13.0
9	Canarium littorale (pulp)	Burseraceae	1.8		75.0												1.8	1.8
43	Santiria rubiginosa	Burseraceae	20.0		40.0				40.0									
33	Myristica maingayi	Myristicaceae	13.8		66.2				1.5		1.5	16.9			1.5		10.6	
8	Bouea oppositifolia	Anacardiaceae			72.7				1.7			13.6	3.3					3.3
38	Pyrenaria acuminata	Theaceae			85.0		1.7						1.5	5.0				
4	Baccaurea parvifolia	Euphorbiaceae			71.6	14.9	2.7				8.1			2.7				
51	Xylopia malayana	Annonaceae			70.0				27.5					2.5				
26	Knema scortechinii	Myristicaceae	12.5		45.8		4.2						29.2		8.3			
40	unknown sp.1	Myristicaceae			46.4	28.6					14.3			10.7				
47	Styrax benzoin	Styracaceae	3.2	0.6	45.2	19.4								3.2	16.1			9.7
34	Ormosia venosa	Fabaceae		0.6	63.7		3.2		0.8		33.8	0.6			0.6			
35	Parartocarpus bracteatus	Moraceae			83.9						4.0		4.8	1.6	3.2			1.6
29	Lithocarpus lucidus	Fagaceae			47.2								52.8					
37	Porterandia anisophylla	Rubiaceae	53.3		40.0	3.3									3.3			
21	Garcinia nigrolineata	Clusiaceae			48.4	41.9			1.6	6.5		1.6						

continued

Table 9.2. *Continued.*

Code	Plant species	Family	Mammals													Birds		Reptiles
			Tupaia	*Presbytis*	*Macaca*	*Lariscus*	*Rhinosciurus*	*Rattus*	*Maxomys* spp.	*M. whiteheadi*	*Leopoldamys*	*Trichys*	*Hystrix*	*Sus*	*Tragulus*	*Chalcophaps*	*Lophura*	*Varanus*
36	*Parkia speciosa*	Mimosaceae			23.8	9.5		20.2	4.8					61.9				
14	*Diospyros singaporensis*	Ebenaceae			34.5	1.2	1.2	7.1			32.1	1.2	3.6		6.0			
12	*Dillenia sumatrana*	Dilleniaceae			1.8					39.3	44.6			7.1				
45	*Sarcotheca monophylla*	Oxalidaceae	18.9		6.6	31.1	0.8		3.3	2.5	9.0	0.8	4.1			2.5	17.2	3.3
50	*Xerospermum noronhianum*	Sapindaceae	44.1		2.9	20.6			26.5									
18	*Endospermum malaccense*	Euphorbiaceae	51.3			12.8			2.6		5.9							
44	*Sapium baccatum*	Euphorbiaceae	4.2		7.3	11.5	1.0		1.0		33.3					75.0		
17	*Elaeocarpus stipularis*	Elaeocarpaceae				90.9						6.1			3.0			
46	*Shorea maxima*	Dipterocarpaceae	6.8		2.7	51.4	1.4		2.7		5.4				28.4			1.4
Mean			7.6	0.0	51.4	11.1	0.5	0.9	3.9	1.5	7.0	1.6	3.1	3.3	3.4	2.4	0.9	1.3
Cluster *Leopoldamys*																		
1	*Alangium ebenaceum*	Alangiaceae	4.9		1.6	23.0			8.2		60.7							
10	*Canarium littorale* (cotyledon)	Burseraceae				37.5			8.7	1.0	52.9							1.6
49	*Terminalia citrina* (cotyledon)	Combretaceae				3.6					96.4							
15	*Dipterocarpus crinitus*	Dipterocarpaceae	16.2		13.1				24.6		49.2	11.5						
42	*Ryparosa kunstleri*	Flacourtiaceae		2.7	10.8				8.1		51.4		5.4	5.4			1.6	
Mean			4.2	0.5	5.1	12.8	0.0	0.0	9.9	0.2	62.1	2.3	1.1	1.1	0.0	0.0	0.3	0.3

Cluster *Presbytis–Hystrix*

#	Species	Family	1	2	3	4	5	6	7	8	9	10	11	12	13	14	15	16
11	Castanopsis megacarpa	Fagaceae		59.1	1.5	2.9			8.8	9.1	16.1		11.7		0.0	0.0	0.0	0.0
23	Horsfieldia sucosa	Myristicaceae		95.0									5.0					
28	Lithocarpus ewyckii	Fagaceae			18.2	9.1			18.2				45.5		0.0	0.0	0.0	0.0
	Mean		0.0	51.4	6.5	4.0	0.0	0.0	9.0	3.0	5.4	0.0	20.7	0.0	0.0	0.0	0.0	0.0

Cluster *Trichys*

#	Species	Family	1	2	3	4	5	6	7	8	9	10	11	12	13	14	15	16
7	unknown sp. 2											95.7						4.3
20	Garcinia nervosa	Clusiaceae			19.0	2.4			3.6			76.2			2.4			
6	Beilschmiedia madang	Lauraceae			1.8		0.5		8.1	1.8		90.9		1.8			17.3	
2	Pithecellobium bubalium	Fabaceae			3.0	0.5			1.4		2.7	31.0	34.5					1.5
32	Myristica elliptica	Myristicaceae			4.1	1.4			1.4		2.7	85.1	1.4					2.7
19	Eugenia sp.	Myrtaceae			10.0	5.0						35.0	15.0	25.0				
41	Ryparosa acuminata	Flacourtiaceae			20.0								20.0				60.0	
30	Lithocarpus sp.	Fagaceae	26.3			41.7			58.3								60.0	
31	Myristica cinnamonea	Myristicaceae			23.7	2.6			2.6			34.2	5.3		1.5			5.3
25	Knema hookeriana	Myristicaceae			23.1	1.5			36.9			33.8			2.0			1.0
16	Dysoxylum acutangulum	Meliaceae			5.6					0.2		91.2						
	Mean		2.4	0.0	10.0	4.2	0.6	0.0	10.1	0.4	1.9	52.1	6.9	0.6	2.4	0.0	7.0	1.3

Table 9.3. Groups of plant species recognized by TWINSPAN. Numbers represent the percentage of photos showing the animal species. Species are sorted in the order of the classification of TWINSPAN.

Code	Plant species	Family	Mammals													Birds		Reptiles
			Tupaia	Presbytis	Macaca	Lariscus	Rhinosciurus	Rattus	Maxomys spp.	Maxomys w.	Leopoldamys	Trichys	Hystrix	Sus	Tragulus	Chalcophaps	Lophura	Varanus
Group Lariscus																		
44	Sapium baccatum	Euphorbiaceae	4.2		7.3	11.5	1.0		1.0									
4	Baccaurea parvifolia	Euphorbiaceae			71.6	14.9	2.7				8.1			2.7		75.0		
47	Styrax benzoin	Styracaceae	3.2		45.2	19.4	3.2							3.2	16.1			9.7
21	Garcinia nigrolineata	Clusiaceae			48.4	41.9			1.6	6.5		1.6						
36	Parkia speciosa	Mimosaceae			23.8	9.5			4.8					61.9				
40	unknown sp.1				46.4	28.6					14.3			10.7				
22	Garcinia parvifolia	Clusiaceae	14.3		71.4	7.1					7.1							
37	Porterandia anisophylla	Rubiaceae	53.3		40.0	3.3									3.3			
43	Santiria rubiginosa	Burseraceae	20.0		40.0				40.0									
18	Endospermum malaccense	Euphorbiaceae	51.3			12.8			2.6		33.3							
50	Xerospermum noronhianum	Sapindaceae	44.1		2.9	20.6			26.5		5.9							
17	Elaeocarpus stipularis	Elaeocarpaceae				90.9						6.1			3.0			
45	Sarcotheca monophylla	Oxalidaceae	18.9		6.6	31.1	0.8		3.3	2.5	9.0	0.8	4.1			2.5	17.2	3.3
46	Shorea maxima	Dipterocarpaceae	6.8		2.7	51.4	1.4		2.7		5.4							1.4
1	Alangium ebenaceum	Alangiaceae	4.9		1.6	23.0			8.2		60.7				28.4			1.6
10	Canarium littorale (cotyledon)	Burseraceae				37.5			8.7	1.0	52.9							
30	Lithocarpus sp.	Fagaceae				41.7			58.3									
	Mean		13.0	0.0	24.0	26.2	0.5	0.0	9.3	0.6	11.6	0.5	0.2	4.6	3.0	4.6	1.0	0.9

Group Leopoldamys

Code	Species	Family	1	2	3	4	5	6	7	8	9	10	11	12	13	14	15	16
14	*Diospyros singaporensis*	Ebenaceae	0.6		34.5	1.2	1.2		20.2		32.1		1.2	3.6	6.0		0.0	
34	*Ormosia venosa*	Fabaceae	0.6	0.6	63.7						33.8		0.6		0.6			
39	rattans	Areacaceae			52.7	4.5	0.9	3.6			20.0		1.8	2.7	10.9			2.7
49	*Terminalia citrina* (cotyledon)	Combretaceae				3.6												
42	*Ryparosa kunstleri*	Flacourtiaceae	16.2	2.7	10.8				8.1	96.4	51.4		5.4	5.4				
12	*Dillenia sumatrana*	Dilleniaceae			1.8		7.1			39.3	44.6			7.1				
	Mean		2.8	0.6	27.3	1.6	0.3	4.6	2.0	6.5	46.4	0.6	1.5	2.5	2.9		0.0	0.5

Group Macaca

Subgroup Macaca

Code	Species	Family	1	2	3	4	5	6	7	8	9	10	11	12	13	14	15	16
26	*Knema scortechinii*	Myristicaceae	12.5		45.8	4.2							29.2		8.3			
28	*Lithocarpus ewyckii*	Fagaceae			18.2	9.1			18.2	9.1			45.5					
29	*Lithocarpus lucidus*	Fagaceae			47.2								52.8					
41	*Ryparosa acuminata*	Flacourtiaceae			20.0								20.0				60.0	
35	*Parartocarpus bracteatus*	Moraceae			83.9				0.8		4.0		4.8	1.6	3.2			1.6
38	*Pyrenaria acuminata*	Theaceae			85.0		1.7		1.7				3.3	5.0				3.3
51	*Xylopia malayana*	Annonaceae			70.0				27.5					2.5				
9	*Canarium littorale* (pulp)	Burseraceae	1.8		75.0										19.6			1.8
24	*Horsfieldia superba*	Myristicaceae			82.2	2.2			6.7									4.4
48	*Terminalia citrina* (pulp)	Combretaceae			71.7									6.5	8.7			13.0
3	*Artocarpus nitidus*	Moraceae			100.0													
5	*Baccaurea reticulata*	Euphorbiaceae			96.0	4.0												
13	*Diospyros cauliflora*	Ebenaceae			91.3						8.7		1.5					
27	*Lithocarpus curtisii*	Fagaceae			100.0						13.6							
8	*Bouea opositifolia*	Anacardiaceae			72.7				1.5								10.6	
	Mean		1.0	0.0	70.6	1.0	0.4	0.0	3.8	0.6	0.6	1.5	10.5	1.0	2.7	0.0	4.8	1.6

continued

Table 9.3. *Continued.*

Code	Plant species	Family	Mammals													Birds		Reptiles
			Tupaia	Presbytis	Macaca	Lariscus	Rhinosciurus	Rattus	Maxomys spp.	Maxomys w.	Leopoldamys	Trichys	Hystrix	Sus	Tragulus	Chalcophaps	Lophura	Varanus
Subgroup *Trichys*																		
31	*Myristica cinnamonea*	Myristicaceae	26.3		23.7	2.6			2.6			34.2	5.3					5.3
33	*Myristica maingayi*	Myristicaceae	13.8		66.2							16.9						
2	*Pithecellobium bubalium*	Mimosaceae			3.0	0.5					1.5	31.0	34.5	0.5	1.5		17.3	1.5
15	*Dipterocarpus crinitus*	Dipterocarpaceae			13.1				8.1		3.6	11.5					1.6	
19	*Eugenia* sp.	Myrtaceae			10.0		5.0		24.6		49.2	35.0	15.0		25.0			
25	*Knema hookeriana*	Myristicaceae			23.1	1.5			36.9		10.0	33.8		1.5				
6	*Beilschmiedia madang*	Lauraceae			1.8				3.6	1.8	3.1	90.9			1.8			
32	*Myristica elliptica*	Myristicaceae			4.1		1.4		1.4	2.7	1.4	85.1	1.4					2.7
7	unknown sp. 2	Leguminosae										95.7						4.3
16	*Dysoxylum acutangulum*	Meliaceae			5.6						0.2	91.2		2.0				1.0
20	*Garcinia nervosa*	Clusiaceae			19.0						2.4	76.2		2.4				
	Mean		3.7	0.0	15.4	0.4	0.6	0.0	7.0	0.4	6.5	54.7	5.1	0.6	2.6	0.0	1.7	1.3
Group *Presbytis*																		
11	*Castanopsis megacarpa*	Fagaceae		59.1	1.5	2.9			8.8		16.1		11.7					
23	*Horsfieldia sucosa*	Myristicaceae		95.0	0.7	1.5			4.4		8.0		5.0					
	Mean		0.0	77.1	0.7	1.5	0.0	0.0	4.4	0.0	8.0	0.0	8.3	0.0	0.0	0.0	0.0	0.0

the most dominant species of each group or subgroup in photos on average. *Macaca nemestrina* was the second most dominant species of the groups *Lariscus* and *Leopoldamys* and of the subgroup *Trichys*. The cotyledons of *Canarium littorale* and *Terminalia citrina* were classified into the groups *Lariscus* and *Leopoldamys*, respectively, while both pulps fell into the same subgroup, *Macaca*.

Comparison between the two classification methods

Figure 9.4 summarizes the results of the classifications using Morisita's index and TWINSPAN. In general, there was good agreement on classification. Four categories and five groups or subgroups were recognized under the name of the same

representative species: *Macaca, Lariscus, Leopoldamys, Presbytis* and *Trichys*, although the number of plant species belonging to each category or group differed. Forty (78.4% of 49 species plus two edible parts) were exactly categorized into corresponding categories or groups by the different methods. Plant species in the cluster *Macaca–Lariscus* of Morisita's similarity index (32 species) were divided into the group *Lariscus* (14), subgroup *Macaca* (13), group *Leopoldamys* (4) and subgroup *Trichys* (1) of TWINSPAN. The former two were considered to be the corresponding categories or groups.

The long-tailed porcupine (*Trichys fasciculata*) showed a high preference for some fruits. *Knema hookeriana, Myristica cinnamomea* and *M. elliptica* (Myristicaceae), *Pithecellobium bubalium* and an unknown species of legume, *Eugenia* sp. (Myrtaceae), *Garcinia nervosa* (Clusiaceae),

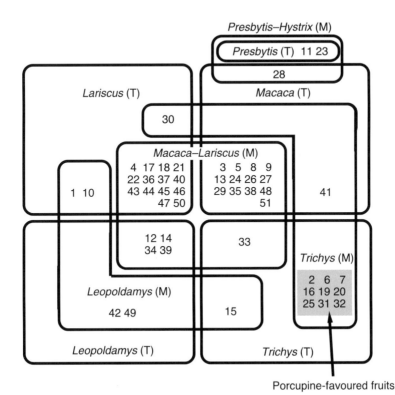

Fig. 9.4. Comparison of classifications by Morisita's similarity index (M) and TWINSPAN (T). Numbers represent the species codes in Table 9.3.

Dysoxylum acutangulum (Meliaceae) and *Beilschmiedia madang* (Lauraceae) were categorized into corresponding categories of the cluster *Trichys* by Morisita's similarity index and the subgroup *Trichys* by TWINSPAN (Fig. 9.4). We call these porcupine-favoured fruits. The porcupine was the first or the second most dominant consumer of these fruits, accounting for 26.2 to 95.7% of the photos.

Nutrient contents of fruits

Six edible parts from five fruit species of the porcupine-favoured fruits and other fruits consumed by other animals were analysed for nutrient content. Nutrient content and gross energy of a total of 19 edible parts from 17 fruits are shown on a fresh matter basis (Table 9.4). Nutrient contents varied widely among fruits: moisture 8.8–90.5%, protein 1.6–13.6%, lipid 0.03–55.0% and NFE 0.8–52.0% on a fresh matter basis. The gross energy was estimated in the range of 1.8–28.7 kJ/g.

The porcupine-favoured fruits were rich in nutrients and energy (Table 9.4). *Pithecellobium bubalium* contained 7.0% protein on a fresh matter basis, which is the second highest value among the fruits analysed. The other five edible parts of the four porcupine-favoured fruits were rich in lipid (6.7–55.0%). The lipid content of the porcupine-favoured fruits was significantly higher than that of the other fruits (Mann–Whitney's U-test, two-tailed, $P = 0.011$, $n = 19$). There was no significant difference in the other nutrients or moisture between the porcupine-favoured fruits and the other fruits (Mann–Whitney's U-test, two-tailed, $P > 0.1$, $n = 19$). The mean gross energy content of the porcupine-favoured fruits was 13.8 kJ/g on a fresh matter basis, while that of the other fruits was 7.8 kJ /g. The porcupine-favoured fruits contained significantly higher energy per unit weight than the others (Mann–Whitney's U-test, two-tailed, $P = 0.043$, $n = 19$).

Fruit colour and animal preferences

No frugivore showed a significant preference for either green or coloured fruits ($P > 0.05$ for all animal species).

Discussion

Fruit characters and fruit preference of frugivores

Fruit colour

Although fruit colour has been considered to be an important factor in determining frugivore food preference and in organizing frugivore communities (Leighton and Leighton, 1983; Gautier-Hion *et al.*, 1985; Lambert and Marshall, 1991), our results suggest that fruit colour is not important in the fruit–frugivore relationship on the forest floor in a Malaysian lowland rainforest. In this study, no vertebrate frugivore showed a significant preference for either green or coloured fruits. For example, the pig-tailed macaque (*Macaca nemestrina*), the most dominant frugivore on the forest floor in the Pasoh Forest Reserve, consumed both green and coloured fruits, and used 89.8% of surveyed fruits in the forest (Table 9.1). The macaque accounted for 20% or more of the photos taken among some green fruits, for example, *Canarium littorale*, *Diospyros cauliflora*, *Lithocarpus* spp., *Parkia speciosa*, *Porterandia anisophylla*, *Styrax benzoin* and *Terminalia citrina* (Table 9.2). The macaque voraciously ate acorns of *Lithocarpus* and *Quercus* (Fagaceae) before the fruit colour turned from green to brown.

Many authors have argued that frugivores in tropical forests, especially primates, prefer coloured fruits. Gautier-Hion *et al.* (1985) suggested that the fruits of the bird–monkey syndrome in an African tropical forest were brightly coloured. Terborgh (1983) found that the only common feature among the fruits consumed by primates in a Peruvian forest was a yellow to orange colour. Julliot (1994) revealed that yellow, orange or red fruits accounted for 64.7% of

Table 9.4. Nutrient contents of porcupine-favoured fruits and some other fruits.

Plant species	Family	Edible part	Moisture	Protein	Lipid	NFE[a]	Fibre	Ash	Energy (kJ/g)
Porcupine-favoured fruits									
Pithecellobium bubalium	Fabaceae	Seed	56.4	7	0.2	34.8	1.1	0.6	8.3
Garcinia nervosa	Clusiaceae	Seed	55.5	1.8	6.7	28.7	2.4	4.9	8.9
Knema hookeriana	Myristicaceae	Seed	37.2	3.8	28.4	24.4	5.9	0.4	18.1
Myristica cinnamomea	Myristicaceae	Seed	46	1.4	8.5	37.2	6.2	0.7	11.9
M. elliptica	Myristicaceae	Seed	29.8	5	55	0.8	8.8	0.6	25.4
M. elliptica	Myristicaceae	Aril	64.1	2.5	15.8	12.7	4.3	0.7	10.2
Mean			48.2	3.6	19.1	23.1	4.8	1.3	13.8
SE			5.3	0.9	8.2	5.7	1.1	0.7	2.7
Fruits consumed by other animals									
Bouea oppositifolia	Anacardiaceae	Pulp	90.1	0.2	0	6.2	3.2	0.3	1.9
Canarium littorale	Burseraceae	Pulp	77.7	0.7	1.4	14.7	4.4	1.1	4.4
C. littorale	Burseraceae	Cotyledon	8.8	13.6	52.3	6.8	14.8	3.8	28.7
Castanopsis megacarpa	Fagaceae	Seed	52.5	2.6	0.1	43.2	1	0.6	8.8
Diospyros singaporensis	Ebenaceae	Seed	49.4	5.3	0.1	23.5	21.3	0.5	10.1
Ormosia venosa	Fabaceae	Seed	33.8	2.8	1	51.8	9.6	1	12.6
Parkia speciosa	Mimosaceae	Seed	80.7	6.5	1.5	8.1	2.2	1	4.1
Pyrenaria acuminata	Theaceae	Pulp	86.4	0.5	0.1	11.4	1.1	0.4	2.5
Sarcotheca monophylla	Oxalidaceae	Pulp	90.5	0.6	0	5.2	3.2	0.4	1.8
Shorea maxima	Dipterocarpaceae	Seed	44	1.8	0.2	52	1.5	0.6	10.3
Styrax benzoin	Styracaceae	Pulp	70.7	1.4	0	19.6	7.7	0.6	5.6
Terminalia citrina	Combretaceae	Pulp	70.9	2.8	0.2	24.2	0.6	1.4	5.3
Xerospermum noronhianum	Sapindaceae	Seed coat	73.4	1.1	0.1	20.4	4.6	0.4	5
Mean			63.8	3.1	4.4	22.1	5.8	0.9	7.8
SE			6.8	1	4	4.7	1.7	0.3	2.9

[a]NFE, nitrogen-free extract (represents carbohydrates).

the fruit eaten by red howler monkeys in French Guiana. Sourd and Gautier-Hion (1986) found that the forest *Cercopithecus* spp. preferred brightly coloured fruits and avoided dull-coloured fruits in a Gabonese forest. In a Bornean forest, Leighton and Leighton (1983) found that primates frequently consumed brightly coloured fruits. It has been suggested that trichromatic vision in primates evolved as an adaptation to frugivory, making it easier to detect fruits against a background of leaves (Osario and Vorobyev, 1996). These primates studied are canopy-dwellers, while *Macaca nemestrina* is more terrestrial. *Macaca nemestrina* usually moves in groups on the forest floor to search for foods, including fallen fruits and animal materials (Bernstein, 1967; Medway, 1978). The foraging behaviour of this species may be the reason why *Macaca nemestrina* does not show a preference for coloured fruits over green fruits. Against the generally brown background for fallen fruits on the ground of shadowy forest, the importance of the contrast between fruits and the background appears to be different for ground foraging primates to search for foods even though they have human-like trichromatic vision.

Summarizing the available knowledge of frugivory in the Oriental region, Corlett (1998) concluded that the fruits eaten largely by mammals are typically duller than fruits taken by birds, which may correspond to the ruminant–rodent–elephant syndrome in an African tropical forest proposed by Gautier-Hion *et al.* (1985). Except for *Macaca nemestrina*, all terrestrial mammals in the present study were likely to be dichromats or, if nocturnal, colour-blind. Therefore, fruit colour appears to be less important for the terrestrial frugivores in determining fruit choice.

From a Thai seasonal tropical forest, Kitamura *et al.* (2002) reported that, with the exception of elephants, all observed vertebrate frugivores including birds, primates, civets, bears and deer, tended to consume black or red fruits, though the latter three mammalian groups are nocturnal. This implies that there are other unidentified essential factors, which may indirectly associate with fruit colour, for mammalian frugivores in determining fruit choice.

Fruit chemistry

Nutrients in edible parts appear to be an important factor for the long-tailed porcupine (*Trichys fasciculata*) in determining fruit choice. The results suggest that *T. fasciculata* prefers fruits that are rich in lipid or protein and therefore contain more energy per unit weight than others (Table 9.4). The porcupine-favoured fruits belong to a variety of plant families, although three of the nine species belong to the family Myristicaceae. We have no information to explain why only the porcupine shows such a tendency, because nutrient content is likely to be important for the other frugivores as well. It is noteworthy to point out that the common treeshrew (*Tupaia glis*) often associated with porcupines (*Trichys* and *Hystrix*) for some dehiscent fruits belonging to the genera *Myristica* and *Knema* (Myristicaceae). The treeshrew preferred to feed on the red oily aril of the fruits on the ground.

Fruit morphology and oral processing mechanism of animals

Patterns of fruit–frugivore relationships may strongly depend on the oral processing mechanism of the animals. Rodents, including rats, squirrels and porcupines in the present study, have well-developed incisors as an adaptation to using morphologically protected fruits (Corbet and Hill, 1992). Fruits in the family Fagaceae, which bear thick-walled acorns, or fruits with a spiny cupule were chiefly consumed by monkeys and rodents (Table 9.2), but never by the common treeshrew (*Tupaia glis*), which has undeveloped teeth (Emmons, 1991). Although *T. glis* is considered to be insectivorous, fruits comprise a considerable proportion of its food resources (Kawamichi and Kawamichi, 1979; Langham, 1982, 1983). In the present study, *T. glis* consumed 38.8% of surveyed fruits in the forest (Table 9.1), and showed a preference for small, soft, juicy fruits (Table 9.2), including the red juicy fruit of *Sarcotheca monophylla* (Oxalidaceae), which is rich in moisture and carbohydrates (Table 9.4), and *Santiria rubiginosa* (Burseraceae), which has a seed (1 cm diameter) that is covered with thin, edible, purple mesocarp. These observations are coincident with the finding of Emmons (1991) that *T. glis* concentrates on small, soft, bird-dispersed fruits.

In general, the relative size of fruit/seed to the animal's mouth and the softness of the edible fractions of the fruit are important factors in fruit choice, particularly for small vertebrate frugivores without specialized teeth (e.g. treeshrews). However, there were some exceptions to morphological limitation in frugivory. *Tupaia glis* accounted for a high proportion of photos in *Porterandia anisophylla* (Rubiaceae) and *Xerospermum noronhianum* (Sapindaceae) (Table 9.2). Soft edible parts of these fruits were covered with a protective tissue and not accessible to the treeshrew when the fruits are intact. *Tupaia glis* consumed the edible parts from the fruit fractions that had been attacked and discarded by other species, such as monkeys and rodents. The significance of this kind of frugivory, as a consequence of indirect interactions between frugivores, on seed fate is unknown.

Influence of frugivores on plants

As discussed above, the fruit nutrient content and morphology are important factors in determining frugivore food preferences. Here we discuss whether the digestive system and foraging behaviour of frugivores define the consequences of the fruit–frugivore interactions.

Macaques

The pig-tailed macaque, *Macaca nemestrina*, is one of the most polyphagous fruit consumers in Malaysian forests. Fruits comprise about three-quarters of its diet (Caldecott, 1986; Feeroz *et al.*, 1994). Owing to its high population density and wide range of distribution in Malaysia (Medway, 1978), *M. nemestrina* could

have a significant role in seed dispersal in Malaysian forests.

We have few data on seed dispersal by *M. nemestrina*, because of the difficulty of following them in forest. Considering results from other macaque species, *Macaca nemestrina* is likely to disperse seeds by two different ways: defecation and spitting. Detailed studies of the long-tailed macaque (*Macaca fascicularis*) in Singapore showed that the fate of seeds in ripe, fleshy fruits was determined by seed size; only the smallest seeds (< 3 mm diameter) were regularly swallowed, and larger seeds were spat out or dropped (Corlett and Lucas, 1990; Lucas and Corlett, 1998). A similar seed-size-dependent range of seed fates has been observed in other macaque species (Corlett, 1998). A variety of seeds was found in the faeces of *M. nemestrina* in the Pasoh forest (M. Yasuda, personal observation). Next, seed spitting will only lead to effective seed dispersal if the seeds have been carried away from the parent plant. The likelihood of this happening in macaques is increased by the presence of cheek pouches (Corlett, 1998).

In this study, *M. nemestrina* consumed the edible part of virtually all surveyed fruit species (Table 9.1) and thus was a dominant component of the species assemblage of terrestrial frugivores in the Pasoh forest (Fig. 9.4). Considering their relatively large home ranges (*c.* 3 km^2; Bennett, 1991), *M. nemestrina* is considered to be a significant long-range seed dispersal agent in the faunally depleted forest of the Pasoh Forest Reserve.

Rodentia

In general, rats, squirrels and porcupines are considered to be seed predators. Our frugivory experiments suggest that those rodents have a different preference for fruits to each other (Fig. 9.4). Although rats and squirrels consumed a variety of fruits, as more than half the fruit species surveyed were eaten by them (Table 9.1), frugivory overlap between *Leopoldamys sabanus* and *Lariscus insignis* was only 44.4% (16 out of 36 edible parts, Table 9.2). *Leopoldamys sabanus* preferred seeds as well as other edible parts, while *Lariscus insignis* often associated with *Macaca nemestrina* and showed a preference for some soft, fleshy fruits (Fig. 9.4 and Table 9.3). Porcupines concentrated on some particular fruits as described previously.

Animals that hoard food are effective dispersers if they hoard fruits near the soil surface (Vander Wall, 1990; several chapters in this volume). Recently, food-hoarding behaviour was observed in Malaysian forest rodents, *Leopoldamys sabanus* and *Lariscus insignis* (Yasuda *et al.*, 2000) – a behaviour that had previously been believed absent in the forests of the region (Forget and Vander Wall, 2001). Caches made by the rodents were usually on the ground under fallen leaves or in soil about 1 cm in depth. Since both species are common in the forest (Yasuda *et al.*, 2003), they might also act as post-dispersal seed dispersers.

On the other hand, thus far no food-hoarding behaviour has been reported for porcupines in South-east Asia. Field observations revealed that porcupines voraciously ate seeds as well as other edible parts, such as arils. Considering their high dominance of the species among frugivores on the porcupine-favoured fruits (Table 9.2), porcupines could act as destructive seed predators.

Ungulates

Mouse-deer are highly frugivorous, small ungulates that live in Asian and African tropical forests. They are considered to be the most primitive ruminants, and have a forestomach enlarged into a rumen as a chamber for cellulose fermentation (Janis, 1984). The lesser mouse-deer (*Tragulus javanicus*) fed on 28.6% of surveyed fruits in the Pasoh forest (Table 9.1), and showed a high preference for *Canarium littorale* (Burseraceae) and *Terminalia citrina* (Combretaceae) (Table 9.2). These plants belong to different taxa, but both fruits are similar in morphology: a hard oblong kernel covered with leathery, fibrous, thick pulp, which *T. javanicus* used as food. Hard and larger seeds of such fleshy fruits eaten by *T. javanicus* are likely to be regurgitated from the mouth in rumination and dispersed at a

distance from the mother tree, as African forest ruminants are considered to be important spit-type dispersers of fleshy fruits, as well as being endozoochoric (through-the-gut) dispersers (Feer, 1995). Harrison (1961) found large seeds in all three stomachs of *T. javanicus* that he examined, showing that seeds can avoid destruction in the mouth and may, therefore, be dispersed intact. Small intact seeds were observed in faeces of mouse-deer in Bornean forests (H. Matsubayashi, 2004, personal communication). Seed dispersal by mouse-deer may be restricted within a home range of individuals, of which size is 3.5–6.2 ha (Matsubayashi *et al.*, 2003). In addition, mouse-deer act as seed predators as well, for instance, they crunched on acorn-like fruits of *Shorea maxima* (Dipterocarpaceae) intensively (Table 9.2).

Scandentia

Fruit is an important part of the natural diet of treeshrews, but their importance in seed dispersal is unclear. Treeshrews can deal with only soft fruits, as discussed previously. They seem ill-adapted for frugivory, as they have undeveloped teeth, a small, simple stomach, a long, narrow small intestine and a narrow, smooth large intestine (Emmons, 1991). Fruits are processed in the mouth in a similar way to fruit bats. Only the juice and soft pulp are swallowed, while the skin, fibres and seeds are spat out in wads. Only tiny seeds could pass through the gut. Langham (1982) and Emmons (1991) suggested that treeshrews are likely to act as seed dispersers of plants bearing small juicy fruits with tiny seeds, e.g. *Ficus* spp., while Corlett (1998) mentioned that treeshrews are unlikely to be significant dispersers for these bird-fruits in habitats with an intact vertebrate fauna.

Notes for further frugivory studies using remote cameras

As shown in this chapter, remote cameras are powerful tools to investigate the interactions between fallen fruits and terrestrial frugivores in a forest landscape. They allow us to conduct a 24-h observation of fruit consumption and provide quantitative data in the number of photos. The results can lead us to a deep understanding of the frugivory and seed fate by use of better statistical methods to describe the patterns of fruit–frugivore interactions. This study needs to be replicated at other sites in the region, including forests with a more intact fauna.

We have to be careful to generalize the fruit–frugivore relationships of the region from the results of this study for three reasons. First, the present study was carried out in a faunally depleted forest where most medium- to large-sized frugivores, such as elephant, tapir, deer and civets, have been nearly wiped out by human activities. These larger frugivores are considered to be essential components of the frugivore community and effective seed dispersers in tropical rain forests of the Oriental region. Some traits of particular fruits may be evolutionarily related to the larger frugivores. For example, frugivory of *Irvingia malayana* (Irvingiaceae) was not observed on the forest floor in this study, in spite of the high abundance of fruits and the long observation period. Local people named the fruit 'Pau kijang', which means 'fruit of barking deer'. We suppose that the low density of barking deer in the Pasoh forest caused such a result. Second, fruit–frugivore relationships may differ in Peninsular Malaysia, Sumatra, Java and Borneo to some extent because the frugivore assemblage differs among the locations. Borneo, for example, lacks tapir, and the Asian elephant found there is considered to be an introduced species (Payne *et al.*, 1985). Information regarding fruit–frugivore relationships in the region is too limited to draw a picture at the community level. Third, frugivory experiments of this study were carried out in non-mast fruiting years. The drastic change in food abundance in a mast fruiting event (Fig. 9.1) is expected to have a great impact on the population dynamics of frugivores and may alter their foraging behaviour. Some plants can recruit only in mast fruiting

events when seed predation satiation is achieved. More research, both during mast and non-mast fruiting events, is needed before we can understand the ultimate fate of seeds in South-east Asian rainforests.

Note

[1]Even though South-east Asian tropics have little seasonal weather variation, flowering trees across many families have a unique rhythm of reproductive phenology known as mass flowering or general flowering, followed by mast fruiting. The phenomenon represents supra-annual synchronization of flowering at irregular intervals of 2–10 years. It results in a massive number of fruits in the forests during a period of 5–6 months. The pollination efficiency hypothesis and predator satiation hypothesis have been proposed for the adaptive significance of the phenomenon (Ashton *et al.*, 1988; Sakai, 2002; Numata *et al.*, 2003).

References

Ashton, P.S., Givinish, T.J. and Appanah, S. (1988) Staggered flowering in the Dipterocarpaceae: new insights into floral induction and the evolution of mast fruiting in the aseasonal tropics. *American Naturalist* 132, 44–66.

Bennett, E.L. (1991) Diurnal primates. In: Kiew, R. (ed.) *The State of Nature Conservation in Malaysia*. Malayan Nature Society, Kuala Lumpur, pp. 150–172.

Bernstein, I.S. (1967) A field study of the pigtail monkey. *Primates* 8, 217–228.

Caldecott, J.O. (1986) An ecological and behavioural study of the pig-tailed macaque. *Contributions to Primatology* 21, 1–259.

Corbet, G.B. and Hill, J.E. (1992) *The Mammals of the Indomalayan Region: a Systematic Review*. Oxford University Press, New York, 488 pp.

Corlett, R.T. (1998) Frugivory and seed dispersal by vertebrates in the Oriental (Indomalayan) Region. *Biological Review* 73, 413–448.

Corlett, R.T. and Lucas, P.W. (1990) Alternative seed-handling strategies in primates: seed-spitting by long-tailed macaques (*Macaca fascicularis*). *Oecologia* 82, 166–171.

Emmons, L.H. (1991) Frugivory in treeshrews (*Tupaia*). *American Naturalist* 138, 642–649.

Feer, F. (1995) Seed dispersal in African forest ruminants. *Journal of Tropical Ecology* 11, 683–689.

Feeroz, M.M., Islam, M.A. and Kabir, M.M. (1994) Food and feeding behaviour of hoolock gibbon (*Hylobates hoolock*), capped langur (*Presbytis pileata*) and pigtailed macaque (*Macaca nemestrina*) of Lawachara. *Bangladesh Journal of Zoology* 22, 123–132.

Fleming, T.H. (1979) Do tropical frugivores compete for food? *American Zoology* 19, 1157–1172.

Forget, P.-M. (1990) Seed dispersal of *Vouacapoua americana* (*Caesalpiniaceae*) by cavimorph rodents in French Guiana. *Oecologia* 85, 434–439.

Forget, P.-M. (1994) Recruitment pattern of *Vouacapoua americana* (*Caesalpiniaceae*), a rodent-dispersed tree species in French Guiana. *Biotropica* 26, 408–419.

Forget, P.-M. and Vander Wall, S.B. (2001) Scatter-hoarding rodents and marsupials: convergent evolution on diverging continents. *Trends in Ecology and Evolution* 16, 65–67.

Forget, P.-M., Moonz, E. and Leigh, E.G. Jr (1994) Predation by rodents and bruchid beetles on seeds of *Scheelea* palms on Barro Colorado Island, Panama. *Biotropica* 26, 420–426.

Gauch, H.G. (1982) *Multivariate Analysis in Community Ecology.* Cambridge University Press, Cambridge, 298 pp.

Gautier-Hion, A., Duplantier, J.M., Quris, R., Feer, F., Sourd, C., Decoux, J.P., Dubost, G., Emmons, L., Erard, C., Hecketsweiler, P., Moungazi, C. and Thiollay, J.M. (1985) Fruit characters as a basis of fruit choice and seed dispersal in a tropical forest vertebrate community. *Oecologia* 65, 324–337.

Gorchov, D.L., Cornejo, F., Ascorra, C.F. and Jaramillo, M. (1995) Dietary overlap between frugivorous birds and bats in the Peruvian Amazon. *Oikos* 74, 235–250.

Grassel, J.F. and Smith, W. (1976) A similarity measure sensitive to the contribution of rare species and its use in investigation of variation in marine benthic communities. *Oecologia* 25, 13–22.

Harrison, J.L. (1954) The natural food of some rats and other mammals. *Bulletin of the Raffles Museum (Singapore)* 25, 157–165.

Harrison, J.L. (1961) The natural food of some Malayan mammals. *Bulletin of the Natural Museum (Singapore)* 30, 5–18.

Hill, M.O. (1979) *TWINSPAN, a FORTRAN Program for Arranging Multivariate Data in an Ordered Two-way Table by Classification of the Individuals and Attributes.* Cornell University Press, New York, 90 pp.

Janis, C (1984) Tragulids as living fossils. In: Eldredge, N. and Stanley, S.M. (eds) *Living Fossils.* Springer-Verlag, New York, 291 pp.

Jongman, R.H.G., ter Braak, C.J.F. and van Tongeren, O.F.R. (1995) *Data Analysis in Community and Landscape Ecology.* Cambridge University Press, Cambridge, 299 pp.

Julliot, C. (1994) Frugivory and seed dispersal by red howler monkeys: evolutionary aspect. *Revue d'Ecologie (La Terre et la Vie)* 49, 331–341.

Karanth, K.U. (1995) Estimating tiger *Panthera tigris* populations from camera-trap data using capture–recapture models. *Biological Conservation* 71, 333–338.

Karanth, K.U. and Nichols, J.D. (1998) Estimation of tiger densities in India using photographic captures and recaptures. *Ecology* 79, 2852–2862.

Kawamichi, T. and Kawamichi, M. (1979) Spatial organization and territory of tree shrews (*Tupaia glis*). *Animal Behaviour* 27, 381–393.

Kitamura, S., Yumoto, T., Poonswad, P., Chuailua, P., Plongmai, K., Maruhashi, T. and Noma, N. (2002) Interactions between fleshy fruits and frugivores in a tropical seasonal forest in Thailand. *Oecologia* 133, 559–572.

Kobayashi, S. (1995) *Multivariate Analyses of Biological Communities.* Soju Shobo, Tokyo, 194 pp. (In Japanese.)

Krebs, C.J. (1998) *Ecological Methodology,* 2nd edn. Addison Wesley Longman, Menlo Park, California, 620 pp.

Lambert, F.R. and Marshall, A.G. (1991) Keystone characteristics of bird-dispersed *Ficus* in a Malaysian lowland rain forest. *Journal of Ecology* 79, 793–809.

Langham, N.P.E. (1982) The ecology of the common tree shrew, *Tupaia glis*, in peninsular Malaysia. *Journal of Zoology, London* 197, 323–344.

Langham, N.P.E. (1983) Distribution and ecology of small mammals in three rain forest localities of peninsular Malaysia with particular references to Kedah Peak. *Biotropica* 15, 199–206.

Leighton, M. and Leighton, D.R. (1983) Vertebrate responses to fruiting seasonality within a Bornean rain forest. In: Sutton, S.L., Whitmore, T.C. and Chadwick, A.C. (eds) *Tropical Rain Forest: Ecology and Management.* Special publications series of the British Ecological Society No. 2, Blackwell Scientific Publications, Oxford, 498 pp.

Lim, B.L. (1970) Distribution, relative abundance, food habits, and parasite patterns of giant rats (*Rattus*) in west Malaysia. *Journal of Mammalogy* 51, 730–740.

Lucas, P.W. and Corlett, R.T. (1998) Seed dispersal by long-tailed macaques. *American Journal of Primatology* 45, 29–44.

MacKinnon, K.S. (1978) Stratification and feeding differences among Malayan squirrels. *Malayan Nature Journal* 30, 593–608.

Magurran, A.E. (1988) *Ecological Diversity and Its Measurement.* Chapman & Hall, London, 179 pp.

Manokaran, N., LaFrankie, J.V., Kochummen, K.M., Quah, E.S., Klahn, J.E., Ashton, P.S. and Hubbell, S.P. (1992) *Stand Table and Distribution of Species in the 50-ha Research Plot at Pasoh Forest Reserve.* FRIM Research Data No. 1. Forest Research Institute Malaysia, Kuala Lumpur, 454 pp.

Matsubayashi, H., Bosi, E. and Kohshima, S. (2003) Activity and habitat use of lesser mouse-deer (*Tragulus javanicus*). *Journal of Mammalogy* 84, 234–242.

McClure, H.E. (1966) Flowering, fruiting and animals in the canopy of a tropical rain forest. *Malayan Forester* 29, 192–203.

McCullough, D.R., Pei, K.C.J. and Wang, Y. (2000) Home range, activity patterns, and habitat relations of Reeves' muntjacs in Taiwan. *Journal of Wildlife Management* 64, 430–441.

Medway, L. (1978) *The Wild Mammals of Malaya (Peninsular Malaysia) and Singapore* (2nd edn., reprinted with corrections in 1983). Oxford University Press, Kuala Lumpur, 131 pp.

Miura, S., Yasuda, M. and Ratnam, L. (1997) Who steals the fruit? *Malayan Nature Journal* 50, 183–193.

Morisita, M. (1959) Measuring of interspecific association and similarity between communities. *Memoirs of Faculty of Science, Kyushu University, Series E (Biol.)* 3, 65–80.

Morisita, M. (1971) Composition of the I_δ-index. *Researches on Population Ecology* 13, 1–27.

Numata, S., Yasuda, M., Okuda, T., Kachi, N. and Nur Supandi, Md.N. (2003) Temporal and spatial patterns of mass flowerings on the Malay Peninsula. *American Journal of Botany* 90, 1025–1031.

Numata, S., Okuda, T., Sugimoto, T., Nishimura, S., Yoshida, K., Quah, E.S., Yasuda, M., Muangkhum, K., Nur Supardi, Md.N. and Nor Azman, H. (2005) Camera trapping: a non-invasive approach as an additional tool in the study of mammals in Pasoh Forest Reserve and adjacent fragmented areas in Peninsular Malaysia. *Malaysian Nature Journal* 57, (in press).

O'Brien, T.G., Kinnaird, M.F. and Wibisono, H.T. (2003) Crouching tigers, hidden prey: Sumatran tiger and prey populations in a tropical forest landscape. *Animal Conservation* 6, 131–139.

Okuda, T., Manokaran, N., Matsumoto, Y., Niiyama, K., Thomas, S.C. and Ashton, P.S. (eds) (2003) *Pasoh: Ecology of Lowland Rainforest in Southeast Asia.* Springer-Verlag, Tokyo, 628 pp.

Osario, D. and Vorobyev, M. (1996) Colour vision as an adaptation to frugivory in primates. *Proceedings of the Royal Society of London B* 263, 593–599.

Otani, T. (2001) Measuring fig foraging frequency of the Yakushima macaque by using automatic cameras. *Ecological Research* 16, 49–54.

Otani, T. (2002) Seed dispersal by Japanese marten *Martes melampus* in the subalpine shrubland of northern Japan. *Ecological Research* 17, 29–38.

Payne, J.B. (1980) Competitors. In: Chivers, D.J. (ed.) *Malayan Forest Primates. Ten Years' Study in Tropical Rain Forest.* Plenum Press, New York, 364 pp.

Payne, J., Francis, C.M. and Phillipps, K. (1985) *A Field Guide to the Mammals of Borneo.* The Sabah Society, Kota Kinabalu, 332 pp.

Sakai, S. (2002) General flowering in lowland mixed dipterocarp forests of South-East Asia. *Biological Journal of the Linnean Society* 75, 233–247.

Sourd, C. and Gautier-Hion, A. (1986) Fruit selection by a forest guenon. *Journal of Animal Ecology* 55, 235–244.

Terborgh, J. (1983) *Five New World Primates: a Study in Comparative Ecology.* Princeton University Press, New Jersey, 260 pp.

Vander Wall, S.B. (1990) *Food Hoarding in Animals.* University of Chicago Press, Chicago, Illinois, 445 pp.

Watt, A.D., Stork, N.E. and Bolton B. (2003) The diversity and abundance of ants in relation to forest disturbance and plantation establishment in southern Cameroon. *Journal of Applied Ecology* 39, 18–30.

Wemmer, C., Kunz, T.H., Lundie-Jenkins, G. and McShea, W.J. (1996) Mammalian signs. In: Wilson, D.E., Cole, F.R., Nichols, J.D., Rudran, R. and Foster, M.S. (eds) *Measuring and Monitoring Biological Diversity.*

Standard Methods for Mammals. Smithsonian Institution Press, Washington, DC, pp. 157–176.

Wolda, H. (1981) Similarity indices, sample size and diversity. *Oecologia* 50, 296–302.

Wolda, H. (1983) Diversity, diversity indices and tropical cockroaches. *Oecologia* 58, 290–298.

Yasuda, M., Miura, S. and Nor Azman, H. (2000) Evidence for food hoarding behaviour in terrestrial rodents in Pasoh Forest Reserve, a Malaysian lowland rain forest. *Journal of Tropical Forest Science* 12, 164–173.

Yasuda, M., Ishii, N., Okuda, T. and Nor Azman, H. (2003) Small mammal community: habitat preference and effects after selective logging. In: Okuda, T., Manokaran, N., Matsumoto, Y., Niiyama, K., Thomas, S.C. and Ashton, P.S. (eds) *Pasoh: Ecology of Lowland Rainforest in Southeast Asia.* Springer-Verlag, Tokyo, pp. 533–546.

10 Myrmecochorous Seed Dispersal in Temperate Regions

Veronika Mayer, Silvester Ölzant and Renate C. Fischer
Institute of Botany, Department of Morphology and Reproduction Ecology,
University of Vienna, Rennweg 14, A-1030 Vienna, Austria

Introduction

Ants are the main group among invertebrates to disperse the seeds and fruits of plants. Diaspores with morphological adaptations to ant transport are called 'myrmecochorous' (Sernander, 1906). These adaptations are usually a variously shaped lipid-rich appendage called an 'elaiosome' (*elaion* = oil, *soma* = body). In some cases (e.g. *Puschkinia*, *Ornithogalum*; see Bresinsky, 1963) only a sarcotesta consisting of a layer of lipid-rich cells is provided. Elaiosomes are used as food by ants and this results in the transport of the diaspores to ant nests where the nutritive tissue is eaten. The diaspores themselves are then usually abandoned, intact and viable, either in the nest or outside the nest on waste piles, or close to the ant midden. The attractiveness of elaiosomes for ants has already been experimentally demonstrated by Sernander (1906), and his monograph on Central European and Mediterranean ant-dispersed plants (he recorded approximately 150 species) is still the most comprehensive work for this region. Myrmecochory is fairly common in eastern North America and Europe (e.g. Beattie *et al.*, 1979), and in Mediterranean climate regions of Australia (Berg, 1975) and southern Africa (Milewski and Bond, 1982). According to Beattie and Hughes (2002) myrmecochory is known in more than 80 plant families involving over 3000 species worldwide.

There are several hypotheses as to why ant dispersal might provide a selective advantage for plants. Beattie (1985) hypothesized five advantages to plants: (i) the nutrient-rich environment surrounding ant nests is beneficial for plant growth; (ii) avoidance of interspecific competition through spatial distribution; (iii) dispersal for distance; (iv) predator avoidance as many rodents and birds forage preferentially on seed clusters; and (v) avoidance of diaspore loss in fire-prone habitats (e.g. sclerophyllous shrublands in Australia, or fynbos in South Africa) due to diaspore burial in ant nests.

The chapter presented here will focus on some less frequently discussed points of myrmecochory and will provide an overview on: (i) the convergent origin of elaiosomes, (ii) their chemistry, (iii) the importance of the lipid-rich elaiosomes as food source for ant populations and demography, (iv) to what extent chemical signals by the plants mediate ant-dispersal, (v) the fate of the seeds and benefits for plants, and (vi) the habitat preferences and phenology of temperate myrmecochorous plants.

Elaiosome Origin in Diaspores

Elaiosomes are a striking example of convergence. Myrmecochorous diaspores can be seeds, clusae or small fruits, suggesting that elaiosomes have diverse morphological origins (Bresinsky, 1963). They derive either from the fruit or different parts of the seed, or, less frequently, from floral parts of the plant (Table 10.1; Figs 10.1 and 10.2). Some myrmecochorous diaspores have a lipid-containing sarcotesta (e.g. *Puschkinia scilloides*), or an ant-attractive structure that consists only of some lipid-containing cells (e.g. *Lathraea squamaria*). Bresinsky (1963) referred to them as 'rudimentary elaiosomes'. In other genera there is a progression from structures without clearly visible borders between elaiosome and seed (e.g. *Viola* spp.; Bresinsky, 1963) towards the formation of a distinct elaiosome. In Austria we found 89 species of 42 genera and 22 families with elaiosomes as well-defined separate organs (B. Krückl and V. Mayer, in preparation). This is similar to the number of myrmecochorous plant species known from southern Japan (Nakanishi, 1994) but much lower than the number found in Australia (*c.* 1500 species; Berg, 1975) or South Africa (*c.* 1300 species; Milewski and Bond, 1982).

Table 10.1. Possible morphological origin of various elaiosomes with examples of the Central European flora.

Plant part	Species	Family
Seed		
Funicle	*Chamaecytisus austriacus*	Fabaceae
Exostome	*Euphorbia amygdaloides, E. cyparissias*	Euphorbiaceae
	Polygala amara, P. chamaebuxus	Polygalaceae
	Scilla vindobonensis	Hyacinthaceae
Raphe	*Asarum europaeum*	Aristolochiaceae
	Chelidonium majus	Papaveraceae
	Corydalis cava	Fumariaceae
	Helleborus niger	Ranunculaceae
Raphe and exostome	*Viola hirta, V. odorata*	Violaceae
Chalaza	*Galanthus nivalis, Leucojum vernum*	Amaryllidaceae
	Luzula luzulina, L. pilosa	Juncaceae
Outer integument	*Ornithogalum* sp.	Hyacinthaceae
	Puschkinia scilloides	
Endosperm	*Melampyrum nemorosum*	Scrophulariaceae
	Lathraea sp.	
Fruit		
Pericarp base	*Borago officinalis*	Boraginaceae
	Pulmonaria officinalis	
	Symphytum officinale, S. tuberosum	
	Lamiastrum montanum, Lamium maculatum	Lamiaceae
	Hepatica nobilis	Ranunculaceae
Floral parts		
Corolla	*Parietaria officinalis*	Urticaceae
	Polygonum capitatum	Chenopodiaceae
Style base	*Carduus nutans*	Asteraceae
Sterile carpel	*Fedia cornucopia*	Valerianaceae
Epicalyx	*Knautia arvensis, K. dipsacifolia*	Dipsacaceae
Utricle	*Carex digitata, C. ornithopoda*	Cyperaceae
Sterile spikelet	*Melica nutans*	Poaceae
Capitulum base	*Centaurea cyanus*	Asteraceae

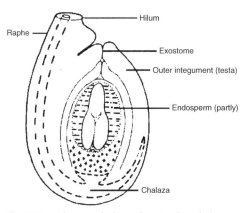

Fig. 10.1. The morphological parts of seeds from where elaiosomes can originate (modified from Boesewinkel and Boumann, 1984). Plant examples of the European flora are given in Table 10.1.

Trends in Morphological Characteristics of Ant-dispersed Diaspores

Consistency of elaiosomes

In contrast to the hard, long-lived elaiosomes of myrmecochorous species in the southern hemisphere (Beattie and Hughes, 2002), elaiosomes of species occurring in temperate regions differ in their consistency. According to B. Krückl and V. Mayer (in preparation) three groups can be distinguished.

● Group 1: Elaiosomes are soft and have large, often elongate or hypertrophic cells (e.g. *Corydalis cava*, Fig. 10.2C;

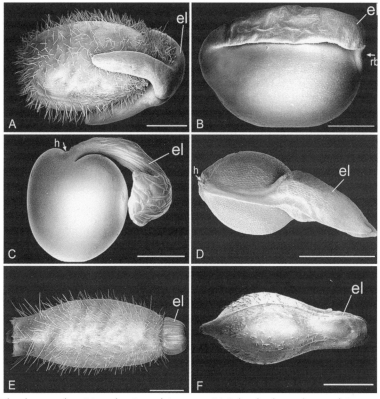

Fig. 10.2. The shape and position of various elaiosomes. (A) *Polygala chamaebuxus*, elaiosome with exostome origin; (B) *Helleborus niger*, raphal elaiosome; (C) *Corydalis cava*, raphal elaiosome; (D) *Luzula luzulina*, elaiosome of chalaza origin; (E) *Knautia arvensis*, elaiosome from the epicalyx; (F) *Carex digitata*, elaiosome from the utricle base. Scale bars: 1 mm; el, elaiosome; h, hilum (from B. Krückl and V. Mayer, in preparation).

Lamium maculatum, Fig. 10.3A; *Scilla vindobonensis*). They contain a lot of water, desiccate quickly and lose their original shape. They become unattractive within a few days. This type of elaiosome occurs in most of the spring-fruiting species.

- Group 2: Elaiosomes are soft and have small (e.g. *Chamaecytisus austriacus*, Fig.10.3B) to medium- (e.g. *Helleborus niger*, Fig. 10.2B) sized cells. Due to an obviously lower water content they rarely collapse and do not lose their original shape.
- Group 3: Elaiosomes are hard and have a prominent cuticula and at least one outer protecting cell layer that consists of small cells (e.g. *Pulmonaria officinalis*, *Knautia*, Fig. 10.2E). They do not visibly collapse. Firm elaiosomes are produced mainly from plants fruiting later in the season.

The consistency of elaiosomes is important for their durability. Storage experiments at 4°C showed that soft elaiosomes are attacked by fungi and decay within a few days if they are kept under moist conditions

(B. Krückl and V. Mayer, in preparation). For hard elaiosomes that are produced mainly from later-fruiting plants fungus attack is not as relevant as for the soft elaiosomes from other species. In contrast to soft elaiosomes, it takes weeks until the firm ones decay or break apart under moist storage conditions (e.g. *Euphorbia* spp.).

Protection of seed and embryo

The seeds, as well as the elaiosomes, of myrmecochorous species are rich in nutrients (Bewley and Black, 1978) and they have to be protected to avoid being eaten by the dispersing ants. Mechanical strength of diaspores is very common among myrmecochores in Australia (Rodgerson, 1998) but evidence for the co-evolution of stronger seed coats and ant dispersal has not been given. As well as stronger diaspores, a number of other protective features could be found in ant-dispersed seeds: (i) the surface of the fruit or seed can be very smooth as well as being firm (e.g. *Corydalis cava*, Fig. 10.2C) and therefore cannot be

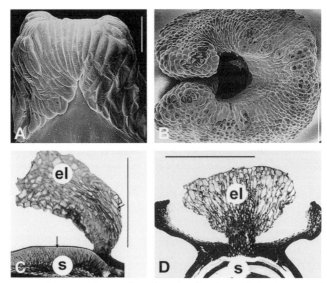

Fig. 10.3. Elaiosome consistency and protection of seed and embryo. (A) *Lamium maculatum*, elaiosome with large, often elongate or hypertrophic cells; (B) *Chamaecytisus austriacus*, elaiosome with soft but small cells; (C) *Chamaecytisus austriacus*, testa with a prominent protective layer of sclereids (arrow); (D) *Pulmonaria officinalis*, the pericarp of the clusa is strongly lignified. Scale bars: A, B 300 μm, C 500 μm, D 1 mm. el, elaiosome; s, seed (from B. Krückl and V. Mayer, in preparation).

cracked by the mandibles of ants. This phenomenon has also been observed by Pulliam and Brand (1975) for smooth, shiny diaspores. (ii) Many diaspores are hairy (Fig. 10.2A,E), which may have a protective function (Janzen, 1971). In *Knautia* spp., for example, the hairy epicalyx may prevent the wall from being cracked (Mayer and Svoma, 1998). (iii) In many myrmecochorous diaspores the testa has a layer of strong sclereids (e.g. *Chamaecytisus austriacus*, Fig. 10.3C), and in fruits the pericarp is strongly lignified (e.g. *Pulmonaria officinalis*, Fig. 10.3D). Also, (iv) some ant-dispersed seeds (e.g. *Luzula* spp., Fig. 10.2D) produce mucilage, which could also fulfil a protective function (Werker, 1997), since myxospermy is not very likely for woodland plants.

Nothing is known about the chemical defence of seed coats against ant predation. Some species lack any mechanical strength or other observable protective structures (e.g. in *Scilla vindobonensis*; B. Krückl and V. Mayer, in preparation). In this case the question remains as to whether their means of protection is still unknown, or whether the benefits of ant dispersal weigh up against the costs of increased predation.

Elaiosome/diaspore mass ratio

Many studies have demonstrated that the presence of elaiosomes on seeds or fruits consistently increases removal rates (e.g. Oostermeijer, 1989). This earlier work indicates that an increase in removal rates is not only due to an increased ease of handling. Diaspores with an artificial handle were not removed faster than those lacking an elaiosome (Hughes and Westoby, 1992), suggesting that there is no selection for structures which only enhance diaspore handling.

B. Krückl and V. Mayer (in preparation) investigated 21 elaiosome-bearing species of the temperate region and found that the diaspores have a defined size range. Diaspore mass ranges from 1 to 15 mg and elaiosome size from 1.2% (*Centaurea cyanus*) up to 26.1% (*Scilla vindobonensis*) of the diaspore mass (Fig. 10.4). The majority of elaiosomes, however, contribute to around 5% of the diaspore mass, and this is also true for other species of temperate regions (*Trillium* spp., Gunther and Lanza, 1989; *Viola* spp., Ohkawara and Higashi, 1994; Lisci *et al.*, 1996, different species). Though it is often reported that removal rates by ants correlate with elaiosome/diaspore mass ratios and that diaspores with large elaiosomes are more attractive to ants (Gunther and Lanza, 1989; Oostermeijer, 1989; Hughes and Westoby, 1992; Mark and Olesen, 1996), only a few myrmecochorous diaspores have larger elaiosomes.

In laboratory experiments with *Hepatica nobilis*, *Myrmica ruginodis* ants collected diaspores with the largest elaiosomes first (Mark and Olesen, 1996). On the other hand, Gunther and Lanza (1989) found, in three species of *Trillium* (*T. erectum*, *T. grandiflorum*, *T. undulatum*), that the species with the smallest elaiosome/

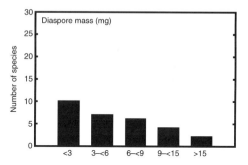

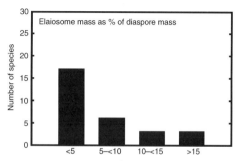

Fig. 10.4. The elaiosome/diaspore mass ratio in Central European myrmecochorous species. In Europe elaiosomes had a mean proportion of 4–10% of the diaspore mass.

diaspore mass ratio (*T. undulatum*) was removed at an intermediate stage and was not the diaspore with the lowest attractiveness. As Mark and Olesen (1996) pointed out, directional selection of larger elaiosomes is probably counterbalanced by the fact that larger elaiosomes (and therefore larger diaspores) have a much higher risk of being predated by rodents and birds than smaller diaspores.

Diaspore characteristics other than size may also play a role in ant dispersal. For example, removal responses to elaiosome/diaspore ratio were found to be species-specific in Australian ants (Hughes and Westoby, 1992). Experiments with an artificially increased elaiosome/mass ratio (elaiosomes being 4% to 125% of the mass of the seed) showed a reduced response if the ratios greatly exceeded those of natural elaiosome-bearing species (Hughes and Westoby, 1992). This suggests that there is no linear correlation between elaiosome size and attractiveness of elaiosomes. Though larger elaiosomes provide greater net food reward per ant trip relative to the increased costs of transport associated with carrying heavier objects (Davidson and Morton, 1981), ants generally seem to prefer diaspores that are compatible with the optimal working mandible span (Gorb and Gorb, 1995). Different ant species, therefore, prefer different diaspore sizes (Gorb and Gorb, 1999a,b), and, in field studies with *Helleborus foetidus* diaspores, the ants' preference was affected by their body size (Garrido *et al.*, 2002). This clearly shows that there is no foraging preference for the diaspore with the largest elaiosome. Dispersal success is correlated with the ant community structure of the locality where the myrmecochorous plants grow and ant size may explain why a certain elaiosome size is usually not exceeded. Interestingly, in a study on the geographical variation in diaspore traits of *Helleborus foetidus*, seed size and mass varied between localities, but elaiosome mass did not. Instead, variability in elaiosome mass and elaiosome/seed mass ratio occurred mostly among plants (Garrido *et al.*, 2002). This suggests that dispersal is favoured over seed mass and resources for

further growth, indicating the importance of successful dispersal and the strength of the selective pressure for that event.

Furthermore, the elaiosome serves to attract seed predators (Culver and Beattie, 1978; Heithaus, 1981; Boyd, 2001), probably through olfactory cues that help predators to find seeds. In only one case (i.e. *Corydalis aurea*, Fumariaceae), the elaiosome acted as repellent for rodent predators, presumably because of alkaloids which are common in Fumariaceae and may also occur in the elaiosomes of *C. aurea* (Hanzawa *et al.*, 1988). Increasing elaiosome/diaspore size can cause increased predation by vertebrates and a higher risk of seed loss. Both diaspore weight and elaiosome proportion to diaspore weight are adaptations to a complex mutualistic system with ant and plant partners. In temperate regions, elaiosomes with a mean proportion of 4–10% of the diaspore mass seem to be an appropriate energy investment for both partners to guarantee ant dispersal.

Chemical components of the elaiosome

Bresinsky (1963) was the first to analyse the chemistry of elaiosomes of myrmecochorous species. Though his main concerns were sugar content and composition, he also analysed the protein and fat content and demonstrated that elaiosomes represent a high quality food source for ants. The proportions of the respective chemical compounds fluctuate between species. Beattie (1985), who worked on Australian species, found a lipid content varying between 6 and 50% of the dry mass (DM). Lanza *et al.* (1992) also showed the considerable variation among various North American species of *Trillium* in the proportions of fatty acids and amino acids in the elaiosomes. Surprisingly, most studies have only compared elaiosome composition between species and not between seed and elaiosome. Fatty acid composition of elaiosomes and seeds was only compared in two studies based on nine North American (Soukup and Holman, 1987) and three Australian *Trillium* species

(Hughes *et al.*, 1994) (see also Table 10.2). Fischer *et al.* (2005b) investigated which chemical components were typical to elaiosomes compared to seeds (Fig. 10.5). Hence, the content and composition of fat, sugar and amino acids, and the protein content of both seeds and elaiosomes of 15 common European ant-dispersed species from different plant families were analysed (Fischer *et al.*, 2005b). The most important fraction was the lipid fraction, which ranged between 14.5 and 398.1 mg/g DM

Table 10.2. Comparison of major fatty acid composition (% of total detected lipids) of elaiosomes (E) and seeds (S). Data for Central European species are from Fischer *et al.* (2005b), for North American *Trillium* species from Soukup and Holman (1987), for the Australian species from Hughes *et al.* (1994) and for Hymenoptera haemolymph from Thompson (1973). Bold numbers indicate higher percentage of either elaiosome or seed.

	Palmitic C16:0		Palmitoleic C16:1		Stearic C18:0		Oleic C18:1		Linoleic C18:2		Linolenic C18:3	
	E	S	E	S	E	S	E	S	E	S	E	S
C European species (n = 15)	**20.0**	9.4	**2.2**	0.1	**2.5**	1.6	**41.3**	15.9	23.1	38.5	7.6	11.8
NA *Trillium* species (n = 9)	**17.6**	5.8	**9.7**	0.5	**1.8**	1.0	36.0	45.3	**26.4**	18.2	**2.0**	0.3
Australian species (n = 3)	**30.0**	12.0	**1.4**	0	**5.8**	3.0	**51.4**	27.3	10.0	45.3	2.3	11.0
Mean	22.5	9.3	4.3	0.2	3.5	1.9	42.9	29.5	19.8	34.0	1.6	4.5
Hymenoptera haemolymph	15.5		4.5		4.3		45.8		10.7		15.5	

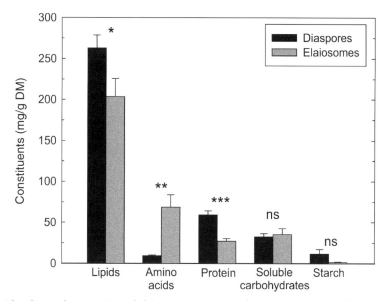

Fig. 10.5. The chemical composition of elaiosomes as compared to seeds in 15 Central European myrmecochorous species. Content of lipid, amino acids, protein, soluble carbohydrates (including glycerol and *myo*-inositol) and starch is indicated in mg/g dry mass. Bars signify means ± SE. Asterisks indicate significance levels (***, $P < 0.001$; **, $P < 0.01$; *, $P < 0.05$; ns, not significant; paired *t*-test). Apart from free amino acids the quantity of chemical constituents in elaiosomes and seeds does not differ markedly. Differences are found in their quality (from Fischer *et al.*, 2005a).

(1.5 and 39.8% of DM) in elaiosomes, and between 56.5 and 417.9 mg/g DM (5.7 and 41.8% of DM) in the seeds. The lipid content in most elaiosomes ranged between 10 and 30% of DM, but in the elaiosomes of four species it was less than 5% (*Asarum europaeum*, *Helleborus dumetorum*, *H. niger* and *Leucojum vernum*). Despite this low lipid content, *A. europaeum* is known to be dispersed by *Formica polyctena* (Gorb and Gorb, 1995, 1999a,b) and *H. niger* were observed to be carried into the nest by *Lasius niger* (V. Mayer, personal observation). The soluble sugars (including polyols) ranged between 1.9 and 134.4 mg/g DM (0.2 and 13.4% of DM) in the elaiosomes and between 10.9 and 89.2 mg/g DM (1.1 and 8.9% of DM) in the seeds. For ants, usable soluble sugars ranged between 1.7 and 152.3 mg/g DM in the elaiosomes. The protein content ranged between 9.3 and 74.9 mg/g DM (0.9 and 7.5% of DM) in elaiosomes, compared to 27.5 and 155.9 mg/g DM (2.7 and 15.6% of DM) in the seeds. The content of free amino acids varied between 4.9 and 355.7 mg/g DM (0.5 and 35.6% of DM) in elaiosomes and 0.8 and 29.0 mg/g DM (0.8 and 2.9% of DM) in seeds. Starch occurred only at very low levels.

Fischer *et al.*'s (2005a) results demonstrate that the soluble sugar content (including polyols) was more or less the same or slightly higher in elaiosomes than in seeds. The total concentration of soluble carbohydrates seems to be plant family specific and not a specific pattern of elaiosomes or seeds. A difference was found in sugar composition: (i) elaiosomes tend to accumulate a variety of different sugars while seeds store predominantly sucrose; and (ii) the amount of monosaccharides is higher in elaiosomes, though it was observed that *Lasius niger* workers prefer trisaccharides over mono- and disaccharides in honeydew (Völkl *et al.*, 1999). Hypotheses regarding potential nutritional benefits of these differences in sugar composition are difficult to generate at present. In all species examined (except *Chelidonium majus* and *Borago officinalis*) the protein content in elaiosomes was lower than in seeds. On the other hand,

the content of free amino acids, especially those needed for the synthesis of proteins, was five to 12 times higher in elaiosomes than in seeds (the elaiosome of *Helleborus niger* has the highest free amino acid content with 355.7 mg/g DM (35.6% of DM), *Leucojum vernum* has the highest among seeds with 29.0 mg/g DM). Amino acids can be stored in insect haemolymph, which has a much higher amino acid content than mammalian blood. It is remarkable that many more nitrogen-delivering acids were found in elaiosomes (41.1 ± 7.0 mg/g DM) than in seeds (8.5 ± 7.0 mg/g DM). Elaiosomes therefore seem to be a valuable source of nitrogen for ant colonies.

There were significant differences in fatty acid content between elaiosomes and seeds: most elaiosomes had distinctly higher concentrations of specific fatty acids (Fischer *et al.*, 2005a). Above all, the content of oleic acid (18:1n-9) was considerably higher in elaiosomes (a mean of 41.3% of total detected fatty acids in elaiosomes vs. 15.9% in seeds), but that of palmitic acid (16:0), palmitoleic acid (16:1) and stearic acid (18:0) was also higher in elaiosomes than in seeds. A comparison using data from the available literature on Australian and North American species suggests that the higher concentration of the latter three fatty acids is a general trend in elaiosomes (Table 10.2). In contrast, the polyunsaturated linoleic (linoleate 18:2) and linolenic acids (linolenate 18:3) are found in considerably higher concentrations in nearly all of the seeds examined except for the North American *Trillium* species (Soukup and Holman, 1987). Linoleate and linolenate may play an important role in the initial oxidative processes during seed germination (Bewley and Black, 1985). Though the linoleic acid content was higher in most seeds than in elaiosomes, it is the second highest lipid constituent in all elaiosomes investigated so far.

Little is known about the nutritional requirements of insects in general and ants in particular, but studies on a few insect species show a dietary requirement for polyunsaturated fatty acids. In Hymenoptera, for example, linolenic acid is very important for metamorphosis (Canavoso et al., 2001). A variety of unsaturated fatty acids can be formed from the fatty acids oleate, linoleate and palmitoleate (which can be oxidized from palmitate), and all are essential for insects (Stryer, 1981). In summary, the major chemical difference between elaiosomes and seeds is that elaiosomes contain low-molecular-weight substances which can be quickly metabolized, while seeds mainly contain high-molecular-weight substances such as proteins and starch. Another, very interesting difference is the much higher quantity of free amino acids, above all nitrogen-containing amino acids (e.g. asparagine, glutamine) in elaiosomes. Nitrogen may be a limiting factor for ants, especially in springtime when insect prey is still rare (Culver and Beattie, 1978) and for aphid tenders, whose diet consists of 98% carbohydrates but is nitrogen poor (Völkl et al., 1999). Elaiosomes are, therefore, a valuable nitrogen source for ants.

Benefit for Ants: Elaiosomes as Food Source

Ants favour different food types according to their tasks in the social life. The food of the food-gathering worker ants, for example, is richer in carbohydrates (especially sugar) than the food of the nurse ants that take care of the larvae (Haskins and Haskins, 1950; Gösswald and Kloft, 1956; Lange, 1967; Howard and Tschinkel, 1981). Proteinaceous food is mainly given to the larvae and the queen (Vinson, 1968; Howard and Tschinkel, 1981; Sorensen et al., 1983). Howard and Tschinkel (1981) found that foragers and nurses of the fire ant Solenopsis invicta feed on soybean oil in equal amounts to larvae and workers. Apart from being eaten at once lipids can be stored as reserve materials in the fat body (Arrese et al., 2001). Due to the different food requirements of the different castes, the amounts of certain compounds need to vary within the life-cycle of an ant colony. Handel (1976) showed that, in laboratory colonies of Aphaenogaster rudis, once workers had transported myrmecochorous diaspores into their nests, they carried larvae to the elaiosomes to feed. In laboratory colonies of Myrmica rubra elaiosome-bearing diaspores are carried into the brood chambers of the nest (Ölzant and Mayer, personal observation). Of those diaspores which were discarded into the arena of the experimental nest, the elaiosome was either completely lacking, or remnants of cell walls were still attached (Fig. 10.6). In labelling experiments, ^{15}N-labelled Corydalis cava elaiosomes and nutritionally well-balanced Bhatkar diet (Bhatkar and Whitcomb, 1970) were simultaneously offered to lab colonies of Myrmica rubra (Fischer et al., 2005b). The results clearly showed that elaiosomes were used above all as 'baby-food' for larvae during the experiment (Fig. 10.7), while workers continued to feed mainly on Bhatkar diet. This may be due to the high content of essential fatty acids (Table 10.2), free amino acids and the use of elaiosomes as a nitrogen source (Fischer et al., 2005a).

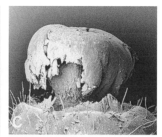

Fig. 10.6. Elaiosomes after being discarded. (A) *Corydalis cava*, the elaiosome is completely lacking; (B) *Chelidonium majus*, the cell content was consumed, only remnants of the cell wall are left; (C) *Knautia dipsacifolia*, parts of the elaiosome were torn out.

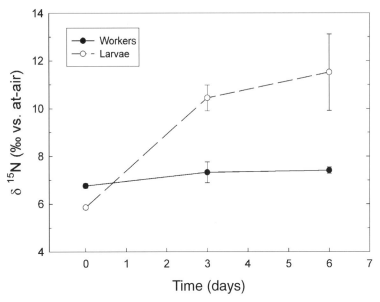

Fig. 10.7. Pulse experiment with *Myrmica rubra* and ^{15}N-labelled *Corydalis cava* elaiosomes. The elaiosomes were offered on days 0 and 1 together with nutritionally well-balanced Bhatkar diet. A significant difference between the ^{15}N incorporation rate of larvae and workers was found (from Fischer *et al.*, 2005a).

Several studies suggest a link between food supply and sex investment in ants (see e.g. Deslippe and Savolainen, 1995). In laboratory experiments, Morales and Heithaus (1998) found that colonies of *Aphaenogaster rudis* increased female sex investment after their diet was supplemented with elaiosomes of *Sanguinaria canadensis* (Papaveraceae), a perennial plant growing in North American woodland habitats. They hypothesized that elaiosomes may have a quantitative effect on larval development because larvae that accumulated more radiolabel from elaiosomes tended to develop into gynes (virgin queens), while other female larvae developed into workers (Morales and Heithaus, 1998). However, later experiments performed both in the laboratory and in the field with radiolabelled elaiosomes of the same species resulted in a discrepancy: lab colonies produced female-biased sex investment ratios, while field colonies mainly invested in males (Bono and Heithaus, 2002). This may be related to differing availabilities of background food

in lab and field. Though the link between elaiosome food and the increased number of females has not yet been satisfactorily clarified, elaiosomes may be an important factor for the reproductive success of the colony.

A recent study on fatty acid composition of the silverleaf whitefly (*Bemisia argentifolii*, Homoptera) showed the largest quantities of oleate, palmitate and linoleate to be in the fourth instar nymphs (Bruckner and Hagen, 2003), all three of them also being the major constituents of elaiosome lipids (Table 10.2). Caution should be exercised in drawing general conclusions, especially when dealing with different insect orders, but the easy availability of these essential fatty acids could be a reason for the investment into gyne development in ant colonies fed with elaiosomes. Additionally, in temperate regions elaiosome-bearing plants are mostly spring flowering species with lots of mature diaspores early in the year when insect prey for ants is still rare (Culver and Beattie, 1978).

How Are Ants Attracted? Do Chemical Signals Act as a Trigger for Ant Dispersal?

Seed-carrying by ants varies depending on the ant species, the respective diaspore, and the context of the interaction. Certain ant species are more likely to transport certain diaspores than others (Davidson and Morton, 1981; Kjellsson, 1985; Ooster-meijer, 1989). The most widely accepted hypothesis as to why ants carry the whole diaspore (elaiosome and fruit/seed) is that there exists a 'chemical behaviour-releaser'. Bresinsky (1963) suggested that ricinoleic acid initiates diaspore transport. 1,2-Diolein, a diglyceride of the oleic acid, acted as trigger for seed dispersal in temperate ants (*Pogonomyrmex rugosus*, *Aphaenogaster rudis*; Marshall *et al.*, 1979; Skidmore and Heithaus, 1988) and in Australian species (*Rhytidoponera victoriae*, *Notoncus ectatommoides*, and others; Brew *et al.*, 1989). Oleic acid is presumed to elicit the so-called 'corpse-carrying' behaviour, a necrophoric response first reported by Wilson *et al.* (1958) for the harvester ant *Pogonomyrmex badius*, and later in *Myrmecia vindex* (Haskins and Haskins, 1974) and in *Solenopsis saevissima* (Blum, 1970). All authors observed that objects treated with oleic acid were carried to the colony's refuse pile or the midden. Gordon (1983), however, found that the response to oleic acid depended on the social context of the colony: when a large percentage was doing midden work or nest maintenance, pieces of paper treated with oleic acid were taken to the midden, and, when a large percentage was foraging or convening, the same pieces were taken into the nest.

Lanza *et al.* (1992) suggest that, apart from the diglyceride 1,2-diolein and other forms of oleic acid (e.g. triolein), substances associated with linoleic acid are also able to function as behavioural triggers; these results stem from data on *Myrmica punctinervis* behaviour and elaiosome fatty acids in three North American *Trillium* species. Hughes *et al.* (1994) offered 1,2-diolein-impregnated artificial diaspores to six Australian ant species and found that their response was dependent on the ant species: *Polyrhachis* sp. and *Camponotus consobrinus*, two nectar-feeding species, did not respond to 1,2-diolein. *Iridomyrmex anceps*, which occasionally collects elaiosomes, did not distinguish between 1,2-diolein-impregnated and 1,2-diolein-free artificial diaspores, and surprisingly collected both with the same frequency. The statement by Carroll and Janzen (1973) that elaiosomes are 'simply a dead insect analogue' is supported by the results of Hughes *et al.* (1994): the Australian ant species that reacted most significantly to the oleic substance (*Aphaenogaster longiceps*, *Rhytidoponera metallica* and *Pheidole* sp.) usually included large amounts of insect material in their diet. Hughes *et al.* (1994) therefore hypothesized that the fatty acid composition of elaiosomes is similar to that of insect prey, indicating that elaiosomes may mimic insects, thereby attracting not only omnivorous but also carnivorous ants that would not normally remove plant material. The comparison of Central European species (Table 10.2) and the data available for Australian (Hughes *et al.*, 1994) and North American species (Soukup and Holman, 1987), together with data for Hymenoptera haemolymph (Thompson, 1973), indeed show a more similar fatty acid composition between elaiosomes and the haemolymph than between seeds and the haemolymph. The observation of Culver and Beattie (1978) that elaiosomes of *Viola* spp. are larger in summer than in spring also supports this hypothesis: in summer elaiosomes have to compete with a surplus of insect prey.

1,2-Diolein is not the only elaiosome component which elicits carrying behaviour (Hughes *et al.*, 1994); *Aphaenogaster* and *Rhytidoponera* species remove elaiosomes faster than diolein piths. Ölzant (personal observation) also did not find an increased carrying response to 1,2-diolein in lab colonies of *Formica fusca* and *Myrmica rubra*. Howard and Tschinkel (1976), however, found that methanol and solvents of dry shellac, or oil-free paints also stimulated corpse-carrying behaviour of dead nest companions, which suggests that, once the behavioural cue is released, 'corpse-carrying behaviour' is different from prey behaviour. The behavioural bases of diaspore dispersal thus remain to be clarified. The hypothesis that diaspore-carrying behaviour is elicited by a single trigger is probably too uni-dimensional. Ants may recognize elaiosomes as an excellent food source like other prey and carry them to their nest for reasons other than just a chemical releaser.

Fate of Diaspores and Benefit for Plants in Temperate Regions

Removal and dropping rates, and the post-dispersal fate of diaspores

Removal of ant-dispersed diaspores has been extensively studied in different habitats with different focuses: influence of season and time of day (Hughes and Westoby, 1990), interactions between one ant and one plant species (Pacini, 1990), one plant and several ant species (Culver and Beattie, 1980; Bond and Stock, 1989; Higashi *et al.*, 1989), one ant and several plant species (Gorb and Gorb, 1995), and several ant and plant species (Sernander, 1906; Oostermeijer, 1989). All authors agree that the primary attractant of the diaspore is the elaiosome: around the world, diaspores with elaiosomes removed have significantly lower removal rates than diaspores with attached elaiosomes (O'Dowd and Hay, 1980; Slingsby and Bond, 1985; Oostermeijer, 1989; Hughes and Westoby, 1992). Ants typically carry the myrmecochorous diaspores into their nests, consume the elaiosome and discard them on aboveground refuse piles or in waste chambers below the soil surface.

For Australian species, Hughes and Westoby (1990) found only a weak correlation between removal rate and diaspore size and shape, in contrast to Gorb and Gorb (1995), who found a clear correlation between removal rate and diaspore size and design in a Ukrainian deciduous forest. In general, ant species with large foragers (e.g.

Formica polyctena) removed large diaspores with significantly higher frequency than ant species with small foragers (e.g. *Leptothorax nylanderi*) (Gorb and Gorb, 1995, 1999a). The behaviour of foragers to a collecting stimulus, however, can differ. Lab colonies of *Myrmica rubra* and *Formica fusca* (S. Ölzant and V. Mayer, unpublished) reacted heterogeneously to intact diaspores, elaiosomes only, and diaspores with artificially removed elaiosome of six different plant species. Apart from differences between *Myrmica* and *Formica* colonies, the ant colonies did not react uniformly from experiment to experiment even for a single plant species. The removal activity as well as where the diaspores were carried differed. In some cases, a single individual removed all offered elaiosomes and intact diaspores. This indicates that the reaction of ants to myrmecochorous diaspores is dependent on the actual constitution of the colony (size of the colony, amount of brood, etc.) (S. Ölzant and V. Mayer, unpublished) and the social context as described by Gordon (1983) for the response to oleic acid as behavioural trigger. Further abiotic factors such as temperature, humidity, time of day and season also play a role: ant activity is known to change with changes in temperature and humidity (Talbot, 1946). Large ant species sometimes also transport diaspores with artificially removed elaiosome (S. Ölzant and V. Mayer, unpublished). The heterogeneity of ant reactions demonstrates the complexity of diaspore removal and removal activity and makes it clear that single observations or short-term experiments with few replicates are not very meaningful in terms of attractiveness of diaspores to ant dispersers.

Not all diaspores are removed to ant nests. In field experiments with *Formica polyctena*, Gorb and Gorb (1999b) found dropping rates of 8–10% for large diaspores (e.g. *Asarum europaeum*) and 20–100% for small diaspores (e.g. *Chelidonium majus*). They correlate this behaviour with the head width of the dispersing ant species, and hypothesize that ants forage for diaspores which are optimal for mandibular span. Repeated removal of the small dropped diaspores was rare; usually they rolled into soil cracks or litter and were no longer detected. The importance of dropping rates of myrmecochorous diaspores has been disregarded until now. A 100% entry of diaspores into a single ant nest and secondary relocation to the colonies' refuse piles can lead to seed clusters, causing intraspecific seedling competition. A good example is provided by Andersen (1988), who graphs *Acacia suaveolens* seeds in relation to ant nests (Fig. 10.8). In this respect diaspore dropping may be an advantage: redispersal by a species of seeds dropped by other ant species avoids secondary clusters within the foraging territory of a single colony and increases the distribution radius of the diaspores. Furthermore, Gorb and Gorb (2000) and

Gorb *et al.* (2000) observed the relocation of diaspores to colony territory borders after elaiosome removal, which increased the distribution area. Lab colonies of *Myrmica rubra* discarded more than 50% of *Corydalis cava* diaspores after 24 h, and after 96 h 100% of all offered diaspores were found in the arena again (Fischer *et al.*, 2005). Storage in underground waste chambers is rare in lab colonies, which may be due to the limited space.

Dispersal Distances of Myrmecochorous Seeds

Dispersal distances, removal rates, and dropping rates are all dependent on ant

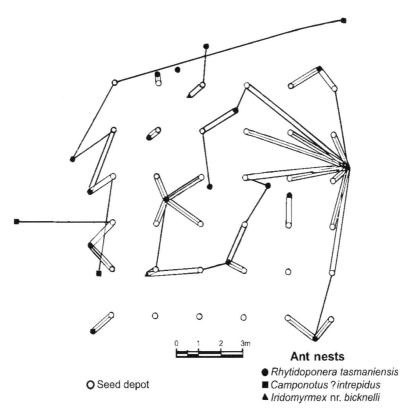

Ant nests
● *Rhytidoponera tasmaniensis*
■ *Camponotus ?intrepidus*
▲ *Iridomyrmex* nr. *bicknelli*

O Seed depot

Fig. 10.8. Transport of *Acacia suaveolens* seeds from seed depots (open circles) to nests of different seed-dispersing ant species. Lines indicate the transport routes. Active colonies with a large foraging territory transport seeds from various depots over different distances into their nests (from Andersen, 1988).

size. A literature survey shows different dispersal distances for the different ant subfamilies: Formicinae and Dolichoderinae disperse at larger mean distances (0.02–70 m and 0.01–77 m) than Myrmicinae and Ponerinae (0.01–21 m) (Gómez and Espadaler, 1998). The mean global dispersal distance estimated from the literature data is 0.96 m, ranging from 0.01 to 77 m (Gómez and Espadaler, 1998), to a new record of 180 m (with a mean of 94 m), reported for *Acacia ligulata* seeds transported from *Iridomyrmex* ants in the arid zone of Australia (Whitney, 2002). Dispersal curves for mesophyllous and sclerophyllous vegetation were significantly different: the mean dispersal distance for mesophyllous vegetation was 0.87 m (0.01–70 m), and for sclerophyllous vegetation it was 1.25 m (0.06–77 m) (Fig. 10.9). In summary, dispersal distances in the northern and southern hemispheres had the same values, which may be due to the fact that studies on myrmecochory have mainly been done in the arid zone of Australia. Global ant dispersal is over short distances; 89% of the registered dispersal distances are between 0 and 2 m (Gómez and Espadaler, 1998). Either the plants

have adapted to this short dispersal distance, or there is no need for distance dispersal as safe and suitable sites for germination are abundant. Alternatively, existing data may still only represent a rough estimation of this complex feature. Dispersal distance is certainly dependent on ant species and the species composition within a specific habitat. Further studies should clarify whether differences in ant and plant species composition in different habitats show different dispersal patterns and distances. Until now only the northern and southern hemispheres have been compared (Gómez and Espadaler, 1998), but not different habitats within a hemisphere.

Elaiosome Removal and Germination

It has been proposed that ant processing of diaspores may stimulate germination (Handel, 1976, 1978; Culver and Beattie, 1980; Horvitz, 1981; Rockwood and Blois, 1986; Lagoa and Pereira, 1987; Pacini, 1990; Mayer and Svoma, 1998). It has been

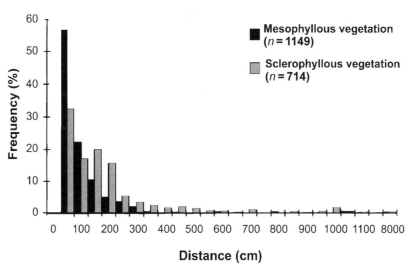

Distance (cm)

Fig. 10.9. Comparison of dispersal distances between mesophyllous and sclerophyllous vegetation. An evaluation of data available in the literature resulted in a mean dispersal distance for mesophyllous vegetation of 0.87 m (0.01–70 m) vs. 1.25 m (0.06–77 m) for sclerophyllous vegetation (from Gómez and Espadaler, 1998).

assumed that elaiosome removal is a trigger for germination or even a suitable dormancy break treatment for diaspores (Pacini, 1990), that it speeds up the germination process (Lagoa and Pereira, 1987; Lisci et al., 1996), or that it is important in processes preceding germination such as dehydration or hydration (Lisci et al., 1996). In contrast, Boyd (2001) observed, in experiments with the central Californian chaparral species *Fremontodon procumbens* (Sterculiaceae), that ant-handled seeds did not germinate better than seeds from dehisced capsules. Germination experiments with 26 common Central European myrmecochorous species showed that hydration was not dependent on the presence or absence of the elaiosome (B. Krückl and V. Mayer, in preparation). Under controlled laboratory conditions, the majority of species did not show any significant differences in germination response with or without elaiosome; in only 15% of the investigated diaspores was germination improved by elaiosome removal. So stimulation of germination due to elaiosome removal seems not to be a benefit of ant-mediated dispersal. Much more important seems to be the fact that ants transport diaspores to different sites, some of which offer suitable conditions for storage until germination and for germination itself. Moist storage is important, especially for seeds that are sensitive to desiccation (e.g. *Scilla*, *Corydalis*, and others). It was observed that fungus attack on the seeds stored under moist conditions was significantly reduced after ant contact. Very few myrmecochorous diaspores needed certain light conditions for germination (only three species out of 26 had light preferences: two germinated better in darkness, one in light). This means that the only trend observed so far in relation to germination requirements is that the majority of ant-dispersed diaspores have no light preferences; they germinate equally well in light and in darkness (B. Krückl and V. Mayer, in preparation). This may be an adaptation to the unpredictable relocation of the diaspores after the elaiosome is consumed (above ground on waste piles, below ground in waste chambers).

Habitat Preferences and Phenology of Temperate Myrmecochorous Plants

Habitat preferences

According to Beattie (1983), myrmecochorous plants comprise up to 30% of the herbaceous flora in temperate forests, and Sernander (1906) reported that ant-dispersed plants are five to ten times more frequent in forest than in open habitats. A more detailed examination of an area with different habitats in western Germany showed that 75% of all 24 ant-dispersed plant species occurred in forest areas, the others in grassland and ruderal sites (Oberrath, 2000). In eastern Austria 40% of 30 ant-dispersed plants are shade-tolerant woodland species, another 40% are light-requiring and occur in more or less open areas, and 20% are found exclusively in grasslands (B. Krückl and V. Mayer, in preparation). A conspicuous characteristic of temperate ant-dispersed diaspores seems to be moisture preference: 67% of the species investigated prefer more or less moist habitats, though only a few occur in markedly wet habitats (e.g. *Leucojum vernum*), and only one-third of the species investigated prefer dry habitats. Moisture conditions of the habitat and flowering and fruiting times were most significantly correlated with diaspore characteristics: species of moist habitats tended to have larger diaspores with larger, soft elaiosomes, whereas species of dry habitats had smaller diaspores with smaller, firm elaiosomes (B. Krückl and V. Mayer, in preparation). The latter elaiosomes are tougher and may persist for a long time, even under dry conditions. Twenty-seven per cent of the species were classified as nitrogen indicators (e.g. *Chelidonium majus*, *Corydalis cava*, *Leucojum vernum*, *Symphytum officinale*, *Viola odorata*) according to Ellenberg et al. (1992), 33% of the species were indifferent and 40% grew in habitats more or less poor in nutrients. It has often been proposed that the surroundings of anthills are suitable sites, regarding nutrients, for seedling recruitment (e.g. Culver and Beattie, 1980). This hypothesis has been rejected for plants of temperate forests, where nutrient deficiency is not an important factor (Higashi *et al.*, 1989), but it could be a benefit for many plants which occur in habitats with poor nutrient availability.

Phenology of temperate myrmecochorous plants

Studies on the synchrony between fruiting time of myrmecochorous plants and dispersing ant species are rare. Oberrath (2000) and Oberrath and Böhning-Gaese (2002) found that, in an area with 24 ant-dispersed and 251 non-ant-dispersed plant species, ant-dispersed plant species flowered on average 5.6 weeks earlier than other plants. The fruiting peak was at the beginning of July and thus up to 7.1 weeks earlier than those with other dispersal modes (Fig. 10.10). This phenological feature could not be explained either by phylogeny or by variation in growth form or habitat. Ant activity was especially high from May to July, during the main fruiting time of myrmecochorous plants (Fig. 10.11). Plant species that fruit from May to July have significantly larger elaiosomes than species that fruit later in the season (Krückl and Mayer, personal observation) and are, therefore, a more profitable food source for foraging ants. Food intake by many ant colonies increases rapidly during April,

with a peak in May (Horstman, 1972) due to brood feeding and the production of gynes and males (e.g. Kirchner, 1964). As elaiosomes are mainly used as food for the brood (Fischer *et al.*, 2005a) it is suggested that the early fruiting time of ant-dispersed plants is an adaptation to ant dispersal. Synchrony of ant activity and diaspore dispersal can exist even at the level of seed release. *Viola nuttallii* synchronized its seed release with the time when most ants and fewest rodents were active (Turnbull and Culver, 1983). This synchrony, however, is not necessarily an adaptation which has evolved on the part of the plant but is possibly the result of several organisms responding to other variables such as temperature.

Future Directions

Since Sernander (1906) wrote his monograph on European myrmecochores, many excellent studies have contributed to the understanding of the complex phenomenon of ant-dispersal. However, many questions still remain. Answering them may contribute decisively to the understanding of the evolution of ant dispersal. Included among these outstanding questions are:

1.   What is the effect of elaiosomes on ant-colony demography? The flow of food has not been studied until now. The finding

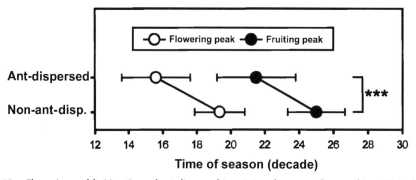

Fig. 10.10.   Flowering and fruiting time of ant-dispersed (*n* = 24) and non-ant-dispersed (*n* = 251) plants. On average, ant-dispersed plants flowered 5.6 weeks earlier than other plants; the fruiting peak was at the beginning of July and thus up to 7.1 weeks earlier than those with other dispersal modes. The study area was observed every 10 days (= decade) between January and December. ***P < 0.001 (from Oberrath, 2000).

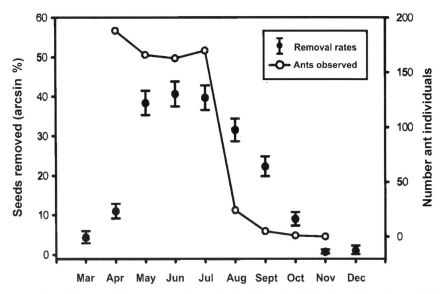

Fig. 10.11. Ant activity and diaspore removal rates. During the main fruiting time of ant-dispersed plants increased ant activity could be observed (from Oberrath, 2000).

by Fischer *et al.* (2005a) that elaiosomes are mainly used as baby-food for larvae gives a new insight into the complexity of the synchronization of plant phenology and colony cycle. The surprising result by Morales and Heithaus (1998) and Bono and Heithaus (2002) that elaiosome food can change the colony sex ratio indicates that much more work still has to be done to understand how ant-dispersed plants contribute to the vitality and growth of ant colonies.

2.   What is the effect of ant-dispersal on plant species composition? Most studies on ant-dispersal have focused on the interactions between one ant and one plant species, one plant and several ant species, one ant and several plant species, or several ant and plant species. Very little is known about ant–plant interactions on the plant communitiy level, changes along habitat gradients and differences among different vegetation types. Reproductive success as well as genetic structure of ant and plant communities are important factors (Folgarait, 1998) that influence habitat diversity.

3.   What is the effect of habitat disturbance and/or invasive species on ant and plant species composition? Habitat disturbance can influence the structure and function of ecosystems. Because of the close relationship between dispersing ants looking for food and seed dispersal, a reduction in or disappearance of one of the interacting species affects the whole system. For example, the invasive Argentine ant (*Linepithema humile*) apparently caused a change in the South African fynbos plant communities because it displaced native seed-dispersing ants (Bond and Slingsby, 1984; Christian, 2001). In California the displacement of native harvester ants by *L. humile* had a negative impact on the population of the myrmecochorous tree poppy *Dendromecon rigida* because of decreased dispersal and increased seed loss (Carney *et al.*, 2003). In the European Mediterranean region, Gomez and Oliveras (2003) demonstrated that the invasion of the Argentine ant decreased seed dispersal from 50% in non-invaded areas to 16.8% in invaded areas; this impact will probably have a dramatic effect on the populations of myrmecochorous plant species. Further studies and monitoring projects are needed to develop sensible conservation strategies.

References

Andersen, A.N. (1988) Dispersal distance as a benefit of myrmecochory. *Oecologia* 75, 507–511.

Arrese, E.L., Canavoso, L.E., Jouni, Z.E., Pennington, J.E., Tsuchida, K. and Wells, M.A. (2001) Lipid storage and mobilization in insects: current status and future directions. *Insect Biochemistry and Molecular Biology* 31, 7–17.

Beattie, A.J. (1983) Distribution of ant-dispersed plants. *Sonderband des Naturwissenschaftlichen Vereins Hamburg* 7, 249–270.

Beattie, A.J. (1985) *The Evolutionary Ecology of Ant–Plant Mutualisms*. Cambridge Studies in Ecology, Cambridge University Press, Cambridge, 182 pp.

Beattie, A.J. and Hughes, L. (2002) Ant–plant interactions. In: Herrera, C.M. and Pellmyr, O. (eds) *Plant–Animal Interactions. An Evolutionary Approach*. Blackwell Science, Oxford, pp. 211–235.

Beattie, A.J., Culver, D.C. and Pudlo, R.J. (1979) Interactions between ants and the diaspores of some common spring flowering herbs. *Castanea* 44, 177–186.

Berg, R.Y. (1975) Myrmecochorous plants in Australia and their dispersal by ants. *Australian Journal of Botany* 23, 475–508.

Bewley, J.D. and Black, M. (1978) *Physiology and Biochemistry of Seeds. Development, Germination and Growth*. Springer Verlag, Berlin, Heidelberg, 306 pp.

Bewley, J.D. and Black, M. (1985) *Seeds: Physiology of Development and Germination*. Plenum Press, New York, 367 pp.

Bhatkar, A.P. and Whitcomb, W.H. (1970) Artificial diet for rearing various species of ants. *Florida Entomologist* 53, 229–232.

Blum, M.S. (1970) the chemical basis of insect sociality. In: Beroza, M. (ed.) *Chemicals Controlling Insect Behaviour*. Academic Press, New York, pp. 61–94.

Boesewinkel, F. and Bouman, F. (1984) The seed: structure. In: Johri, B.M., Ambegaokar, K.B. and Srivastava, P.S. (eds) *Embryology of Angiosperms*. Springer Verlag, Berlin, pp. 567–610.

Bond, W.J. and Slingsby, P. (1984) Collapse of an ant–plant mutualism: the Argentine ant (*Iridomyrmex humilis*) and myrmecochorous Proteaceae. *Ecology* 65, 1031–1037.

Bond, W.J. and Stock, W.D. (1989) The costs of leaving home: ants disperse myrmecochorous seeds to low nutrient sites. *Oecologia* 81, 412–417.

Bono, J.M. and Heithaus, E.R. (2002) Sex ratios and the distribution of elaiosomes in colonies of the ant, *Aphaenogaster rudis*. *Insectes Sociaux* 49, 320–325.

Boyd, R.S. (2001) Ecological benefits of myrmecochory for the endangered chaparral shrub *Fremontodendron decumbens* (Sterculiaceae). *American Journal of Botany* 88, 234–241.

Bresinsky, A. (1963) Bau, Entwicklungsgeschichte und Inhaltsstoffe der Elaiosomen. Studien zur myrmekochoren Verbreitung von Samen und Früchten. *Bibliotheca Botanica* 126, 1–54.

Brew, G.R., O'Dowd, D.J. and Rae, J.D. (1989) Seed dispersal by ants: behaviour releasing compounds in elaiosomes. *Oecologia* 80, 490–497.

Bruckner, J.S. and Hagen, M.M. (2003) Triacylglycerol and phospholipid fatty acids of the silverleaf whitefly: composition and biosynthesis. *Archives of Insect Biochemistry and Physiology* 53, 66–79.

Canavoso, L.E., Jouni, Z.E., Karnas, K.J., Pennington, J.E. and Wells, M.A. (2001) Fat metabolism in insects. *Annual Review of Nutrition* 21, 23–46.

Carney, S.E., Byerley, M.B. and Holway, D.A. (2003) Invasive ants (*Linepithema humile*) do not replace native ants as seed dispersers of *Dendromecon rigida* (Papaveraceae) in California, USA. *Oecologia* 135, 576–582.

Carroll, C.R. and Janzen, D.H. (1973) The ecology of foraging by ants. *Annual Review of Ecology and Systematics* 4, 231–258.

Christian, C.E. (2001) Consequences of a biological invasion reveal the importance of mutualism for plant communities. *Nature* 413, 635–639.

Culver, D.C. and Beattie, A.J. (1978) Myrmecochory in *Viola*: dynamics of seed–ant interactions in some West Virginia species. *Journal of Ecology* 66, 53–72.

Culver, D.C. and Beattie, A.J. (1980) The fate of *Viola* seeds dispersed by ants. *American Journal of Botany* 67, 710–714.

Davidson, D.W. and Morton, S.R. (1981) Myrmecochory in some plants (Chenopodiaceae) of the Australian arid zone. *Oecologia* 50, 357–366.

Deslippe, R.J. and Savolainen R. (1995) Sex investment in a social insect: the proximate role of food. *Ecology* 76, 375–382.

Ellenberg, H., Weber, H.E., Düll, R., Wirth, V., Werner, W. and Paulissen, D. (1992) Zeigerwerte von Pflanzen in Mitteleuropa. *Scripta Geobotanica* 18, 2nd edn.

Fischer, R.C., Ölzant, S.M., Wanek, W. and Mayer, V. (2005a) The fate of *Corydalis cava* elaiosomes within an ant colony of *Myrmica rubra*: elaiosomes are preferentially fed to larvae. *Insectes Sociaux* 52 (in press).

Fischer, R.C., Richter, A. and Mayer, V. (2005b) Chemical composition of ant-dispersed diaspores and their elaiosome appendages: are elaiosomes adapted to nutritional needs of ants? *Functional Ecology* (in review).

Folgarait, P.J. (1998) Ant biodiversity and its relationship to ecosystem functioning: a review. *Biodiversity and Conservation* 7, 1221–1244.

Garrido, J.L., Rey, P.J., Cerda, X. and Herrera, C.M. (2002) Geographical variation in diaspore traits of an ant-dispersed plant (*Helleborus foetidus*): are ant community composition and diaspore traits correlated? *Journal of Ecology* 90, 446–455.

Gómez, C. and Espadaler, X. (1998) Myrmecochorous dispersal distances: a world survey. *Journal of Biogeography* 25, 573–580.

Gomez, C. and Oliveras, J. (2003) Can the Argentine ant (*Linepithema humile* Mayr) replace native ants in myrmecochory? *Acta Oecologia – International Journal of Ecology* 24, 47–53.

Gorb, S.N. and Gorb, E.V. (1995) Removal rates of seeds of five myrmecochorous plants by the ant *Formica polyctena* (Hymenoptera: Formicidae). *Oikos* 73, 367–374.

Gorb, S.N. and Gorb, E.V. (1999a) Dropping rates of elaiosome-bearing seeds during transport by ants (*Formica polyctena* Foerst.): implications for distance dispersal. *Acta Oecologica – International Journal of Ecology* 20, 509–518.

Gorb, S.N. and Gorb, E.V. (1999b) Effects of ant species composition on seed removal in deciduous forest in Eastern Europe. *Oikos* 84, 110–118.

Gorb, E.V. and Gorb, S.N. (2000) Effects of seed aggregation on the removal rates of elaiosome-bearing *Chelidonium majus* and *Viola odorata* seeds carried by *Formica polyctena* ants. *Ecological Research* 15, 187–192.

Gorb, S.N., Gorb, E.V. and Punttila, P. (2000) Effects of redispersal of seeds by ants on the vegetation pattern in a deciduous forest: a case study. *Acta Oecologica – International Journal of Ecology* 21, 293–301.

Gordon, D.M. (1983) Dependence of necrophoric response to oleic acid on social context in the ant *Pogonomyrmex badius*. *Journal of Chemical Ecology* 9, 105–111.

Gösswald, K. and Kloft, W. (1956) Untersuchungen über die Verteilung von radioaktiv markiertem Futter im Volk der Klienen roten Waldameise (*Formica rufopratensis minor*). *Waldhygiene* 1, 200–202.

Gunther, R.W. and Lanza, J. (1989) Variation in attractiveness of *Trillium* diaspores to a seed-dispersing ant. *American Midland Naturalist* 122, 321–328.

Handel, S.N. (1976) Dispersal ecology of *Carex pedunculata* (Cyperaceae), a new North American myrmecochore. *American Journal of Botany* 63, 1071–1079.

Handel, S.N. (1978) The competitive relationship of three woodland sedges and its bearing on the evolution of ant-dispersal of *Carex pedunculata*. *Evolution* 32, 151–163.

Hanzawa, F.M., Beattie, A.J. and Culver, D.C. (1988) Directed dispersal: demographic analysis of an ant–seed mutualism. *American Naturalist* 131, 1–13.

Haskins, C.P. and Haskins, E.F. (1950) Notes on the biology and social behaviour of the archaic Ponerine ants of the genera *Myrmecia* and *Promyrmecia*. *Annals of the Entomological Society of America* 43, 461–491.

Haskins, C.P. and Haskins, E.F. (1974) Notes on necrophoric behaviour in the archaic ant *Myrmecia vindex* (Formicidae: Myrmeciinae). *Psyche* 81, 258–267.

Heithaus, E.R. (1981) Seed predation by rodents on three ant-dispersed plants. *Ecology* 62, 136–145.

Higashi, S., Tsuyuzaki, M., Ohara, M. and Ito, F. (1989) Adaptive advantages of ant-dispersed seeds in the myrmecochorous plant *Trillium tschonoskii* (Liliaceae). *Oikos* 54, 389–394.

Horstman, K. (1972) Untersuchungen über den Nahrungserwerb der Waldameisen (*Formica polyctena* Foerster) im Eichenwald. *Oecologia* 8, 371–390.

Horvitz, C.C. (1981) Analysis of how ant behaviours affect germination in a tropical myrmecochore *Calathea microcephala* (P. and E.) Koernicke (Marantaceae): microsite selection and aril removal by neotropical ants, *Odontomachus*, *Pachyhondyla* and *Solenopsis* (Formicidae). *Oecologia* 51, 47–52.

Howard, D.F. and Tschinkel, W.R. (1976) Aspects of necrophoric behavior in the red imported fire ant, *Solenopsis invicta*. *Behaviour* 56, 157–180.

Howard, D.F. and Tschinkel, W.R. (1981) The flow of food in colonies of the fire ant *Solenopsis invicta*: a multifactorial study. *Physiological Entomology* 6, 297–306.

Hughes, L. and Westoby, M. (1990) Removal rates of seeds adapted for dispersal by ants. *Ecology* 71, 138–148.

Hughes, L. and Westoby, M. (1992) Fate of seeds adapted for dispersal by ants in an Australian sclerophyll vegetation. *Ecology* 73, 1285–1299.

Hughes, L., Westoby, M. and Jurado, E. (1994) Convergence of elaiosomes and insect prey: evidence from ant foraging behaviour and fatty acid composition. *Functional Ecology* 8, 358–365.

Janzen, D.H. (1971) Seed predation by animals. *Annual Review of Ecology and Systematics* 2, 465–492.

Kirchner, W. (1964) Jahreszyklische Unter-suchungen zur Reservestoffspeicherung und Überlebensfähigkeit adulter Waldameisenar-beiterinnen (Gen. Formica). *Zoologisches Jahrbücher, Abteilung für Allgemeine Zoologie und Physiologie der Tiere* 71, 1–72.

Kjellsson, G. (1985) Seed fate in a population of *Carex pilulifera* L. I. Seed dispersal and ant–seed mutualism. *Oecologia* 67, 416–423.

Lagoa, A.M.M.A. and Pereira, M.A. (1987) The role of the caruncle in the germination of seeds of *Ricinus communis*. *Plant Physiology and Biochemistry* 25, 125–128.

Lange, R. (1967) Die Nahrungsverteilung unter den Arbeiterinnen des Waldameisen-staates. *Zeitschrift für Tierpsychologie* 24, 513–545.

Lanza, J., Schmitt, M.A. and Awad, A.B. (1992) Comparative chemistry of elaiosomes of 3 species of *Trillium*. *Journal of Chemical Ecology* 18, 209–221.

Lisci, M., Bianchini, M. and Pacini, E. (1996) Structure and function of the elaiosome in some angiosperm species. *Flora* 191, 131–141.

Mark, S. and Olesen, J.M. (1996) Importance of elaiosome size to removal of ant-dispersed seeds. *Oecologia* 107, 95–101.

Marshall, D.L., Beattie, A.J. and Bollenbacher, W.E. (1979) Evidence for diglycerides as attractants in an ant–seed interaction. *Journal of Chemical Ecology* 5, 335–344.

Mayer, V. and Svoma, E. (1998) Development and function of the elaiosome in *Knautia* (Dipsacaceae). *Botanica Acta* 111, 402–410.

Milewski, A.V. and Bond, W.J. (1982) Conver-gence of myrmecochory in mediterranean Australia and South Africa. In: Buckley, R.C. (ed.) *Ant–Plant Interactions in Australia*. Geobotany 4, Dr W. Junk, The Hague, pp. 89–98.

Morales, M.A. and Heithaus, E.R. (1998) Food from seed-dispersal mutualism shifts sex ratios in colonies of the ant *Aphaenogaster rudis*. *Ecology* 79, 734–739.

Nakanishi, H. (1994) Myrmecochorous adapta-tions of *Corydalis* species (Papaveraceae) in southern Japan. *Ecological Research* 9, 1–8.

Oberrath, R. (2000) Seed dispersal by ants and its consequences for the phenology of plants. PhD thesis, Rheinisch Westfälische Technische Hochschule Aachen, Aachen, Germany.

Oberrath, R. and Böhning-Gaese, K. (2002) Pheno-logical adaptation of ant-dispersed plants to seasonal variation in ant activity. *Ecology* 83, 1412–1420.

O'Dowd, D.J. and Hay, E. (1980) Mutualism between harvester ants and a desert ephemeral seed escape from rodents. *Ecology* 61, 531–540.

Ohkawara, K. and Higashi, S. (1994) Relative importance of ballistic and ant dispersal in two diplochorous *Viola* species (Violaceae). *Oecologia* 100, 135–140.

Oostermeijer, J.G.B. (1989) Myrmecochory in *Polygala vulgaris* L., *Luzula campestris* (L.) DC. and *Viola curtisii* Forster in a Dutch dune area. *Oecologia* 78, 302–311.

Pacini, E. (1990) *Mercurialis annua* L. (Euphor-biaceae) seed interactions with the ant *Messor structor* (Latr.), Hymenoptera: Formicidae. *Acta Botanica Neerlandica* 39, 253–262.

Pulliam, H.R. and Brand, M.R. (1975) The production and utilization of seeds in plains grasslands of southeastern Arizona. *Ecology* 56, 1158–1166.

Rockwood, L.L. and Blois, M.C. (1986) Effects of elaiosome removal on germination of two ant-dispersed plants. *American Journal of Botany* 73, 675 [abstract].

Rodgerson, L. (1998) Mechanical defence in seeds adapted for ant dispersal. *Ecology* 79, 1669–1677.

Sernander, R. (1906) Entwurf einer Mono-graphie der europäischen Myrmekochoren. *Kungliche Svenska Vetenskapsakademiens Handlingar* 41, 1–409.

Skidmore, B.A. and Heithaus, E.R. (1988) Lipid cues for seed-carrying by ants in *Hepatica americana*. *Journal of Chemical Ecology* 14, 2185–2196.

Slingsby, P. and Bond, W.J. (1985) The influence of ants on the dispersal distance and seedling recruitment of *Leucospermum conocarpo-dendron* (L.) Bueck (Proteaceae). *South African Journal of Botany* 51, 30–34.

Sorensen, A.A., Busch, T.M. and Vinson, S.B. (1983) Behaviour of worker subcastes in the fire ant *Solenopsis invicta*, in response to proteinaceous food. *Physiological Entomology* 8, 83–92.

Soukup, V.G. and Holman, R.T. (1987) Fatty acids of seeds of North American pedicillate *Trillium*-species. *Phytochemistry* 26, 1015–1018.

Stryer, L. (1981) *Biochemistry*. Freeman & Company, New York.

Talbot, M. (1946) Daily fluctuations in above ground activity of three species of ants. *Ecology* 27, 65–70.

Thompson, S.N. (1973) A review and comparative characterization of the fatty acid composition of seven insect orders. *Comparative Biochemistry and Physiology* 45B, 467–482.

Turnbull, C.L. and Culver, D.C. (1983) The timing of seed dispersal in *Viola nuttallii*: attraction of dispersers and avoidance of predators. *Oecologia* 59, 360–365.

Vinson, S.B. (1968) The distribution of an oil, carbohydrate and protein food source to members of the imported fire ant colony. *Journal of Economic Entomology* 61, 712–714.

Völkl, W., Woodring, J., Fischer, M., Lorenz, M.W. and Hoffmann, K.H. (1999) Ant–aphid mutualisms: the impact of honeydew production and honeydew sugar composition on ant preferences. *Oecologia* 118, 483–491.

Werker, E. (1997) *Seed Anatomy. Encyclopedia of Plant Anatomy*, Bd. 10, Teil 3. Gebrueder Borntraeger, Berlin, 424 pp.

Whitney, K.D. (2002) Dispersal for distance? *Acacia ligulata* seeds and meat ants *Iridomyrmex viridiaeneus*. *Australian Ecology* 27, 589–595.

Wilson, E.O., Durlach, N.J. and Roth, L.M. (1958) Chemical releasers of necrophoric behaviour in ants. *Psyche* 65, 108–114.

11 Scatterhoarding in Mediterranean Shrublands of the SW Cape, South Africa

Jeremy J. Midgley and Bruce Anderson

Botany Department, University of Cape Town, P. Bag Rondebosch 7701, South Africa

Introduction

Scatterhoarding refers to the behaviour of animals who bury food items in many small depots for later recovery (Vander Wall, 1990) and it is increasingly being shown to be a widespread behaviour among small mammals (Forget and Vander Wall, 2001). It is a newly discovered phenomenon in the fire-prone Mediterranean shrublands ('fynbos') of the SW Cape, South Africa (Midgley *et al.*, 2002). Cape fynbos is restricted to the low nutrient soils of the SW Cape (Fig. 11.1). Physiognomically, fynbos is a shrubland (from 0.5 to 2 m tall) which burns naturally every 5–25 years (Cowling *et al.*, 1997a). The flora of the fynbos constitutes a global regional hotspot of plant species richness (Cowling *et al.*, 1997b).

Our aim in this chapter is to provide further data concerning the phenomenon of scatterhoarding and to discuss the implications of this for ecological and evolutionary processes in the Cape. For example, the occurrence of scatterhoarding has evolutionary implications for understanding the widespread evolution of serotiny (canopy seed storage; occurs in 200+ species) and myrmecochory (dispersal by ants which are attracted by the oily elaiosome on the seeds, which they subsequently bury in their nests; occurs in 1000+ species) (see Mayer *et al.*, Chapter 10, this volume) in Cape shrublands. Serotiny and myrmecochory are both considered to have evolved as anti-rodent predation traits (Bond, 1984; Bond and Slingsby, 1984; Bond and Breytenbach, 1985; Le Maitre and Midgley, 1992). However, if some rodents disperse seeds then it becomes unclear whether serotiny and myrmecochory evolved to avoid rodents. The experimental part of this chapter aimed to discover whether scatterhoarding was more widespread than reported in Midgley *et al.* (2002). Also we wished to discover whether the currently known scatterhoarder, *Acomys subspinosus*, is able to relocate experimentally buried seeds, and other relevant details of the behaviour of this rodent. Finally, we reconsider previous studies on seed predation, seed traits and rodent distribution patterns in the light of the discovery of scatterhoarding. To date only one species of rodent, the fynbos endemic Cape spiny mouse (*Acomys subspinosus*), and one plant species, the nut-fruited *Leucadendron sessile* (Proteaceae), have been confirmed (both in the field and in the laboratory) to be involved in scatterhoarding (Midgley *et al.*, 2002). Since there are many nut-fruited (we consider a nut to be a large seed > 5 mm in length, with a coat > 2 mm) fynbos species, much remains to be discovered about this interaction.

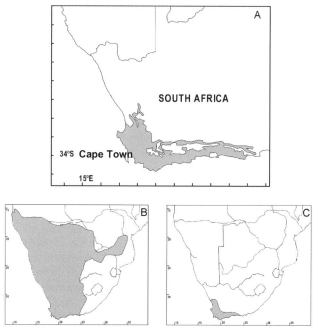

Fig. 11.1. Maps of the distribution of the fynbos biome (A), *G. paeba* (B) and *A. subspinosus* (C).

Methods

How widespread is scatterhoarding?

We detected scatterhoarding by placing 10–20 groups of ten seeds in habitats which had nut-fruited species and observed whether any were eaten *in situ* (i.e. remains of seed-coat could be found) and/or whether any nuts apparently disappeared. Once sites were found where seeds disappeared, we then placed out seeds (10–20 piles of ten seeds per pile) on to which a 10 cm length of colourful thin dacron thread (fly-fishing line) had been attached using quick-setting glue. Buried seeds were located by finding the threads. We used the seeds of the following species: *Leucadendron sessile*, *L. pubescens* and *L. concavum* (Proteaceae), *Ceratocaryum argenteum* and *Willdenowia incurvata* (Restionaceae) as well as occasionally using commercial sunflower seeds. Finally, laboratory observations were made on suspected scatterhoarders to confirm whether they buried seeds. Rodents were placed in glass tanks with a 10-cm deep layer of sand.

They were observed in the evening to determine whether they buried or retrieved buried seeds.

We sampled 11 sites spread across several hundred kilometres of the western and southern Cape (three sites at Potberg (34°26'S, 20°44'E with 3 nights of observation repeated over 2 years), two sites at Fernkloof (34°23'S, 19°16'E with 3 nights of observation), one site at Bains Kloof (33°35'S, 19°05'E with 5 nights of observation) and one site at Jonaskop (33°58'S, 19°30'E with 3 nights of observation)). At all of these sites, despite the presence of plants with nut-fruits, we obtained negative results (no seeds were removed or buried). In contrast, we noted burial at a site near Clanwilliam (32°09'S, 18°31'E), two sites on Pakhuis Pass (32°09'S, 18°57'E; a site in mid-south slope and a site at top of pass) and a site near Salmonsdam (34°27'S, 19°34'E). We also trapped rodents at Pakhuis and Clanwilliam. We used seeds as bait and used transects of 30–50 live-traps each 10 m apart. At two sites (Pakhuis Pass Top and Salmonsdam) we followed seeds over a 3-week period (see Table 11.1).

Can *Acomys* relocate experimentally buried seeds and, if so, from what burial depth?

Single seeds were spaced 5 cm apart in lines of 5×2 and pushed down 1 cm or 5 cm into the ground using a small stick and then covered over with soil. We repeated this with three plant species (commercial sunflowers, *L. sessile* and *W. incurvata*) and observed which were relocated out of a sample size of 20 seeds per species. This experiment was located in the field over 1 night at Sir Lowrys Pass site where previously *Acomys* was confirmed as the disperser. These experiments were performed at night-time as *Acomys* is the only granivorous nocturnal rodent at the site (Midgley *et al.*, 2002).

Results

At Clanwilliam some seeds were scatter-hoarded (Table 11.2) and at this site we trapped 11 *G. paeba* and a single *M. minutoides* and *R. pumillio* (using 30 traps over a single trapping night). *G. paeba* was the only nocturnal rodent at the Clanwilliam site (seeds were placed out in late afternoon) and was thus assumed to bury the seeds of *Willdenowia incurvata* (Table 11.2). Dispersal distances (mean of 2.4 to 4.5 m) are similar to *Acomys* although *G. paeba* appears to bury multiple seeds in a hole more frequently (1.1–2.5; Table 11.2) than *Acomys* (Midgley *et al.*, 2002). The scatterhoarding behaviour of *G. paeba* was later confirmed in the laboratory, where it was observed to be an avid digger and was observed carrying and burying seeds.

At the mid-slope site on Pakhuis Pass we found some burial (Table 11.2). Here we trapped six *Aethomys namaquensis* individuals, and a single *A. subspinosus*, *G. paeba* and *O. irroratus* (using 30 traps and 1 night of trapping). Subsequent laboratory observations suggested that *A. namaquensis* does not bury seeds. Unlike the gerbil, it was not observed to carry seeds or to dig much. Presumably *A. subspinosus* or *G. paeba* buried the seeds at this site (*O. irroratus* is a herbivore). At the top site on Pakhuis we found signs of burial (Table 11.1). Here we trapped eight *A. subspinosus* individuals (using 50 traps over 3 trapping nights). At Salmonsdam we found signs of burial of *C. argenteum* (Table 11.1). At both of the latter two sites, where the fate of seeds was followed over 3 weeks, there appeared to be a gradual increase in seed removal or seed predation with time (Table 11.1). It was clear that rodents did not bury or consume a whole pile once they had encountered the

Table 11.1. Mean numbers of removed (subsequently eaten or buried) seeds per pile (20 piles of ten each); *Leucadendron concavum* at Pakhuis Pass Top and *C. argenteum* at Salmonsdam over a 3-week period. Number refers to seeds buried within 2 cm of others.

Site	Date	Removed	Buried	Eaten	Number
Pakhuis Pass Top	30/11	1.5	1.3	5.3	1.1 (1–3)
	21/12	1.5	1.4	8.3	1.1 (1–3)
Salmonsdam	22/12	0.2	0.2	0	1.0 (1)
	14/01	2.0	2.0	0	1.0 (1)

Table 11.2. Numbers of seeds (out of ten groups of ten labelled seeds) removed, buried or eaten *in situ*. Mean distance moved (m and range) and mean numbers buried together (within 2 cm of another seed and range).

Site	Date	Removed	Buried	Eaten	Distance moved	Number buried
Clanwilliam	06/02	6.4	5.3	0.2	4.5 (0.8–14.0)	2.2 (1–4)
	07/02	8.6	5.7	0.1	4.2 (0.5–12.0)	1.8 (1–3)
Pakhuis Pass Mid	07/02	10	3.8	0.3	2.4 (0.2–6.0)	1.1 (1–2)

pile. This suggests the burying behaviour is only partially in response to encountering more food items than can be immediately consumed.

Acomys was found to be able to locate all seeds experimentally buried 1 cm deep (all 20 seeds of each of three species were located) and attempted to retrieve about 65% of the seeds buried at 5 cm. It abandoned all attempts at the deeper buried seeds before actually reaching the seeds. These results suggest that seeds are relocated using a strong sense of smell (at least to 5 cm soil depth). Acomys was attracted to all species buried 1 cm (i.e. sunflower seeds as well as the indigenous W. involucrata and L. sessile), suggesting a general, non-specific, choice of food items.

Discussion

Distribution of scatterhoarding

We have extended the scatterhoarding phenomenon geographically (west coast lowland fynbos and south coast lowland fynbos) and to a new plant (Restionaceae) and animal (Cricetidae) family. Both G. paeba and W. incurvata have a wide distribution in the more arid and sandy areas of the Cape (Linder, 1985; Skinner and Smithers, 1990). At this stage it appears that there are probably 60–100 plant species with nuts and without elaiosomes in these two families alone, which are probably primarily dispersed by rodents. It seems, therefore, that scatterhoarding may be fairly widespread in fynbos, although highly localized (as evidenced by the lack of burial at many sites with nut-fruited species). The Restionaceae and Proteaceae should be added to Vander Wall's (2001) list of nut-fruited families. At many sites we found all nut seeds to be untouched, being neither dispersed nor consumed in situ. Presumably, the absence of scatterhoarding rodent species will mean that unburied seeds will be incinerated in fires. In other words, and in the absence of interfire establishment, the mutualism is crucial for the nut-seeded

plants. One reason that we did not always detect seed burial is that seed storing behaviour may be seasonal. We were not always able to experimentally place seeds out during seed-fall of local nut-fruited species. For example, it remains to be seen whether seeds placed out in winter are also buried, given that summer is the period of seed release.

A re-evaluation of granivory of large nuts in fynbos

In this section we focus our discussion on large seeds (> 5 mm) and on post-dispersal predation. This is because most of the information concerning seed dispersal and predation in the Cape is from the Proteaceae. We investigate whether experiments, results and conclusions are reliable in the light of the discovery of scatterhoarding. Johnson (1992) provides a review of the limited information available on other aspects of seed-predation such as by small granivores (ants) on small seeds and of pre-dispersal predation (such as by lepidopteran larvae of developing seeds). Thus the main granivores we are concerned with are small mammals, mainly rodents but also some shrews, and the main plant family for which information is known: the Proteaceae.

Granivory by birds in fynbos, for example, in post-fire stands where seed resources are high due to massive seed release by serotinous species, has been poorly studied. In part this is because the itinerant opportunistic nature of the phenomenon makes it difficult to study. It may, however, be important ecologically and evolutionarily. For example, Midgley (2000) noted a common colour polymorphism (black versus mottled) in serotinous seeds of the genus Leucadendron (Proteaceae). As is common for other serotinous species (such as pines), most have black seeds and presumably black seeds are cryptic in the blackened post-fire environment. The mottled seeds match the pale soils of some sub-habitats, such as limestone areas. Scatterhoarded seeds are brown and match litter in the pre-fire environment. The point here is that a visually acute selective agent needs to be invoked to explain the tremendous variation in seed colour in Cape Proteaceae. Since most small mammal granivores are nocturnal and do not have colour vision, this implies a role for birds such as canaries and doves. However, bird granivory is not well known in fynbos (see discussion in Fraser, 1990).

The occurrence of scatterhoarding has evolutionary implications for re-evaluating the experimental evidence for granivory and for understanding the reasons for widespread evolution of serotiny and myrmecochory in fynbos. It has been hypothesized that the evolution of serotiny in fynbos and the impact of fires in different seasons are due to high levels of rodent seed-predation (Bond, 1984). Germination of serotinous seeds occurs in the first winter after fire. Bond (1984) noted that higher levels of seedling regeneration occur after autumn fires rather than summer fires. He argued this is because a shorter period during which seed predation can occur favours autumn over summer fires. Midgley and Clayton (1990) questioned this because rodent numbers crash after fires and this is the main period of seed release. Also, seed mortality could explain variation in establishment after fires in different seasons (Midgley *et al.*, 1989).

In the most recent analysis, van Hensbergen *et al.* (1992) performed a detailed experiment which showed that high levels of seed predation by small mammals essentially prevent any establishment in mature fynbos. Mean levels of seedling establishment of the serotinous species *Protea neriifolia* inside exclosures were about 70%, whereas establishment outside exclosures ranged from 33% to nil, with a mean of 16%. However, after fire, the rodent impact was only a 20% decrease in establishment outside exclosures compared to inside exclosures. This was because small mammal numbers in their study site were shown to have declined strongly after the fire. This suggests that post-fire seed predation is not as intense as suggested by Bond (1984). Furthermore, it is not yet known whether scatterhoarders (*Acomys* occurred in the study site of van Hensbergen *et al.*, 1992) removed and buried some experimental

Protea seeds, i.e. some seeds removed from outside the exclosures may have been buried. At the same general study area, Fraser (1990) noted low levels of granivory (using commercial seeds) and almost no difference in percentage of visits to experimental seed depots placed out in burned (27%) versus unburned tall fynbos (21%). Given that some sites indicate high levels of granivory in mature fynbos and others do not, our assessment (also based on our own observations) is that both granivory and seed burial are variable in time (e.g. season) and space.

It has also been argued that myrmecochory evolved as a consequence of intense granivory by small mammals. For example, Bond and Breytenbach (1985) noted no differences in rates of removal of myrmecochorous seeds exposed to ants only, or to small mammals only. Both levels of removal were nearly 100%. Because they considered all seed removal by small mammals to be equivalent to predation, and all seeds taken by ants to be protected against granivory, they argued for the importance of seed burial by ants in preventing granivory. Indigenous fynbos seed-dispersing ants can be displaced by an exotic ant that is too small to carry large nuts (Bond and Slingsby, 1984). This ant merely removes the elaiosome. Thus, myrmecochorous plant species may also be eliminated by ant invasion because seed predation by rodents of unburied seeds is considered to be extreme (Bond and Slingsby, 1984). More recently, and following the same reasoning, Christian (2001) argued that invasion by exotic ants may lead to community-wide loss of large-seeded plants due to excessive rodent predation. The problem with these studies concerning myrmecochory is that further data are needed to demonstrate that: (i) all seeds that are removed from experimental depots are removed only by ants (and not scatterhoarding rodents); and (ii) all seeds buried by ants are safe (e.g. escape rodent detection). Christian (2001) explicitly assumed that rodents only eat seeds *in situ*. In light of the discovery of scatterhoarding it is clear that this assumption is erroneous in some situations. *Acomys* clearly can locate seeds experimentally buried up to 5 cm deep, although in the field it does not persist with digging beyond about 2–3 cm. Seeds buried by ants (typically < 2 cm deep; Brits, 1987) are well within the scope of retrieval of *Acomys*. Thus the interaction between myrmecochory and scatterhoarding needs to be further investigated. For example, research may show that myrmecochorous plants are rare where scatterhoarders occur.

There have been few studies in the Cape fynbos of seed predation of species with large seeds which are neither serotinous nor myrmecochorous (and thus they are presumably scatterhoarded). Botha (1989) noted 30% establishment of seeds of *Widdringtonia cederbergensis* inside double-mesh exclosures as opposed to only 8% outside. He attributed this difference to the dominant granivores he trapped in the area: *Acomys subspinosus* and *Aethomys namaquensis*. Again, the discovery that *Acomys* is a disperser suggests that some of the seeds removed from outside exclosures may eventually establish as seedlings.

The scatterhoarding rodents: distribution and succession

Most studies of the distribution of small mammals have used vegetation structure (e.g. vegetation height and density) and environmental variables (e.g. rockiness and altitude) as correlates. However, this approach has failed for some species, notably the scatterhoarder *Acomys* (Bond *et al.*, 1980; van Hensbergen *et al.*, 1992; Fleming and Nicholson, 2002). One reason for this could be that distribution patterns have not been considered relative to food resources. Thus it is possible that the distribution of *Acomys* is dependent on the occurrence of nut-fruited species. Also, since *Acomys* is an important pollinator (Fleming and Nicholson, 2002), it may also be dependent on nectar resources during winter as well (nut production generally occurs in summer).

We have few data on the fine-scale distribution of *G. paeba* in the Cape, other than their affinity for sandy arid sites (e.g.

Skinner and Smithers, 1990). Interestingly, this species is also an important pollinator (Johnson *et al.*, 2001) and therefore both it and *Acomys* may be considered important mutualistic species.

Evolution of scatterhoarded plants

The Restionaceae is one of the Cape families with large nuts and for which good phylogenies exist. On the basis of phylogenetic evidence and seedling growth rate data, Caddick and Linder (2002) speculate that large nut-fruited Restionaceae species evolved 5–7 million years ago in response to the occurrence of dry summers. They argue that large nuts facilitate large seedlings, which have lower summer mortality rates, and note that nut-fruited Restionaceae predominate in arid, rather than mesic, regions. Nut-fruited *Leucadendron* species also appear to predominate in more arid areas (Le Maitre and Midgley, 1992). It thus seems that scatterhoarding is more important in arid than in mesic areas of the Cape. In *Ceratocaryum* (Restionaceae) the scatterhoarded species (i.e. the large nut-seeded species without elaiosome in Linder, 2001) are restricted to the more arid sandy lowland habitats whereas species with elaiosomes occur in more mesic and rocky, mountain habitats. In part this may be because digging rodents are more common in arid, sandy habitats and nut-fruited species depend on the occurrence of the rodent-disperser. Evidence for this is that *Acomys subspinosus* is more common in arid areas than in mesic areas. It is rare in moist coastal mountains (Midgley and Clayton, 1990) compared with drier inland mountains (Breytenbach, 1987; Botha, 1989). *G. paeba* is also an arid-adapted species (Skinner and Smithers, 1990). Clearly further work is needed on the distribution patterns of scatterhoarders and nut-fruited plant species to determine relative degrees of dependence.

The evolution of scatterhoarding may also have led to the evolution of large nuts, rather than this being only a response to aridification 5–7 million years ago. Small nuts would simply have been uneconomical food-sources to rodents (i.e. ignored) or may have been eaten *in situ*. The inability of ants to physically move large seeds provides an upper limit on the size of myrmecochorous seeds and thus scatterhoarding may be associated with large seeds. Presumably, serotiny is associated with small winged seeds because dispersal is an important selective agent. Another benefit of scatter-hoarding is reduced clumping. Serotinous species are often extremely clumped (Midgley and von Maltitz, 1990) and myr-mecochores can also be extremely clumped (e.g. Brits, 1987). The present data (Table 11.2) and those from Midgley *et al.* (2002) suggest that scatterhoarders rarely bury seeds close together.

The possible costs of scatterhoarding from the plants' perspective are first that emerging from depth implies an energetic cost. This, together with the impact of delayed germination with increasing depth of burial, may influence relative seedling size. Most scatterhoarded species have thick seed coats, which will also represent a cost. Thus the positive influence of seed size on seedling size (Caddick and Linder, 2002) cannot simply be determined by nursery experiments where germination is induced and germinants are grown on the soil sur-face. In *Ceratocaryum* loss of the elaiosome and large seededness are derived conditions (Linder, 2001). Since the *Ceratocaryum* clade is derived in the family Restionaceae, this suggests that scatterhoarding is a recently evolved and convergent trait in this family.

In conclusion, we see the following as outstanding questions:

1.   What is the importance of nut seeds to scatterhoarders? For example, do breeding episodes and migratory events coincide with periods of mass seed release? Are distribution patterns of scatterhoarders determined by nut-seeded plants (and vice versa)?
2.   Does the loss of either partner in this mutualism have an impact on the other? For example, does fragmentation of fynbos affect scatterhoarding mammals and thus nut-seeded plants?
3.   What fire mosaic is needed to maintain the presence of scatterhoarding mammals?

Acomys declines strongly after fire (Midgley and Clayton, 1990; van Hens-bergen *et al.*, 1992), is rare in firebreaks (van Hensbergen *et al.*, 1992) and therefore may be disadvantaged by a frequent fire regime.

The limited information to date suggests that scatterhoarding is widespread geographically and also occurs in several plant families. The discovery of scatter-hoarding has implications for seed preda-tion experiments and understanding the evolution of serotiny and myrmecochory and it has conservation implications. To conserve not only species, but also inter-actions, we need further information on the biology of scatterhoarding rodents such as *Acomys subspinosus*. Serotiny and myrmecochory are widespread plant traits globally in Mediterranean systems and a more comprehensive view of their evolution is now needed. This is especially so if further research shows scatterhoarding occurs in other Mediterranean systems, notably the analogous kwongan vegetation of Western Australia.

Acknowledgements

We thank Claude Midgley, Jasper Slingsby, Ben Wrigley and Gareth Hempson for assistance in the field and J. Lambert and P.-M. Forget for comments on this chapter.

References

Bond, W.J. (1984) Fire survival of Cape Proteaceae – influence of fire season and seed predators. *Vegetatio* 56, 65–74.

Bond, W.J. and Breytenbach, G.J. (1985) Ants, rodents and seed predation in Proteaceae. *South African Journal of Zoology* 20, 150–154.

Bond, W.J. and Slingsby, P. (1984) Collapse of an ant–plant mutualism: the Argentine ant (*Iridomyrmex humilis*) and myrmecochorous Proteaceae. *Ecology* 65, 1031–1037.

Bond, W.J., Ferguson, M. and Forsyth, G. (1980) Small mammals and habitat structure along altitudinal gradients in the southern Cape mountains. *South African Journal of Zoology* 15, 34–43.

Botha, S.A. (1989) Effects of sowing season and granivory on the establishment of the Clanwilliam cedar. *South African Journal of Wildlife Research* 19, 112–117.

Breytenbach, G.J. (1987) Small mammal dynamics in relation to fire. In: Cowling, R.M., Le Maitre, D.C., McKenzie, B., Prys-Jones, R.P. and van Wilgen, B.W. (eds) *Disturbance and the Dynamics of Fynbos Biome Communities.* South African National Scientific Programme Report 135, CSIR, Pretoria, pp. 56–58.

Brits, G.J. (1987) Germination depth versus temperature requirements in naturally dispersed seeds of *Leucospermum cordifolium* and *L. cuneiforme* (Proteaceae). *South African Journal of Botany* 53, 119–124.

Caddick, L.R. and Linder, H.P. (2002) Evolutionary strategies for reproduction and dispersal in African Restionaceae. *Australian Journal of Botany* 50, 339–355.

Christian, C.E. (2001) Consequences of a biological invasion reveal the importance of mutualism for plant communities. *Nature* 413, 635–636.

Cowling, R.M., Richardson, D.M. and Mustart, P.J. (1997a) Fynbos. In: Cowling, R.M., Richardson, D.M. and Pierce, S.M. (eds) *Vegetation of Southern Africa.* Cambridge University Press, Cambridge, pp. 99–129.

Cowling, R.M., Richardson, D.M., Schulze, R.E., Hoffman, M.T., Midgley, J.J. and Hilton-Taylor, C. (1997b) Species diversity at the regional scale. In: Cowling, R.M., Richardson, D.M. and Pierce, S.M. (eds) *Vegetation of Southern Africa.* Cambridge University Press, Cambridge, pp. 447–472.

Fleming, P.A. and Nicholson S.W. (2002) Opportunistic breeding in the Cape spiny mouse (*Acomys subspinosus*), *African Zoology* 37, 101–105.

Forget, P.-M. and Vander Wall, S.B. (2001) Scatter-hoarding rodents and marsupials: convergent evolution on diverging continents. *TREE* 16, 65–67.

Fraser, M.W. (1990) Small mammals, birds and ants as seed predators in post-fire mountain fynbos. *South African Journal of Wildlife Research* 20, 52–55.

Johnson, S.D. (1992) Plant–animal relationships. In: Cowling, R. (ed.) *The Ecology of Fynbos. Nutrients, Fire and Diversity.* Oxford University Press, Oxford, pp. 175–205.

Johnson, S.D., Pauw, A. and Midgley, J.J. (2001) Rodent pollination in the African lily *Massonia depressa* (Hyacinthaceae). *American Journal of Botany* 88, 1768–1773.

Le Maitre, D. and Midgley J.J. (1992) Plant reproductive ecology. In: Cowling, R. (ed.) *The Ecology of Fynbos. Nutrients, Fire and Diversity.* Oxford University Press, Oxford, pp. 135–174.

Linder, H.P. (1985) A conspectus of the African species of Restionaceae. *Bothalia* 15, 387–503.

Linder, H.P.L. (2001) Two new species of *Ceratocaryum* (Restionaceae). *Kew Bulletin* 56, 465–477.

Midgley, J.J. (2000) Seed colour in serotinous Cape proteas. *Protea Atlas Newsletter* 48, 8.

Midgley, J.J. and Clayton, P. (1990) Short-term effects of an autumn fire on small mammal populations in Southern Cape Mountain Fynbos. *South African Forestry Journal* 153, 27–30.

Midgley, J.J. and von Maltitz, G. (1990) Comparison of seedling distribution patterns of wind and ant-dispersed Proteaceae. *South African Journal of Ecology* 1, 60–62.

Midgley, J.J., Hoekstra, T. and Bartholomew, R. (1989) Implications of a field germination trial of serotinous fynbos Proteaceae for season of burn. *Vegetatio* 79, 185–192.

Midgley, J.J., Anderson, B.C., Anderson, B., Bok, A. and Fleming, T. (2002) Scatter-hoarding of Cape Proteaceae nutlets. *Evolutionary Ecology Research* 4, 623–626.

Skinner, J.D. and Smithers, R.H.N. (1990) *The Mammals of the Southern African Subregion.* University of Pretoria, Pretoria, 736 pp.

van Hensbergen, H.J., Botha, S.A., Forsyth, G.G. and Le Maitre, D.C. (1992) Do small mammals govern vegetation recovery after fire in fynbos? In: van Wilgen, B.W., Richardson, D.M., Kruger, F.J. and van Hensbergen, H.J. (eds) *Fire in South African Mountain Fynbos.* Springer-Verlag, Berlin, pp. 182–202.

Vander Wall, S.B. (1990) *Food Hoarding in Animals.* University of Chicago Press, Chicago, Illinois, 445 pp.

Vander Wall, S.B. (2001) The evolutionary ecology of nut dispersal. *Botanical Review* 67, 74–117.

12 Selection, Predation and Dispersal of Seeds by Tree Squirrels in Temperate and Boreal Forests: are Tree Squirrels Keystone Granivores?

M. Steele,[1] L.A. Wauters[2] and K.W. Larsen[3]

[1]Department of Biology, Wilkes University, Wilkes-Barre, PA 18766, USA;
[2]Department of Structural and Functional Biology, University of Insubria, Varese,
I-21100 Varese, Italy; [3]Department of Natural Resource Sciences, University College
of the Cariboo, Kamloops, British Columbia V2C 5N3, Canada

Introduction

Since the classic work by C.C. Smith (1968, 1970) on the co-evolution of conifers and squirrels, numerous studies have addressed one or more aspects of the ecological and evolutionary interactions between seed trees and squirrels. This review summarizes these interactions for seed trees and Holarctic tree squirrels. We review squirrel–seed interactions prior to dispersal and hoarding of seeds as well as those in which squirrels exert a significant positive effect on dispersal and establishment of seeds. Across all Holarctic systems, we recognize three primary selective pressures that tree squirrels exert on tree seeds: two as seed predator and one as seed disperser. And we consider the close evolutionary relationship between several species of tree squirrels and the tree species on which they feed, including the influence of tree squirrels on seed and tree characteristics and the effects of seeds on the demography, behaviour and social system of the squirrels. Finally, we suggest that in some systems tree squirrels might be considered keystone consumers as

a result of their disproportionate influence on seed fates.

The squirrel family (Sciuridae) is a diverse group of rodents, ranging in size from chipmunks (e.g. *Tamias, Eutamias*) to large marmots (*Marmota*) and prairie dogs (*Cynomys*). Herein we limit our discussion to the so-called 'tree squirrels' found in North America and Eurasia. These animals belong to two genera, *Sciurus* and *Tamiasciurus*, that contain eight and three species, respectively (Gurnell, 1987; Hoffman *et al.*, 1993), although arguments recently have been made for consolidating the latter into one species (Arbogast *et al.*, 2001). Flying squirrels (e.g. *Glaucomys, Pteromys*) constitute another group of Holarctic sciurids well-adapted for an arboreal existence, but they belong to a separate subfamily and are not included here. For general scientific treatises on tree squirrels, consult Gurnell (1987) and Steele and Koprowski (2001).

As arboreal granivorous rodents, tree squirrels occupy a unique niche in the Holarctic. They represent one of the few mammals that harvest, consume, and/or disperse tree seeds before seed fall. Moreover, in many systems some seed types are not consumed or dispersed by any other species of vertebrates (e.g. *Carya*, *Juglans*). Consequently, tree squirrels can significantly influence seed fates and are likely to exert significant ecological and evolutionary effects on seeds, fruit morphology, tree demography, and forest structure. The objectives of this review are to:

1. Summarize these effects on seed fates across the Holarctic.
2. Characterize the primary role of tree squirrels in different biomes and forest types.
3. Review the close ecological and evolutionary relationship between seeds and tree squirrels in select systems.

Tree Squirrels as Seed Predators

Coniferous forests

In many temperate or boreal forests dominated by conifers, tree squirrels have a significant impact on seed survival. Many tree squirrels, including several species of *Sciurus* (*S. aberti*, *S. niger* and *S. vulgaris*) and *Tamiasciurus* are residents of predominantly coniferous forests and are significant predators of conifer seeds.

In coniferous forests, tree squirrels influence seed fates in three important ways. Squirrels: (i) cause the indirect loss of cones and their seeds as a result of other feeding activities such as bark stripping or twig clipping (Allred *et al.*, 1994; Snyder, 1998); (ii) consume seeds in the tree often before they are mature (Elliot, 1974, 1988; Moller, 1986; Wauters and Dhondt, 1987; Steele, 1988; Steele and Weigl, 1992; Steele and Koprowski, 2001); and (iii) store seeds and cones in central larders (i.e. territorial *Tamiasciurus*) where successful recruitment of seeds is unlikely (C.C. Smith, 1968, 1970; Gurnell, 1984, 1987). Despite reports that some tree squirrels and similar species (e.g. chipmunks) scatterhoard and disperse conifer seeds, which may increase seed establishment of conifer seeds (Hall, 1981; Kato, 1985; Miyaki, 1987; Vander Wall,

1992; Wauters and Casale, 1996), most tree squirrels function instead as significant seed predators in coniferous forests. Here we briefly review the primary causes of conifer seed predation.

Indirect effects of feeding activity

Tree squirrels can cause notable reductions in tree reproduction as the indirect result of consumption of flower buds and flowers of pine and spruce species (Lampio, 1967; Grönwall, 1982; Wauters and Dhondt, 1987; Wauters *et al.*, 1992, 2001; Allred *et al.*, 1994). Moreover, it has been shown that Abert's squirrel, which relies on the inner cambium of ponderosa pine (*Pinus ponderosa*) as a primary food source during much of the year, can significantly affect tree fitness including reproductive output. Abert's squirrels show a strong preference for the cambium of individual trees (i.e. those with higher sodium and specific xylem and carbohydrate properties) and selectively feed in the same trees year after year (see review by Snyder, 1998). The result is a change in tree architecture, reduced growth and significantly lower cone and seed production, and, ultimately, strong selection on many of these characters which are under tight genetic control (Snyder, 1998). Although these few studies strongly suggest that indirect feeding activities of tree squirrels can exert a significant impact on tree reproduction, the impact in most systems (e.g. deciduous forests) and in various tree species has not been studied.

Consumption of immature cones

In many conifer forests, several species of *Sciurus* consume the immature female cones of conifer species, immediately in the trees at the time of harvest, often weeks or months before cone maturation. As with larderhoarding, these feeding activities may result in significant losses to cone crops and may, in fact, drive the evolution of conifer cones in some forests (e.g. Elliot, 1974;

Benkman, 1995; Parchman and Benkman, 2002). Although consumption of conifer cones is common in nearly all species of *Sciurus* where they reside in conifer stands (Gurnell, 1987; Wauters and Dhondt, 1987; Wauters *et al.*, 1992; Steele and Koprowski, 2001), the behaviour of cone consumption and its impact on tree reproduction has been studied most extensively in three systems. These systems include Abert's squirrels of ponderosa pine forests of southwestern USA, south-eastern fox squirrels (*S. niger niger*) of the longleaf pine (*P. palustris*) forests of south-eastern USA and Eurasian red squirrels (*S. vulgaris*) in various conifer forests of Europe and Asia.

Consumption of green immature conifer cones usually occurs in trees from which they are harvested. Squirrels first cut the cones, often carry them to a position near the trunk and then systematically remove one bract at a time from the basal end of the cone. Seeds are consumed immediately as encountered at the base of the bracts (Steele and Koprowski, 2001). The remaining cone rachus (cone core) is then often dropped to the ground providing an obvious record of cone consumption that several authors have used to follow patterns of cone selection and use (Moller, 1986; Steele, 1988). The manner

and efficiency with which these immature cones are harvested results in heavy pre-dispersal predation. The proportion of cores and cones collected on quadrats under sampling trees over an entire year (Wauters and Lens, 1995) can be used to estimate total annual conifer cone (seed) production and the proportion of the cone crop consumed by squirrels (Table 12.1).

Evidence from several of these studies indicates that tree squirrels are extremely efficient when harvesting cones, either for immediate consumption (Moller, 1986; Steele, 1988; Steele and Weigl, 1992) or even when hoarding for subsequent use (C.C. Smith, 1965, 1968, 1970). Evidence from mixed stands suggests that, when harvesting cones, some tree squirrels first selectively remove cones from the tree species with the most profitable cones (i.e. those providing the highest energy return). They then shift to the tree species with cones of the next highest profitability. Selection between individual trees of one species or individual cones occurs only after the cones of all but the least profitable species are depleted (C.C. Smith, 1965, 1968, 1970; Moller, 1986). Detailed patterns of cone and tree selectivity are evident from studies of *S. vulgaris* in monospecific stands of Scots pine

Table 12.1. Seed production and predation in a coniferous and a mixed deciduous woodland in Belgium (data modified from Wauters and Lens, 1995). Annual total tree seed production (in MJ/ha and in the number of cones, acorns and beechnuts/ha). Cone numbers indicate pine cone consumption by red squirrels (between June of the year seeds are produced and end of May the following year). Acorn and beechnut consumption by birds and mammals (from July to December the year seeds are produced), including those previously cached by squirrels. Red squirrel autumn densities were between 1.0 and 1.8 squirrels/ha in the coniferous woodland and between 0.8 and 1.6 squirrels/ha in the deciduous woodland (Wauters and Lens, 1995).

Year	Total seed crop (MJ/ha) Coniferous	Deciduous	Cones/ha produced (% consumed) Coniferous	Deciduous	Acorns/ha produced (% consumed) Deciduous	Beechnuts/ha produced (% consumed) Deciduous
1984	966	3,121	226,000 (61%)	294,000 (60%)	No data	No data
1985	668	7,374	156,000 (51%)	533,000 (91%)	184,400 (13%)	10,000 (60%)
1986	496	8,666	116,000 (81%)	231,000 (81%)	227,800 (6%)	486,000 (45%)
1987	1,722	14,662	403,000 (26%)	562,000 (48%)	336,700 (15%)	424,000 (28%)
1988	1,159	6,256	271,000 (26%)	321,000 (67%)	154,400 (37%)	112,000 (14%)
1989	1,483	16,089	347,000 (40%)	497,000 (73%)	307,800 (19%)	636,000 (30%)
1990	1,008	20,000	236,000 (40%)	184,000 (71%)	116,700 (28%)	12,781,000 (3%)
1991	1,457	2,386	341,000 (41%)	187,000 (50%)	55,500 (60%)	8,000 (50%)

(*P. sylvestris*) in Scotland (Moller, 1986) and from studies of south-eastern fox squirrels in forests of longleaf pine in the southern USA. In both of these systems tree squirrels frequently feed in those trees with cones of greatest profitability.

South-eastern fox squirrels seem to have fine-tuned the efficiency of seed harvest with a rather involved method of sampling behaviour that allows them to maximize their rate of cone harvest (Steele and Weigl, 1992). They begin in early July – a full 70–100 days before cones dehisce and release their seeds – by sampling cones to first determine when cones are ripe enough for harvest. Regular harvesting of cones begins when the caloric value of the seeds allows the squirrels to obtain a positive energy balance. Squirrels continue to sample one or a few cones from > 40% of the trees they visit every few weeks, while feeding intensively in a few preferred trees. Preferred trees often are those with the greatest number of cones or those with the most profitable cones (largest cones). However, Steele and Weigl (1992) also found at one site, where cone numbers per tree were negatively correlated with cone size, that squirrels selected trees of intermediate cone size and number in which the feeding rate per tree was highest (Steele and Weigl, 1992). In other words, fox squirrels based tree preferences on the overall profitability of the tree, rather than other simple measures of cone quality (cone size or number). By the end of the season, before cones ever open, fox squirrels can have a devastating impact on seed crops (> 85% in some stands). Even in moderate or high mast years this feeding pattern results in those individual trees investing the most in reproduction (female cones) sustaining the greatest loss (90–100%). Such patterns of cone and seed predation suggest that tree squirrels may exert strong selection on cone morphology and patterns of cone production.

The amount of seed predation due to consumption of immature cones, at the population level, was estimated by Wauters and Dhondt (1987) in temperate mixed conifer woodlands in Belgium, dominated by Scots pine and Corsican pine (*P. nigra*) with some oak and beech. Eurasian red squirrels (*S. vulgaris*) harvested and consumed 80% of pine cones when seed crops were poor, 51–61% in years of medium seed crops, and about 26–41% in years with heavy seed crops. In woodlands dominated by oak and beech with only 10% of Scots and Corsican pine, pine seed predation was even higher, ranging from 48% (highest seed crop) to 91% of seed crops (Table 12.1). These data, together with the extremely high proportion of cached pine cones that were retrieved and consumed (99–100%; Wauters and Casale, 1996), confirm that red squirrels prefer pine seeds during most of the year (e.g. Wauters and Dhondt, 1987; Wauters *et al.*, 1992). In boreal and subalpine forests, tree squirrels, like crossbills (*Loxia* spp.), feed heavily on Norway spruce (*Picea abies*; Lampio, 1967; Pulliainen, 1973; Degn, 1974; Grönwall, 1982; Gurnell, 1987; Andrén and Lemnell, 1992), but studies quantifying rates of seed predation in these habitats are lacking.

The effects of larderhoarding

Larderhoarding, or the caching of multiple food items in one or a few locations, is performed primarily by species of the genus *Tamiasciurus* (Smith and Reichman, 1984) throughout montane and boreal forests of North America (Steele, 1998, 1999). Throughout this range, where conifer species are the dominant forest cover, these squirrels cut female cones that they store in a central 'midden', where discarded cone debris accumulates from year to year. Cone cutting often begins in late summer and continues through the autumn, in some cases allowing squirrels to larderhoard the cones needed to survive one or two harsh winters (M.C. Smith, 1968). Located in the centre of the squirrel's territory, middens are aggressively and efficiently defended against conspecifics.

Several attempts have been made to determine the number of cones larderhoarded by individual *Tamiasciurus* within their middens. M.C. Smith (1968) excavated middens in a spruce forest in Alaska and found an average of 3057 cones per midden ($n = 7$). Gurnell (1984) found an average of 1458 cones in four middens in a Colorado lodgepole pine stand, and Zimmerling (1990) reported an average of 2708 cones from eight middens in a jack pine forest in central Alberta, although even higher estimates have been reported by M.C. Smith (1968). These numbers probably represent conservative estimates of the cones hoarded by individuals, as, even in larderhoarding populations, smaller, auxiliary larders may exist some distance from the central midden (Hurly and Lourie, 1997). Conversely, M.C. Smith (1968) argued that, in his northern, cold location, surplus cones from past years may accumulate in the midden.

Gurnell (1984) and Becker et al. (1998) discuss how larderhoarding strategies and cone predation may differ in serotinous pine forests, as compared with non-serotinous conifers (e.g. northern spruce forests). In the former, competition for the seed of the hard serotinous cones may be relatively less critical, and hoarding the cones on the ground may not provide long-term benefits of cone preservation. In the latter, conditions in the midden provide a relatively cool and moist environment, preventing the cones from opening and sustaining pilferage by other small granivores. Sullivan et al. (1984) also discuss the presence in cone hoards of a fungus that prevents deterioration of the seeds. In summary, it seems that the number of cones cached per animal, as well as how that number relates to the overwinter energy demands of the squirrel, vary tremendously both spatially and temporally.

The impact of seed predation and larderhoarding by Tamiasciurus on North American boreal forest stands and individual trees is not clear. Cone losses resulting from the provisioning of middens by Tamiasciurus can be quite high, ranging from 60 to 100% of the cone crops of individual trees, species or single- and mixed-species stands (Flyger and Gates, 1982). Seed losses may occur over extensive tracts of coniferous forest, and may significantly retard reforestation in many species of conifers

(e.g. pine, spruce and fir; see review by Flyger and Gates, 1982; Peters et al., 2003). Conversely, Larsen et al. (1997) determined that only approximately one-third of the current year's crop of jack pine (Pinus banksiana) cones were harvested off the territories of female Tamiasciurus. Jack pine, like many populations of lodgepole pine (P. contorta), are serotinous, retaining their cones from year to year. Thus, the large number of older cones often visible in stands of these trees bears testimony to the fact that the cones harvested by squirrels in many years is only a small portion of the total crop. However, Larsen et al. (1997) reported high levels of cone harvesting from certain individual trees, and C.C. Smith (1965, 1968, 1970) showed that Tamiasciurus would preferentially harvest from trees that were more profitable. Estimates of the overall impact of cone harvesting by Tamiasciurus (and other squirrels) will require measures of cone selection within and between trees and across stands. These sorts of data are necessary if we are to better understand how spatial patterns of cone productions influence patterns of cone use by squirrels.

In some situations, Tamiasciurus is known to scatterhoard rather than larderhoard cones (Hurly and Robertson, 1987; Dempsey and Keppie, 1993; Hurly and Lourie, 1997). Hurly and Robertson (1987, 1990), for example, studied mixed-conifer plantations in Ontario, Canada, where over 71% of all Tamiasciurus cone caches contained only a single cone, accounting for 50% of all cones cached (up to 13,000 Scots pine cones). Intuitively, this behaviour might be more conducive to the germination of seeds, compared to larderhoarding (middens) where abiotic (temperature) and biotic (competition) factors would make successful seedling establishment unlikely. Dempsey and Keppie (1993), however, found the average number of cones scatterhoarded by Tamiasciurus was small, providing only enough food to last a few weeks (< 37 days). To date, the relationship between scatterhoarding by Tamiasciurus and tree establishment has not been examined (see Seed Dispersal by Tree Squirrels).

Mixed deciduous forests

Few studies have rigorously monitored both seed production and seed losses in temperate, mixed deciduous woodlands. Gurnell (1993) studied seed production and seed fate of oak (*Quercus robur*), sweet chestnut (*Castanea sativa*), beech (*Fagus sylvatica*) and hazel (*Corylus avellana*) in an oak-dominated wood in southern England between 1975 and 1988. The percentage of seeds eaten before dispersal by woodland birds (mainly jays, *Garrulus glandarius*, woodpigeons, *Columbia palumbus*, and nuthatches, *Sitta europaea*) and rodents (mainly bank voles, *Clethrionomys glareolus*, wood mice, *Apodemus sylvaticus*, and introduced grey squirrels, *Sciurus carolinensis*) averaged 38% for oak, 80% for hazel, 33% for beech and 40% for chestnut (Gurnell, 1993). However, predation rates were highly variable between years. Only 3–10% of acorns and < 1–25% of beechnuts were eaten prior to dispersal during mast years, while in poor seed years this increased to 32–87% and 43–98%, respectively. In 5 years, post-dispersal losses accounted for 100% of the seeds by the end of winter, and, in 2 years with the poorest seed production, post-dispersal losses reached 100% by mid-November (Gurnell, 1993). In the remaining 6 years of the 13-year study, with medium to heavy seed crops, seeds disappeared quickly during autumn and predation rates slowed thereafter. In only 3 years of mast crops, < 3% of acorns survived until the following spring or summer when seeds began to germinate (estimated from Gurnell, 1993).

In oak–beech woods in Belgium, Eurasian red squirrels feed heavily in the canopy on beechnuts, and to a lesser extent on acorns, from August (unripe seeds) to October (Wauters *et al.*, 1992). During autumn and winter, interspecific competition with other rodents, ungulates and birds is intense, especially in non-mast years (Moller, 1983; Gurnell, 1987, 1993; Wauters *et al.*, 1992). Hence, the availability of high-energy tree seeds is much more limited in time than in coniferous forests. In poor seed years, about 50–60% of acorns and

beechnuts were consumed by late December (Table 12.1) and all seeds disappeared from the forest floor by late winter (L. Wauters, unpublished data). In years with a medium to good seed crop, seed predation amounted to 13–28% for oak and 14–45% for beech (Table 12.1). Only after the occurrence of a mast crop of beech seeds were these predation rates considerably lower. This was followed by relatively high levels of seedling production in the following summer (L. Wauters, unpublished data).

Seed Dispersal by Tree Squirrels

The significance of scatterhoarding

In contrast to *Tamisciurus*, many non-territorial species of *Sciurus* (eastern grey squirrels, fox squirrels, Japanese squirrels (*S. lis*), Eurasian red squirrels) are scatterhoarders, burying seeds individually in widely dispersed sites (Kato, 1985; Gurnell, 1987; Wauters and Dhondt, 1987; Hayashida, 1988, 1989; Vander Wall, 1990; Wauters *et al.*, 1992, 1995, 2002; Wauters and Casale, 1996; Steele and Koprowski, 2001; Lee, 2002). In these species, home ranges of neighbouring animals overlap considerably (e.g. Wauters and Dhondt, 1992) and scatterhoarding allows individuals to store seeds rapidly in a spatial pattern that reduces pilferage by competitors (Stapanian and Smith, 1978; Jenkins and Peters, 1992; Wauters and Casale, 1996). The spatial distribution of seeds that results from scatterhoarding and its influence on the probability that seeds are recovered or pilfered are the two most significant factors determining whether seed dispersal and seedling establishment result from scatterhoarding.

The ways that tree squirrels space their caches appears to be due to several factors. Earlier studies suggest that the dispersion of scatterhoards represent a trade-off between the costs of storage (and retrieval) of caches when items are widely spaced and the increased risk of pilferage at high densities (Stapanian and Smith, 1978, 1984; Hurly and Lourie, 1997). However, more recent studies suggest that optimal spacing of caches may also be related to habitat structure, risks of predation to the hoarder and the probability of pilfering (M.A. Steele, unpublished results). Steele (2004, unpublished results), for example, found that eastern grey squirrels cache larger more profitable items farther from cover than less preferred items (see also Stapanian and Smith, 1986), which are frequently cached under tree canopies. He suggests that these squirrels trade off a slightly higher risk of predation so they can store more valuable seeds in sites that are more secure and less susceptible to pilfering. Most scatterhoarders, including the tree squirrels, are well adapted for recovering their caches, relying on spatial memory and olfactory cues (McQuade et al., 1986; Jacobs and Liman, 1991; Lavenex et al., 1998). Estimates of cache recovery of seeds collectively scatterhoarded by squirrels and other small mammals often exceed 90% (Wauters and Casale, 1996; Steele and Smallwood, 2001). However, more quantitative studies are needed to determine the specific factors and conditions that result in seedling establishment as a consequence of scatterhoarding by tree squirrels.

In Eurasian red squirrels (S. vulgaris), the amount of time spent hoarding seeds (and the amount cached) varied greatly among individuals within and between populations (Wauters et al., 1992, 1995, 2002; Wauters and Casale, 1996). In Scots pine-dominated woodlands, red squirrels recovered and consumed cached seeds from September to June; about 80% of cached seeds were eaten during the following winter and spring (Wauters and Casale, 1996; Wauters et al., 2002). Seeds and nut shells near recovered caches suggested that the vast majority of scatterhoarded seeds were retrieved either by the animal that had created them or by 'naive' conspecifics (L. Wauters, personal observation). Autumn seed dispersal in deciduous woods causes seed supplies to become depleted more rapidly than in conifer woods, where most cones retain seeds until the next spring (Wauters and Dhondt, 1987; Wauters et al.,

1992; Gurnell, 1993). Red squirrels respond to this decrease of food in the canopy by hoarding more and consuming a larger proportion of caches earlier (60% by the end of March) in deciduous than in coniferous woodlands (Wauters and Casale, 1996). Consequently, the proportion of the total daily energy intake attributed to cached seeds (January–April) is much higher in the deciduous than in coniferous woodlands. Japanese squirrels (Sciurus lis) recovered most of their underground food caches (walnuts, Juglans mandshurica, and Japanese red pine cones, P. densiflora) from March to May (Kato, 1985). Thus it appears that the extent of scatterhoarding and the timing of cache consumption by squirrels depend largely on the availability of seeds during the pre-dispersal period.

In mixed woodlands in Belgium, Eurasian red squirrels retrieved most caches and consumed 99–100% of cached pinecones, 83–92% of cached beechnuts and 54–62% of cached acorns (Wauters and Casale, 1996). Thus, the impact of seed dispersal by tree squirrels on seed fate will probably vary among tree species. However, the proportions of seeds of Scots and Corsican pine that were scatterhoarded in mixed woods in Belgium (0.1–1.4% of cone crop), as well as those of acorns (< 1%) and beechnuts (0.3–9%; L. Wauters, unpublished results), were quite low compared to pre-dispersal seed losses due to direct seed predation (Table 12.1; see also Wauters and Casale, 1996). This was confirmed by estimates of seed hoarding in mixed temperate forests in northern Italy (Wauters et al., 2001). Red squirrels there cached a nearly insignificant proportion of the seed crop of sweet chestnuts (average 576 per squirrel or 0.2–0.7% of seed crop; Wauters et al., 2001, 2002). However, nearly all nuts of walnut trees (J. regia) were eaten in autumn, or cached (Wauters et al., 2001, 2002; L. Wauters, unpublished results). Hazelnuts (Corylus avellana) were cached at a density of 94–134/ha (an average 134 nuts per squirrel; Wauters et al., 2002), corresponding with 4.5–13% of total seed crop (1020–2070 nuts/ha, estimated from Wauters et al., 2001). Thus, most quantitative data on scatterhoarding by some species (e.g. Eurasian red squirrels) show that only a portion of total seed crop is cached, and its impact on seed losses is trivial compared to pre- and post-dispersal seed losses by direct predation. In terms of energy intake, scatterhoarding mainly allows squirrels to: (i) prolong the period that seeds are available, and (ii) increase rates of energy intake in periods of food shortage.

Despite the ability of tree squirrels to find cached seeds, seedling establishment from scatterhoards can be common, especially when seeds are abundant (Silvertown, 1980; Crawley and Long, 1995) and squirrels die or disperse before seeds are recovered (Steele and Koprowski, 2001; Steele and Smallwood, 2001). In fact, the scatter-hoarding of acorns by tree squirrels probably increases the probability of seed germination and establishment compared with seeds that are not cached (Barnett, 1977; Steele and Smallwood, 2001). And, unlike the seeds and nuts moved by many avian dispersal agents, those scatterhoarded by tree squirrels are generally dispersed for relatively shorter distances (< 150 m), usually within forest patches, influencing the structure and composition of woodlands (Stapanian and Smith, 1978; Steele and Smallwood, 2001). Tree squirrels, like other hoarding species (e.g. jays and small granivorous rodents), often scatterhoard seeds in cache sites where the probability of seed and seedling mortality, desiccation and competition is reduced (Barnett, 1977; Vander Wall, 1990; Steele and Smallwood, 2001). The eastern grey squirrel in fact may select cache sites, because of their suitability for storage, that are also optimal for germination (Barnett, 1977).

In many systems, the impact of tree squirrels on the dispersal of seeds and establishment of seedlings has been inferred from only a limited number of mostly observational investigations. However, a few experimental studies have more clearly delineated the impact of tree squirrels. Stapanian and Smith (1978, 1984, 1986), for example, have documented patterns of scatterhoarding in fox squirrels (Stapanian and Smith, 1978, 1984), patterns of seed recovery from

artificial caches, and at least strong circumstantial evidence that some of these caches may result in tree establishment (Stapanian and Smith, 1984, 1986). Stapanian and Smith (1986) argue that scatterhoarding fox squirrels have greatly influenced the establishment of nut-bearing trees such as black walnut (*Juglans nigra*) in the prairies of the Midwestern USA. More recently, Goheen and Swihart (2003) modelled the effects of walnut dispersal by grey squirrels and red squirrels in Indiana, where forest fragmentation over the last century has favoured the spread of the non-native red squirrel at the expense of the grey squirrel. They argue that this range expansion by the red squirrels, which are far less likely to scatterhoard walnuts than grey squirrels, has resulted in a decline in dispersal and establishment of walnut trees in portions of the central hardwoods region.

Despite these few studies that link seed dispersal to the scatterhoarding activity of squirrels, there currently exist few quantitative data on the frequency with which seedling establishment results specifically from the caches of tree squirrels. We call for more quantitative studies investigating the link between seedling establishment and the caching activity of different species of tree squirrels.

Oak dispersal by tree squirrels

Several studies, primarily in eastern deciduous forests of North America, strongly suggest that tree squirrels may influence the structure of oak forests (Steele *et al.*, 2004). Although many species rely heavily on oak mast for autumn and winter food supplies and provisioning of their caches, tree squirrels are highly selective with respect to the acorns they choose to eat or cache (Smallwood and Peters, 1986; Steele and Smallwood, 2001), and much of the dispersal and establishment of oaks may follow from these patterns of acorn selection. Patterns of acorn selectivity, in turn, follow from the physical and chemical characteristics of these fruits (Steele *et al.*,

2004). Most oaks in North America belong to two subgenera. These include the red oaks (*Erythrobalanus*), which produce acorns that are generally high in lipids (18–25% by mass) and tannins (6–10%) and remain dormant in winter before they germinate, and the white oaks (*Quercus*), which produce acorns with lower levels of both lipid (5–10%) and tannin (2%) that lack dormancy and germinate within days of maturation in autumn.

The acorns selected for consumption by tree squirrels have been attributed to the influence of both of these compounds, which together present a significant trade-off for squirrels (Smith and Follmer, 1972; Lewis, 1980, 1982). Tannins bind to salivary and digestive enzymes and reduce both palatability and digestibility of acorns (Smallwood and Peters, 1986; Chung-MacCoubrey et al., 1997), whereas lipids are the primary energy sources in the cotyledon of acorns. Although tree squirrels may exhibit some physiological adaptations for processing tannins, the acorn preferences of squirrels and other acorn consumers have long been attributed to the combined effects of both tannins and lipids (Smith and Follmer, 1972; Smallwood and Peters, 1986). However, Smallwood and Peters (1986) systematically separated the effects of these two compounds by presenting free-ranging grey squirrels with artificial acorns (dough balls) constructed of ground white oak cotyledon to which varying amounts of tannin and lipid were added. In the autumn when energy demands were lower, animals selected artificial acorns lower in tannin even when they had lower lipid levels. Later in winter, however, squirrels showed a preference for acorns with higher lipids regardless of tannin levels (Smallwood and Peters, 1986).

Although lipids and tannins clearly influence patterns of acorn consumption by tree squirrels, acorn caching decisions are influenced more directly by differences in the germination schedules between the two acorn groups (Smallwood et al., 2001; Steele et al., 2004). Hadj-Chikh et al. (1996) demonstrated that eastern grey squirrels selectively cache dormant acorns of red oak over those of white oak because of the reduced perishability resulting from the delayed germination in red oak acorns. By presenting free-ranging squirrels with pairs of acorns that varied with respect to germination schedules as well as several other factors (size, handling time, tannins and lipids), Hadj-Chikh et al. (1996) found seed perishability (due to germination schedules) to be the sole proximate determinant of acorn caching decisions.

More recent investigations demonstrate that tree squirrels selectively cache dormant red oak acorns over those that have broken dormancy, even when the latter exhibit no physical signs of germination (Steele et al., 2001a). Moreover, manipulation of seed chemistry suggests that tree squirrels will cache modified acorns constructed of the pericarps of red oak acorns, even when they contain white oak acorn cotyledon (Steele et al., 2001a). Thus it appears that grey squirrels rely on a specific cue in the pericarp (e.g. appearance, odour) of dormant acorns to determine which acorns to cache.

The selective caching of dormant acorns by grey squirrels underscores the importance of seed perishability in caching decisions and is consistent with other studies showing that tree squirrels selectively cache sound red oak acorns over those infested with weevils (Steele et al., 1996). It is also consistent with reports that tree squirrels often prevent germination by excising the embryos of white oak acorns prior to caching them (Fox, 1974, 1982; Pigott et al., 1991; Steele et al., 2001b). In field experiments, in which the fate of metal-tagged seeds was monitored, grey squirrels and other small mammals selectively cached red oak acorns but consistently ate white oak acorns or excised their embryos before caching them (Steele et al., 2001b).

Embryo excision in particular points to a keen sensitivity to seed perishability in tree squirrels. The behaviour, which involves a few quick scrapes of the incisors on the apical end of the acorn, results in the removal of the embryo and allows tree squirrels to store what is otherwise a highly perishable food for up to 6 months (Steele et al., 2001b). This behaviour, now thought to be unique to the tree squirrels (*Sciurus*;

Steele *et al.*, 2001b), is considered a counter-adaptation to the early autumn germination of white oaks, which involves the rapid transfer of the seed energy into a taproot that is indigestible for squirrels and other granivores. Moreover, in experiments with captive-raised grey squirrels with no previous experience with acorns, they appear to possess an innate tendency to perform embryo excision when caching white oak acorns (M.A. Steele, 2004, unpublished results).

Finally, there is strong evidence that chemical gradients within acorn cotyledons (Steele *et al.*, 1993) as well as the shape of the nut (Steele *et al.*, 1998) promote partial consumption of the basal portion of these acorns in at least nine species of red oaks. These partially eaten acorns can successfully germinate and establish seedlings (Steele *et al.*, 1993; Steele, unpublished results), suggesting a possible adaptation by oaks for escaping partial predation.

Collectively, results of the above studies strongly suggest that tree squirrels and other small mammals influence establishment of red and white oak species differently (Smallwood *et al.*, 1998; Steele *et al.*, 2001b). Steele *et al.* (2001b, 2004) predict that red oaks should establish in a variety of sites at some distance from maternal trees, whereas white oaks will be dispersal limited and may establish closer to parent trees. Steele, Smallwood, and others are currently testing this differential dispersal hypothesis and its implications for forests by means of DNA fingerprinting.

Effects of Seed Production on Squirrel Demography and Behaviour

There are few long-term studies on the dynamics of tree squirrels, particularly those that examine correlates with seed production by trees and other plants. In North America, Kemp and Keith (1970) found significant positive correlations between white spruce (*Picea glauca*) cone crops and densities of North American red squirrels (*Tamiasciurus hudsonicus*) over a

large area of Canadian boreal forest (1946–1962). More recently, Boonstra *et al.* (2001) reported on a 10-year study of the same species. Although the squirrel populations showed little variation in relation to the hare cycle, larger cone crops of white spruce were correlated with higher squirrel numbers in the following year. Adult survival of squirrels did not appear related to cone supply, whereas more offspring were produced when the cone crops were relatively high. Wheatley *et al.* (2002) detected almost a 50% drop in squirrel densities in spruce habitat 1 year after a cone failure, and M.C. Smith (1968) found a 67% decrease in squirrel densities 2 years after a cone failure, attributing the lag effect to a surplus of cached food. Densities of the Eurasian red squirrel (*Sciurus*) fluctuate strongly (five- to 14-fold) in relation to tree seed crop, often with a 1-year time-lag. Rusch and Reeder (1978), however, found high densities of North American red squirrel during several years of poor cone production.

The proximate and ultimate factors that may cause squirrel populations to respond to fluctuating food levels have also been investigated more directly in Eurasian red squirrel populations. Tree seed availability affects fitness, through a direct effect on squirrel body condition (Wauters and Dhondt, 1989, 1995). For example, rich seed supplies increase winter–spring body mass of red squirrels; subadults in good condition are more likely to survive winter and subsequently emigrate and establish a home range (Wauters and Dhondt, 1989, 1993). Likewise, the probability of recruitment of locally born juveniles is higher when body mass is higher and they are weaned early in the breeding season (Wauters *et al.*, 1993). Also, body mass of adult females, which is positively related to their probability of successful reproduction, increases with home range quality, and is determined by spatio-temporal availability of tree seeds (Wauters and Dhondt, 1989, 1995; Wauters *et al.*, 1995). Finally, an important component of variation in lifetime reproductive success among female Eurasian red squirrels is the number of successful breeding events (litters), which strongly depends on the number of good seed years encountered during the 'reproductive' lifetime (Wauters and Dhondt, 1995).

In North America, the mechanisms by which changing food levels affect populations have been explored through food addition studies. Densities of *Tamiasciurus* have been shown to increase two to four times when provided with supplemental food, most commonly sunflower seeds (Sullivan and Sullivan, 1982; Sullivan, 1990; Ransomme and Sullivan, 1997). Klenner and Krebs (1991) found similar results using sunflower seed additions, and also showed that the increase was brought about by a significant decrease in squirrel territory size in Douglas fir (*Pseudotsuga menziesii*). Larsen *et al.* (1997) added both natural (cones) and unnatural food (sunflower seeds) to the midden larders of individual female squirrels (*Tamiasicurus*) living in jack pine forests. They also depleted the natural larders of some of the animals, by removing hoarded cones. They reported no significant effects on the survival of the females, their body mass, or reproductive success. Parturition date, however, was significantly earlier (2–11 days) for female squirrels receiving the seed additions.

Finally, we close with a brief reference to C.C. Smith's arguments that, for some species of tree squirrels, both their hoarding behaviour and social structure have evolved in direct response to tree reproduction (Smith, 1998; Smith and Stapanian, 2002). In boreal forests of North America, the territorial behaviour of *Tamiasciurus douglasii* and *T. hudsconicus* (and the social behaviour that results from it) allows for the efficient harvest and storage of cones and the ability to protect larders from competitors (Larsen and Boutin, 1994). In temperate deciduous and mixed forests, however, most food cannot be larderhoarded efficiently. Here, species of *Sciurus* have adopted scatterhoarding as a means of storing and protecting seeds from competitors. As a consequence stored food cannot be defended and these species are not territorial but instead exhibit overlapping home ranges and social hierarchies (Smith and Stapanian, 2002).

Conclusion

The relationship between tree squirrels and the tree seeds on which they feed is a close and complex one, leading many authors to argue that the two are co-evolved (C.C. Smith, 1970; Elliot, 1974; Smith and Balda, 1979). Masting patterns of seed trees greatly influence the survival and reproduction of tree squirrels, influencing a variety of demographic, behavioural and morphological characteristics of squirrels. In contrast, tree squirrels can act as seed predators, often significantly reducing seed crops, limiting the distribution and regeneration of many species, and shaping the physical and chemical characteristics of seeds. But, as dispersal agents, squirrels can also greatly affect establishment of trees, thereby influencing forest structure and composition (Fig. 12.1).

The reproductive success of plants that are adapted for wind dispersal (e.g. light, winged seeds of conifers) are affected differently by squirrel predation than are plant species adapted for animal dispersal (heavy, wingless seeds). Although tree squirrels act primarily as seed predators on conifer seeds and often agents of dispersal for many deciduous tree species, this dichotomy is not as simple as it may seem. Some conifers have not evolved protective measures, indicating that selection for seed protection has not been particularly strong (see Smith and Balda, 1979). In others, however, such as the North American lodgepole pine, co-evolution between the tree and its principal predator, the red squirrel (*Tamiasciurus*), appears so strong that the seeds now only constitute approximately 2% of the weight of the heavily armoured cone. In yet other pines that produce large, wingless seeds such as the Korean pine (*P. koraiensis*; Hayashida, 1989) and the Swiss stone pine (*Pinus cembra*; L. Wauters, unpublished results) tree squirrels may be critical for successful dispersal and even germination although much study remains to be done.

In deciduous tree species, predation of seeds can also be significant, but for many of these species (e.g. *Quercus, Fagus, Carya, Juglans, Castanea*) tree squirrels appear to

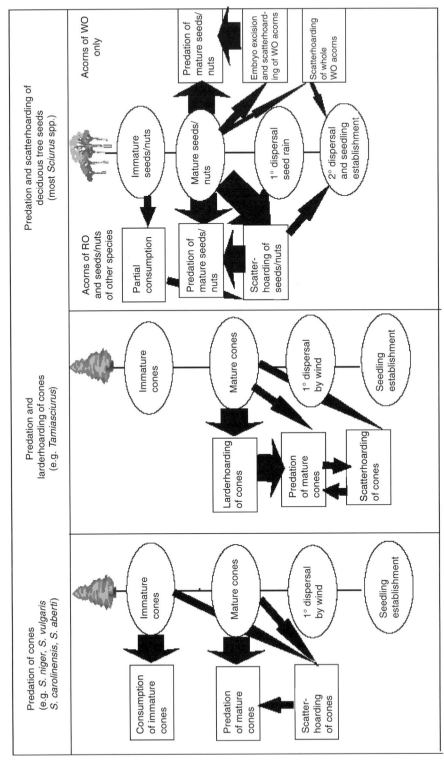

Fig. 12.1. Overview of the three most significant ways in which tree squirrels influence seed fates in temperate and boreal forests. Ellipses indicate primary stages in the development and establishment of seeds or cones; the path from seed development to seedling establishment progresses from top to bottom. Rectangles and arrows indicate the influence of tree squirrels on seed fates. Arrow width is proportional to effect. RO, red oak species (subgenus *Erythrobalanus*); WO, white oak species (subgenus *Quercus*).

serve as a major agent of dispersal (Fig. 12.1), although more quantitative estimates of seed fates from squirrel caches are needed. Scatterhoarding by many species of *Sciurus* results in the placement of seeds in individual cache sites often a considerable distance from sources where the probability of establishment and recruitment is high. Indeed, it appears that some of these seeds are well adapted for dispersal, residence in cache sites, and even partial predation by tree squirrels following dispersal. However, for at least one group (the white oaks) the tree squirrels may have evolved the upper hand (via embryo excision), which may ultimately limit the dispersal and establishment of several white oak species across North America (Fig. 12.1). Only after careful examination of the specific physical and chemical characteristics of seeds, the masting patterns of trees, the responses of tree squirrels to such factors and, most importantly, rigorous quantitative measures of seed fates will we begin to uncover the true evolutionary relationship between tree squirrels and their seeds. Much remains to be done.

References

Allred, S.W., Gaud, W.S. and States, J.S. (1994) Effects of herbivory by Abert squirrels (*Sciurus aberti*) on cone crops of ponderosa pine. *Journal of Mammalogy* 75, 700–703.

Andrén, H. and Lemnell, P.A. (1992) Population fluctuations and habitat selection in the Eurasian red squirrel *Sciurus vulgaris*. *Ecography* 15, 303–307.

Arbogast, B.S., Browne, R.A. and Weigl, P.D. (2001) Evolutionary genetics and Pleistocene biogeography of North American tree squirrels (*Tamiasciurus*). *Journal of Mammalogy* 82, 302–319.

Barnett, R.J. (1977) The effect of burial by squirrels on germination and survival of oak and hickory nuts. *The American Midland Naturalist* 98, 19–30.

Becker, C.D., Boutin, S. and Larsen, K.W. (1998) Constraints on first reproduction in North American red squirrels. *Oikos* 81, 81–92.

Benkman, C.W. (1995) The impact of tree squirrels (*Tamiasciurus*) on limber pine seed dispersal adaptations. *Evolution* 49, 585–592.

Boonstra, R., Boutin, S., Byrom, A., Karels, T., Hubbs, A., Stuart-Smith, K., Blower, M. and Antoehler, S. (2001) The role of red squirrels and Arctic ground squirrels. In: Krebs, C., Boutin, S. and Boonstra, R. (eds) *Ecosystem Dynamics of the Boreal Forest: the Kluane Project*. Oxford University Press, New York, pp. 180–214.

Chung-MacCoubrey, A.L., Hagerman, E.E. and Kirkpatrick, R.L. (1997) Effects of tannin on digestion and detoxification activity in gray squirrels. *Physiological Zoology* 70, 270–277.

Crawley, M.J. and Long, C.R. (1995) Alternate bearing, predator satiation and seedling recruitment in *Quercus robur* L. *Journal of Ecology* 83, 683–696.

Degn, H.J. (1974) Feeding activity in the red squirrel (*Sciurus vulgaris*). *Journal of Zoology, London* 174, 516–520.

Dempsey, J.A and Keppie, D.M. (1993) Foraging patterns of eastern red squirrels. *Journal of Mammalogy* 74, 1007–1013.

Elliot, P.F. (1974) Evolutionary responses of plants to seed-eaters: pine squirrel predation on lodgepole pine. *Evolution* 28, 221–231.

Elliot, P.F. (1988) Foraging behavior of a central-place forager: field tests of theoretical predictions. *American Naturalist* 131, 159–174.

Flyger, V. and Gates, J.E. (1982) Pine squirrels: *Tamiasciurus hudsonicus, T. douglasii*. In: Chapman, J.A. and Feldhamer, G.A. (eds) *Wild Mammals of North America*. Johns Hopkins University Press, Baltimore, Maryland, pp. 230–235.

Fox, J.F. (1974) Coevolution of white oak and its seed predators. PhD thesis, The University of Chicago, Illinois, 90 pp.

Fox, J.F. (1982) Adaptation of gray squirrel behavior to autumn germination by white oak acorns. *Evolution* 36, 800–809.

Goheen, J.R. and Swihart, R.K. (2003) Food-hoarding behavior of gray squirrels and North American red squirrels in the central hardwoods region: implications for forest regeneration. *Canadian Journal of Zoology* 81, 1636–1639.

Grönwall, O. (1982) Aspects of the food ecology of the red squirrel (*Sciurus vulgaris* L.). PhD thesis, University of Stockholm, Sweden.

Gurnell, J.C. (1984) Home range, territoriality, caching behaviour and food supply of the red squirrel (*Tamiasciurus hudsonicus fremonti*) in a subalpine forest. *Animal Behaviour* 32, 1119–1131.

Gurnell, J. (1987) *The Natural History of Squirrels*. Academic Press, London.

Gurnell, J. (1993) Tree seed production and food conditions for rodents in an oak wood in southern England. *Forestry* 66, 291–315.

Hadj-Chikh, L.Z., Steele, M.A. and Smallwood, P.D. (1996) Caching decisions by grey squirrels: a test of the handling-time and perishability hypotheses. *Animal Behaviour* 52, 941–948.

Hall, J.G. (1981) A field study of the Kaibab squirrel in Grand Canyon National Park. *Wildlife Monographs* 75, 1–54.

Hayashida, M. (1988) The influence of social interactions on the pattern on scatter-hoarding in red squirrels. *Research Bulletin Faculty of Agriculture, Hokkaido University* 45, 267–278.

Hayashida, M. (1989) Seed dispersal by red squirrels and subsequent establishment of Korean pine. *Forest Ecology and Management* 28, 115–129.

Hoffman, R.S., Anderson, C.G., Thorington, R.W. and Heaney, L.R. (1993) Family Sciuridae. In: Wilson, D.E. and Reeder, D.M. (eds) *Mammal Species of the World*. Smithsonian Institution Press, Washington, DC, pp. 419–465.

Hurly, T.A. and Lourie, S.A. (1997) Scatter-hoarding and larder-hoarding by red squirrels: size, dispersion, and allocation of hoards. *Journal of Mammalogy* 78, 528–537.

Hurly, T.A. and Robertson, R.J. (1987) Scatter-hoarding by territorial red squirrels: a test of the optimal density model. *Canadian Journal of Zoology* 65, 1247–1252.

Hurly, T.A. and Robertson, R.J. (1990) Variation in the food hoarding behaviour of red squirrels. *Behavioral Ecology and Sociobiology* 26, 91–97.

Jacobs, L.F. and Liman, E.R. (1991) Grey squirrels remember the locations of buried nuts. *Animal Behaviour* 41, 103–110.

Jenkins, S.H. and Peters, R.A. (1992) Spatial patterns of food storage by Merriam's kangaroo rats. *Behavioral Ecology* 3, 60–65.

Kato, J. (1985) Food and hoarding behavior of Japanese squirrels. *Japanese Journal of Ecology* 35, 13–20.

Kemp, G.A. and Keith, L.B. (1970) Dynamics and regulation of red squirrel (*Tamiasciurus hudsonicus*) populations. *Ecology* 51, 763–779.

Klenner, W. and Krebs, C.J. (1991) Red squirrel population dynamics. I. The effect of supplemental food on demography. *Journal of Animal Ecology* 60, 961–978.

Lampio, T. (1967) Sex ratios and the factors contributing to them in the squirrel, *Sciurus vulgaris*, in Finland, II. *Finnish Game Research* 29, 1–69.

Larsen, K.W. and Boutin, S. (1994) Movements, survival, and settlement of red squirrel (*Tamiasciurus hudsonicus*) offspring. *Ecology* 75, 214–223.

Larsen, K.W., Becker, C.D., Boutin, S. and Blower, M. (1997) Effects of hoard manipulations on life history and reproductive success of female red squirrels (*Tamiasciurus hudsonicus*). *Journal of Mammalogy* 78, 192–203.

Lavenex, P.M., Shiflett, R., Lee, R.K. and Jacobs, L.F. (1998) Spatial versus nonspatial relational learning in free-ranging fox squirrels (*Sciurus niger*). *Journal of Comparative Psychology* 112, 1–10.

Lee, T.H. (2002) Feeding and hoarding behaviour of the Eurasian red squirrel *Sciurus vulgaris* during autumn in Hokkaido, Japan. *Acta Theriologica* 47, 459–470.

Lewis, A.R. (1980) Patch use by gray squirrels and optimal foraging. *Ecology* 61, 1371–1379.

Lewis, A.R. (1982) Selection of nuts by gray squirrels and optimal foraging theory. *American Midland Naturalist* 107, 250–257.

McQuade, D.B., Williams, E.H. and Eichenbaum, H. (1986) Cues used for localizing food by the gray squirrel (*Sciurus carolinensis*). *Ethology* 72, 22–30.

Miyaki, M. (1987) Seed dispersal of the Korean pine, *Pinus koraiensis*, by the red squirrel, *Sciurus vulgais*. *Ecological Research* 2, 147–157.

Moller, H. (1983) Foods and foraging behaviour of red (*Sciurus vulgaris*) and grey (*Sciurus carolinensis*) squirrels. *Mammal Review* 13, 81–98.

Moller, H. (1986) Red squirrels (*Sciurus vulgaris*) feeding in a Scots pine plantation in Scotland. *Journal of Zoology, London* 209, 61–83.

Parchman, T.L. and Benkman, C.W. (2002) Diversifying coevolution between crossbills and black spruce on Newfoundland. *Evolution* 56, 1663–1672.

Peters, S., Boutin, S. and Macdonald, E. (2003) Pre-dispersal seed predation of white spruce cones in logged boreal mixedwood forests. *Canadian Journal of Forest Research* 33, 33–40.

Pigott, C.D., Newton, A.C. and Zammitt, S. (1991) Predation of acorns and oak seedlings by gray squirrels. *Quarterly Journal of Forestry* 85, 173–178.

Pulliainen, E. (1973) Winter ecology of the red squirrel (*Sciurus vulgaris* L.) in northeastern

Lapland. *Annales Zoologici Fennici* 10, 487–494.

Ransomme, D.B. and Sullivan, T.P. (1997) Food limitation and habitat preference of *Glaucomys sabrinus* and *Tamiasciurus hudsonicus*. *Journal of Mammalogy* 78, 538–549.

Rusch, D.A. and Reeder, W.G. (1978) Population ecology of Alberta red squirrels. *Ecology* 59, 400–420.

Silvertown, J.W. (1980) The evolutionary ecology of mast seeding in trees. *Biological Journal of the Linnean Society* 14, 235–250.

Smallwood, P.D. and Peters, W.D. (1986) Grey squirrel food preferences: the effects of tannin and fat concentration. *Ecology* 67, 168–174.

Smallwood, P.D., Steele, M.A., Ribbens, E. and McShea, W.J. (1998) Detecting the effects of seed hoarders on the distribution of tree species: gray squirrels (*Sciurus carolinensis*) and oaks (*Quercus*) as a model system. In: Steele, M.A., Merritt, J.F. and Zegers, D.A. (eds) *Ecology and Evolutionary Biology of Tree Squirrels*. Virginia Museum of Natural History, Charlottesville, Virginia, Special Publication 6, pp. 211–222.

Smallwood, P.D., Steele, M.A. and Faeth, S. (2001) The ultimate basis of the caching preferences of rodents and the oak-dispersal syndrome: tannins, insects, and seed germination. *American Zoologist* 41, 840–851.

Smith, C.C. (1965) Interspecific competition in the genus of tree squirrels, *Tamiasciurus*. PhD thesis, University of Washington, Seattle, Washington, 269 pp.

Smith, C.C. (1968) The adaptive nature of social organization in the genus of tree squirrel *Tamiasciurus*. *Ecological Monographs* 38, 30–63.

Smith, C.C. (1970) The coevolution of pine squirrels (*Tamiasciurus*) and conifers. *Ecological Monographs* 40, 349–371.

Smith, C.C. (1998) The evolution of reproduction in trees: its effect on ecology and behavior of squirrels. In: Steele, M.A., Merritt, J.F. and Zegers, D.A. (eds) *Ecology and Evolutionary Biology of Tree Squirrels*. Virginia Museum of Natural History, Charlottesville, Virginia, Special Publication 6, pp. 203–210.

Smith, C.C. and Balda, R.P. (1979) Competition between insects, birds, and mammals for conifer seeds. *American Zoologist* 19, 1065–1083.

Smith, C.C. and Follmer, D. (1972) Food preferences of squirrels. *Ecology* 53, 82–91.

Smith, C.C. and Reichman, O.J. (1984) The evolution of food caching by birds and mammals. *Annual Review of Ecology and Systematics* 15, 329–351.

Smith, C.C. and Stapanian, M.A. (2002) Squirrels and oaks. In: McShea, W.J. and Healy, W.M. (eds) *Oak Forest Ecosystems: Ecology and Management for Wildlife*. Johns Hopkins University Press, Baltimore, Maryland, pp. 256–268.

Smith, M.C. (1968) Red Squirrel responses to spruce cone failure. *Journal of Wildlife Management* 32, 305–316.

Snyder, M.A. (1998) Abert's squirrels (*Sciurus aberti*) in ponderosa pine (*Pinus ponderosa*) forests: directional selection, diversifying selection. In: Steele, M.A., Merritt, J.F. and Zegers, D.A. (eds) *Ecology and Evolutionary Biology of Tree Squirrels*. Virginia Museum of Natural History, Charlottesville, Virginia, Special Publication 6, pp. 195–201.

Stapanian, M.A. and Smith, C.C. (1978) A model for seed scatterhoarding: coevolution of fox squirrels and black walnuts. *Ecology* 59, 884–896.

Stapanian, M.A. and Smith, C.C. (1984) Density-dependent survival of scatterhoarded nuts: an experimental approach. *Ecology* 65, 1387–1396.

Stapanian, M.A. and Smith, C.C. (1986) How fox squirrels influence the invasion of prairies by nut-bearing trees. *Journal of Mammalogy* 67, 326–332.

Steele, M.A. (1988) Patch use and foraging behavior by the fox squirrel (*Sciurus niger*): tests of theoretical predictions. PhD thesis, Wake Forest University, Winston-Salem, North Carolina, 220 pp.

Steele, M.A. (1998) *Tamiasciurus hudsonicus*. *Mammalian Species* 586, 1–9.

Steele, M.A. (1999) *Tamiasciurus douglasii*. *Mammalian Species* 630, 1–8.

Steele, M.A. and Koprowski, J.L. (2001) *North American Tree Squirrels*. Smithsonian Institution Press, Washington, DC, 201 pp.

Steele, M.A. and Smallwood, P.D. (2001) Acorn dispersal by birds and mammals. In: McShea, W.J. and Healy, W.M. (eds) *Oak Forest Ecosystems: Ecology and Management for Wildlife*. Johns Hopkins University Press, Baltimore, Maryland, pp. 182–195.

Steele, M.A. and Weigl, P.D. (1992) Energetics and patch use in the fox squirrel *Sciurus niger*: responses to variation in prey profitability and patch density. *The American Midland Naturalist* 128, 156–167.

Steele, M.A., Knowles, T., Bridle, K. and Simms, E. (1993) Tannins and partial consumption of acorns: implications for dispersal of oaks by seed predators. *American Midland Naturalist* 130, 229–238.

Steele, M.A., Hadj-Chikh, L.Z. and Hazeltine, J. (1996) Caching and feeding decisions by *Sciurus carolinensis*: responses to weevil-infested acorns. *Journal of Mammalogy* 77, 305–314.

Steele, M.A., Gavel, K. and Bachman, W. (1998) Dispersal of half-eaten acorns by gray squirrels: effects of physical and chemical seed characteristics. In: Steele, M.A., Merritt, J.F. and Zegers, D.A. (eds) *Ecology and Evolutionary Biology of Tree Squirrels*. Virginia Museum of Natural History, Charlottesville, Virginia, Special Publication 6, pp. 223–232.

Steele, M.A., Smallwood, P.D., Spunar, A. and Nelsen, E. (2001a) The proximate basis of food-hoarding decisions and the oak dispersal syndrome: detection of seed dormancy by rodents. *American Zoologist* 41, 852–864.

Steele, M.A., Turner, G., Smallwood, P.D., Wolff, J.O. and Radillo, J. (2001b) Cache management by small mammals: experimental evidence for the significance of acorn embryo excision. *Journal of Mammalogy* 82, 35–42.

Steele, M.A., Smallwood, P.D., Terzaghi, W.B., Carlson, J.E., Contreras, T. and McEuen, A. (2004) Oak dispersal syndromes: do red oaks and white oaks exhibit different dispersal strategies? *US Forest Service Technical Publication* (in press).

Sullivan, T.P. (1990) Responses of red squirrel (*Tamiasciurus hudsonicus*) populations to supplemental food. *Journal of Mammalogy* 71, 579–590.

Sullivan, T.P. and Sullivan, D.S. (1982) Population dynamics and regulation of the Douglas Squirrel (*Tamiasciurus douglasii*) with supplemental food. *Oecologia* 53, 264–279.

Sullivan, T.P., Sutherland, J.R., Woods, T.A.D. and Sullivan, D.S. (1984) Dissemination of the conifer seed fungus *Caloscypha fulgens* by small mammals. *Canadian Journal of Forest Research* 14, 134–137.

Vander Wall, S.B. (1990) *Food Hoarding in Animals*. University of Chicago Press, Chicago, Illinois, 445 pp.

Vander Wall, S.B. (1992) The role of animals in dispersing a 'wind-dispersed' pine. *Ecology* 73, 614–621.

Wauters, L.A. and Casale, P. (1996) Long-term scatterhoarding by Eurasian red squirrels (*Sciurus vulgaris*). *Journal of Zoology, London* 238, 195–207.

Wauters, L.A. and Dhondt, A.A. (1987) Activity budget and foraging-behaviour of the red squirrel (*Sciurus vulgaris*, Linnaeus, 1758) in a coniferous habitat. *Zeitschrift für Säugetierkunde* 52, 341–352.

Wauters, L.A. and Dhondt, A.A. (1989) Variation in length and body weight of the red squirrel (*Sciurus vulgaris*) in two different habitats. *Journal of Zoology, London* 217, 93–106.

Wauters, L.A. and Dhondt, A.A. (1992) Spacing behaviour of the red squirrel, *Sciurus vulgaris*: variation between habitats and the sexes. *Animal Behaviour* 43, 297–311.

Wauters, L.A. and Dhondt, A.A. (1993) Immigration patterns and success in red squirrels. *Behavioral Ecology and Sociobiology* 33, 159–167.

Wauters, L.A. and Dhondt, A.A. (1995) Lifetime reproductive success and its correlates in female Eurasian red squirrels. *Oikos* 72, 402–410.

Wauters, L.A. and Lens, L. (1995) Effects of food availability and density on red squirrel (*Sciurus vulgaris*) reproduction. *Ecology* 76, 2460–2469.

Wauters, L.A., Swinnen, C. and Dhondt, A.A. (1992) Activity budget and foraging behaviour of red squirrels (*Sciurus vulgaris* Linnaeus, 1758) in coniferous and deciduous habitats. *Journal of Zoology, London* 227, 71–86.

Wauters, L.A., Bijnens, L. and Dhondt, A.A. (1993) Body mass at weaning and juvenile recruitment in the red squirrel. *Journal of Animal Ecology* 62, 280–286.

Wauters, L.A., Suhonen, J. and Dhondt, A.A. (1995) Fitness consequences of hoarding behaviour in the Eurasian red squirrel. *Proceedings of the Royal Society, London, Series B* 262, 277–281.

Wauters, L.A., Gurnell, J., Preatoni, D. and Tosi, G. (2001) Effects of spatial variation in food availability on spacing behaviour and demography of Eurasian red squirrels. *Ecography* 24, 525–538.

Wauters, L.A., Tosi, G. and Gurnell, J. (2002) Interspecific competition in tree squirrels: do introduced grey squirrels (*Sciurus carolinensis*) deplete tree seeds hoarded by red squirrels (*S. vulgaris*)? *Behavioral Ecology and Sociobiology* 51, 360–367.

Wheatley, M., Larsen, K.W. and Boutin, S. (2002) Does density reflect habitat quality for North American red squirrels during a spruce-cone failure? *Journal of Mammalogy* 83, 716–727.

Zimmerling, T. (1990) Caching behaviour differences between breeding and non-breeding female red squirrels (*Tamiasciurus hudsonicus*). BSc Honours thesis, Department of Zoology, University of Alberta, Edmonton, Alberta, Canada.

13 Jays, Mice and Oaks: Predation and Dispersal of *Quercus robur* and *Q. petraea* in North-western Europe

Jan den Ouden, Patrick A. Jansen and Ruben Smit

Centre for Ecosystem Studies, Wageningen University, PO Box 47, 6700 AA, Wageningen, The Netherlands

Introduction

Acorns of the North European oaks (*Quercus robur* and *Q. petraea*) are an important food source for forest vertebrates in north-western Europe. This is not surprising because acorns are highly nutritious (Gurnell, 1993) and abundant. Seed production can amount to over 600 kg/ha or 60 acorns/m² (Jones, 1959; Ovington and Murray, 1964; Tanton, 1965; Shaw, 1968a; Gurnell, 1993; Crawley and Long, 1995). Consumers of acorns in north-western Europe include mammals such as wild boar (*Sus scrofa*), roe deer (*Capreolus capreolus*), red deer (*Cervus elaphus*), rabbit (*Oryctolagus cuniculus*), hare (*Lepus europaeus*), bank vole (*Clethrionomys glareolus*), wood mouse (*Apodemus sylvaticus*), yellow-necked mouse (*A. flavicollis*), red squirrel (*Sciurus vulgaris*) and the introduced grey squirrel (*S. carolinensis*), and birds such as Eurasian jay (*Garrulus glandarius*), wood pigeon (*Columba palumbus*), crow (*Corvus corone*) and magpie (*Pica pica*) (Tanton, 1965; Corbet, 1974; Bossema, 1979; Kenward and Holm, 1993; Groot Bruinderink and Hazebroek, 1995; Gomez *et al.*, 2003). The number of vertebrate species feeding on acorns is also large in other regions

containing *Quercus* species (e.g. McShea and Healy, 2002).

Some species continue to consume acorns well outside the crop season. These acorns often come from food caches that have been made by hoarding animals during the period of seed ripening and seed fall. Notably jay and wood mouse hide large numbers of acorns in the top soil layer. Because a high proportion of oak recruitment comes from cached seeds that are not retrieved or predated (Bossema, 1979; Crawley and Long, 1995; Kollmann and Schill, 1996), jays and wood mice are the most important dispersers and hoarders of acorns in Europe (Bossema, 1979; Jensen and Nielsen, 1986; Gomez, 2003).

Most other species that forage on acorns are mainly seed predators or only occasional dispersers. Wild boar forage for acorn reserves made by jays and wood mice by rooting the topsoil. They crack the seed husk and swallow the acorns, effectively killing the seed (Groot Bruinderink *et al.*, 1994). Wood pigeons also collect acorns and grind these in their gizzard. Although they may accidentally drop viable acorns elsewhere (Mellanby, 1968), we do not consider pigeons important dispersers of oak. Corvids like crows and magpies also hoard food, but these species exploit acorns opportunistically (Clarkson et al., 1986; Goodwin, 1986), in contrast to the specialist jays, which actively search for acorns throughout the period of seed ripening (Chettleburgh, 1952; Bossema, 1979). Finally, several rodent species besides wood mouse store acorns as a winter food source, but, in contrast to wood mouse, mostly hoard seeds in larders or up in trees (Corbet and Harris, 1991), which makes them less effective dispersers (Vander Wall, 1990).

In this chapter, we review the role of European jay and wood mouse as predators and dispersers of *Quercus robur* and *Q. petraea* acorns. Our aim is to show the similarities and differences in hoarding behaviour between the two species, and to compare seed fate from the moment acorns ripen on the tree until the shoot appears from the germinated seed. We also evaluate both species' impact on oak populations in different habitats and landscapes. For the jay, we heavily draw from the work of Bossema (1979). His dissertation, entitled 'Jays and oaks', is one of the first exhaustive studies on food caching in animals. Bossema's numerous experiments with captured and free-ranging birds cover almost all stages in the seed fate pathway, and we consider his work as a benchmark in the field of plant–animal interactions. For the wood mouse, we present data from a number of field experiments that investigated effects of structural characteristics of the habitat on seed dispersal and predation by rodents (den Ouden, 2000; Jansen and den Ouden, Chapter 22, this volume; J. den Ouden and P.A. Jansen, unpublished data).

Acorn Abundance

Oaks produce seeds almost every year, yet their crop size varies greatly between years, and does so synchronously within populations (Crawley and Long, 1995). The occurrence of good crops – also referred to as mast, although oak is not a strict masting species – is determined by reproductive investments in the previous years, combined with climatic conditions (Sork et al.,

1993; Kelly and Sork, 2002). The resulting alternation of rich and poor acorn years has important consequences for wild boar, wood mice and jays. Winter survival rates for these species are much greater in high acorn years than in low acorn years (Watts, 1969; Gurnell, 1981; Jensen, 1982; Groot Bruinderink et al., 1994).

The size of a seed crop affects the ratio between immediate predation and hoarding. Several authors (Theimer, 2001; Vander Wall 2002, 2003; Jansen et al., 2004) have shown that a greater proportion of hoarded seeds escapes predation in high seed years, in accordance with the predator satiation hypothesis for mast seeding (Janzen, 1970), while dispersal distances are smaller. The proportion of hoarded *Q. robur* and *Q. petraea* acorns surviving until germination in high versus low seed years is not known, but our own experiments do suggest that dispersal distances are smaller in high seed years than in low seed years (den Ouden, 2000; see section 'Cache Properties'). Crawley and Long (1995), however, found evidence for satiation of *Q. robur* acorn parasites – the larger the crop, the lower the proportion of acorns infested by weevils – but not for satiation of vertebrate predators.

Seed Removal

Seed removal in oak is difficult to quantify because acorns are both plucked from the tree by birds and arboreal mammals and gathered below trees after shedding by many different species. The proportion of acorns directly taken from individual trees can be high. By comparing the numbers of cupules and acorns that dropped below trees, Tanton (1965) estimated that as much as 92.5% of the acorns were plucked from oak crowns in Monks Wood (UK), and only 8.4% in Yarner Woods (UK). Gurnell (1993) found that on average 38% of the acorns were plucked from the tree. The proportion of acorns taken directly from tree crowns determines the amount of acorns available for terrestrial foragers. Although rates and proportions of seed removal by different agents vary widely, no viable seeds are generally remaining on the forest floor at the onset of winter. So, effectively, all seeds are either eaten, infested by insects or hoarded within the period of seed ripening and seed fall.

Jay

Seed removal by jays starts early in September and continues until there are no more seeds on the trees. Jays often aggregate when feeding, forming social groups that collectively use the same trees but cache seeds in their own territories (Chettleburgh, 1952; Bossema, 1979). A single bird can remove and store many thousands of acorns: estimates range from 2200 to 5700 acorns per bird per season (Schuster, 1950; Chettleburgh, 1952; Wadewitz, 1976). Gomez (2003) reports an average of 13 transport flights per hour by jays collecting *Quercus ilex* in Spain. Individual birds usually transport several acorns at once (Bossema, 1979; Gomez, 2003). Trees on forest edges and isolated trees are visited far more frequently than trees inside forest stands. There may be two important reasons for this. First, trees inside the forest produce less seed than trees with well-lit crowns in the forest edge (Jones, 1959; Ovington and Murray, 1964). Second, jays tend to avoid enclosed conditions (Bossema, 1979; Kollmann and Schill, 1996; Gomez, 2003) and preferentially cache seeds in open areas (see below).

Wood mouse

Wood mice, in contrast to jays, forage for fallen acorns (Fig. 13.1). Wood mouse activity, and consequently the chance of wood mice detecting and removing acorns strongly depends on the vegetation structure. Wood mice prefer the protection of vegetation cover, yet they can also be found foraging in structurally less complex habitats such as grasslands and heath lands, and may venture several hundred metres away from their burrows (Kikkawa, 1964;

Fig. 13.1. Wood mouse (*Apodemus sylvaticus*). Photo: R. Smit/OUTDOORVISION.NL.

Tattersall *et al.*, 2001). Other rodents, such as bank voles, strictly depend on structures such as high vegetation, dead wood and deep litter that provide protective cover (Hansson, 1978; Van Apeldoorn *et al.*, 1992).

We investigated the activity of wood mice and bank voles in the forest floor vegetation and compared their habitat use (see den Ouden (2000) for details). In five different oak–birch (*Betula pendula*) woodlands, we used pairs of trapping grids: one placed inside bracken vegetation and another placed in a short and patchy vegetation of wavy hair-grass (*Deschampsia flexuosa*) and blueberry (*Vaccinium myrtillus*). Bank voles were found exclusively underneath bracken, whereas wood mice were trapped both inside and outside bracken vegetation (Fig. 13.2). We also studied seed removal from seed plots in the same areas inside and outside the bracken stand, and found only minor differences in seed removal: nearly all seeds were removed within a month (Fig. 13.3).

We also compared the habitat use of rodents between an oak–birch woodland and a beech (*Fagus sylvatica*) forest with scattered oaks (*Quercus robur* and *Q. petraea*). These forests contrasted sharply in understorey cover. While the oak forest had bracken swards and patches with wavy hair-grass and blueberry (Fig. 13.4), the surface of the beech forest was covered with just litter. The only structure-rich places in the beech forest were a large thicket of holly (*Ilex aquifolium*) and a canopy gap occupied by bracken and some low (< 2.5 m) beech

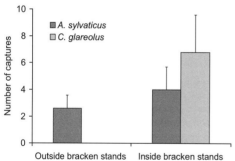

Fig. 13.2. Effect of habitat structure on the activity – expressed as number of captures per 100 trapping days – of wood mouse (*Apodemus sylvaticus*) and bank vole (*Clethrionomys glareolus*). Significantly more wood mice were caught inside the bracken stands (sign-test, $P = 0.032$). Rodents were caught in five different *Quercus robur/Betula pendula* forests either inside or outside stands of bracken (*Pteridium aquilinum*) in the understorey. Two trapping grids were laid out containing 25 Longworth life traps in a 10 m × 10 m grid. A 15 m buffer through the edge of the bracken stand separated the trapping grids. Traps were baited with rolled oats and apple, set for 4 days and checked before sunset and after sunrise. Data are means over the five sites (± 1 SE).

regeneration. Rodents were trapped and marked in four weekly sessions in autumn, winter and early and late spring. Again, results indicated that bank voles were mainly restricted to structure-rich habitats, where wood mice also used more exposed areas, although they rarely ventured on to the bare litter underneath beech canopies (Fig. 13.5). Consequently, rodents removed only 15% of 150 seeds presented in the bare litter area in the oak–beech forest.

Seed removal rates

Seed removal rates by rodents from a given area underneath an oak canopy or from artificial caches are the composite of two rates: the time it takes for rodents to locate the food source and the time to select and cache seeds. The detection rate of seed sources depends on the activity of animals, which is related to population density, food availability (see Jansen and den Ouden, Chapter 22, this volume) and habitat structure. We found that seeds were removed from

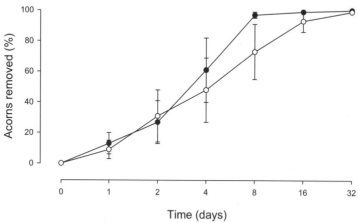

Fig. 13.3. Removal rates of acorns (mean ± SE) from seed plots inside (closed symbols) or outside (open symbols) bracken stands in five mixed *Quercus robur/Betula pendula* woodlands. Seed plots (see below) were paired within forest stands with a 15 m buffer in between. Overall removal rates did not significantly differ between bracken and non-bracken habitat. When analysed per forest stand, three out of five plot pairs showed significantly higher removal rates from underneath bracken (log-rank tests, df = 1, $P < 0.001$). Seed plots contained 200 acorns on an area of 400 m². Each plot consisted of a 20 m baseline. At 2 m intervals, ten 10 m lines ran perpendicular to the baseline, alternating to the left or right. Along these lines, acorns were placed at 50 cm intervals. Plots were checked daily until all acorns were removed. The experiment was conducted during the period of seed fall, and the natural acorn crop was negligible. See den Ouden (2000) for detailed description of site and methods.

Fig. 13.4. Mixed beech (*Fagus sylvatica*) and oak (*Quercus robur* and *Q. petraea*) forest in Speulderbos. Photo: J. den Ouden.

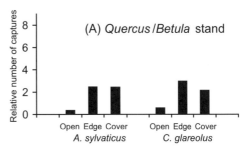

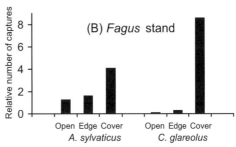

Fig. 13.5. Relative amount of captures of wood mice (*Apodemus sylvaticus*) and bank voles (*Clethrionomys glareolus*) in two forest types and in relation to vegetation structure. For all species, significantly fewer rodents were captured in open areas (G-test: all $G_{adj} > 15$, df = 2, all $P < 0.001$). Rodents were captured in baited Longworth life traps, set up in a 10 m × 10 m trapping grid. Total number of traps was 46 in site A and 42 in site B. Data are pooled over four trapping sessions in late summer, early and late autumn 1995, and spring 1996. Each session was conducted over 4 days, checking traps before sunset and after sunrise. Plots were classified as being located in open area (site A: sparse vegetation of *Deschampsia flexuosa* and *Vaccinium myrtillus*; site B: bare litter), under cover (site A: mainly bracken stands and a small oak thicket; site B: bracken stands and small thickets of *Ilex aquifolium*, *Rubus* sp. and beech) or at the edge (within 5 m of edge between cover and open area). Relative number of catches is calculated by multiplying the proportion of captures of a species in one of the three habitat types by the ratio of total trap number and number of traps in that habitat type (see den Ouden (2000) for details).

experimental caches (0.36 seeds/m²) significantly faster underneath downed tree crowns and significantly slower from low vegetation underneath the canopies of oak woodland (Table 13.1). This is consistent with the higher number of captured animals, and thus activity, in structurally more complex habitat (see Figs 13.2, 13.3).

We also carried out five experiments in which we presented 49 acorns on a 1 m² plot to rodents and video-recorded seed removal (see Jansen and den Ouden, Chapter 22, this volume). Acorns were tagged with small neodymium magnets and a numbered metal tag, inserted through a drilled hole in the blunt end of the acorn and plugged with grafting wax. Combined weight of magnet, tag and wax compensated for the removed cotyledon tissue, so original seed fresh weight was maintained. We searched the surrounding area the day after removal to relocate acorns and record their fate. We found that on each seed plot a single individual wood mouse removed all seeds within 1 night. On average, it took a mouse 3.6 h to remove all seeds, which corresponds to a removal rate of 13.9 ± 1.7 acorns/hour (mean ± SE, $n = 5$).

Infested acorns

Acorn-feeding insects may infest a large part of the acorn crop (Gurnell, 1993; Crawley and Long, 1995). Foragers clearly discriminate against insect-infested acorns, which have lower nutritional value and are perishable. Crawley and Long (1995), for example, found that rodents removed only 0–18% of acorns infested by weevils, whereas they removed 45–100% of sound acorns. In our own seed removal

Table 13.1. Median discovery times (in days) by rodents of experimental caches in mixed *Quercus robur/Betula pendula* woodland. Plot sizes were 10 m × 10 m containing 36 *Q. robur* acorns in an evenly spaced grid (1.67 m between acorns) and buried to simulate shallow caches. All treatments were replicated five times. Plots were part of an experiment on seed removal with different seed densities, plot sizes and different cover grouped within five different boar-exclosures (clustered plots), or distributed within stands (scattered).

Presence of vegetation cover on seed plot	Distribution of seed plots in experiment		Mean rank
	Scattered plots	Clustered plots	
No cover	10	3	20.5
Cover by dead bracken	6	2	17.2
Cover by branches	2	1	8.8
Mean rank[a]	19.7	11.3	

[a] Mean ranks are calculated for Kruskal Wallis test to examine differences in discovery times. For the contrast scattered vs. clumped: $\chi^2 = 7.2$, df = 1, $P = 0.007$; for the effect of cover: $\chi^2 = 9.7$, df = 2, $P = 0.008$.

experiments, all acorns that rodents ignored turned out to be infested or rotten upon inspection. Bossema (1979) experimentally showed that jays can visually distinguish between infested and non-infested seeds: acorns with artificial holes simulating infestation were rejected. Other jay species, like the Mexican jay (*Aphelocoma ultramarina*), show similar preference for sound, non-infested acorns (Hubbard and MacPherson, 1997). Hence, even though infested seeds may still be able to germinate (Vázquez Pardo, 1998), pre-dispersal infestation of acorns is at the cost of dispersal and burial by animals.

Dispersal Patterns

Jays and wood mice differ greatly in body size and activity range. Both species are territorial, but, while wood mice have territory sizes up to 1 ha (Corbet and Harris, 1991; Tattersall *et al.*, 2001), jays can have territory sizes up to 10–100 ha (Andren, 1990; Lensink, 1993; Rolando, 1998). This causes considerable differences in dispersal distances of acorns.

Jay

Jays are pure scatterhoarders, as caches typically contain one acorn only (Bossema,

1979; Gomez, 2003). Jays, unlike wood mice, often carry several seeds at the same time, but even then they will hide each acorn at a different spot, albeit relatively close to each other. Bossema (1979) recorded inter-cache distances between 0.2 and 15 m, with 0.5–1 m predominating. In general, dispersal distances of acorns by jays are in the order of several hundred metres (Chettleburgh, 1952; Bossema, 1979; Kollmann and Schill, 1996; Gomez, 2003). However, jays often feed outside their territories, and transport acorns back into their own territory for caching. Thus, acorns can be dispersed over distances up to several kilometres (Schuster, 1950).

Bossema (1979) found that dispersal distance was related to the amount of acorns simultaneously transported by the birds. Flights with only one acorn in the bill tended to be much shorter (< 20 m) than flights with three or more seeds (> 100 m). Gomez (2003) found a mean dispersal distance of 263 m in jays collecting *Quercus ilex* acorns in a heterogeneous landscape in Spain. The distance distribution clearly had two peaks: one close to the trees where the jays collected acorns and another several hundred metres away from the tree. Dispersal distance was clearly determined by the distance of favourable habitats for acorn caching. Bossema (1979) found similar results for jays caching *Q. robur* seeds. The resulting seed shadows produced by jays are clearly skewed to the right and demonstrate the importance of jays for the long distance dispersal of oak (cf. Darley-Hill and Johnson, 1981; Johnson and Webb, 1989).

Wood mouse

Wood mice are also scatterhoarders, though less purely than jays. We found that about 20% of the caches in our experiments contained more than one seed (Figs 13.6, 13.7). Dispersal by wood mice results in relatively modest dispersal distances. We conducted several experiments to investigate the dispersal patterns produced by wood mice. In our video monitoring experiment (see section 'Seed removal rates') we retrieved the

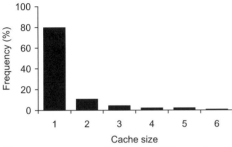

Fig. 13.6. Frequency distribution of the number of acorns in caches made by wood mice (*Apodemus sylvaticus*), measured directly after seed removal.

Fig. 13.7. A seed cache made by wood mice containing six acorns. Photo: J. den Ouden.

cached seeds immediately after they were removed – one by one – from the seed plot. Most seeds were initially cached close to the seed plot, and number of caches strongly declined with increasing distance (Fig. 13.8). Median dispersal distance ranged between 2.7 and 9.2 m for the five seed plots, and was 7.5 m for the pooled data. In a random search, we found one acorn 70 m away from the source. So, despite the mean short dispersal distances, the seed shadow may have a tail of considerable length. Other experiments involving *Apodemus* or closely related mice resulted in similar estimates of dispersal distance (Sork, 1984; Iida, 1994; Sone and Kohno, 1996; den Ouden, 2000).

Our video recordings of seed removal showed that wood mice placed additional acorns in existing caches only after having cached other seeds in between, rather than directly after another as we expected. This suggests that the animals re-use caches because they run out of suitable caching sites. It also appears that wood mice rapidly sequester seeds and hide at least part of the seeds in provisional caches that they may re-cache later to hide them better (cf. Jenkins and Peters, 1992). In new caches, for example, almost 65% of the caches were very shallow (< 2 cm deep: see Fig. 13.9), while in older caches – located 1 month after seed removal from seed plots in another experiment – shallow caches were few, and the proportion of deep caches was significantly higher (Fig. 13.9). It also appears that the distribution of dispersal distances becomes flatter over time (Jensen and Nielsen, 1986; den Ouden, 2000; see Fig. 13.8). The shift in the distribution of cache distances can be the result of re-caching, but also of higher cache use (by either the disperser or food competitors) close to the seed source, as seed retrieval is related to cache density (Stapanian and Smith, 1984; Clarkson *et al.*, 1986).

Cache Properties

Contrasting habitat preferences also produce differences in cache locations between

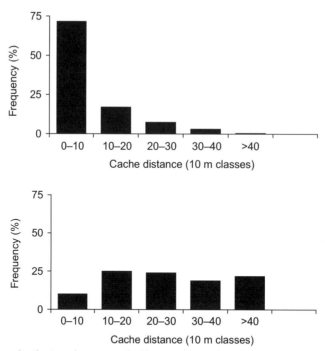

Fig. 13.8. Distance distribution of acorns cached by wood mice directly after dispersal (top) and 1 month later (bottom). Top: Pooled data from five seed plots in oak–beech woodland where 49 tagged acorns were surveyed by camera and seen taken by wood mice. Seeds were relocated directly after dispersal. Bottom: Data from another experiment in the same woodland. In this case, 200 tagged acorns were offered, not surveyed by camera and relocated 1 month after dispersal. Only intact caches are shown. Data from den Ouden (2000).

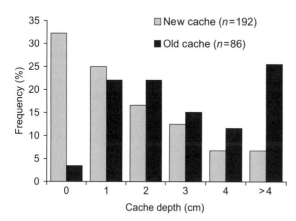

Fig. 13.9. Frequency distribution of depth of acorn caches made by wood mice (*Apodemus sylvaticus*). Cache depths of new caches were determined directly after removal from five seed plots in the video monitoring experiment, containing 49 acorns each. Depths of old caches are from caches relocated 1 month after removal (latter data from den Ouden, 2000). These can be original caches or re-cached seeds. Caches were significantly deeper 1 month after seed removal (Mann–Whitney $U = 4692$, $P < 0.001$).

jay and wood mouse. Jays spend much time looking for good cache locations and tend to store seeds in open and exposed habitat with no or low ground vegetation. Particularly when they transport several acorns at once, they are highly selective with respect to such areas (Bossema, 1979; Gomez, 2003). Wood mice, on the other hand, are mostly active in more closed habitat where they are protected by the cover of the vegetation. Hence, jay and wood mouse bring acorns into completely different habitats.

Jay

While caching, jays make a hole in the soil surface and push the acorn into the soil. Caches are typically shallow. Gomez (2003) found an average cache depth of 1.5 cm (range 0.5–4 cm). Kollmann and Schill (1996) report that jays may additionally cover the hoard with 1–3 cm of litter or moss. This depth is ideal for acorn germination and seedling establishment (Shaw, 1968b; Crawley and Long, 1995). Jays preferentially cache seeds in transitions between different structures or near distinct objects, which helps them to memorize cache locations (Bossema, 1979; Gomez, 2003). Visual markers, such as little sticks placed next to caches, serve as memory cues for later cache retrieval. Jays clearly prefer short grassland and other open habitats for seed storage (Fig. 13.10), while areas covered by dense tree or shrub canopies or high grasses and herbs tend to be avoided (Chettleburgh, 1952; Bossema, 1979; Kollmann and Schill, 1996; Gomez, 2003). The birds have also been found to cache seeds in open pine forests with a scarce understorey vegetation (Kollmann and Schill, 1996; Gomez, 2003). Effectively, this means that jays mainly contribute to the dispersal of acorns into open habitats and light forests. This is favourable to early seedling establishment and survival

Fig. 13.10. Eurasian jay (*Garrulus glandarius*) caching an acorn in grassland. Photo: R. Smit/ OUTDOORVISION.NL.

because oak is a light-demanding species (Jones, 1959; Ziegenhagen and Kausch, 1995; Bakker *et al.*, 2004).

Wood mouse

Wood mice, in contrast, prefer storing seeds in closed microhabitats. We looked at the dispersal of acorns that were presented in open and dense understorey habitat in oak coppice woodland (consisting of mainly *Deschampsia flexuosa* and *Vaccinium myrtillus*, surrounded by a more closed structure built by swards of bracken fern (*Pteridium aquilinum*)) and an oak–beech forest with bare litter, bracken swards and a thicket of holly (*Ilex aquifolium*) (den Ouden, 2000; see 'Seed Removal' above). In the oak–beech forest, rodents cached acorns under the canopies of bracken and holly, or under a dead tree in the open area, but never in open and exposed areas (Fig. 13.11). In the oak woodland, dispersal distances in a year with acorn mast were half the distance in the subsequent year with a failed seed crop (Fig. 13.11). Seeds were not preferentially cached in areas under the bracken canopy, probably because of the relatively low contrast in habitat structure between bracken and grass/blueberry, as compared to the bracken–litter contrast in the oak–beech forest (Fig. 13.11). However, seed dispersal was directed towards the centres of rodent activity as determined in prior rodent trapping (den Ouden, 2000). Wood mice in our experiments made half the caches near objects such as branches and clumps of grass and moss. Stopka and Macdonald (2003) showed that wood mice use objects for way-marking, so the distinct objects near caches are likely to serve as memory cues to the position of caches, helping wood mice to retrieve food, at least in the short run.

These results are in line with those of earlier studies investigating the importance of the structural characteristics of the habitat for the caching of seeds by rodents. Jensen and Nielsen (1986) found that, in a mosaic of

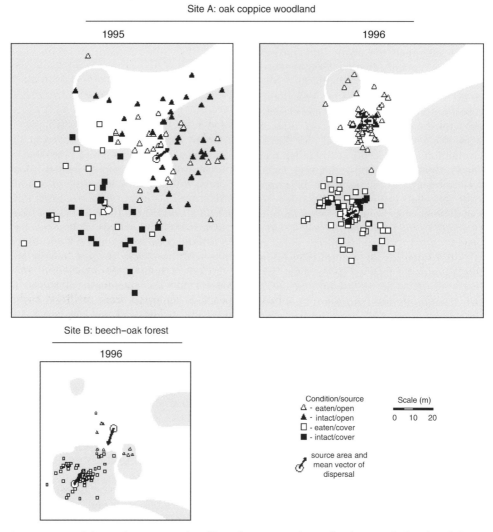

Fig. 13.11. Seed dispersal patterns produced by rodents in an oak woodland (top) and a beech–oak forest (bottom) in 1995 (2 × 200 tagged acorns presented) and 1996 (2 × 150 tagged acorns presented). Acorns were retrieved 1 month (1995) or 5 months after removal from the seed plots that were located in structure-rich microhabitat (squares) and open microhabitat (triangles). Arrows represent the mean vector of dispersal direction and distance. In 1995, there was a rich seed crop, whereas in 1996 the seed crop was negligible (from den Ouden, 2000).

Calluna, Empetrum and *Deschampsia* heath land, acorns were cached mostly in the areas dominated by *Empetrum*, not in the more open *Calluna* heath. In forested habitat, potentially favourable sites to which wood mice may disperse seeds are tree fall gaps and forest edges, which are often rich in structure as well as light. Such sites are also less accessible to foraging boar that pilfer caches. Here, the caching wood mouse faces a dilemma: caching seeds in structurally complex patches makes the cache vulnerable to pilferage by other rodents. Caching seeds in open habitat decreases the chance of pilferage (Table 13.1) but at the same time exposes the animal to predators.

Cache Survival

Although acorns can be very abundant, animals exploit them so heavily that, in general, no viable acorns remain on the forest floor at the onset of winter. Effectively, all seeds are eaten, infested or hoarded within the period of seed ripening and seed fall. Seedling establishment from non-buried seeds is rare even in the absence of seed predators. Acorns are highly sensitive to desiccation (Finch-Savage, 1992). Frost also damages acorns severely, and a high percentage will rot before they can germinate (Shaw, 1968b). This implies that oak recruitment must largely come from hoarded seeds.

Once hoarded, there are several escape routes possible for the cached seed to survive until spring (Jansen and Forget, 2001). The hoarding animal may die, but animals can also forget their cache locations. Observations by Bossema (1979) and Goodwin (1986) suggest that the long-term spatial memory of jays is as well developed as in related corvids like Clark's nutcracker, *Nucifraga columbiana* (Vander Wall, 1982; Balda and Kamil, 1992). Jays can even retrieve seeds from snow-covered caches (Goodwin, 1986). We have no information on long-term memory capacity of the wood mouse, but we assume that wood mice only have a short-term memory of cache locations, as have squirrels (Macdonald, 1997), which may be enhanced by using markers or specific cues (Stopka and Macdonald, 2003). This implies that wood mice eventually lose their memory of cache sites, and become naïve with respect to their own caches, depending on olfactory cues for cache retrieval. This will increase survival chances of acorns cached by wood mice, whereas caches from jays remain subject to retrieval by the caching animal year-round (Bossema, 1979).

The survival percentage of caches until spring, when acorns germinate, has hardly been quantified for jays or mice. We found only one seedling from a primary cache in the spring from 147 tagged acorns in three seed plots, but more acorns may have germinated in secondary or higher-order caches. Gomez *et al.* (2003) found that in a *Quercus pyrenaica* forest less that 4% of experimental acorns and less than 2% of the experimental seedlings survived. Wood mice and wild boar were the main seed predators and the authors state that, in southern European oak ecosystems, wood mice act mainly as predators rather than dispersers of acorns.

The potential survival of acorns in jay-made caches is probably higher than the survival of rodent caches because of the large differences in cache environment. The preference of jays to cache acorns in open areas, outside areas with dense tree or shrub canopy cover or in dense herb and grass vegetation, puts seeds into an environment were predation pressure by rodents is relatively low (Hulme and Kollmann, Chapter 2, this volume). Consequently, many more seedlings appeared from artificial caches in open habitat like short grasslands than from caches in woodlands, where few seedlings emerged (Crawley and Long, 1995; Kollmann and Schill, 1996). However, seed predation pressure by wild boar is higher in open areas (Gomez *et al.*, 2003) (Fig. 13.12), so composition and density of the different predator populations will ultimately determine the surviving proportions of jay caches. Also, in open areas herbivory pressure on seedlings is high. Hence, the emerging seedling has higher survival probability near or under shrubs and other vegetation than in open areas (Gomez *et al.*, 2003; Bakker *et al.*, 2004).

The emerged seedlings, however, are still exposed to damage by granivorous animals, since the emergence of the primary shoot acts as a flag to attract seed predators. Jays, wood mice and wild boar will all feed on the remaining cotyledons. In the case of wild boar, this almost always leads to the death of the seedlings since they are completely pulled out of the ground (J. den Ouden, personal observation). Jays and wood mice generally damage only a fraction of the seedlings when they remove the cotyledons, and the seedling is well able to survive since seedling growth is not affected when remaining cotyledons are removed (Sonesson, 1994).

Fig. 13.12.   Wild boar (*Sus scrofa*) foraging in grassland. Photo: R. Smit/OUTDOORVISION.NL.

Jays, Wood Mice and Oak Population Dynamics

The positive role of jays in the oak life cycle is widely acknowledged. Wood mice, in contrast, have been treated mainly as seed predators and were often considered a main cause of regeneration failure (cf. Gomez *et al.*, 2003). A striking example of this misappreciation is the extensive application of rodenticides in forests to enhance oak regeneration (cf. Margaletic *et al.*, 2002). We believe, however, that wood mice also have a net positive effect on recruitment of oak, despite the fact that they may consume most of the seeds they hide. While jays provide long distance transportation of seeds into open habitats, wood mice provide short distance dispersal into structurally more complex habitat patches, mostly under the dense canopies of trees, shrubs, high ground vegetation and woody debris. Thus, jays and wood mice are complementary in their seed dispersal and caching.

Oak recruitment in a landscape context

Considering the fact that a multitude of animal species feed on acorns (Tanton, 1965; Corbet, 1974; McShea, 2000; McShea and Healy, 2002), merely burying seeds will already enhance seed survival (Shaw, 1968b; Crawley and Long, 1995). Indeed,

most oak recruitment comes from caches (Bossema, 1979; Crawley and Long, 1995; Kollmann and Schill, 1996). However, oak seedling survival and growth will also be determined by the location of caches with respect to the biotic and abiotic environment (Herrera, 1995; Gomez, 2003; Bakker *et al.*, 2004), which differs greatly between jay and wood mouse. The ultimate contribution of jays and mice to oak population dynamics depends on whether caches are safe sites.

The two main factors that regulate oak survival and growth within its natural habitat are light and herbivory (Jones, 1959; Vera, 2000). Jays preferentially cache seeds in open, low vegetation near transitions between vegetation types or in open pine forest (Chettleburgh, 1952; Bossema, 1979; Kollmann and Schill, 1996; Gomez, 2003). Such sites provide ample light for seedling growth, but simultaneously expose the seedlings to herbivores, which may cause high seedling mortality (Herrera, 1995; Gomez, 2003). Wood mice are mainly active in microhabitats that provide cover and will cache most seeds under the canopy of trees, under shrubs or in other dense vegetation. This places most emerging seedlings in shady conditions, but simultaneously reduces the exposure to herbivores. So, when protection by shrubs is required for successful establishment and survival (associational resistance, cf. Bakker *et al.*, 2004), wood mice rather than jays seem ideal vectors for carrying acorns into such structure-rich places. Therefore, jays and wood mice work in concert with respect to the dispersal of oak in heterogeneous landscapes.

The dispersal of oak within dense woodland or shrub takes place almost entirely via scatterhoarding by wood mice, since jays avoid this habitat when caching food (Bossema, 1979; Gomez, 2003). Wood mice not only hoard seeds under the canopy, but also in gaps and among large woody debris. Germinating seeds from caches maintain a seedling population that is able to sustain itself in the shade for a few years on the food reserves drawn from the cotyledons (Ziegenhagen and Kausch, 1995). After 2 years, seedlings require higher light levels, and survival then depends on the occurrence of a disturbance event to release the seedling. The potential for oak seedlings to grow up in these conditions subsequently depends on herbivory pressure and competition from other vegetation (Vera, 2000; Smit, 2002).

Post-glacial colonization

An important argument behind the idea that jays are the principal dispersers of oak is the high rate of post-glacial colonization of *Quercus* into the northern hemisphere. Oaks achieved a migration rate of as much as 350–500 m/year on continental Europe, and 150–500 m/year in the British Isles (Huntley and Birks, 1983), which is similar to many wind-dispersed species (Birks, 1989). Reasoning that the age of first acorn production ranges between 15 and 40 years (Jones, 1959), this implies that oak migrated in the order of 5–20 km per generation. *Quercus* in North America, in comparison, achieved a post-glacial migration rate at the low end of this estimate: approximately 7 km per generation (Johnson and Webb, 1989). Johnson and Webb concluded that the blue jay (*Cyanocitta cristata*) must have been the principal dispersal agent for North American oaks, discarding other hypotheses such as ultra-long dispersal by

passenger pigeons. But is the Eurasian jay the main candidate for the role of main vector in the rapid post-glacial colonization of northern Europe by oak?

If jays were the main post-glacial dispersers of oak, this would imply that there has been a constant long right-tailed distribution of dispersal distances that results in effective dispersal over distances greater than 5 km. We are not convinced that the underlying assumption is met that dispersal distances are continuously exceeding many kilometres. Schuster (1950) indeed saw birds travelling over 4–5 km distances between trees from which they harvested acorns and their territories in which they cached them, but Chettleburgh (1952), Bossema (1979) and Gomez (2003) only occasionally observed jays caching acorns further away than 1 km. Also, the dispersal should be directed towards oak-free habitat, which in turn requires that jays occupy territories without oaks. Indeed, the current distribution range of the jay extends to central Scandinavia and Finland (Hagemeijer *et al.*, 1997), which is several hundreds of kilometres more north than the range of oak (Jahn, 1991). Also, a large proportion of the forest ecosystem that oak colonized consisted of *Pinus sylvestris*, a forest type preferably used by jays to cache acorns (Kollmann and Schill, 1996; Mosandl and Kleinert, 1998; Gomez, 2003).

Nevertheless, the absence of other (avian) long distance dispersers does not automatically imply that jays are the sole candidates for explaining the fast rate of both European and North American oak migration. In this respect other mechanisms are still needed to sufficiently explain oak migration rates (cf. Birks, 1989). Oaks in Europe could have been transported northward over several hundred kilometres by water (Ridley, 1930). Also, we believe we should not disregard the possibility that migrating prehistoric humans have been taking acorns along as food and could have dispersed oaks. Indeed, acorns were found as food remnant in prehistoric settlements (Kubiak-Martens, 1999), and are still a food source for humans today (Luna-José *et al.*, 2003).

Conclusion

Bossema (1979) gave his thesis the subtitle: 'An eco-ethological study of a symbiosis'. He argued that the jay depended on oak as food source, and oak depends on the jay for dispersal and reproduction. We argue that a similar symbiosis exists between wood mice and oaks. In terms of dispersal, the jay acts on the scale of the landscape with home ranges of the order 10–100 ha (Andren, 1990; Lensink, 1993; Rolando, 1998), while wood mice act on the level of habitat patches in home ranges of the order 0.1–1 ha (Corbet and Harris, 1991; Tattersall *et al.*, 2001). Dispersal distances are in the range of 10–50 m for wood mice and 100–1000 m for jays. The amount of seeds dispersed by wood mice is unknown, but, assuming a single wood mouse caches 10 seeds per night during an acorn crop season of 50 days, an individual may cache 500 seeds. A wood mouse population density of ten animals per hectare would thus be able to disperse and cache 5000 acorns/ha. This is comparable to the amount of acorns cached by an individual jay. Note, however, that population densities are ten times higher in wood mouse than in jay, so, in absolute figures, wood mice disperse more seeds than jays.

The relative contributions to oak reproduction by wood mice and jays are largely unknown, and will strongly depend on the context of the habitat. Wood mice are the principal dispersers in dense woodland and shrubs, while jays are the main vectors for reproduction of isolated trees and recruitment in open landscapes. Thus, wood mice seem as important to oak regeneration as jays.

Acknowledgements

We thank Pascalle Jacobs, Willem Loonen, Jaco van Altena, Melchert Meijer zu Schlochtern and Miranda Aronds for their assistance in the fieldwork, and H. Hees of the Dutch State Forest Service for permission to work at Speulderbos. Comments by Joanna Lambert, Pierre-Michel Forget, José Gomez, Frans Vera, Markus Feijen and one anonymous reviewer greatly improved earlier versions of the manuscript.

References

Andren, H. (1990) Despotic distribution, unequal reproductive success, and population regulation in the jay *Garrulus glandarius* L. *Ecology* 71, 1796–1803.

Bakker, E.S., Olff, H., Vandenberghe, C., de Maeyer, K., Smit, R., Gleichman, J.M. and Vera, F.W.M. (2004) Ecological anachronisms in the recruitment of temperate light-demanding tree species in wooded pastures. *Journal of Applied Ecology* 41, 571–582.

Balda, R.P. and Kamil, A.C. (1992) Long-term spatial memory in Clark's nutcracker, *Nucifraga columbiana. Animal Behaviour* 44, 761–769.

Birks, H.J.B. (1989) Holocene isochrone maps and patterns of tree spreading in the British Isles. *Journal of Biogeography* 16, 503–540.

Bossema, I. (1979) Jays and oaks: an eco-ethological study of a symbiosis. *Behaviour* 70, 1–117.

Chettleburgh, M.R. (1952) Observations on the collection and burial of acorns by jays in Hainault Forest. *British Birds* 45, 359–364.

Clarkson, K.S., Eden, S.F., Sutherland, W.J. and Houston, A.I. (1986) Density dependence and magpie food hoarding. *Journal of Animal Ecology* 55, 111–121.

Corbet, G.B. (1974) The importance of oak to mammals. In: Morris, M.G. and Perring, F.H. (eds) *The British Oak: its History and Natural History.* Classey, Faringdon, UK, pp. 312–323.

Corbet, G.B. and Harris, S. (1991) *The Handbook of British Mammals,* 3rd edn. Blackwell Science, Oxford, 588 pp.

Crawley, M.J. and Long, C.R. (1995) Alternate bearing, predator satiation and seedling recruitment in *Quercus robur* L. *Journal of Ecology* 83, 683–696.

Darley-Hill, S. and Johnson, W.C. (1981) Acorn dispersal by the blue jay (*Cyanocitta cristata*). *Oecologica* 50, 231–232.

den Ouden, J. (2000) The role of bracken (*Pteridium aquilinum*) in forest dynamics. Dissertation, Wageningen University, Wageningen, The Netherlands, 218 pp.

Finch-Savage, W.E. (1992) Seed development in the recalcitrant species *Quercus robur* L.

Germinability and desiccation tolerance. *Seed Science Research* 2, 17–22.

Gomez, J.M. (2003) Spatial patterns in long-distance dispersal of *Quercus ilex* acorns by jays in a heterogeneous landscape. *Ecography* 26, 573–584.

Gomez, J.M., Garcia, D. and Zamora, R. (2003) Impact of vertebrate acorn- and seedling-predators on a Mediterranean *Quercus pyrenaica* forest. *Forest Ecology and Management* 180, 125–134.

Goodwin, D. (1986) *Crows of the World*, 2nd edn. British Museum (Natural History), London, 299 pp.

Groot Bruinderink, G.W.T.A. and Hazebroek, E. (1995) Ingestion and diet composition of red deer (*Cervus elaphus* L.) in the Netherlands from 1954 till 1992. *Mammalia* 59, 187–195.

Groot Bruinderink, G.W.T.A., Hazebroek, E. and van der Voet, H. (1994) Diet and condition of wild boar, *Sus scrofa scrofa*, without supplementary feeding. *Journal of Zoology, London* 233, 631–648.

Gurnell, J. (1981) Woodland rodents and tree seed supplies. In: Chapman, J.A. and Pursley, D. (eds) *The Worldwide Furbearer Conference Proceedings*. R.R. Donnelly & Sons Co., Virginia, pp. 1191–1214.

Gurnell, J. (1993) Tree seed production and food conditions for rodents in an oak wood in southern England. *Forestry* 66, 291–315.

Hagemeijer, W.J.M., Blair, M.J., Van Turnhout, C., Bekhuis, J. and Bijlsma, R. (1997) *EBCC Atlas of European Breeding Birds: their Distribution and Abundance*. Poyser, London, 903 pp.

Hansson, L. (1978) Small mammal abundance in relation to environmental variables in three Swedish forest phases. *Studia Forestalia Suecica* 147, 5–38.

Herrera, J. (1995) Acorn predation and seedling production in a low-density population of cork oak (*Quercus suber* L.). *Forest Ecology and Management* 76, 197–201.

Hubbard, J.A. and MacPherson, G.R. (1997) Acorn selection by Mexican jays: a test of a tri-trophic symbiotic relationship hypothesis. *Oecologia* 110, 143–146.

Huntley, B. and Birks, H.J.B. (1983) *An Atlas of Past and Present Pollen Maps for Europe 0–13,000 Years Ago*. Cambridge University Press, Cambridge, 667 pp.

Iida, S. (1994) Quantitative analysis of acorn transportation by rodents using magnetic locator. *Vegetatio* 124, 39–43.

Jahn, G. (1991) Temperate deciduous forests of Europe. In: Röhrig, E. and Ulrich, B. (eds) *Ecosystems of the World, 7. Temperate Deciduous Forests*. Elsevier, Amsterdam, pp. 377–503.

Jansen, P.A. and Forget, P.-M. (2001) Scatter-hoarding and tree regeneration. In: Bongers, F., Charles-Dominique, P., Forget, P.-M. and Théry, M. (eds) *Nouragues: Dynamics and plant–animal Interactions in a Neotropical Rainforest*. Kluwer, Dordrecht, The Netherlands, pp. 275–288.

Jansen, P.A., Hemerik, L. and Bongers, F. (2004) Seed mass and mast seeding enhance dispersal by scatter-hoarding rodents in a Neotropical rainforest tree. *Ecological Monographs* 4 (in press).

Janzen, D.H. (1970) Herbivores and the number of tree species in tropical forests. *American Naturalist* 104, 501–528.

Jenkins, S.H. and Peters, R.A. (1992) Spatial patterns of food storage by Merriam's kangaroo rats. *Behavioral Ecology* 3, 60–65.

Jensen, T.S. (1982) Seed production and outbreaks on non-cyclic rodent populations in deciduous forests. *Oecologia* 54, 184–192.

Jensen, T.S. and Nielsen, O.F. (1986) Rodents as seed dispersers in a heath–oak wood succession. *Oecologia* 70, 214–221.

Johnson, W.C. and Webb, T. III (1989) The role of blue jays (*Cyanocitta cristata* L.) in the postglacial dispersal of fagaceous trees in eastern North America. *Journal of Biogeography* 16, 561–571.

Jones, E.W. (1959) Biological flora of the British Isles. *Quercus* L. *Journal of Ecology* 47, 169–222.

Kelly, D. and Sork, V.L. (2002) Mast seeding in perennial plants: why, how, where? *Annual Review of Ecology and Systematics* 33, 427–447.

Kenward, R.E. and Holm, J.L. (1993) On the replacement of the red squirrel in Britain: a phytotoxic explanation. *Proceedings of the Royal Society London Series B* 251, 187–194.

Kikkawa, J. (1964) Movement, activity and distribution of the small rodents *Clethrionomys glareolus* and *Apodemus sylvaticus* in woodland. *Journal of Animal Ecology* 33, 259–299.

Kollmann, J. and Schill, H.-P. (1996) Spatial patterns of dispersal, seed predation and germination during colonization of abandoned grasslands by *Quercus petraea* and *Corylus avellana*. *Vegetatio* 125, 193–205.

Kubiak-Martens, L. (1999) The plant food component of the diet at the late Mesolithic (Ertebølle) settlement at Tybrind Vig,

Denmark. *Vegetation History and Archeo-botany* 8, 117–127.

Lensink, R. (1993) *Vogels in het Hart van Gelderland.* KNNV/SOVON, Utrecht, 561 pp.

Luna-José, A.L., Montalvo-Espinosa, L. and Rendón-Aguilar, B. (2003) Los usos no leñosos de los encinos en Mexico. *Boletín de la Sociedad Botánica de México* 73, 107–117.

Macdonald, I.M.V. (1997) Field experiments on duration and precision of grey and red squirrel spatial memory. *Animal Behaviour* 54, 879–891.

Margaletic, J., Glavaš, M. and Bäumler, W. (2002) The development of mice and voles in an oak forest with a surplus of acorns. *Journal of Pest Science* 75, 95–98.

Mellanby, K. (1968) The effects of some mammals and birds on regeneration of oak. *Journal of Applied Ecology* 5, 359–366.

McShea, W.J. (2000) The influence of acorn crops on annual variation in rodent and bird populations. *Ecology* 81, 228–238.

McShea, W.J. and Healy, W.M. (eds) (2002) *Oak Forest Ecosystems. Ecology and Management for Wildlife.* Johns Hopkins University Press, Baltimore, Maryland, 432 pp.

Mosandl, R. and Kleinert, A. (1998) Development of oak (*Quercus petraea* (Matt.) Liebl.) emerged from bird-dispersed seeds under old-growth pine (*Pinus sylvestris* L.) stands. *Forest Ecology and Management* 106, 35–44.

Ovington, J.D. and Murray, G. (1964) Determination of acorn fall. *Quarterly Journal of Forestry* 58, 152–159.

Ridley, H.N. (1930) *The Dispersal of Plants Throughout the World.* L. Reeve & Co, Ashfort, 744 pp.

Rolando, A. (1998) Factors affecting movements and home ranges in the jay (*Garrulus glandarius*). *Journal of Zoology, London* 246, 249–257.

Schuster, L. (1950) Über den Sammeltrieb des Eichelhähers (*Garrulus glandarius*). *Vogelwelt* 71, 9–17.

Shaw, M.W. (1968a) Factors affecting the natural regeneration of sessile oak (*Quercus petraea*) in North Wales. I. A preliminary study of acorn production, viability and losses. *Journal of Ecology* 56, 565–583.

Shaw, M.W. (1968b) Factors affecting the natural regeneration of sessile oak (*Quercus petraea*) in North Wales. II. Acorn losses and germination under field conditions. *Journal of Ecology* 56, 647–660.

Smit, R. (2002) The secret life of woody species. A study on woody species establishment, interactions with herbivores and vegetation succession. Dissertation, Wageningen University, Wageningen, The Netherlands.

Sone, K. and Kohno, A. (1996) Application of radiotelemetry to the survey of acorn dispersal by *Apodemus* mice. *Ecological Research* 11, 187–192.

Sonesson, L.K. (1994) Growth and survival after cotyledon removal in *Quercus robur* seedlings, grown in different natural soil types. *Oikos* 69, 65–70.

Sork, V.L. (1984) Examination of seed dispersal and survival in red oak, *Quercus rubra* (Fagacaea), using metal-tagged acorns. *Ecology* 65, 1020–1022.

Sork, V.L., Bramble, J. and Sexton, O. (1993) Ecology of mast-fruiting in three species of North American deciduous oaks. *Ecology* 74, 528–541.

Stapanian, M.A. and Smith, C.C. (1984) Density-dependent survival of scatterhoarded nuts: an experimental approach. *Ecology* 65, 1387–1396.

Stopka, P. and Macdonald, D.W. (2003) Way-marking behaviour: an aid to spatial navigation in the wood mouse (*Apodemus sylvaticus*). *BMC Ecology* 3. Available at: www.biomedcentral.com/1472-6785/3/3

Tanton, M.T. (1965) Acorn destruction potential of small mammals and birds in British woodland. *Quarterly Journal of Forestry* 59, 1–5.

Tattersall, F.H., Macdonald, D.W., Hart, B.J., Manley, W.J. and Feber, R.E. (2001) Habitat use by wood mice (*Apodemus sylvaticus*) in a changeable arable landscape. *Journal of Zoology, London* 255, 487–494.

Theimer, T.C. (2001) Seed scatterhoarding by white-tailed rats: consequences for seedling recruitment by an Australian rain forest tree. *Journal of Tropical Ecology* 17, 177–189.

Van Apeldoorn, R.C., Oostenbrink, W.T., Van Winden, A. and Van Der Zee, F.F. (1992) Effect of habitat fragmentation on the bank vole (*Clethrionomys glareolus*) in an agricultural landscape. *Oikos* 65, 265–274.

Vander Wall, S.B. (1982) An experimental analysis of cache recovery in Clark's nutcracker. *Animal Behaviour* 30, 84–94.

Vander Wall, S.B. (1990) *Food Hoarding in Animals.* University of Chicago Press, Chicago, Illinois, 445 pp.

Vander Wall, S.B. (2002) Secondary dispersal of Jeffrey pine seeds by rodent scatterhoarders: the roles of pilfering, recaching, and a variable environment. In: Levey, D., Silva, W.R. and Galetti, M. (eds) *Seed Dispersal and Frugivory: Ecology, Evolution* and *Conservation*. CAB International, Wallingford, UK, pp. 193–208.

Vander Wall, S.B. (2003) Effects of seed size of wind dispersed pines (*Pinus*) on secondary seed dispersal and the caching behavior of rodents. *Oikos* 100, 25–34.

Vázquez Pardo, F.M. (1998) *Semillas del género Quercus L. (Biología, Ecología y Manejo).* Publicaciones de la Secretaría General Técnica, Junta de Extremadura, 234 pp.

Vera, F.W.M. (2000) *Grazing Ecology and Forest History.* CAB International, Wallingford, UK, 506 pp.

Wadewitz, O. (1976) Die Sammelflüge des Eichelhähers. *Falke* 23, 160–164.

Watts, C.H.S. (1969) The regulation of wood mouse (*Apodemus sylvaticus*) numbers in Wytham Woods, Berkshire. *Journal of Animal Ecology* 38, 285–304.

Ziegenhagen, B. and Kausch, W. (1995) Productivity of young shaded oaks (*Quercus robur* L.) as corresponding to shoot morphology and leaf anatomy. *Forest Ecology and Management* 72, 97–108.

14 Walnut Seed Dispersal: Mixed Effects of Tree Squirrels and Field Mice with Different Hoarding Ability

Noriko Tamura,[1] Toshio Katsuki[1] and Fumio Hayashi[2]

[1]Tama Forest Science Garden, Forestry and Forest Product Research Institute, Todori, Hachioji, Tokyo 193-0843, Japan; [2]Department of Biology, Tokyo Metropolitan University, Minamiosawa 1-1, Hachioji, Tokyo 192-0397, Japan

Introduction

Many plant species depend on scatter-hoarding animals for seed dispersal. In particular, nut bearing trees such as walnuts, hickories, beeches, oaks and several tropical species are thought to have evolved in response to selection by rodents (Vander Wall, 1990). Such species produce larger, more nutritious seeds with woody husks that are fewer in number than plants with other dispersal modes (Vander Wall, 1990). For example, in Japan, at least 21 species of woody plants are reported to depend on rodent seed dispersal (Table 14.1). Most of these seeds were large in size, nutritious and covered by hard shell.

Four study types have been conducted in Japan to investigate rodent seed dispersal:

1. Direct observation of rodent hoarding behaviour in the field (Hayashida, 1988).
2. Observation of rodent removal of experimentally marked seeds placed on the forest floor (Kanazawa and Nishikata, 1976; Kikuzawa, 1988; Miyaki and Kikuzawa, 1988; Isaji and Sugita, 1997).
3. Analysis of seedling distribution relative to crowns of parent trees (Isaji and Sugita, 1997; Goto and Hayashida, 2002).

4. Monitoring of foraging rodents via traceable seeds marked by spools, magnets and radio transmitters (Yasuda *et al.*, 1991; Iida, 1996; Nagahama, 1996; Soné and Kohno, 1996; Tamura and Shibasaki, 1996). This earlier work indicates that some seeds are transported by rodents, escape predation and eventually germinate. However, we have few quantitative field data regarding plant fitness (Hoshizaki *et al.*, 1997).

Co-evolution between a single species of plant and vertebrate in seed dispersal systems is very rare (Herrera, 1985). The concept of 'diffuse co-evolution' may be more applicable in seed dispersal, wherein plants with similar seed characteristics rely on the seed dispersal behaviour of several species of animals belonging to the same guild. As indicated (Table 14.1), in Japan several plant species (*Quercus, Aesculus, Juglans*) depend on seed dispersal by both tree squirrels and field mice. It is possible that one plant species is dispersed by several sympatric species of animals as well as several allopatric species throughout their geographic distribution. However, interactions between sympatric species in seed hoarding behaviour have not been fully investigated (e.g. Vander Wall, 2000; Leaver and Daly, 2001).

Table 14.1.	Woody plants in Japan depending on seed dispersal by rodents.

Plant species	Animal species (seed hoarder)	Reference
Abies homolepis	*Sciurus vulgaris*	Lee (1999)
A. sachalinensis	*Sciurus vulgaris*	Lee (1999)
Aesculus turbinata	*S. vulgaris, Apodemus speciosus*	Isaji and Sugita (1997); Lee (1999)
Castanea creanata	*S. vulgaris*	Lee (1999)
Castanopsis cuspidata	*A. argenteus*	Shimada (2001)
Corylus sieboldiana	*S. lis*	Kato (1985)
Fagus crenata	*A. speciosus*	Miguchi (1996)
Juglans ailanthifolia	*A. speciosus, S. lis, S. vulgaris*	Hayashida (1988); Tamura and Shibasaki (1996)
Kalopanax pictus	*Tamias sibiricus*	Kawamichi (1980)
Larix leptolepis	*S. lis*	Kato (1985)
Lithocarpus edulis	*A. speciosus, A. argenteus*	Nagahama (1996); Soné and Kohno (1996)
Pinus densiflora	*S. lis/ S. vulgaris*	Kato (1985); Lee (1999)
P. koraiensis	*S. vulgaris*	Miyaki (1987); Hayashida (1988)
P. parviflora	*S. lis/ S. vulgaris*	Kato (1985); Hayashida (1988)
P. thunbergii	*S. vulgaris*	Lee (1999)
Prunus sp.	*T. sibiricus*	Kawamichi (1980)
Quercus acutissima	*A. speciosus, A. argenteus*	Iida (1996)
Q. crispula	*A. speciosus, A. argenteus, T. sibiricus*	Kanazawa and Nishikata (1976); Kawamichi (1980)
Q. dentata	*T. sibiricus*	Kawamichi (1980)
Q. serrata	*A. speciosus/A. argenteus*	Miguchi (1994); Iida (1996); Shimada (2001)
Q. variabilis	*A. speciosus*	Miguchi (1994)

In this chapter, we compare food selection by two sympatric species of rodents. One of the most preferred foods, walnuts (*Juglans ailanthifolia*), was counted in the field to investigate predation rates by the two species. We next compare hoarding rates, hoarding sites and retrieval rates of cache by both rodent species. Finally, we present field data on seedling distribution and survival rates. Our purpose in this study is to evaluate the effects of these two rodent species; we thus conclude with an analysis of their effects by employing conceptual models.

Study Sites

The study was conducted in suburban forests at Hachioji City, Western Tokyo, Central Honshu Island, Japan (35°39′N, 139°16′E). The average temperature was from 2.2°C in January to 24.6°C in August. The average precipitation varied from 33.5 mm in December to 258.1 mm in

September, with a total of 1628 mm precipitation for a year (Toyoda and Tanimoto, 2000). Study sites 1 and 2 are parts of the experimental forests of the Tama Forest Science Garden (TFSG: 57 ha). This forest consists of six types of mosaic forests (Tamura, 1998): (i) natural forests (23.0% in area) dominated by *Abies firma* and *Q. glauca*; (ii) secondary forest (18.2% in area) dominated by *Q. glauca* and *Q. serrata*; (iii) artificial conifer forest (14.0% in area) of *Cryptomeria japonica* and *Chamaecyparis obtusa*; (iv) artificial deciduous forests (7.8% in area) of *J. ailanthifolia, Zelkova serrata*, and so on; (v) arboretum (18.2% in area); and (vi) shrub and grassland (18.8% in area). The two study sites are walnut groves in the artificial deciduous forests (forest type iv). At site 1, five walnut trees were planted in a 0.12 ha patch at 210 m in elevation, while, at site 2, *c*. 50 trees were planted in a 0.39 ha patch along the stream at 230 m in elevation. These trees were *c*. 80 years old, with heights ranging between 15 and 20 m. More than half of the understorey consisted of *Aucuba japonica* Thunb.

Study site 3 is located in the northern part of the protected forests as the Tama Mausoleum (TM: 47 ha). This forest has not been logged for the past 70 years and access is restricted. The predominant species there are: *Q. serrata* and *Q. acutissima*, with *Pinus densiflora* on the ridge (Azegami *et al.*, 1991). We set the study plot of 80 m × 160 m in size at 180 to 210 m in elevation. The dominant species in the study plot were deciduous *Q. serrata* in the secondary forest (76% in area) and *C. japonica* in the plantation (23% in area).

Materials

The Japanese walnut, *J. ailanthifolia*, is distributed on Hokkaido, Honshu, Shikoku, Kyushu and several other small Japanese islands (Horikawa, 1976). This species' habitat is characterized by flat or mild slope areas of riparian forests with humid soils (Tozawa and Akasaka, 1980). According to Fjellstrom and Parfitt (1994), there has been minimal genetic differentiation between *J. ailanthifolia* and *J. mandshurica* or *J. cathayensis* distributed in China and Formosa. The seeds of the Japanese walnuts are 1.5–4.3 cm in shell length and 4.3 to 14.8 g in fresh weight (Tamura, 1994).

Two species of tree squirrels occur in Japan: *Sciurus vulgaris orientalis* found only on Hokkaido Island; and *S. lis*, which occurs on Honshu, Shikoku and Awaji islands. The body weight of *S. vulgaris* is 300–410 g, while that of *S. lis* is 250–310 g (Abe, 1994). Both species are known to eat and hoard walnuts (Table 14.1). Two species of field mice, *Apodemus argenteus* and *A. speciosus*, are distributed on Hokkaido, Honshu, Shikoku, Kyushu and several small islands. *A. argenteus* is smaller (10–20 g) than *A. speciosus* (20–60 g) (Abe, 1994) and has not been reported to use the walnuts seed as food. Another field mouse species, *A. peninsulae*, lives on Hokkaido Island, although its food habits have not been investigated. The chipmunk, *Eutamias sibiricus*, is also distributed on Hokkaido Island, but there are no reports of it using walnuts as food.

Walnuts have extremely hard shells and are rarely preyed upon by animals other than rodents. Black bears (Hashimoto and Takatsuki, 1997) and crows (Nihei, 1995) are known as potential predators in some localities on Honshu Island, but not at the present study site, where *S. lis* and *A. speciosus* were thought to be the main predators of walnut seeds. Both rodent species are known to scatterhoard seeds (Tamura, 1997, 2001; Goto and Hayashida, 2002). We determine whether a seed was eaten by squirrels or field mice by the feeding marks on the shell; squirrels shave along the raphe of the shell and crack the halves with their teeth, while mice shave and make holes on two sides of the shell, and then pick the meat, piece by piece, from the holes (Fig. 14.1).

Food Selection

We selected 12 species of large seeds that are distributed in the lowland forests of central Honshu, Japan, including: *Q. glauca*, *Q. myrsinaefolia*, *Q. acutissima*, *Q. serrata*, *Custanopsis cuspidata*, *Castanea crenata*, *Lithocarpus edulis*, *Torreya nucifera*, *Ginkgo biloba*, *Aesculus turbinata*, *Camellia japonica* and *J. ailanthifolia*. Seeds used for the following experiment were collected in the forests of Tama Forest Science Garden. We selected six species, and set 30 seeds (five seeds from each species) on the feeding stand (1.5 m in height) at study site 1. Seeds were maintained on the stand for 4 days and we recorded the process of animals removing seeds by using a video-camera set 1.5 m from the feeding stand. These trials were repeated 16 times in different combinations of plant species at 5–7 day intervals from September to November 1999. A total of 480 seeds, 40 seeds from each species, was provided during the study period.

Sixteen per cent of the 480 seeds were removed by squirrels; 6% were taken by the various tits (*Parus varius*) and jays (*Garrulus glandarius*); and 1% were removed by nuthatches (*Sitta europaea*) (Table 14.2). Squirrels preferentially selected walnuts and chestnuts, followed by the larger acorns, but never the other six species. The avian species used smaller seeds than the squirrels.

We set five feeding boxes on the ground at study site 1, separated by more than 10 m, to conduct the same experiment for field mice. The entrance of each feeding box (10 cm × 15 cm × 20 cm) was covered with a wire net that allowed mice to enter the box, but not squirrels. Thirty seeds (five seeds from each of the six species) were placed in the feeding boxes, and seed transportation was recorded for 2 days by an automatic camera (September–October 1999). A total of 40 experiments were conducted using 1200 seeds (100 seeds from each species); each trial was replicated eight times per

Fig. 14.1. Japanese walnuts with feeding signs of the Japanese squirrel (left) and the large Japanese field mouse (right). Scale bars: 1 mm in the minimum scale.

Table 14.2. The number of seeds taken from the feeders by Japanese squirrels and large Japanese field mice. The number of provided seeds was 40 for each species at the feeding stand and 100 for each species at the feeding box. Only the data for mice are from a feeding box set on the floor, the rest being from a feeding stand on the branch (refer to text).

Plant species	Seed weight (g) (mean ± SD)	Number of seeds removed by				
		Squirrels	Tits	Jays	Nuthatches	Mice
Castanea crenata	18.2 ± 5.0	22	0	5	0	86
Aesculus turbinata	11.7 ± 1.7	0	0	0	0	65
Juglans ailanthifolia	8.1 ± 1.3	28	0	0	0	58
Quercus acutissima	6.7 ± 1.9	5	0	11	0	96
Lithocarpus edulis	3.0 ± 0.5	3	0	4	1	94
Torreya nucifera	2.0 ± 0.3	16	18	0	2	97
Quercus serrata	1.9 ± 0.4	0	0	5	0	66
Ginkgo biloba	1.7 ± 0.3	0	1	1	0	100
Camellia japonica	1.6 ± 0.4	0	0	0	0	92
Castanopsis cuspidata	0.8 ± 0.1	1	10	2	3	98
Quercus myrsinaefolia	0.7 ± 0.1	0	0	0	0	41
Quercus glauca	0.7 ± 0.2	0	0	0	0	53
Total		75 (15.6%)	29 (6.0%)	28 (5.8%)	6 (1.3%)	946 (78.8%)

feeding box in different species combinations. Seventy-nine per cent of the 1200 seeds were removed by the field mice (Table 14.2). To varying degrees, the field mice used all 12 seed species. The most frequently selected species were *G. biloba*, *C. cuspidata*, *T. nucifera*, *Q. acutissima*, *L. edulis* and *C. japonica*. More than 90% of provided seeds were taken away for these selected species, while only 58% of provided walnuts were used. As a result, compared with other species at our study site, the walnut seeds were preferentially used by tree squirrels, but not field mice.

Predation Rates of Walnuts by Two Rodent Species

To determine how much and when the walnuts were used by the two rodent species under natural conditions, six plots (3 m × 3 m) were set on the floor under walnut canopies at study site 2. The seeds found in the plots were counted at 1- or 2-week intervals from 11 August to 30 November 1992. To prevent double

counting, the seeds were marked with a red dot on each side of the shells with nail polish and kept at the place they were found. The species eating the seeds could be easily identified by the feeding sign on the shell (Fig. 14.1). Because squirrels break the walnuts into two shells to eat, each shell was counted as 0.5.

Walnuts fell from August to November, but mainly in September and October (Fig. 14.2). During the whole period, 64 seeds fell and were marked per 9 m² plot on the floor of walnut forests. Seeds with feeding signs by squirrels and mice increased from October to November, and an average of 5.9 and 18.2 seeds per plot were eaten on the plot by squirrels and mice, respectively. On 30 November, only 0.2 seeds per plot were left without eating.

Although we cannot deny annual differences in these numbers, most seeds on the floor were eaten or transported by rodents before winter, the same as in previous Japanese studies (Kanazawa and Nishikata, 1976; Kikuzawa, 1988; Miyaki and Kikuzawa, 1988; Isaji and Sugita, 1997). The number of seeds with evidence of mouse feeding was three times that of squirrels.

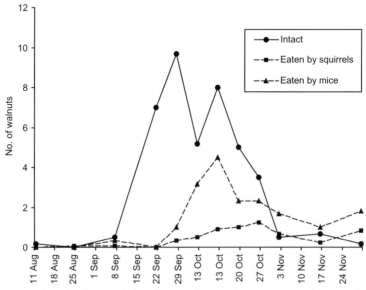

Fig. 14.2. Mean number of walnut seeds found on six plots (each 3 m × 3 m quadrat set under walnut tree canopy) from 11 August to 30 November 1992. Three types of walnuts, intact ones, and seeds with feeding signs of squirrels or mice, were discriminated and counted.

These differences may result partly from differences in population density at this study site; the density of mice was 5–63 individuals/ha (Soné and Kohno, 1999), while that of squirrels was estimated as 0.06–0.13 individuals/ha (N. Tamura, unpublished).

However, the abundance of feeding signs indicates only the number of seeds eaten at the site, and not the number of seeds transported and then eaten/hoarded by the rodents. In the next section, we analyse the number of seeds transported and hoarded by squirrels and mice, respectively.

Hoarding Behaviour by Two Species of Rodents

Two feeding stands (1.5 m in height) were set under the walnut trees at study site 2. Ten walnut seeds with radio-transmitters (2 g in weight, 50 MHz band, SR-43 SW, Nakane Studio) were placed on the feeding stand. When the walnuts were removed from the feeding stand, their locations were traced by a receiver (FT-690, Yaesu Co., Ltd) within 1 or 2 days. When the seeds were stored on the ground or among twigs by squirrels, the seeds were kept as they were. The locations of hoarded seeds were checked every week until they were retrieved. These procedures were conducted five times for each feeding stand from September to December in 1992, so that a total of 100 seeds were provided (Tamura and Shibasaki, 1996).

Sixty per cent of the 100 provided seeds were originally scatterhoarded by squirrels. The mean distance from the feeding stand to cache sites was 18.3 m ($n = 60$, range 1–62) (Tamura and Shibasaki, 1996). Twenty-five per cent of the hoarded seeds were pilfered by field mice, 63% were retrieved by squirrels and 12% were left until germination in spring. Similar experiments were conducted at sites 1 and 2 in different years and seasons (Tamura *et al.*, 1999). The transport distance differed between seasons and sites (ranging from 0 to 168 m), and the mean value ranged from 10.3 m to 19.9 m. The hoarding rate was lowest in winter (28%) and highest in

autumn (68%). A total of 720 walnuts were provided during the course of the study; of these, 51% were scatterhoarded and 33% were eaten immediately. Thus, the number of scatterhoarded walnuts versus eaten walnuts by squirrels in this study site was 1:0.65.

To investigate the hoarding behaviour of field mice, we set up a wooden feeding box (10 cm × 15 cm × 20 cm) on the ground at study site 2. The entrance of the box was covered with a wire net that allowed mice to enter the box, but not squirrels. Five walnuts with the same radio-transmitters as in the experiment for squirrels were placed in the feeding box. The seed weight used for this experiment was 6–10 g. Nineteen trials ($n = 95$ seeds) were conducted from September to February in 1996 and 16 trials ($n = 80$ seeds) were done from September to February in 2000 (Tamura, 2001).

In 1996, only 30% of the seeds were scatterhoarded by the field mice, 2% were larderhoarded under the ground and 68% were eaten immediately by the mice. In 2000, 36% were scatterhoarded, 32% were larderhoarded under the ground and 32% were eaten immediately. Because larderhoarding, which was deep in burrows containing at least two walnuts in the same place, seemed not to play an important role in the regeneration of walnuts, we considered here only the role of scatterhoarding. The number of scatterhoarded walnuts versus eaten walnuts by the field mice in this study site was 1:1.58 on the average of 1996 and 2000. Scatterhoarding was never observed in the winter of 1996 (December 1996–February 1997) and only 17% in the winter of 2000 (December 2000–February 2001). The transport distance for hoarding was not different between years and was 6.2 m on the average (range: 0–21 m, $n = 66$). Out of 30 hoarded seeds by mice in 1996, 3% was pilfered by squirrels, 73% was retrieved by mice and 23% remained intact by May (Tamura, 2001).

As Janzen (1970) pointed out, fruiting trees that are both spatially and temporally patchy will attract many seed predators; seed survival hence increases with distance from parents. Survival rates of seeds do not increase linearly, but reach a constant level at some distance from the source. In fact, disappearance rates of artificially hoarded walnut seeds are generally higher near the source trees, and decreased rapidly with distance (Tamura et al., 1999).

Differences in body size seem to affect the differences in transport distance between the two species; the body weights of squirrels and mice are c. 250 g and 31.5 g, respectively (Tamura, 2001). The mice transported the acorns of Q. acutissima (3.4–6.6 g) 22.1 m on the average from the source (Iida, 1996), while they transported the large seeds of horse chestnuts, A. turbinata (13–27 g) only 0.6–2.1 m from the feeder (Isaji and Sugita, 1997). The field mice in the present study site transported the acorns of L. edulis (6 g including a transmitter) farther than the walnut seeds (Soné and Kohno, 1996). Therefore, the seed size may affect the transport distance by the field mice.

Seedling Distribution

Seedling distributions were surveyed in a study plot at study site 3, where squirrels and mice are sympatric. There were two walnut trees producing seeds at both the northern and western edges (79 m apart from each other). The elevation was lowest at the northern and western sides (180 m above sea level) and increased gradually towards the hill (210 m) at the south-eastern edge. Therefore, walnut seeds did not roll down by themselves within the study plots. Locations of the seedlings were recorded on a map in July 1994, 1995 and 1996, and in September 1997. All the seedlings were identified with a number tape and aged based on their leaf scars formed after wintering. Presence/absence of seedlings was recorded yearly to monitor survival rates. The nearer one of the two walnut trees was assumed to be the parent of each seedling.

Seedling distance from trees is shown in Fig. 14.3. The mean distance of the current year seedlings was 33.9 m ($n = 34$, SD = 20.0,

range 7–74 m) in 1994, 27.7 m ($n = 60$, SD = 17.0, range 5–93 m) in 1995, 23.6 m ($n = 88$, SD = 16.7, range 4–111 m) in 1996 and 32.4 m ($n = 29$, SD = 18.2, range 7–70 m) in 1997. The seedlings at least 1 year old (saplings) were much fewer and distributed farther from the trees than the current year seedlings. The mean distance of the saplings was 46.3 m ($n = 19$, SD = 32.7, range 10–106 m) in 1994, 52.0 m ($n = 15$, SD = 31.8, range 12–106 m) in 1995, 49.6 m ($n = 10$, SD = 31.0, range 13–105 m) in 1996 and 38.6 m ($n = 14$, SD = 37.5, range 9–111 m) in 1997. The distances (log-transformed values) differed significantly between current year seedlings and saplings significantly in 1995 ($t = 2.93$, $P < 0.01$) and 1996 ($t = 3.32$, $P < 0.01$), but did not differ in 1994 ($t = 0.94$, $P = 0.35$) and 1997 ($t = 0.15$, $P = 0.88$). In 1994 and 1997, the number of current year seedlings was lower than in the other 2 years. In 1997, seedling data were not taken in July but in September, so that some current year seedlings may have been lost during these 2 months. Although we did not gather quantitative data on seed production, there is a possibility that the seed amount in 1993 (and probably 1996) may have been lower than normal. In such years, the squirrels may harvest most of the walnuts from branches before falling on the ground, so that the field mice handle fewer walnuts than in a normal year and the mean transport distance increases as a result.

Figure 14.4 shows the number of each year-class seedlings surviving after marking, which are categorized into the following three distance groups from the source trees: less than 10 m, 10–20 m, and more than 20 m. For all combined data of 1994 to 1996, 6% of 26 current year seedlings survived for only 1 year within the distance of 10 m. In the distance of 10–20 m, 15% of 53 current year seedlings survived to 1 year, and 22% of nine 1-year-old seedlings survived to 2 years, but no individual survived to 3 years. In the distant seedlings (more than 20 m from the tree), survival rates were higher; 10% of individuals ($n = 103$) survived in the first year, and eight and five individuals survived in the second and third years, respectively. Thus, farther transportation

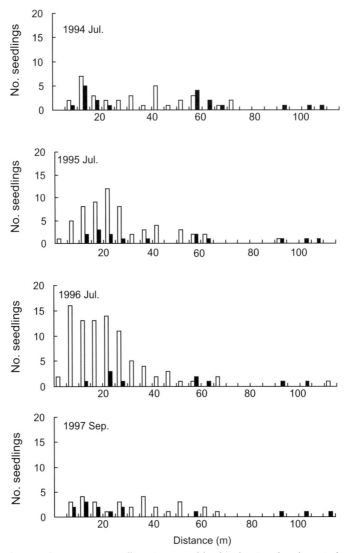

Fig. 14.3. Distribution of current year seedlings (0 years old, white bars) and saplings (at least 1-year-old seedlings, black bars) of the Japanese walnut in relation to the distance from the parent trees in 1994–1997.

from the parent trees seems to be advantageous for seedlings to increase their survival rate.

Distance effects to increase seedling survival, known as the 'Janzen–Connell model', have been well examined for various tree species in tropical forests (e.g. Clark and Clark, 1984; Wenny, 2000). In temperate forests of Japan, some species showed a similar phenomenon; for example, in *Q. crispula* serious leaf damage to seedlings occurred within 10 m from the parent tree (Wada *et al.*, 2000) and the survival rates of current year seedlings in *Carpinus* spp. increased with distance from the parent tree (Shibata and Nakashizuka, 1995). However, the distance from the source tree did not affect the survival rates of seedlings in *Q. serrata* (Nakashizuka *et al.*, 1995). In future work we will investigate the cause of mortality in the seedlings of Japanese walnut near the parent trees.

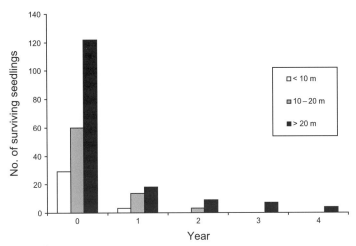

Fig. 14.4. The number of each age-class of seedlings observed in three categories of distance from the trees: 0–10 m, 10–20 m and more than 20 m. The number of seedlings was summarized through study periods from 1994 to 1997.

Conclusion: Factors Affecting the Fitness of Walnuts

A critical factor influencing plant fitness is hoarding frequency. In the present study, the frequency of scatterhoarding by squirrels was twice that of field mice. In *S. vulgaris* on Hokkaido Island, 66% of walnut seeds provided were hoarded, the rest being eaten (Hayashida, 1988). In the field mice, 100% of *L. edulis* and 47% of *Q. serrata* were hoarded in the same study site as the present study (Soné and Kohno, 1996). The frequency of hoarding may be affected by several factors, including: perishability of food items (Fox, 1982; Steele *et al.*, 1996), handling time (Woodrey, 1990; Jacobs, 1992; Hadj-Chikh *et al.*, 1996; Cristol, 2001), fat levels (Lucas and Walter, 1991; Brodin, 2000), social interaction (Brotons, 2000), food availability (Brodin, 2000) and deprivation (Lucas, 1994; Clayton and Dickinson, 1999). Further studies will be necessary to know the variation in range of hoarding rates for squirrels and mice.

The survival rate of seeds which are hoarded but not recovered by animals is also important for increasing the fitness of walnuts. However, this value seems to include more difficult to determine variables, such as hoarder's death, availability of other food resources and competitors. Distance transported by hoarders and density of the hoarder species will also influence plant fitness. We found that, after germination, the survival rate of the seedlings increased with distance and the seedlings of more than 2 years old are rarely found within 20 m from the source trees. The seeds hoarded by mice at 0–21 m from the source may survive less after germination than those hoarded by squirrels at 0–168 m.

Here, we propose a fitness model of walnuts hoarded by rodents and include several of the above factors. The fitness of walnut was generated by adding the number of surviving seedlings derived from the squirrel's hoarding and the field mouse's hoarding. We found that the number of survived seedlings was affected by the number of seeds hoarded, their survival rate until germination, and the survival rate of seedlings after germination (Eqn 1). On the other hand, we assume that the total number of seeds used (including both eaten and hoarded) by dispersers is constant (Eqn 2). We assume that under stable conditions the number of hoarded seeds vs. eaten seeds by each species of rodents is constant, and both the number of seeds hoarded and the number eaten are affected by rodent density (Eqn 2).

$$F = Hs \times Gs \times Ss + Hm \times Gm \times Sm \qquad (1)$$

$$K = Es + Hs + Em + Hm = aHs + bHm \quad (2)$$

by (1) and (2),

$$F = Hs \, (cSs - dSm) + fKSm \quad (3)$$

where F is fitness of walnut, H is total number of seeds hoarded by rodents (Hs: by squirrel; Hm: by mouse), G is survival rate of seeds until germination (Gs: survival rate of seeds hoarded by squirrel; Gm: survival rate of seeds hoarded by mouse), S is survival rate of seedlings after germination (Ss: seedlings surviving from squirrel's hoarding; Sm: seedlings surviving from mouse's hoarding), E is total number of seeds eaten without hoarding by rodents (Es: by squirrel; Em: by mouse), K is a constant: total number of seeds hoarded and eaten by rodents, a, b are constants: factors to know the total number of seeds used by rodents (including eaten without hoarding) estimated from hoarding vs. eating rates of squirrel and mouse, respectively, and c, d and f are constants calculated from a, b, Gs and Gm.

For example, the value estimated from the present study was put in this conceptual model. Values were as follows: Gs = 0.12, Gm = 0.23, a = 1.65 (because eaten vs. hoarded = 0.65:1.00), and b = 2.58 (because eaten vs. hoarded = 1.58:1.00).

$$F = 0.12 \, HsSs + 0.23 \, HmSm \quad (4)$$

$$K = 1.65Hs + 2.58Hm,$$
$$Hm = 0.39K - 0.64Hs \quad (5)$$

by (4) and (5),

$$F = Hs \, (0.12Ss - 0.15Sm) + 0.09KSm \quad (6)$$

Under condition Ss > Sm, the fitness of walnut tree will increase linearly with Hs (Eqn 6). This fitness formula suggests that the number of survived seedlings increases as squirrels take seeds more frequently than mice.

We have presented data on hoarding and retrieval behaviour in the field; however, these numbers vary depending upon site, time and environment. The assumption that the number of hoarded seeds vs. the number of eaten seeds is constant for each rodent species is uncertain, especially when the food availability, food quality and/or density of competitors change drastically.

Unfortunately, the plasticity of hoarding rate, retrieval rate and transport distance has not been adequately investigated, although our data suggest that walnut fitness may depend on the proportion of the two rodent species. To best understand the evolution of walnut seeds that rely on two hoarder species, we need to improve the fitness model, including variation in the factors mentioned above.

Acknowledgements

We thank K. Soné and T. Yokoyama for giving valuable suggestions to start this study, Y. Hashimoto and E. Shibasaki for assisting with the fieldwork, and P.-M. Forget, K. Hoshizaki, J.E. Lambert, R. Tabuchi and anonymous reviewers for their critical reading of this manuscript. We also greatly appreciate the staff of Tama Mausoleum for allowing us to conduct field studies there.

References

Abe, H. (1994) *A Pictorial Guide to the Mammals of Japan.* Tokai University, Tokyo, 195 pp.

Azegami, C., Tanimoto, T. and Toyoda, T. (1991) A list of the higher plants in Tama Mausoleum, Tokyo Metropolis. *Science Report of Takao Museum Natural History* 14, 1–20.

Brodin, A. (2000) Why do hoarding birds gain fat in winter in the wrong way? Suggestions from a dynamic model. *Behavioral Ecology* 11, 27–39.

Brotons, L. (2000) Individual food-hoarding decisions in a nonterritorial coal tit population: the role of social context. *Animal Behaviour* 60, 395–402.

Clark, D.A. and Clark, D.B. (1984) Spacing dynamics of a tropical rain forest tree: evaluation of the Janzen–Connell model. *The American Naturalist* 124, 769–788.

Clayton, N.S. and Dickinson, A. (1999) Motivational control of caching behaviour in the scrub jay, *Aphelocoma coerulescens. Animal Behaviour* 57, 435–444.

Cristol, D.A. (2001) American crows cache less preferred walnuts. *Animal Behaviour* 62, 331–336.

Fjellstrom, R.G. and Parfitt, D.E. (1994) Walnut genetic diversity determined by restriction fragment length polymorphisms. *Genome* 37, 690–700.

Fox, J.E. (1982) Adaptation of gray squirrel behavior to autumn germination by white oak acorns. *Evolution* 36, 800–809.

Goto, S. and Hayashida, M. (2002) Seed dispersal by rodents and seedling establishment of walnut trees in a riparian forest. *Journal of the Japanese Forestry Society* 84, 1–8. (In Japanese with English summary.)

Hadj-Chikh, L.Z., Steele, M.A. and Smallwood, P.D. (1996) Caching decisions by grey squirrels: a test of the handling time and perishability hypotheses. *Animal Behaviour* 52, 941–948.

Hashimoto, Y. and Takatsuki, S. (1997) Food habits of Japanese black bears: a review. *Honyurui Kagaku* (Mammalian Science) 37, 1–19. (In Japanese with English abstract.)

Hayashida, M. (1988) The influence of social interactions on the pattern of scatterhoarding in red squirrels. *Bulletin of Hokkaido University, Faculty of Agriculture* 45, 267–278. (In Japanese with English summary.)

Herrera, C.M. (1985) Determinants of plant–animal coevolution: the case of mutualistic dispersal of seeds by vertebrates. *Oikos* 44, 132–141.

Horikawa, Y. (1976) *Atlas of the Japanese Flora*, 2. Gakken, Tokyo, 518 pp.

Hoshizaki, K., Suzuki, W. and Sasaki, S. (1997) Impacts of secondary seed dispersal and herbivory on seedling survival in *Aesculus turbinata*. *Journal of Vegetation Science* 8, 735–791.

Iida, S. (1996) Quantitative analysis of acorn transportation by rodents using magnetic locator. *Vegetatio* 124, 39–43.

Isaji, H. and Sugita, H. (1997) Removal of fallen *Aesculus turbinate* seeds by small mammals. *Japanese Journal of Ecology* 47, 121–129. (In Japanese with English summary.)

Jacobs, L.F. (1992) The effect of handling time on the decision to cache by grey squirrels. *Animal Behaviour* 43, 522–524.

Janzen, D.H. (1970) Herbivores and the number of tree species in tropical forests. *The American Naturalist* 104, 501–528.

Kanazawa, Y. and Nishikata, S. (1976) Disappearance of acorns from the floor in *Quercus crispula* forests. *Journal of Japanese Forest Society* 58, 52–56.

Kato, J. (1985) Food and hoarding behavior of Japanese squirrels. *Japanese Journal of Ecology* 35, 13–20.

Kawamichi, M. (1980) Food, food hoarding and seasonal changes of Siberian chipmunks. *Japanese Journal of Ecology* 30, 211–220.

Kikuzawa, K. (1988) Dispersal of *Quercus mongolica* acorns in a broadleaved deciduous forest. *Forest Ecology and Management* 25, 1–8.

Leaver, L.A. and Daly, M. (2001) Food caching and differential cache pilferage: a field study of coexistence of sympatric kangaroo rat and pocket mice. *Oecologia* 128, 577–584.

Lee, T.H. (1999) Habitat use, food habits and mating ecology of the Eurasian red squirrel (*Sciurus vulgaris* L.). Doctoral dissertation, Environment Earth Science, Hokkaido University, Sapporo, Hokkaido, Japon, 72 pp.

Lucas, J.R. (1994) Regulation of cache stores and body mass in Carolina chickadees (*Parus carolinensis*). *Behavioral Ecology* 5, 171–181.

Lucas, J.R. and Walter, L.R. (1991) When should chickadees hoard food? Theory and experimental results. *Animal Behaviour* 41, 579–601.

Miguchi, H. (1994) Role of wood mice on regeneration of cool temperate forest. In: Kobayashi, S., Nishikawa, K., Danilin, I.M., Matsuzaki, T., Abe, N., Kamitani T. and Nakashizuka, T. (eds) *Proceeding of NAFRO Seminar of Sustainable Forestry and its Biological Mechanisms*. NAFRO, pp. 115–121.

Miguchi, H. (1996) Dynamics of beech forest from the view point of rodents ecology. *Japanese Journal of Ecology* 46, 185–189 (in Japanese).

Miyaki, M. (1987) Seed dispersal of the Korean pine, *Pinus koraiensis*, by the red squirrel, *Sciurus vulgaris*. *Ecological Research* 2, 147–157.

Miyaki, M. and Kikuzawa, K. (1988) Dispersal of *Quercus mongolica* acorns in a broadleaved deciduous forest. 2. Scatterhoarding by mice. *Forest Ecology and Management* 25, 9–16.

Nagahama, D. (1996) Acorn dispersal by rodents. *Ringyo Gijyutu* 652, 30–31 (in Japanese).

Nakashizuka, T., Iida, S., Masaki, T., Shibata, M. and Tanaka, H. (1995) Evaluating increased fitness through dispersal: a comparative study on tree population in a temperate forest, Japan. *Ecoscience* 2, 245–251.

Nihei, Y. (1995) Variations of behaviour of carrion crows *Corvus corone* using automobiles as nutcrackers. *Japanese Journal of Ornithology* 44, 21–35. (In Japanese with English abstract.)

Shibata, M. and Nakashizuka, T. (1995) Seed and seedling demography of co-occurring *Carpinus* species in a temperate deciduous forest. *Ecology* 76, 1099–1108.

Shimada, T. (2001) Hoarding behaviors of two wood mouse species: different preference for acorns of two Fagaceae species. *Ecological Research* 16, 127–133.

Soné, K. and Kohno, A. (1996) Handling of acorns by *Apodemus* mice. *Research Bulletin of the Kagoshima University Forests* 24, 89–94.

Soné, K. and Kohno, A. (1999) Acorn hoarding by the field mouse, *Apodemus speciosus* Temminck (Rodentia: Muridae). *Journal of Forest Research* 4, 167–175.

Steele, M.A., Hadj-Chikh, L.Z. and Hazeltine, J. (1996) Caching and feeding decisions by *Sciurus carolinensis*: responses to weevil-infested acorns. *Journal of Mammalogy* 77, 301–314.

Tamura, N. (1994) Application of radio-transmitter for studying seed dispersion by animals. *Journal of the Japanese Forestry Society* 76, 607–610.

Tamura, N. (1997) Japanese squirrels as a seed disperser of walnuts. *Primate Research* 13, 129–135 (in Japanese).

Tamura, N. (1998) Forest type selection by the Japanese squirrel, *Sciurus lis*. *Japanese Journal of Ecology* 48, 123–127. (In Japanese with English abstract.)

Tamura, N. (2001) Walnut hoarding by the Japanese field mouse, *Apodemus speciosus* Temminck. *Journal of Forest Research* 6, 187–190.

Tamura, N. and Shibasaki, E. (1996) Fate of walnut seeds, *Juglans ailanthifolia*, hoarded by Japanese squirrels, *Sciurus lis*. *Journal of Forest Research* 1, 219–222.

Tamura, N., Hashimoto, Y. and Hayashi, F. (1999) Optimal distances for squirrels to transport and hoard walnuts. *Animal Behaviour* 58, 635–642.

Toyoda, T. and Tanimoto, T. (2000) Forest succession in the Tama Forest Science Garden (formerly Asakawa Experiment Forest), Hachioji, Central Japan. *Bulletin of the Forestry and Forest Products Research Institute* 377, 1–60 (in Japanese).

Tozawa, T. and Akasaka, Y. (1980) Ecological studies of the secondary forest (3): the walnuts forest in Ryugamori. *Japan Forest Society Tohoku Branch* 32, 142–145 (in Japanese).

Vander Wall, S.B. (1990) *Food Hoarding in Animals*. University of Chicago, Illinois, Press, Chicago, 445 pp.

Vander Wall, S.B. (2000) The influence of environmental conditions on cache recovery and cache pilferage by yellow pine chipmunks (*Tamias amoenus*) and deer mice (*Peromyscus maniculatus*). *Behavioral Ecology* 11, 544–549.

Wada, N., Murakami, M. and Yoshida, K. (2000) Effects of herbivore-bearing adult trees of the oak *Quercus crispula* on the survival of their seedlings. *Ecological Research* 15, 219–227.

Wenny, D.G. (2000) Seed dispersal, seed predation, and seedling recruitment of a neotropical montane tree. *Ecological Monographs* 70, 331–351.

Woodrey, M.S. (1990) Economics of caching versus immediate consumption by white-breasted nuthatches: the effect of handling time. *The Condor* 92, 621–624.

Yasuda, M., Nakagoshi, N. and Takahashi, F. (1991) Examination of the spool-and-line method as quantitative technique to investigate seed dispersal by rodents. *Japanese Journal of Ecology* 41, 257–262. (In Japanese with English abstract.)

15 Influence of Forest Composition on Tree Seed Predation and Rodent Responses: a Comparison of Monodominant and Mixed Temperate Forests in Japan

Kazuhiko Hoshizaki[1] and Hideo Miguchi[2]

[1]Department of Biological Environment, Faculty of Bioresource Sciences, Akita Prefectural University, Akita 010-0195, Japan; [2]Faculty of Agriculture, Niigata University, Niigata 950–2181, Japan

Introduction

Plants often suffer heavy mortality in their seed and seedling stages. While some vertebrates play a significant role in seed dispersal, others are predators of seeds and seedlings. In many large-seeded, nut-bearing trees, rodents are responsible both for effective seed dispersal by their vigorous activity of seed hoarding (Vander Wall, 2001), and for early-stage mortality through seed and seedling predation (Crawley, 2000; Kitajima and Fenner, 2000; Hulme and Benkman, 2002). Rodents are pivotal in the regeneration and demography of large-seeded trees.

Conversely, large-seeded trees can affect rodent populations. Large seeds are a high quality food resource (Vander Wall, 2001), but they are not reliable foods for rodents because the trees often exhibit large interannual variations in seed production (Kelly, 1994; Vander Wall, 2001). The large annual variation in seed crops (e.g. masting followed by low seed crops in subsequent years) has a major influence on the population size of predators, such that the population increases after masting and decreases during low seed-crop years (Wolff, 1996b; Wright et al., 1999; Ostfeld and Keesing, 2000). Variation in rodent numbers may in turn affect seed predation and dispersal in some trees (Vander Wall, 2002; Zhang et al., Chapter 16, this volume). Ingestion of large seeds may have negative effects on rodent physiology; many large seeds, especially Quercus acorns and Aesculus seeds, contain secondary compounds which operate as a chemical defence against seed predators (e.g. Vander Wall, 2001; Shimada and Saitoh, 2003).

The interaction between large seeds and rodent populations has been intensively studied (e.g. Crawley, 2000; Vander Wall, 2001; Hulme, 2002; Hulme and Kollmann, Chapter 2, this volume, and references therein). The ecological significance of masting is supposedly that synchronous seed production satiates seed predators and so produces a disproportionately large cohort of seedlings following masting (Janzen, 1971; Kelly, 1994). However, because rodents are polyphagous and generalist seed predators (Hulme and

Benkman, 2002), their response to variation in seed crops is likely to be complex in multi-species forest systems (Hoshizaki and Hulme, 2002). In fact, there appears to be considerable variation in rates of pre- and post-dispersal seed predation and subsequent seedling recruitment (Hulme, 1998, 2002). For example, the rate of predation on temperate tree seeds varies with stand locality, habitats, seasons and years (e.g. Willson and Whelan, 1990; Kollmann et al., 1998; Hulme and Borelli, 1999; Hoshizaki and Hulme, 2002; Hulme and Kollmann, Chapter 2, this volume).

An explanation for the variability in seed predation patterns is that indirect interactions within a plant guild in which seed predators are shared may play a role in mixed forest stands (Ostfeld et al., 1996; Hoshizaki and Hulme, 2002). For example, rodents show different responses when the relative abundance of seed species changes (Hulme and Hunt, 1999; Hoshizaki and Hulme, 2002), generating inconsistent rates of seed predation. A number of studies have examined the relationship between abundance of one species of seed and seed predation (e.g. Jensen, 1985; Schupp, 1990; Crawley and Long, 1995), but only a few have attempted to examine predator abundance with reference to community-wide seed availability (Wright et al., 1999; Curran and Webb, 2000; Hoshizaki and Hulme, 2002; Schnurr et al., 2002).

In cool-temperate Japan, the dominant vegetation is beech forests (Nakashizuka, 1987). Forest types differ over relatively small geographical distances but share common component species. Forest stands show marked variation in species composition; there are beech-monodominant stands and mixed stands of large-seeded trees (Nakashizuka and Iida, 1995). Patterns of interaction may vary among stand types. In this chapter, we review the role of rodents in the recruitment of several large-seeded tree species in Japan, with emphasis on the early demography across several years and sites. We explain that the early demography of a single tree population can be affected by the early-stage dynamics of other tree

species within the same forest community. In addition we discuss how the variations in tree community structure and those in seed attributes (e.g. nutritional values) of large-seed producing plants may cause variations in their interaction with rodents. Finally we also attempt to compare case histories for temperate trees and rodents in Japan with findings from similar temperate forests in Western Europe and eastern North America.

Selected Species and Areas

The large-seeded trees

The tree species focused upon in this review are the horse chestnut (*Aesculus turbinata* Blume), beeches (*Fagus crenata* Blume and *F. japonica* Maxim.) and oaks (*Quercus crispula* Blume and *Q. serrata* Thumb. ex Murray). These taxa were chosen because the demography of their seeds and seedlings in relation to community structure is comparatively well understood (e.g. Nakashizuka et al., 1995; Hoshizaki et al., 1997; Homma et al., 1999; Nakashizuka, 2001). Species in warm-temperate, evergreen forests are excluded because less is shown of their interaction with rodents.

Aesculus turbinata (*Hippocastanaceae*) is one of the dominant species in cool-temperate riparian forests, co-occurring with *F. crenata*, *Q. crispula* and several wind-dispersed trees (Suzuki et al., 2002; Table 15.1). It bears extremely large seeds (21 g fresh weight, 6.2 g dry weight; Hoshizaki et al., 1997). Seeds are carbohydrate- and sugar-rich, but also contain high levels of secondary compounds, especially saponins (Hoshizaki, 1999; Shimada, 2001). Seed dispersal is exclusively by rodents (Isaji and Sugita, 1997). Seeds of *A. turbinata* show hypogeal germination, which enhances seedling resistance to herbivory (Hoshizaki et al., 1997).

In Japan there are two *Fagus* species (*F. crenata* and *F. japonica*) and 15 *Quercus* species. Both the *Fagus* and two *Quercus* species, *Q. crispula* (synonymous to *Q. mongolica* Fischer var. *grosseserrata* (Blume) Rehd. et Wilis.) and *Q. serrata*, are abundant in cool-temperate old-growth forests. In particular, *F. crenata* is dominant in cool-temperate montane zones and is recognized as a keystone species (see next section). Seeds are removed mainly by rodents (e.g. Miguchi and Maruyama, 1984; Kikuzawa, 1988) and to a lesser extent by birds (passerines and corvids). Fresh weight of the nut or acorn of these Fagaceae species is, in ascending order, 0.16 g for *F. japonica* (S. Taniguchi, unpublished data), 0.20 g for *F. crenata* (Miguchi, 1994), 2.8 g for *Q. serrata* (Iida, 1996) and 3.5 g for *Q. crispula* (Hoshizaki, 1999).

Stand structures

There are two contrasting types of beech forest, differing in both stand structure and geographical distribution. One is a mono-dominant forest of *F. crenata* and is typically distributed along the Sea of Japan (Japan Sea type). Here, *F. crenata* occupies 70 to nearly 100% of the stand basal area (Table 15.1). In contrast, the Pacific Ocean type is a mixed-species forest in which the dominant species are *F. crenata*, *Q. crispula* (or *Q. serrata*) and in some locations *F. japonica*, distributed along the Pacific Ocean. The dominance of each of these species ranges from 10 to 40% of the stand basal area (Table 15.1).

Table 15.1. Stand characteristics for seven study sites described in this chapter. Sites with long-term data or with data sets for both rodents and trees in Japan are selected. The seven sites are chosen from cool-temperate old-growth forests. Sites are sorted by location, within the same stand type, from north to south.

Stand type[a] and site	Location	Max. snow depth (m)	Dominant tree species[b]	Rodent investigation	Duration (yr)	Ref.[c]
Japan Sea type beech forest						
Utasai	42°38′N, 140°19′E	≈ 3	*F. crenata* (> 90)	Habitat	1	6
Lake Towada	40°24′N, 140°53′E	≈ 2	*F. crenata* (69)	Abundance and habitat	6+	1, 8
Nukumi-daira	37°55′N, 139°41′E	3–4+	*F. crenata* (99)	Abundance and habitat	11+	2, 7, 8
Buna-daira	36°58′N, 139°16′E	2–2.5	*F. crenata* (94)	Behaviour	1	11
Riparian mixed forest						
Kanumazawa	39°06′N, 140°52′E	≈ 2	*Cercidiphyllum japonicum*[d] (26) *A. turbinata* (19) *F. crenata* (15) *Q. crispula* (13)	Abundance	10+	3, 4, 12
Pacific Ocean type beech forest						
Ogawa Forest Reserve	36°56′N, 140°35′E	0–0.5[e]	*Q. serrata* (27) *F. japonica* (21) *F. crenata* (9)	None	16+	2, 9, 10
Mount Mito	35°45′N, 139°00′E	0–0.3[e]	*F. crenata* (36) *F. japonica* (25)	Behaviour	1	5, 11

[a]See text for definition and references for detailed stand characteristics.
[b]The value in parentheses represents relative basal area (%) for each species in each site.
[c]References: (1) Abe *et al.*, 2001; (2) Homma *et al.*, 1999; (3) Hoshizaki and Hulme, 2002; (4) Hoshizaki *et al.*, 1997; (5) Irie *et al.*, 1998; (6) Kitabatake and Wada, 2001; (7) Miguchi and Maruyama, 1984; (8) H. Miguchi *et al.*, unpublished; (9) Nakashizuka and Matsumoto, 2002; (10) Shibata *et al.*, 2002; (11) Shimano and Masuzawa, 1998; (12) Suzuki *et al.*, 2002.
[d]Seeds of this species are extremely small (0.8 mg), wind-dispersed and not damaged by rodents.
[e]Winter climate is generally dry and snow cover is not continuous throughout winter.

The geographical distribution of the two beech forest types corresponds well to distinct climatic regions with contrasting snowfall regimes (Nakashizuka and Iida, 1995; Homma *et al.*, 1999). The Japan Sea type forests are distributed in areas of humid climate throughout the year and with heavy winter snow (2–4 m or more maximum depth), and the Pacific Ocean type forests are distributed in areas with dry winter and without continuous snow cover throughout the winter.

Riparian forests in cool-temperate Japan are mixtures of various species and occur in both areas along the Pacific Ocean and the Sea of Japan (Sakio, 1997; Suzuki *et al.*, 2002). Riparian stands include *A. turbinata*, *F. crenata*, *Q. crispula* and several wind-dispersed trees and are distinct from the two types of beech forest in Japan (Suzuki *et al.*, 2002; Table 15.1).

We selected seven study sites, covering both Japan Sea and Pacific Ocean beech forests and a cool-temperate riparian mixed forest (Table 15.1). These sites were chosen to provide data on both seedfall and seed fate within temperate Japan in order to review demographic or behavioural interrelationships between large seeds and rodents, and to examine the response of rodents to interannual variation in seed crops in monodominant and mixed forests. The distance between most distant sites (Utasai to Mt Mito) is ≈ 800 km.

Post-dispersal seed predators

In the temperate forests in Japan, the most important seed dispersers and predators after seedfall are rodents. Two murid and three microtine species are well documented for their impacts on tree regeneration. The dominant murid rodents in most Japanese temperate forests are *Apodemus speciosus* Temminck and *A. argenteus* Temminck (Miguchi, 1988, 1996a,b). Among the microtines, *Microtus montebelli* Milne-Edwards and/or *Eothenomys andersoni* Thomas occur in most temperate forests except Hokkaido, northern Japan.

These species are usually fewer in number than murid rodents, but can have a substantial negative impact on the tree regeneration (Miguchi, 1988; Kitabatake and Wada, 2001). *Clethrionomys rufocanus* Sundevall is another microtine vole dominant in Hokkaido, but its geographical distribution only overlaps the northernmost populations of *F. crenata*.

Patterns in Seed and Seedling Demography

Negative and positive impacts of rodents on *A. turbinata* regeneration

Seed predation, seedling herbivory and seedling resistance

Seeds of *A. turbinata* are intensively removed (≈ 100%) and scatterhoarded by rodents. Caches usually contain a single seed (Hoshizaki and Hulme, 2002). Rodents typically transport the seeds several times from cache site to cache site. They cache and secondarily disperse seeds within 1–5 m per movement (Isaji and Sugita, 1997). The ultimate fate of each seed is usually to be eaten, but seedling emergence does occur from caches (Hoshizaki *et al.*, 1997; Hoshizaki and Hulme, 2002). A 10-year trend (1992–2001) in overall seed survival (proportion of fallen seeds that emerge) shows a large range of 0.8–27.0% with a mean of 7.7% (K. Hoshizaki, unpublished data). Annual patterns of seed production and seedling emergence are distinct. Based on a 7-year observation period (1992–1999; Hoshizaki and Hulme, 2002), seed production does not vary greatly, with max./min. < 3 (coefficient of variation (CV) = 38%). On the other hand, seedling emergence ranges from 97 to 1828 seedlings/ha. Seed success therefore varies greatly from year to year (CV = 83%). Seedling emergence for *A. turbinata* varies considerably from year to year but is not seed-limited.

Hoshizaki *et al.* (1997) undertook detailed observations of herbivory and of the fate of seedlings. Seedlings experienced diverse types of herbivory, including shoot clipping and cotyledon removal by rodents. Many seedlings suffered damage several times and from differing types of herbivory. Seedlings had differing experiences during the first-year growing season. In the case of a large cohort (1248 seedlings/ha) at Kanumazawa riparian forest in 1993, 39% of the seedlings had their cotyledons removed, 40% experienced shoot clipping and many seedlings (n = 207) experienced both. When seedlings suffer clipping, they are likely to die; clipping had the strongest effect on seedling mortality, followed by cotyledon removal. In contrast, seedlings not suffering herbivory survived well; in this same cohort the survival rate of intact seedlings reached 74%. Herbivory is therefore the major factor determining seedling demography in this species (Hoshizaki *et al.*, 1997).

Although herbivory decreased seedling survivorship, it never led to 100% mortality. This resistance was associated with the large seed size and hypogeal germination such that the seedlings whose hypogeal cotyledons were intact showed higher survivorship than those whose cotyledons were removed by rodents. Even 1 month after emergence of seedlings of *A. turbinata*, a considerable fraction of seed reserves remains in hypogeal cotyledons (Fig. 15.1A). By drawing on these resources, seedlings were able to produce a callus and resprout after the shoot has been clipped (Fig. 15.1B). Resprouting corresponded significantly to retention of hypogeal cotyledons (Hoshizaki *et al.*, 1997). Since removal of hypogeal cotyledons had no effect on seedling size, these results indicate that the large reserve resources in cotyledons act to reduce the mortality risks from herbivory, as has been found for tropical trees (Forget, 1992; Harms and Dalling, 1997; Kitajima and Fenner, 2000). Regeneration in *A. turbinata* is therefore herbivore-limited (*sensu* Crawley and Long, 1995; Crawley, 2000). The seedlings, however, show high resistance associated with the large seed size, mitigating to some extent the negative impact of rodents on the survival of seedlings (Hoshizaki *et al.*, 1997).

Fig. 15.1. The role of seed reserves stored in hypogeal cotyledons in *Aesculus turbinata* seedlings. (A) A large amount of reserves is stored in hypogeal cotyledons after seedling emergence. Seedlings 3 weeks and 5 weeks after emergence are shown, when their shoots no longer elongate. (B) A seedling resprouted after shoot clipping by rodents. The reserves in hypogeal cotyledons play a significant role in the resprouting (Hoshizaki *et al.*, 1997). Photo by: K. Hoshizaki.

Demographic consequence of seed dispersal by rodents

The role of rodent caching in regeneration can be evaluated from the survival and growth of seedlings (e.g. Nakashizuka et al., 1995). In Japan, dispersal effectiveness has been evaluated by these means in A. turbinata, Q. serrata and several other species (Nakashizuka et al., 1995; Iida, 1996; Hoshizaki et al., 1997, 1999). In A. turbinata, three hypotheses concerning the selective significance of seed dispersal have been tested, namely the colonization, escape, and directed-dispersal hypotheses (Howe and Smallwood, 1982). Hoshizaki et al. (1997, 1999) have proposed that the colonization and escape hypotheses apply to A. turbinata, with more support for the former.

The spatial distribution patterns of fallen seeds and emergent seedlings are distinct. Seedlings are widely distributed, but fallen seeds are highly aggregated beneath the canopy of mother trees (Hoshizaki et al., 1997). Rodents play a significant role in enlarging the seed–seedling shadow by gathering seeds and transporting them to cache sites; the tail of the seed–seedling shadow distribution reaches up to 42–115 m from the parent tree (12.2–44.7 m for the means of 3 years; Hoshizaki et al., 1999). These values are comparable to other large-seeded Fagaceae (Jensen and Nielsen, 1986; Miguchi, 1994; Iida, 1996). Since seedlings of A. turbinata showed higher rates of survival and growth in sites with higher light levels, such as canopy gaps, these results support the colonization hypothesis. Similar effects were found for Q. serrata (Nakashizuka et al., 1995; Iida, 1996). Second, seedling survival rate was slightly density-dependent; in sites with a sufficiently low density of seedlings the survival rate was higher, supporting the escape hypothesis (Hoshizaki et al., 1997). However, its role in seedling regeneration may be weak, since seedling herbivory is not density-dependent and the light level was a more important determinant of seedling survival than the seedling density (Hoshizaki et al., 1997, 1999). These results suggest that the escape hypothesis is less important for A. turbinata.

At the other extreme, the seed shadow of A. turbinata cannot extend beyond the canopy in the absence of rodents. In autumn 2002, the density of A. speciosus was unusually sparse (≈ 2 animals/ha). The spatial distribution of A. turbinata seedlings in 2003 was highly aggregated; $\approx 85\%$ of the emerging seedlings were beneath the canopy of A. turbinata, even though the seedling density was much higher (≈ 3700 seedlings/ ha) than in other years of study (K. Hoshizaki, unpublished data). Although E. andersoni was relatively abundant in this period (≈ 27 animals/ha), they are larder-hoarders and have a smaller home range than A. speciosus (Miguchi, 1996a). These results suggest that E. andersoni plays only a minor role, if any, on seed-shadow enlargement. Aggregation of seedlings beneath conspecifics leads to an increase in attacks on roots and shoots by E. andersoni, and a high rate of infestation by shoot-mining moth larvae (Hepialidae sp.), whose density is low in normal years. When the density of this minor herbivore is high, it reduces performance of second-year seedlings by damaging the apical shoot (K. Hoshizaki, unpublished data). Thus, A. turbinata cannot regenerate successfully without seed-dispersing rodents, especially A. speciosus. Although rodents have a strong negative impact on A. turbinata, the net effect is positive and the interaction between them may be mutualistic; the size of the rodent populations is important, because too few or too many rodents can reduce regeneration of A. turbinata.

Geographical variation in F. crenata nut predation

Several explanations have been proposed for the geographical variation in beech dominance between the Japan Sea and Pacific Ocean types of beech stand. Homma et al. (1999) summarized these explanations and set out the following four hypotheses concerning beech regeneration:

1.   Nut production of F. crenata is lower in regions where it is less dominant (the production hypothesis).

2.   Pre-dispersal predation is severe in regions of lower snowfall (pre-dispersal predation hypothesis).

3. Because the snow pack constrains rodent activity for a longer period, post-dispersal nut predation is less severe in snow-rich regions (post-dispersal predation hypothesis).

4. A relatively dry winter climate in regions of lower snowfall may cause higher mortality of nuts by desiccation (desiccation hypothesis).

Homma *et al.* (1999) tested these hypotheses by simultaneously monitoring the early demography of *F. crenata* for a mast event in 1993, when this species fruited synchronously across its range. They monitored the density of *F. crenata* nuts after nutfall and the following spring in 15 locations in *F. crenata* forests, covering the distribution range of *F. crenata* forests. Production of *F. crenata* nuts varied among the locations but was not correlated with snow depth (no support for the production hypothesis). The proportion of viable nuts, however, was correlated with snow depth, and the proportion of nuts damaged by invertebrates was negatively correlated with snow depth. These results support the pre-dispersal predation hypothesis. Seedling emergence among the locations showed a contrasting pattern. The number of nuts surviving in the following spring was $\approx$ 120 nuts/m^2 at the heavy snow (> 2 m deep) sites, but was only 17 nuts/m^2 at sites having less snow (< 1 m deep). The seedling emergence ratio (density of emerged seedlings to that for the preceding nutfall) was positively correlated with snow depth; at deep snow sites, 15–50% of fallen nuts produced seedlings, but only 0–20% did so at sites with less snow. These results partly support the post-dispersal predation hypothesis. Nut mortality from causes other than vertebrate predation varied widely (0–58%) among sites, but was dominated by microbial decay. Death from desiccation was not observed in most of the studied sites. Maruta *et al.* (1997) also reports low nut mortality from desiccation in a Pacific Ocean type beech forest, Mount Fuji. These findings suggest that desiccation may not be critical for *F. crenata* nuts even under dry winter condition in the Pacific Ocean type forests. Regeneration of *F. crenata*, therefore,

varies geographically and this variation is correlated with snow depth, and pre- and post-dispersal predation are major factors relating to the variation.

The clear contrast in snowfall regime between the stand types (especially for maximum snow depth and duration of snow cover) may be a key to understanding the variation in *F. crenata* regeneration. Shimano and Masuzawa (1998) studied *F. crenata* nut predation in a Pacific Ocean type forest and a Japan Sea type *F. crenata* forest (Mounts Mito and Buna-daira, respectively; Table 15.1) and found a marked difference in nut removal rates between the forests (100% in Mount Mito vs. 30% in Buna-daira). However, in Mount Mito, rodents did not find nuts which were wrapped in a combination of a paper envelope (to remove visibility) and a zippered polyethylene bag (to reduce smell). The detection rate was modest when the nuts were bagged by either the envelope or the polyethylene bag. These results imply that thick snow accumulation could interfere with rodent feeding activity and ability to detect nuts, through reduction in olfactory and visual cues, leading to higher nut survival in snow-rich regions.

Response of Rodents to Variable Seed Resource

Assumptions for predator satiation

The ecological significance of masting is generally taken to be that synchronous seed production leads to satiation of seed predators, both specialists and generalists, resulting in a disproportionately large cohort of seedlings. Additionally, in inter-mast years seed production is often much reduced, reducing the density of seed predators (Kelly, 1994; Ostfeld and Keesing, 2000; and references therein). Several assumptions are involved in these scenarios. However, these patterns may not hold in complex communities, such as multi-species mixed forests with generalist seed predators, for the following reasons.

First, the abundance of seed predators decreases during periods of reduced resources. However, this may be problematic when the predators are generalist, and other tree species may influence the predator population between mast events. For example, we can expect that rodents maintain a high population level even in inter-mast years by consuming alternative resources (Hoshizaki and Hulme, 2002).

Second, seed predators are assumed to respond to *single* species of seed of a mast-fruiting plant. However, the foraging pattern of generalist predators depends on resource availability (e.g. Hulme and Hunt, 1999). Vander Wall (2002) found that synchronous fruiting of sympatric *Pinus* species facilitates seed dispersal via active caching of seeds by rodents, resulting in increased dispersal distance. Furthermore, most studies have evaluated seed supply to predators by the total number of fallen seeds (e.g. Pucek *et al.*, 1993; Choquenot and Ruscoe, 2000). This may be an inappropriate measure if the seedfall of different species is not equivalent (Schnurr *et al.*, 2002). Although seeds vary in nutritional value and in secondary compounds (Grodziñski and Sawicka-Kapusta, 1970; Shimada, 2001; Vander Wall, 2001), differences in the resource quality of the mast to predators has received little quantitative attention.

Finally, the numerical response of mammalian seed-consumers often shows a time-lag after masting events (Curran and Leighton, 2000; Ostfeld and Keesing, 2000). This point is important for rodents, because a numerical response of rodents in the spring following a mast crop could increase their seedling predation, and the beneficial effect of masting should therefore be tested using both seed and seedling censuses (Hoshizaki *et al.*, 1997; Hoshizaki and Hulme, 2002).

Annual seed production and synchrony

In evaluating patterns of annual seed production, the degree of among-species synchrony is important. Here, we selected patterns in annual production and synchrony of large-seeded trees in three different forest stand types: a Japan Sea (Nukumi-daira), a Pacific Ocean (Ogawa Forest Reserve) and a riparian mixed forest (Kanumazawa). In Nukumi-daira, the pattern is the simplest and clearest. During 11 years of monitoring, *F. crenata* fruited in 7 years, with three very good crops (1990, 1993 and 1999) and four bad crops (1992, 1997, 1999 and 2000). The nut supply is highly pulsed in this monodominant forest (Fig. 15.2A).

In mixed-species forests (Kanumazawa and Ogawa), the patterns of among-year variation in the community-wide seed abundance are more complex (Fig. 15.2B). In Kanumazawa, *F. crenata* and *Q. crispula* showed large year-to-year variation in seed production. Masting of *F. crenata* occurred in 1993 (small crop), 1995 (large) and 2000 (moderate), and *Q. crispula* showed an 'alternate bearing' pattern (*sensu* Crawley and Long, 1995). The annual seed production of *A. turbinata* was less variable (CV = 45%, 1992–2000). The patterns of annual seed production among the three species were not synchronized (Fig. 15.2B), as reported in England (Gurnell, 1993) and California (Koenig *et al.*, 1994). Since *A. turbinata* seeds are much larger in size and have similar energy per gram to *Q. crispula*, the community-wide seed energy varied relatively little from year to year (Hoshizaki and Hulme, 2002), except for 1999. In Ogawa, seed production of major component species has been monitored since 1987 (e.g. Nakashizuka and Matsumoto, 2002). Shibata *et al.* (2002) has analysed seed crop patterns over a 9 year period for 16 tree species. Many species shared mast years, with a high degree of synchrony among species (in 1988, 1993 and 1995), but *Q. serrata* and *Q. crispula* showed little between-year variability (Shibata *et al.*, 2002). The annual fruiting patterns differed among major large-seeded trees in Ogawa forest (*F. crenata*, *F. japonica*, *Q. serrata* and *Q. crispula*). Both *Fagus* species showed large inter-annual fluctuations in nut production, but their fruiting was not correlated with each other. In contrast, the fruiting patterns of *Quercus*

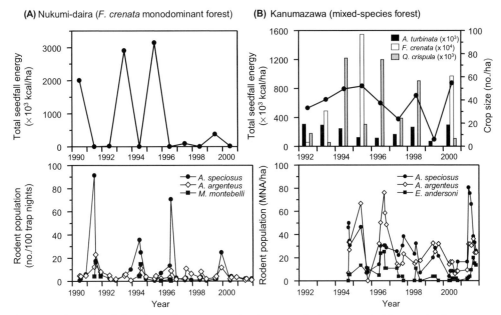

Fig. 15.2. Annual variations in seed resources and rodent numerical responses in (A) monodominant forest of *Fagus crenata* (Nukumi-daira) and (B) a mixed-species forest (Kanumazawa). Seed resources are represented by numbers of fallen seeds (bars) and community-wide seedfall energy (lines). Rodent populations are monitored using Sherman live traps placed in a matrix. Rodent abundance was represented by number of animals per 100 trap-nights (Nukumi-daira) or minimum number of animals alive (MNA; Kanumazawa). Data sources: (A) Miguchi (1995, 1996a, and unpublished); (B) Hoshizaki and Hulme (2002) and K. Hoshizaki (unpublished).

spp. were less variable and more synchronized than *Fagus* spp. Thus, community-wide seed availability in mixed-species forests is less variable than in monodominant forest, but includes a pulse of nutritious nuts of *Fagus* spp. In North America, acorn production of *Quercus* spp. is not highly synchronized among species (Koenig *et al.*, 1994; Schnurr *et al.*, 2002), implying a less variable community-wide seed resource between different years.

Numerical response after masting

We examine the numerical response of rodents using data for Japan Sea type beech stands (Nukumi-daira and Lake Towada) and a mixed-species stand (Kanumazawa). Although the seed-resource availability differs between the Japan Sea and the mixed-species stands, rodent numbers in

these forests show similar patterns of response (Fig. 15.2). In *F. crenata* monodominant forests, rodent abundance clearly follows the masting cycle of *F. crenata*; in Nukumi-daira, the number of rodents, especially for *A. speciosus*, increases dramatically every spring after mast events, but such high densities are maintained for only a few months (Fig. 15.2A). *M. montebelli*, which usually inhabits pastures and cultivated fields, is found in beech forests only in the year following *F. crenata* mast events. Similar patterns are observed in Lake Towada (M. Abe *et al.*, unpublished). Jensen (1982) also documents clear eruptions of the *Clethrionomys glareolus* population after masting in a *Fagus sylvatica* monodominant forest in Denmark. Miguchi (1988) investigated changes in the population structure of rodents in relation to masting at Nukumi-daira, and found that recruitment of young individuals makes a large contribution to the population-increase phase. Thus, the large pulse of *Fagus* mast causes large fluctuations in rodent populations in monodominant forests, with a high degree of predictability.

Population fluctuations also occur in mixed forests but are not easy to explain. In Kanumazawa, rodent populations showed large fluctuations from year to year, as in Nukumi-daira, but maintained on average greater total biomass (Miguchi, 1988; K. Hoshizaki, unpublished). Similar results have been reported for *Nothofagus*-monodominant stands and *Podocarpus*–hardwood mixed stands in New Zealand (Choquenot and Ruscoe, 2000). These findings support the idea that rodent abundance should be buffered in mixed forests, because of asynchronous seed production (Hoshizaki and Hulme, 2002). Nevertheless, the effects of *F. crenata* masting are clear in Kanumazawa. The amplitude of density fluctuations in rodent populations varied from year to year, and there were distinct density peaks after *F. crenata* masting, especially in summer 2001 (Fig. 15.2B). In North America, patterns of population increase differ between species; *Peromyscus* species numerically respond to red oak (*Q. rubra*) masting, but the population of *C. gapperi* is correlated with the masting of *Acer rubrum* and *Tsuga canadensis* and not with red oak (Schnurr et al., 2002). There appear to be species-specific responses to seed crops, resulting in complex ecological patterns in mixed forests.

Thus, in addition to the effect of less snow cover, the greater seed-predation pressure in Pacific Ocean beech forests (Homma et al., 1999) may also be associated with higher rodent densities, which may be buffered by another seed pulse of *Q. crispula* and/or *F. japonica* in non-mast years of *F. crenata*. In fact, fruiting of *F. crenata* and *F. japonica* is not synchronized (Shibata et al., 2002), but this idea should nevertheless be tested empirically.

In Nukumi-daira and Kanumazawa, the peak rodent density did not persist for long. Densities, especially for *A. speciosus* and microtine species, decrease rapidly within the year. Several authors have also reported population declines in generalist seed predators after the mast is depleted (Miguchi, 1988; Wright et al., 1999; Ostfeld and Keesing, 2000). However, the population fluctuation in Kanumazawa cannot be fully explained under an assumption that rodents similarly respond to different seed species, since the rodent fluctuations showed much larger variation between years than the large-seed availability and because higher total seedfall energy in the preceding autumn does not always lead to a subsequent population increase of rodents in some years (e.g. 1998 seedfall) and vice versa (e.g. 1996 seedfall) (Fig. 15.2B). In 1996, *A. turbinata* made a larger contribution to the total seedfall energy than *Q. crispula*, and the reverse is true in 1998. *F. crenata* did not fruit in those years. This suggests that the response of rodent populations to seed supply is not a simple function of total seedfall energy and, in addition, that the effect of the larger seed crop of *Q. crispula* on the rodent population is unclear (Miguchi, 1996b). Jensen (1982) argues that one factor stimulating the rapid increase in the *C. glareolus* population is seed quality, since *F. sylvatica* seeds have the highest energy content of the tree seeds there. The pattern in mixed stands implies different mechanisms than simple numerical response, such as changes in reproductive schedule (Wolff, 1996b) and winter survival (Jensen, 1982; Wolff, 1996a). The reproductive schedule of Japanese rodents differs from that in Europe and North America. In Europe and North America, rodents often begin to reproduce in late winter following masting (Jensen, 1982; Pucek et al., 1993; Wolff, 1996b). However, winter reproduction has not been observed in Nukumi-daira or Kanumazawa even after good seed crops (Miguchi, 1988; K. Hoshizaki, unpublished data).

Influence of *F. crenata* mast on *A. turbinata* with shared predators

Since *A. turbinata* shares seed predators with *F. crenata*, its regeneration may be affected indirectly by *F. crenata* masting.

The discordant annual patterns in seeds and seedlings found in a 7-year study of *A. turbinata* seed dynamics in Kanumazawa (Hoshizaki and Hulme, 2002) strongly suggest that the regeneration cannot be explained solely by direct interaction between *A. turbinata* and its seed predators and dispersers.

The overall seed survival rate in *A. turbinata* was not correlated with its seed production, but tended to be low in mast years of *F. crenata* and was unaffected by *Q. crispula* seed crops. This long-term trend suggest that there is little evidence for predator satiation by *A. turbinata*, and also shows that *A. turbinata* seed survival may be indirectly affected via rodent responses to *F. crenata* masting. However, the mechanisms of these indirect effects are complex. In the 1995 *F. crenata* mast year and the 2 subsequent non-mast years, Hoshizaki and Hulme (2002) followed the fate of *A. turbinata* seeds almost daily, and examined whether the rate of *A. turbinata* seed predation is affected by the nutfall pattern of *F. crenata*. The rate of predation on *A. turbinata* seeds varied both within the fruiting period and from year to year. Rodents clearly responded in a frequency-dependent manner to *F. crenata* nutfalls and ignored *A. turbinata* caches. This indicates that rodents preferentially shifted their diet from *A. turbinata* seeds to *F. crenata* nuts (Hoshizaki and Hulme, 2002). However, this change in seed predation played only a minor role in the overall seed demography of *A. turbinata*, because the seeds suffered high mortality just before seedling emergence, resulting in lower overall survival in the *F. crenata* mast year than in the other 2 years. The reason for the change in predation on *A. turbinata* seeds in spring remains unknown. Nevertheless, the data have strong implications for the mechanisms underlying apparently inconsistent seed predation patterns; rodent foraging can vary in response to community-wide resource availability, even on a month-to-month basis.

The *A. turbinata*–*F. crenata* interaction in Kanumazawa seems to contrast with the seed dynamics in *F. crenata* monodominant forests. In *F. crenata* monodominant forests, rodents consume *F. crenata* nuts in a density-dependent manner (direct interaction) in mast years (Abe *et al.*, 2001; Tomita *et al.*, 2002) and thus predator satiation (direct interactions) may be expected (see also Schupp, 1992). It is suggestive that the direct effects on seed survival were apparent in *F. crenata* monodominant forests, but not in mixed forest (Kanumazawa), where indirect effects must be taken into account.

Seed quality and rodent physiology

Why does the response of rodents to resource fluctuation vary with seed species? Differences in seed nutritional composition may be a key factor associated with the population response, since seed nutrition also affects the physiological condition of individual rodents. In particular, levels of lipids have a major influence on seed preference and energy assimilation (Grodziński and Sawicka-Kapusta, 1970; Smith and Follmer, 1972; Smallwood and Peters, 1986; Stapanian, 1986). For instance, Smith and Follmer (1972) found that grey squirrels prefer lipid-rich (20% dry weight) *Q. shumardii* acorns over lipid-poor (4.6% dry weight) *Q. alba* acorns, and that they have a higher rate of energy assimilation of *Q. shumardii* acorns. The predictable numerical response found in Nukumi-daira and the clear preference for *F. crenata* nuts over *A. turbinata* in Kanumazawa (Hoshizaki and Hulme, 2002) suggest that, among *F. crenata*, *Q. crispula* and *A. turbinata*, lipid-rich *F. crenata* nuts are the highest quality food resource for rodents.

Secondary compounds in seeds, such as tannins, can have deterrent effects on rodents (chemical defence). Tannins are known to reduce digestibility of seeds and body mass of rodents (Vander Wall, 2001 and references therein), and in some cases even to cause mortality (Shimada and Saitoh, 2003). Vander Wall (2001) compiled the nutritional values of acorns of various *Quercus* in North America, and noted that black-oak acorns (species of subgenus

Erythrobalanus) have higher levels of both lipids and tannins than white-oak acorns (subgenus *Lepidbalanus*). Some rodent species cannot maintain their body mass when fed tannin-rich acorns of black oaks (Briggs and Smith, 1989). In contrast, white-oak acorns serve as high-quality foods for rodents, because rodent populations increase rapidly after masting of these oak species (e.g. Wolff, 1996b).

In temperate Japan, evidence for a rodent population increase after masting of *Quercus* acorns is not clear (Miguchi, 1996b; Hoshizaki and Hulme, 2002). The reason why masting of Japanese *Quercus* acorns does not lead to an increase in rodent populations may lie in the chemical properties of the seeds. Recently, Shimada and Saitoh (2003) studied the physiology of captive *A. speciosus* individuals fed with *Q. crispula* and *Q. serrata* acorns. They found that: (i) feeding on acorn alone leads to a significant decrease in rodent body mass and daily intake for both acorn species, sometimes resulting in death; (ii) *Q. crispula* acorns, which contain about three times the levels (11.7% dry weight) of tannins in *Q. serrata*, have stronger effects, causing mortality or severe decrease in body mass; and (iii) a tannin-free formula diet that is otherwise similar to natural acorns causes little change in body weight. These results strongly suggest that tannins in *Q. crispula* and *Q. serrata* acorns reduce the quality of these large, carbohydrate-rich seeds. Hoshizaki (1999) made a similar speculation to explain the discordance between the large fluctuation in rodent abundance and the small variation in energy availability (mostly owing to *A. turbinata* seeds) in Kanumazawa (Fig. 15.2B); for rodents, the quality of *A. turbinata* seeds appears to be lower because of the effects of saponins, though this hypothesis needs further experimental testing. Evidently seed attributes, especially lipid contents and chemical defence, may be important in the mechanisms underlying the complex patterns of rodent abundance (e.g. Stapanian, 1986; see also Hoshizaki and Hulme, 2002) in relation to variable seed production in mixed-species forests.

Conclusions

Rodents have significant impacts on tree regeneration. In particular, granivorous, seed-dispersing mice may have a large influence on the early demography of trees. Studies on *A. turbinata* regeneration illustrate that seeds and seedlings suffer large impacts, through seed predation and seedling herbivory. The seedling resistance to herbivory and seed dispersal by rodents, particularly *Apodemus speciosus*, are effective to mitigate these negative effects. However, the effects vary from year to year and between forests with differing stand structures. In temperate Japan, forests differing in structure, but with the same major species, occur in a relatively narrow geographic area. By comparing results of demographic studies having similar methodology between *F. crenata* monodominant forests and multi-species forests, differences in the patterns of interactions between large seeds and rodents are discernible. In monodominant forests, the patterns of regeneration and rodent fluctuation may be predictable, because the annual resource supply is highly pulsed. In contrast, studies of indirect interactions among *F. crenata*, *A. turbinata* and rodents demonstrate that the patterns of interactions are much more complex in multi-species forests. Important points in understanding the regeneration dynamics of multi-species systems are:

1. The degree of asynchrony in seed production among species and years.
2. The inter-species variation in seed quality, which includes energy content and deterrent effects of secondary compounds.
3. Rodents' foraging response to seeds with different quality.
4. The species-specific response of rodents to seed production.

Levels of pre-dispersal seed predation may also be important, which should be reviewed elsewhere. Empirical studies on tree–rodent ecology in multi-species system remain sparse. Further accumulation of data on the dynamics in multi-species forests would increase our understanding of the mechanisms of regeneration and co-existence and of patterns in tree–rodent interactions in temperate forests (Naka-shizuka, 2001), and perhaps in more species-rich systems such as tropical forests.

Acknowledgements

We are grateful to Pierre-Michel Forget and the other co-editors of this volume for their invitation to contribute. Helpful comments by S.B. Vander Wall, P.-M. Forget and P.E. Hulme improved the clarity of the manuscript. We also thank T. Shimada for constructive discussions and T. Masaki and M. Abe for suggestions and encouragement. K.H. is grateful to T. Nakashizuka, W. Suzuki and S. Sasaki for supervision during the course of the study on *A. turbinata*. This study was supported by Grants-in-Aid for Scientific Research (Ministry of Education, Culture, Sports, Science and Technology of Japan; Grant No. 14740423) to K.H.

References

Abe, M., Miguchi, H. and Nakashizuka, T. (2001) An interactive effect of simultaneous death of dwarf bamboo, canopy gap, and predatory rodents on beech regeneration. *Oecologia* 127, 281–286.

Briggs, J.M. and Smith, K.G. (1989) Influence of habitat on acorn selection by *Peromyscus leucopus*. *Journal of Mammalogy* 70, 35–43.

Choquenot, D. and Ruscoe, W.A. (2000) Mouse population eruptions in New Zealand forests: the role of population density and seedfall. *Journal of Animal Ecology* 69, 1058–1070.

Crawley, M.J. (2000) Seed predators and plant population dynamics. In: Fenner, M. (ed.) *Seeds: the Ecology of Regeneration in Plant Communities*, 2nd edn. CAB International, Wallingford, UK, pp. 167–182.

Crawley, M.J. and Long, C.R. (1995) Alternate bearing, predator satiation and seedling recruitment in *Quercus robur* L. *Journal of Ecology* 83, 683–696.

Curran, L.M. and Leighton, M. (2000) Vertebrate responses to spatiotemporal variation in seed production of mast-fruiting Dipterocarpaceae. *Ecological Monographs* 70, 101–128.

Curran, L.M. and Webb, C.O. (2000) Experimental tests of the spatiotemporal scale of seed predation in mast-fruiting Dipterocarpaceae. *Ecological Monographs* 70, 129–148.

Forget, P.-M. (1992) Regeneration ecology of *Eperua grandiflora* (Caesalpiniaceae), a large seeded tree in French Guiana. *Biotropica* 24, 146–156.

Grodziński, W. and Sawicka-Kapusta, K. (1970) Energy values of tree-seeds eaten by small mammals. *Oikos* 21, 52–58.

Gurnell, J. (1993) Tree seed production and food conditions for rodents in an oak wood in Southern England. *Forestry* 66, 291–315.

Harms, K.E. and Dalling, J.W. (1997) Damage and herbivory tolerance through resprouting as an advantage of large seed size in tropical trees and lianas. *Journal of Tropical Ecology* 13, 617–621.

Homma, K., Akashi, N., Abe, T., Hasegawa, M., Harada, K., Hirabuki, Y., Irie, K., Kaji, M., Miguchi, H., Mizoguchi, N., Mizunaga, H., Nakashizuka, T., Natume, S., Niiyama, K., Ohkubo, T., Sawada, S., Sugita, H., Takatsuki, S. and Yamanaka, N. (1999) Geographical variation in the early regeneration process of Siebold's beech (*Fagus crenata* Blume) in Japan. *Plant Ecology* 140, 129–138.

Hoshizaki, K. (1999) Regeneration dynamics of a sub-dominant tree *Aesculus turbinata* in a beech-dominated forest. PhD thesis, Kyoto University, Kyoto, Japan.

Hoshizaki, K. and Hulme, P.E. (2002) Mast seeding and predator-mediated indirect interactions in a forest community: evidence from post-dispersal fate of rodent-generated caches. In: Levey, D.J., Silva, W.R. and Galetti, M. (eds) *Seed Dispersal and Frugivory: Ecology, Evolution and Conservation*. CAB International, Wallingford, UK, pp. 227–239.

Hoshizaki, K., Suzuki, W. and Sasaki, S. (1997) Impacts of secondary seed dispersal and herbivory on seedling survival in *Aesculus turbinata*. *Journal of Vegetation Science* 8, 735–742.

Hoshizaki, K., Suzuki, W. and Nakashizuka, T. (1999) Evaluation of secondary dispersal in a large-seeded tree *Aesculus turbinata*: a test of directed dispersal. *Plant Ecology* 144, 167–176.

Howe, H.F. and Smallwood, J. (1982) Ecology of seed dispersal. *Annual Review of Ecology and Systematics* 13, 201–228.

Hulme, P.E. (1998) Post-dispersal seed predation: consequences for plant demography and evolution. *Perspectives in Plant Ecology, Evolution and Systematics* 1, 32–46.

Hulme, P.E. (2002) Seed eaters: seed dispersal, destruction and demography. In: Levey, D.J., Silva, W.R. and Galetti, M. (eds) *Seed Dispersal and Frugivory: Ecology, Evolution and Conservation*. CAB International, Wallingford, UK, pp. 257–273.

Hulme, P.E. and Benkman, C.W. (2002) Granivory. In: Herrera, C.M. and Pellmyr, O. (eds) *Plant–Animal Interactions*. Blackwell Science, Oxford, pp. 132–154.

Hulme, P.E. and Borelli, T. (1999) Variability in post-dispersal seed predation in deciduous woodland: relative importance of location, seed species, burial and density. *Plant Ecology* 145, 149–156.

Hulme, P.E. and Hunt, M.K. (1999) Rodent post-dispersal seed predation in deciduous woodland: predator response to absolute and relative abundance of prey. *Journal of Animal Ecology* 68, 417–428.

Iida, S. (1996) Quantitative analysis of acorn transportation by rodents using magnetic locator. *Vegetatio* 124, 39–43.

Irie, K., Homma, K., Masuzawa, T., Miguchi, H. and Shimano, K. (1998) Population dynamics of rodents in the beech forests with different snow accumulation and the predation number estimation. *Journal of Phytogeography and Taxonomy* 46, 37–45. (In Japanese with English abstract.)

Isaji, H. and Sugita, H. (1997) Removal of fallen *Aesculus turbinata* seeds by small mammals. *Japanese Journal of Ecology* 47, 121–129. (In Japanese with English abstract.)

Janzen, D.H. (1971) Seed predation by animals. *Annual Review of Ecology and Systematics* 2, 465–492.

Jensen, T.S. (1982) Seed production and outbreaks of non-cyclic rodent populations in deciduous forests. *Oecologia* 54, 184–192.

Jensen, T.S. (1985) Seed–seed predator interactions of European beech, *Fagus sylvatica* and forest rodents, *Clethrionomys glareolus* and *Apodemus flavicollis*. *Oikos* 44, 149–156.

Jensen, T.S. and Nielsen, O.F. (1986) Rodents as seed dispersers in a heath–oak wood succession. *Oecologia* 70, 214–221.

Kelly, D. (1994) The evolutionary ecology of mast seeding. *Trends in Ecology and Evolution* 9, 465–470.

Kikuzawa, K. (1988) Dispersal of *Quercus mongolica* acorns in a broadleaved deciduous forest 1. Disappearance. *Forest Ecology and Management* 25, 1–8.

Kitabatake, T. and Wada, N. (2001) Notes on beech (*Fagus crenata* Blume) seed and seedling mortality due to rodent herbivory in a northernmost beech forest, Utasai, Hokkaido. *Journal of Forest Research* 6, 111–115.

Kitajima, K. and Fenner, M. (2000) Ecology of seedling regeneration. In: Fenner, M. (ed.) *Seeds: The Ecology of Regeneration in Plant Communities*, 2nd edn. CAB International, Wallingford, UK, pp. 331–359.

Koenig, W.D., Mumme, R.L., Carmen, W.J. and Stanback, M.T. (1994) Acorn production by oaks in central coastal California: variation within and among years. *Ecology* 75, 99–109.

Kollmann, J., Coomes, D.A. and White, S.M. (1998) Consistencies in post-dispersal seed predation of temperate fleshy-fruited species among seasons, years and sites. *Functional Ecology* 12, 683–690.

Maruta, E., Kamitani, T., Okabe, M. and Ide, Y. (1997) Desiccation-tolerance of *Fagus crenata* Blume seeds from locations of different snowfall regime in central Japan. *Journal of Forest Research* 2, 45–50.

Miguchi, H. (1988) Two years of community dynamics of murid rodents after a beechnut mastyear. *Journal of Japanese Forestry Society* 70, 472–480. (In Japanese with English abstract.)

Miguchi, H. (1994) Role of wood mice on the regeneration of cool temperate forest. In: Kobayashi, S., Nishikawa, K., Danilin, I.M., Matsuzaki, T., Abe, N., Kamitani, T. and Nakashizuka, T. (eds) *Proceedings of NAFRO Seminar on Sustainable Forestry and its Biological Environment*. Japan Society of Forest Planning Press, Niigata, Japan, pp. 115–121.

Miguchi, H. (1995) Caprice forest mother trees: beech (*Fagus crenata*) masting habit and its effects on the forest ecosystem. *Bulletin of the Society of Population Ecology* 52, 33–40 (in Japanese).

Miguchi, H. (1996a) Dynamics of beech forest from the view point of rodent ecology – ecological interactions of the regeneration characteristics of *Fagus crenata* and rodents. *Japanese Journal of Ecology* 46, 185–189. (In Japanese with English abstract.)

Miguchi, H. (1996b) Study on the ecological interactions of the regeneration characteristics of Fagaceae and the mode of life of wood mice and voles. Dissertation, Niigata University, Niigata, Japan (in Japanese).

Miguchi, H. and Maruyama, K. (1984) Ecological studies on a natural beech forest (XXXVI) Development and dynamics of beechnuts in a mastyear. *Journal of Japanese Forestry Society* 66, 320–327. (In Japanese with English abstract.)

Nakashizuka, T. (1987) Regeneration dynamics of beech forests in Japan. *Vegetatio* 69, 169–175.

Nakashizuka, T. (2001) Species coexistence in temperate, mixed deciduous forests. *Trends in Ecology and Evolution* 16, 205–210.

Nakashizuka, T. and Iida, S. (1995) Composition, dynamics and disturbance regime of temperate deciduous forests in Monsoon Asia. *Vegetatio* 121, 23–30.

Nakashizuka, T. and Matsumoto, Y. (2002) *Diversity and Interaction in a Temperate Forest Community: Ogawa Forest Reserve of Japan.* Springer-Verlag, Tokyo, 319 pp.

Nakashizuka, T., Iida, S., Masaki, T., Shibata, M. and Tanaka, H. (1995) Evaluating increased fitness through dispersal: a comparative study on tree populations in a temperate forest, Japan. *Écoscience* 2, 245–251.

Ostfeld, R.S. and Keesing, F. (2000) Pulsed resources and community dynamics of consumers in terrestrial ecosystems. *Trends in Ecology and Evolution* 15, 232–237.

Ostfeld, R.S., Jones, C.G. and Wolff, J.O. (1996) Of mice and mast: ecological connections in eastern deciduous forests. *BioScience* 46, 323–330.

Pucek, Z., Jedrzejewski, W., Jedrzejewska, B. and Pucek, M. (1993) Rodent population dynamics in a primeval deciduous forest (Bialowieza National Park) in relation to weather, seed crop, and predation. *Acta Theriologica* 38, 199–232.

Sakio, H. (1997) Effects of natural disturbance on the regeneration of riparian forests in a Chichibu Mountains, central Japan. *Plant Ecology* 132, 181–195.

Schnurr, J.L., Ostfeld, R.S. and Canham, C.D. (2002) Direct and indirect effects of masting on rodent populations and tree seed survival. *Oikos* 96, 402–410.

Schupp, E.W. (1990) Annual variation in seedfall, postdispersal predation, and recruitment of a neotropical tree. *Ecology* 71, 504–515.

Schupp, E.W. (1992) The Janzen–Connell model for tropical tree diversity: population implications and the importance of spatial scale. *American Naturalist* 140, 526–530.

Shibata, M., Tanaka, H., Iida, S., Abe, S., Masaki, T., Niiyama, K. and Nakashizuka, T. (2002) Synchronized annual seed production by 16 principal tree species in a temperate deciduous forest, Japan. *Ecology* 83, 1727–1742.

Shimada, T. (2001) Nutrient compositions of acorns and horse chestnuts in relation to seed-hoarding. *Ecological Research* 16, 803–808.

Shimada, T. and Saitoh, T. (2003) Negative effects of acorns on the wood mouse *Apodemus speciosus.* *Population Ecology* 45, 7–17.

Shimano, K. and Masuzawa, T. (1998) Effects of snow accumulation on survival of beech (*Fagus crenata*) seed. *Plant Ecology* 134, 235–241.

Smallwood, P.D. and Peters, W.D. (1986) Grey squirrel food preferences: the effects of tannin and fat concentration. *Ecology* 67, 168–174.

Smith, C.C. and Follmer, D. (1972) Food preferences of squirrels. *Ecology* 53, 82–91.

Stapanian, M.A. (1986) Seed dispersal by birds and squirrels in the deciduous forests of the United States. In: Estrada, A. and Fleming, T.H. (eds) *Frugivores and Seed Dispersal.* Dr W. Junk Publishers, Dordrecht, The Netherlands, pp. 225–236.

Suzuki, W., Osumi, K., Masaki, T., Takahashi, K., Daimaru, H. and Hoshizaki, K. (2002) Disturbance regimes and community structures of a riparian and an adjacent terrace stand in the Kanumazawa Riparian Research Forest, northern Japan. *Forest Ecology and Management* 157, 285–301.

Tomita, M., Hirabuki, Y. and Seiwa, K. (2002) Post-dispersal changes in the spatial distribution of *Fagus crenata* seeds. *Ecology* 83, 1560–1565.

Vander Wall, S.B. (2001) The evolutionary ecology of nut dispersal. *Botanical Review* 67, 74–117.

Vander Wall, S.B. (2002) Masting in animal-dispersed pines facilitates seed dispersal. *Ecology* 83, 3508–3516.

Willson, M.F. and Whelan, C.J. (1990) Variation in post-dispersal survival of vertebrate-dispersed seeds: effects of density, habitat, location, season, and species. *Oikos* 57, 191–198.

Wolff, J.O. (1996a) Coexistence of white-footed mice and deer mice may be mediated by fluctuating environmental conditions. *Oecologia* 108, 529–533.

Wolff, J.O. (1996b) Population fluctuations of mast-eating rodents are correlated with production of acorns. *Journal of Mammalogy* 77, 850–856.

Wright, S.J., Carrasco, C., Calderon, O. and Paton, S. (1999) The El Niño Southern Oscillation, variable fruit production, and famine in a tropical forest. *Ecology* 80, 1632–1647.

16 Impact of Small Rodents on Tree Seeds in Temperate and Subtropical Forests, China

Zhi-Bin Zhang, Zhi-Shu Xiao and Hong-Jun Li

State Key Laboratory of Integrated Management of Pest Insects and Rodents in Agriculture, Institute of Zoology, Chinese Academy of Sciences, Beijing 100080, P.R. China

Introduction

In recent decades, numerous studies have reported that many mammals and birds are important agents of forest regeneration in temperate and tropical regions (e.g. Abbott and Quink, 1970; Miyaki and Kikuzawa, 1988; Forget, 1990, 1991, 1992, 1993; Vander Wall, 1990, 1993, 1994; Forget and Milleron, 1991; Brewer and Rejmánek, 1999; Theimer, 2001; Zhang and Wang, 2001a,b). Previous studies have largely focused on seed or fruit display, seed or fruit removal, seed predation, seed dispersal, seedling establishment and their interactions with animals or animal groups. One of the biggest obstacles to studying seed dispersal by animals is the difficulty of tracking seeds and estimating the shape of seed shadows (Levey and Sargent, 2000). For many forests, there is still a challenging problem of determining how tree seeds move, die or survive, after they have left the parent plant.

Recently, there have been four major trends in the study of seed dispersal and predation by small rodents: (i) tracking individual seeds with coded numbers to characterize seed dispersal and seed fates in animal-dispersed tree species (Xiao and Zhang, 2003; Forget and Wenny, Chapter 23, this volume); (ii) pursuing ultimate seed

fates to re-evaluate the contribution of seed-caching rodents to tree reproductive success (e.g. Vander Wall, 1994; Vander Wall and Joyner, 1998; Jansen et al., 2002); (iii) linking seed dispersal and other plant life history traits to illustrate forest dynamics, tree diversity and distribution (Wang and Smith, 2002); and (iv) exploring the mutual interactions between animal-dispersed plants and scatterhoarding animals to understand the evolution of seed dispersal (e.g. Vander Wall, 1990, 2001, 2002a,b, 2003; Jansen and Forget, 2001; Hoshizaki and Hulme, 2002; Hulme, 2002; Jansen et al., 2002; Theimer, Chapter 17, this volume). However, our knowledge of the role of seed dispersal by scatterhoarding animals in tree reproductive success is still poor because dispersal is often a complicated, multistage process and there is great temporal and spatial variation in forest dynamics (Vander Wall, 2002a; Wang and Smith, 2002).

Generally, large-seeded species vary greatly in traits such as seed size, nutrient content, secondary compounds and seed-coat hardness, which affect the selection, harvest, transportation and scatterhoarding by small rodents, and subsequently influence seed survival, seed shadow and seedling establishment (e.g. Price and Jenkins, 1986; Vander Wall, 1990, 2001; Forget et al.,

©CAB International 2005. *Seed Fate*
(eds P.-M. Forget, J.E. Lambert, P.E. Hulme and S.B. Vander Wall)

1998; Jansen and Forget, 2001). In addition, plant species abundance and distributions often vary across different scales in time and space, which can influence vegetation structure and seed availability, and affect the structure and dynamics of animal communities (e.g. seed-eating rodents and birds) (Forget *et al.*, 1998; Jansen and Forget, 2001; Hoshizaki and Hulme, 2002). However, few studies have examined the interactions between seed species and the animals that feed on and store them in a community context (e.g. Kollmann *et al.*, 1998; Hulme and Hunt, 1999; Vander Wall, 2002b, 2003; see Hoshizaki and Hulme, 2002).

In China, few projects have been carried out on the interactions between seed-caching rodents and forest tree seeds. In 1998, we began a project to assess the impact of rodents on seed regeneration of *Quercus liaotungensis* (Fagaceae) and *Prunus armeniaca* (Rosaceae) in the temperate deciduous forest of the Donglingshan Mountains in the Beijing region. In 2000, we launched another project to study seed fate of nut-bearing tree species (i.e. *Quercus variabilis*, *Quercus serrata*, *Castanopsis fargesii* (Fagaceae) and *Camellia oleifera* (Theaceae)) in a subtropical evergreen broadleaved forest in the Dujiangyan region of Sichuan Province. We have found that seed-eating rodents (e.g. field mice *Apodemus peninsulae* and grey squirrels *Sciurotamias davidianus* in the Donglingshan forest, and Edwards' long-tailed rats *Leopoldamys edwardsi* in the Dujiangyan forest) may be very important for forest regeneration by scatterhoarding seeds in soil. The purpose of this chapter is to summarize our results about interactions between several common nut-bearing trees and their rodent associates in both temperate and subtropical forests in China, including seed predation, scatter-hoarding and seed dispersal, and linking seed production to seedling establishment.

Study Sites and Species

The research was carried out in a temperate and a subtropical forest in China. Both forests have been fragmented by agricultural development, with only small parts remaining as natural forests.

The temperate forest (elevation 1200–2000 m) is located in the Donglingshan Mountains (40°0′N, 115°30′E). The site has a warm temperate continental monsoon climate, with mean annual temperature of 11°C and annual precipitation of 639 mm. In the Donglingshan forest, *Quercus liaotungensis* (Fagaceae), *Juglans mandshurica* (Juglandaceae), *Prunus armeniaca* (Rosaceae), *Vitex negundo* (Vitaceae) and *Prunus davidiana* (Rosaceae) are common trees or shrubs, and *Elymus excelsus*, *Poa* spp., *Elsholtzia stauntoni* are common grasses (Wang *et al.*, 2000; Zhang and Wang, 2001b; Li and Zhang, 2003).

The subtropical forest (elevation 700–1000 m) is in the Dujiangyan region (31°4′N/103°43′E) of Sichuan Province, with a mean annual temperature of 15.2°C and an annual precipitation of 1200–1800 mm. Because of variation in stand age and vegetation structure, we conducted our experiments in two stands: primary stands (80–90 years) and secondary stands (< 50 years). In the primary stand (9 ha), dominant canopy trees are *Castanopsis fargesii* (Fagaceae), *Q. variabilis* (Fagaceae), *Pinus massoniana* (Pinaceae) and *Acer catalpifolium* (Aceraceae), with a small population of other tree species *Quercus serrata* (Fagaceae), *Lithocarpus harlandii* (Fagaceae), *Phoebe zhenman* (Lauraceae) and *Cyclobalanopsis glauca* (Fagaceae). Dominant shrubs are *Camellia oleifera* (Theaceae), *Symplocos stellaris* (Symplocaceae), *S. laurina* (Symplocaceae) and *Pittosporum daphniphylloides* (Pittosporaceae). The ground flora is poorly developed, consisting of small patches of *Dicranopteris pedata*. In secondary stands (20 ha), *Q. variabilis*, *Q. serrata* and *C. fargesii* are dominant canopy trees. The understorey layer is mainly composed of *S. stellaris*, *S. laurina*, *Ilex purpurea* and *Myrsine africana*. The ground flora is dominated by *Dicranopteris pedata*.

We have studied two common tree species (*Q. liaotungensis* and *P. armeniaca*) in the Donglingshan forest, and four tree species (*Q. variabilis*, *Q. serrata*, *C. fargesii* and *C. oleifera*) in the Dujiangyan forest. All these trees and shrubs (Table 16.1) are dependent on seed-caching rodents to disperse their seeds (Xiao *et al.*, 2001; Zhang and Wang, 2001b; Li and Zhang, 2003).

General Methods

From 1999 to 2001 in the Donglingshan forest and from 2000 to 2002 in the Dujiangyan forest, we used wooden snare kill-traps baited with one peanut to monitor rodent species and abundance in the autumn (October). Two or three transects were selected, and 25 trap stations were set at 5 m intervals along each transect for 2 consecutive nights. We determined the number of each species of captured rodents. Seed-eating birds and diurnal mammals were censused by direct observation. We also observed seed hoarding of dominant rodent species in outdoor arenas (2 m × 2 m) at the Dujiangyan forest and (4 m × 3 m) at the Donglingshan forest.

Since 2000, we have recorded annual seed production of *Q. variabilis*, *Q. serrata*, *C. fargesii* and *C. oleifera* in two permanent plots (a primary stand and a secondary stand) in the Dujiangyan forest. We set up 40–60 seed traps (0.5 m² sample area) at intervals of 8–10 m throughout the plots

(3 ha) for estimating seed rain, and we established another 40–60 permanent quadrats (1 m² sample area) 1 metre from the seed traps for estimating intensity of seed removal by small rodents, birds, insects and microbes. Seeds in both seed traps and quadrats were checked at 2- or 4-day intervals. We recorded species and seed fate (i.e. consumed *in situ*, removed, infested by fungi or insects or remaining). We monitored seed removal on the ground in permanent quadrats by marking seeds (Xiao *et al.*, 2001). After counting, all new acorns falling in quadrats were marked by using indelible ink. To estimate seed removal rates, we let M_t and M_{t+1} be the numbers of marked seeds in quadrats when checked at time t and $t+1$, N_t is the number of new seeds falling in quadrats when checked at time t, and the removal rate (R_t) during period t to $t+1$ is calculated as: $R_t = \dfrac{M_t + N_t - M_{t+1}}{M_t + N_t}$. (Note: N_t is the non-marked seeds when checking at time t, but they are marked after counting. Thus, M_{t+1} is the survival number from M_t and N_t.) Similar methods were used to measure seed rain and seed removal of *Q. liaotungensis* in Donglingshan forest, but seed rain and seed removal of *P. armeniaca* was only quantified for permanent quadrats from 2000 to 2001.

To track the fates of seeds after they are removed from parent trees by small rodents,

Table 16.1. Some morphological and ecological characters of rodent-dispersed tree species in the temperate Donglingshan forest and the subtropical Dujiangyan forest.

Species	Habit	Fruiting	Seed mass (g) Fresh	Dry	Seed coat	Tannin (%)[a]	Potential dispersers
Donglingshan forest							
Quercus liaotungensis	Tree	Sep.–Nov.	1.76	1.22	Leathery	8.3	Rodents, birds
Prunus armeniaca	Shrub	June–July	1.61	1.53	Hard	Low	Rodents
Dujiangyan forest							
Q. variabilis	Tree	Sep.–Dec.	2.42	1.77	Leathery	11.7	Rodents, birds
Q. serrata	Tree	Sep.–Dec.	0.97	0.77	Leathery	10.6	Rodents, birds
Castanopsis fargesii	Tree	Oct.–Dec.	0.46	0.31	Leathery	0.2	Rodents, birds
Camellia oleifera[b]	Shrub	Sep.–Nov.	0.90	0.47	Leathery	0.1	Rodents

[a]Tannin is represented in % dry weight.
[b]*Camellia oleifera* exists mainly in primary stands and is rare in secondary stands in the Dujiangyan forest.

we used coded tin tags to label individual seeds (Zhang and Wang, 2001b). A tiny hole 0.5 mm in diameter was drilled near the germinal disc of each seed. Though the cotyledon was partly damaged, the embryo remained intact and was capable of germinating. Seeds were tied with a small, light tin tag (≈ 3.5 cm × 1 cm, < 0.1 g) through the hole by using a fine steel wire 3 cm long (Zhang and Wang, 2001b; Li, 2002; Li and Zhang, 2003) or 8 cm long (Xiao, 2003). Each tag was numbered using a fine-point metal pen to make each seed identifiable. When rodents buried seeds in soil, all the tin tags were often left on the surface, making them easy for us to relocate. Tagging had a negligible effect on seed removal and caching by rodents (Zhang and Wang, 2001b).

For all tree species (except for those of *C. fargesii* seeds, which were unavailable in 2002), sound fresh seeds were collected from the ground or trees in the Donglingshan forest (1998–2001) and in the Dujiangyan forest (2000–2002). We followed individual *Q. liaotungensis* and *P. armeniaca* seeds from 1998 to 2001 in the Donglingshan forest, and we also traced individual *Q. variabilis*, *Q. serrata*, *C. fargesii* and *C. oleifera* seeds in primary stands and in secondary stands in the Dujiangyan forest from 2000 to 2002. We released 2654 acorns and 3920 nuts in the Donglingshan forest, and 4720 seeds in the Dujiangyan forest. Usually we placed 10–40 tagged seeds at each seed station with 15–22 stations spaced at intervals of 10 m along each transect line (2 ha area). We used bamboo sticks (10 cm from the cache sites) coded with the number of the relocated seeds to mark cache locations. The marked sticks might give rodents some cues for pilfering, but we tested whether marked sticks had any effects on cache survival by establishing artificial caches (Z.-S. Xiao and Z.-B. Zhang, unpublished data). We monitored cache sites to determine spatio-temporal changes of the seeds cached by small rodents. Relevant information on relocated seeds was recorded, including seed condition (intact, eaten), cache size (number of seeds per cache), dispersal distance from seed sources and cache microsites.

Results

Potential seed dispersal agents and seed predators

From 1999 to 2001, we trapped 171 individual small nocturnal rodents in 900 trap nights in the Donglingshan forest, including field mice (*Apodemus peninsulae*, n = 118), white-bellied mice (*Niviventer confucianus*, n = 44), rat-like hamsters (*Cricetulus triton*, n = 5), striped field mice (*A. agrarius*, n = 2) and brown-backed voles (*Clethrionomys rutilus*, n = 2) (Fig. 16.1A; Li, 2002). In addition, striped hamsters (*Cricetulus barabensis*), long-tailed hamsters (*Cricetulus longicaudatus*), Siberia chipmunks (*Tamias sibericus*), grey squirrels (*Sciurotamias davidianus*) and house mice (*Mus musculus*) were also found in the Donglingshan Mountains (Meng and Zhang, 1999). These rodent species (especially *A. peninsulae*, *S. davidianus* and *N. confucianus*) are known to harvest and cache seeds of *Q. liaotungensis* and *P. armeniaca* (Meng and Zhang, 1999; Wang and Ma, 1999, 2001; Wang *et al.*, 1999, 2000; Sun and Chen, 2000; Zhang and Wang, 2001a,b; Li, 2002; Li and Zhang, 2003). Through caching experiments in outdoor arenas, we found *A. peninsulae* and *S. davidianus* to be the main scatterhoarding rodent species, while *N. confucianus* is the main seed predator in the Donglingshan forest (L.-Q. Lu and Z.-B. Zhang, unpublished data). Besides small rodents, Eurasian jays (*Garrulus glandarius*) are potential seed dispersers, but brown-eared pheasants (*Crossoptilon mantchuricus*), Koklass pheasants (*Pucrasia macrolopha*) and ring-necked pheasants (*Phasianus colchicus*) are potential seed predators of *Q. liaotungensis* (Meng and Zhang, 1997). Eurasian jays may be less important for short-distance seed dispersal than rodents because of small population numbers, but they are very important for long-distance seed dispersal (see den Ouden *et al.*, Chapter 13, this volume).

From 2000 to 2002, 228 small nocturnal rodents were captured in 2250 trap nights at the Dujiangyan forest, including ten species, i.e. chestnut rats (*Niviventer fulvescens*,

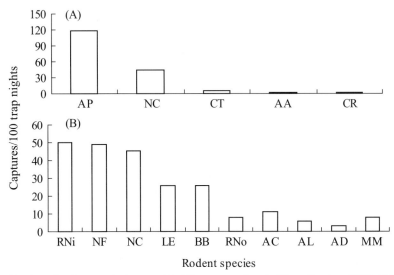

Fig. 16.1. Number of rodent species captured in (A) temperate forest of Donglingshan (1999–2001, n = 171 captures; Li, 2002) and (B) subtropical forest of Dujiangyan (2000–2003, n = 225 captures; Xiao, 2003). AA, *Apodemus agrarius*; AC, *A. chevrieri*; AD, *A. draco*; AL, *A. latronum*; AP, *A. peninsulae*; BB, *Berylmys bowersi*; CR, *Clethrionomys rutilus*; CT, *Cricetulus triton*; LE, *Leopoldamys edwardsi*; MM, *Micromys minutus*; NC, *Niviventer confucianus*; NF, *N. fulvescens*; RNi, *Rattus nitidu*; RNo, *R. norvegicus*.

n = 50), Himalayan rats (*Rattus nitidu*, n = 49), Chinese white-bellied rats (*Niviventer confucianus*, n = 45), Edwards' long-tailed rats (*Leopoldamys edwardsi*, n = 25), Bowers' rats (*Berylmys bowersi*, n = 25), Norway rats (*Rattus norvegicus*), Sichuan field mice (*Apodemus latronum*), Chevrier's field mice (*Apodemus chevrieri*), Chinese field mice (*Apodemus draco*) and harvest mice (*Micromys minutus*) (Fig. 16.1B; Xiao, 2003). All these rodent species consume seeds of *Q. variabilis*, *Q. serrata*, *C. fargesii* and *C. oleifera* in the Dujiangyan forest (Xiao et al., 2003b). *Leopoldamys edwardsi* has been observed to scatterhoard seeds in outdoor arenas (Xiao et al., 2003a). In the Dujiangyan forest, there are other potential seed predators (e.g. masked palm civet (*Paradoxurus larvata*) (Viverridae) and seed dispersers (e.g. Eurasian jays).

Seed rain

In the Dujiangyan forest, seed production of *Q. variabilis*, *Q. serrata*, *C. fargesii* and

C. oleifera varied among years and between primary and secondary stands (Fig. 16.2). At the community level, seed abundance and total energy values were higher in the primary stand than in the secondary stand (Xiao, 2003). In the Donglingshan forest, acorn production of *Q. liaotungensis* varied greatly among years. For example, production was 1.6 ± 2.6 acorns/m² in 1997 (Wang et al., 2000), 8.5 ± 12.4 acorns/m² in 1999 (Yan, 2000), 56.4 ± 81.5 acorns/m² in 2000 (Li, unpublished data) and nearly non-existent in 1998 and 2001. For *P. armeniaca*, the annual seed crop was more consistent at 9.6 ± 14.2 stones/m² in 2000 and 8.2 ± 13.4 stones/m² in 2001 (Li, unpublished data). We did not estimate effects of birds on seed removal in either forest. Based on the very small populations of seed-eating birds, the impact of birds on seeds falling on the ground may be potentially minor, e.g. *Q. liaotungensis* acorns in the Donglingshan forest and acorns of *Q. variabilis*, *Q. serrata* and *C. fargesii* in the Dujiangyan forest, but no impacts of birds on *P. armeniaca* nuts with hard hulls in the Donglingshan forest and *C. oleifera* in the

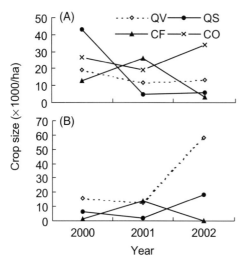

Fig. 16.2. Annual variation in seed crop of four common tree species (QV, *Quercus variabilis*; QS, *Q. serrata*; CF, *Castanopsis fargesii*; and CO, *Camellia oleifera*) in primary (A) and secondary (B) stands in the Dujiangyan forest (2000–2002 data from Xiao, 2003).

Dujiangyan forest. Insects destroyed a large proportion of the seed crop before seed dispersal. Two weevil species (*Curculio dentipes* and *Curculio* sp.) and two moth species (*Cydia kurokoi* and a Pyralidae species) are important seed predators for *Q. liaotungensis* in the Donglingshan forest (Yu *et al.*, 2001), while at least one weevil species (*Curculio* sp.) and another unidentified moth species are seed predators for *Q. variabilis, Q. serrata* and *C. fargesii* in

the Dujiangyan forest (Z.-S. Xiao, personal observation). For *Q. liaotungensis* in the Donglingshan forest, the proportion of insect infestation was 17.8% in 1997 (Wang *et al.*, 2000), 41.9% in 1999 (Yu *et al.*, 2001) and 14.1% in 2000 (Li, 2002), indicating that insect infestation tends to be greater in low crop years. In the Dujiangyan forest, the proportion of insect infestation was $32.6 \pm 12.5\%$ for *Q. variabilis*, $68.1 \pm 22.7\%$ for *Q. serrata*, and $44.0 \pm 24.3\%$ for *C. fargesii*. Insect infestation was much higher in the secondary stand than in the primary stand for these three species during this study (Xiao, 2003). Insects seldom infested the seeds of *P. armeniaca* and *C. oleifera* during our survey. After mature seeds were shed, factors affecting seed fates on the ground included consumption *in situ* and removal by small rodents, infestation by fungi or germination *in situ* (Table 16.2). Seed removal and predation by small rodents was the main factor in both the Dongling-shan and Dujiangyan forests (Table 16.2; Sun and Chen, 2000; Wang *et al.*, 2000; Wang and Ma, 2001; Xiao *et al.*, 2001).

Spatial and temporal dispersal of seeds

Harvest of tagged seeds

Nearly 100% of the tin-tagged seeds were harvested (i.e. removed or consumed *in situ*) by small rodents within 1 or 2 weeks

Table 16.2. Seed fates (removal and consumption *in situ* by rodents, fungi infestation and germination *in situ*) of several tree species from permanent quadrats in both Donglingshan and Dujiangyan forests. Data are means ± 1 SD.

Tree species	Removal (%)	Consumed (%)	Fungi (%)	Germination *in situ*[a] (%)
Dujiangyan forest				
Quercus variabilis[b]	84.5 ± 11.4	13.2 ± 9.6	0.9 ± 2.1	1.5 ± 2.5
Q. serrata[b]	46.6 ± 19.7	23.0 ± 15.5	26.5 ± 22.7	3.8 ± 5.1
Castanopsis fargesii[b]	16.7 ± 1.8	83.3 ± 1.8	0	0
Camellis oleifera[b]	93.9 ± 8.7	6.2 ± 8.7	0	0
Donglingshan forest				
Q. liaotungensis[c]	70	15	15	0

[a]Of seeds that were not removed or eaten and established as seedlings in the next spring.
[b]Data from Xiao (2003).
[c]Data from Sun and Chen (2000).

after release at seed stations. The proportion of seed removal versus consumption *in situ* varied greatly among seed species, seasons, years and stands. In the Dujiangyan forest, the proportion of seed removal was 87.8 ± 5.5% for *Q. variabilis*, 44.0 ± 13.9% for *Q. serrata*, 34.0 ± 15.6% for *C. fargesii* and 95.1% for *C. oleifera* in the primary stand. In the secondary stand, animals removed 84.5 ± 12.0% for *Q. variabilis*, 56.0 ± 14.0% for *Q. serrata*, 48.0 ± 25.5% for *C. fargesii* and 91.6% for *C. oleifera* (Xiao, 2003). These data indicate that rodents prefer to remove large seeds (e.g. *Q. variabilis*) or high value seeds (e.g. *C. oleifera*). In the Donglingshan forest, the proportion of seed removal of *Q. liaotungensis* was 24.4% in 1999 and 48.6% in 2000 (Li and Zhang, 2003), and that of *P. armeniaca* was very high, e.g. 90% in 1998 (Zhang and Wang, 2001b), 94.9% in 2000 and 99.3% in 2001 (Li, 2002).

Seed fate

After removal from seed stations, 44.9 ± 19.0% of the tagged seeds were relocated at least once in the Dujiangyan forest (Xiao, 2003), and 36.6 ± 21.8% in the Donglingshan forest (Zhang and Wang, 2001b; Li, 2002; Li and Zhang, 2003). Post-removal seed fate can be sorted into three categories: (i) cached, including buried intact in the soil by rodents or deposited intact on the surface, (ii) eaten, leaving only tin tags and seed fragments, and (iii) missing with their true fates unknown. In both forests, scatterhoarding by small rodents was very common, but the proportion of the cached seeds varied among seed species, years and stands. In the Dujiangyan forest, the proportion of cached seed in primary caches was the highest for *C. oleifera* (secondary stand: 48.2%; primary stand: 75.7%), intermediate for *Q. variabilis* (secondary stand: 34.3 ± 29.4%; primary stand: 60.5 ± 14.1%), and lowest for *Q. serrata* (secondary stand: 5.7 ± 8.0%; primary stand: 20.6 ± 12.3%) and *C. fargesii* (secondary stand: 6.3 ± 9.0%; primary stand: 13.1 ± 9.8%). These data show that rodents prefer to cache large or high value

seeds. The proportion of seed caching was higher in the primary stand than in the secondary stand for all species (Xiao, 2003). In the Donglingshan forest, the proportion of seed caching of *Q. liaotungensis* was very low, e.g. 6.4% in 1999 and 4.0% in 2000 (Li and Zhang, 2003), but that of *P. armeniaca* with very hard seed cover was much higher, ranging from 26.8 to 58.8% (Zhang and Wang, 2001b; Li, 2002; Li and Zhang, 2003).

Dynamics of caches

Through regular surveys, we documented the timing of cache discovery as well as occurrences of both primary and secondary caches, and their survival conditions. Seeds were often recovered from caches within several days and transported to new cache sites or eaten near cache sites. For example, the cached seeds of *C. oleifera* were much more ephemeral at *Camellia*-poor (mean of 4 days in primary caches, 2 days in secondary caches) and *Camellia*-rich (73 days in primary caches, 95 days in secondary caches) areas in the Dujiangyan forest (Xiao, 2003). We also found differences in the dynamics of the caches of tree species in both forests. In the Dujiangyan forest, a small proportion of the primarily cached seeds of *C. oleifera* (3–13%) and *Q. variabilis* (5–10%) were transported and moved two to three times from one cache site to another, but this was seldom observed in *Q. serrata* and *C. fargesii* (Xiao, 2003). In the Donglingshan forest, we observed that only a few seeds of *Q. liaotungensis* (0.15–0.75%) and *P. armeniaca* (0–1.5%) were repeatedly cached by small rodents (Li, 2002). The amount of seed recaching by rodents was higher in subtropical forest than that in temperate forest. Seeds with a hard hull or higher nutritional value were recached more frequently than seeds with a soft hull or lower value.

In the Dujiangyan forest, 33.5 ± 0.4% (*C. oleifera*), 31.5 ± 0.9% (*Q. variabilis*), 75.5 ± 0.9% (*Q. serrata*) and 87.3 ± 0.9% (*C. fargesii*) of the tagged seeds (including part of the cached seeds) were consumed by small rodents, and the rest (including cached seeds) eventually disappeared and their ultimate fate was unknown (Xiao, 2003). In the Donglingshan forest, $36.9 \pm 10.9\%$ (*P. armeniaca*) and $81.1 \pm 7.36\%$ (*Q. liaotungensis*) of the tagged seeds were consumed by small rodents, and the rest eventually disappeared (Zhang and Wang, 2001b; Li, 2002; Li and Zhang, 2003). Small rodents may have transported the missing seeds into underground burrows or nests, rock caves and dense shrubs (for larderhoarding), where we could not find them. Some seeds were larderhoarded underground (Z.-S. Xiao, personal observation). In addition, small rodents may also have transported some seeds, especially the larger ones, outside the surveyed area because some seeds were cached or eaten more than 60 m away from seed stations.

Cache size

In both forests, small rodents made caches with similar numbers of seeds. For all species in both forests, 76–100% of primary caches and nearly 100% of higher order caches (i.e. secondary and tertiary caches) contained only one seed. We only found one cache ($n = 580$ caches, pooled data for both species) in the Donglingshan forest and 89 caches ($n = 616$ caches, pooled data for all four species) in the Dujiangyan forest that contained two or three seeds (Li, 2002; Z.-S. Xiao and Z.-B. Zhang, unpublished data). In addition, we found two big caches. In the primary stand of Dujiangyan forest in spring, we found ten germinating nuts of *C. fargesii* and a tagged seed of *C. oleifera*, and another cache had three tagged seeds of *C. oleifera* germinating and 12 germinating nuts of *C. fargesii* (Xiao, 2003).

Dispersal distances

In both forests, over 80% of cached seeds for all seed species were dispersed less than 20 m from our seed stations. In the Dujiangyan forest, mean dispersal distances for the 3-year data set (2000–2002) of seeds in primary caches were 4.7 ± 2.1 m (max:

38.1 m) for *Q. variabilis*, 3.6 ± 2.4 m (max: 12.5 m) for *Q. serrata*, 4.4 ± 3.2 m (max: 7 m) for *C. fargesii*, and 8.1 ± 0.2 m (mean for 2000 and 2001; max: 62.0 m) for *C. oleifera* (Xiao, 2003). In the Donglingshan forest, mean dispersal distances (1999 and 2000) of seeds in primary caches were 4.3 ± 5.0 m (max: 20 m) for *Q. liaotungensis* and 6.0 ± 5.1 m (max: 35 m) for *P. armeniaca* (Zhang and Wang, 2001b; Li, 2002; Li and Zhang, 2003). Mean dispersal distances gradually increased as seeds were recached at new cache sites. For example, in a primary forest, primary caches of *C. oleifera* were 8.2 m ($n = 238$), secondary caches were 10.2 m ($n = 33$) and tertiary caches were 28.8 m ($n = 2$) from the source (Xiao, 2003).

Microsites around the caches

More than 70% and 60% of all cached *Q. liaotungensis* and other tree species seeds, respectively, were deposited under shrubs or near the edge of shrubs in the Donglingshan and the Dujiangyang forests. More than 80% of the tagged seeds were eaten under shrubs, at the edge of shrubs and rock caves. We thus hypothesized that predation risk determines the microsite where animals choose to eat or cache seeds. However, it is unclear whether the microsites where rodents bury seeds benefit germination. Field observations suggest that seeds buried at the edge of shrubs or under shrubs could establish seedlings (Z.-S. Xiao, personal observation).

Seedling establishment

A small part of the cached seeds in Dujiangyan forest (*C. oleifera*, 0–3.2%; *Q. variabilis*, 0–1.0%; *Q. serrata*, 0–1.5%; *C. fargesii*, 0%) and the Donglingshan forest (*P. armeniaca*, 0–0.6%; *Q. liaotungensis*, 0%) survived to establish seedlings in the next spring (Zhang and Wang, 2001b; Li, 2002; Xiao, 2003). In the Donglingshan forest, natural seedlings of *Q. liaotungensis* are very rare in the field. We did not observe a single seedling after releasing 2654 acorns. In both forests, tagged seeds with a hard hull (e.g. *P. armeniaca*) or high caloric value (e.g. *C. oleifera* and *Q. variabilis*) were more likely to survive and establish seedlings than tagged acorns of other studied tree species.

Discussion

We found that the proportions of seed removal, consumption near the source plant, caching and survival varied greatly among tree species, stands, seasons, years and regions. There are several general patterns:

1. Larger seeds and harder seeds were more likely to be transported greater distances and cached than consumed *in situ* by rodents.
2. Seeds in secondary forest suffered higher insect infestation and predation by rodents.
3. Seeds were mostly cached under shrubs or at the edge of shrubs.
4. Seeds were more likely to be recached and established as seedlings in subtropical regions than in temperate regions.

Large seeds have higher nutritional values than small ones, and thus are more attractive to scatterhoarding animals (Vander Wall, 1990, 2001, 2003; Jansen and Forget, 2001; Jansen et al., 2002). There may exist a general trend that larger seeds are cached more often than smaller ones among different plant species (Hurly and Robertson, 1987; Vander Wall, 1995a; Forget et al., 1998) and within the same species (Hallwachs, 1994; Jansen et al., 2002; Jansen, 2003), though there are some exceptions (e.g. Brewer, 2001). Besides seed size, other seed traits may also have important impacts on seed caching and consumption. Scatterhoarding animals often prefer seeds with higher fat content (Jansen and Forget, 2001). Since long handling time can increase predation risk (Jacobs, 1992; Hadj-Chikh et al., 1996) and seeds with hard seed hulls can be stored for a long time (Jansen and

Forget, 2001), seeds with hard seed hulls may increase caching and reduce consumption *in situ* (e.g. *P. armeniaca*, compared to the acorns of *Q. liaotungensis* in Dongling-shan forest). In the Dujiangyan forest, small rodents prefer to consume *C. fargesii* seeds, which have low tannin content (0.24%) and small seed mass, and seldom cache them (Xiao, 2003).

Seed dispersal distances were also influenced by seed size and other seed traits. Large seeds are often carried further than small ones (Hurly and Robertson, 1987; Hallwachs, 1994; Vander Wall, 1995a; Forget et al., 1998; this study). *P. armeniaca* seeds, with a hard seed hull and high nutritional value, had longer dispersal distances than *Q. liaotungensis* seeds in the Dongling-shan forest, and *C. oleifera*, with a high fat content, had the longest dispersal distances in the Dujiangyan forest. Jansen et al. (2002) found that seed dispersal distances of *Carapa procera* increased with increased seed mass.

A growing body of evidence indicates that secondary caching of seeds by scatter-hoarding animals seems very common in many forests all over the world, e.g. North America (e.g. Vander Wall, 1994, 1995b, 2002a,b, 2003; Vander Wall and Joyner, 1998; Longland et al., 2001), South America (e.g. Forget and Milleron, 1991; Jansen et al., 2002; Jansen, 2003), Australia (e.g. Theimer, 2001) and Asia (e.g. Iida, 1996; Hoshizaki and Hulme, 2002; this study). Large seeds may facilitate secondary caching (Jansen et al., 2002; Xiao, 2003) and mast seeding may reduce secondary caching (Vander Wall, 2002b). Secondary caching may greatly change the final pattern of seed dispersal and seed fates (Xiao et al., 2004). First, the fate of seeds in the initial cache site is often not the ultimate seed fate, indicating that part of seeds can survive to become seedlings in higher order cache sites (e.g. secondary caches, tertiary caches) (e.g. Vander Wall and Joyner, 1998; Vander Wall, 2002b, 2003; Xiao, 2003). Second, seed number per cache seems to become smaller as rodents repeatedly move the seeds from primary cache sites into higher order cache sites, which may potentially increase seedling establishment and subsequently reduce seedling competition (Vander Wall, 2002a, 2003; Xiao, 2003). Third, dispersal distances gradually increase and the distribution of caches becomes more even (less clumped), which may reduce seed predation and increase seed survival and seedling establishment (e.g. Vander Wall, 2002a, 2003; Xiao, 2003). Fourth, the initial microsites of the cached seeds may not be the final microsites. At present, the potential contributions of secondary caching by scatterhoarding animals to seed dispersal and forest dynamics are poorly known (Vander Wall and Joyner, 1998; Xiao *et al.*, 2003a,b, 2004). Future studies are needed to better understand how secondary caching by rodents influences forest dynamics.

Seed dispersal and seed fate are often influenced by mast seeding (Vander Wall, 2001, 2002b; Jansen, 2003), local seed availability at the community level (Forget and Milleron, 1991; Forget, 1992; Hoshizaki and Hulme, 2002) and rodent abundance (Ostfeld *et al.*, 1996; Wolff, 1996). Mast seeding often results in predator saturation, which reduces seed predation by rodents. The reason that the seed-poor secondary forests have higher seed predation by rodents than in seed-rich primary forests may be because of predator saturation. But mast seeding may not be synchronized at the community level (Ims, 1990), which may have some important consequences for scatterhoarder-dispersed tree species. For example, the less preferred species might experience reduced seed predation when the seed abundance of other more preferred species increases (Hoshizaki and Hulme, 2002). In the Dujiangyan forest, the most-preferred species is *C. fargesii*. During our 3-year study, *C. fargesii* had a large seed crop in 2001 and a poor crop in 2002 (Fig. 16.2). We expected that seed consumption rate of the less preferred seed species (e.g. *Q. variabilis* and *Q. serrata*) should be lower in 2001 when mast seeding occur in *C. fargesii*. Seed consumption of *Q. variabilis* was lower in 2001 (9%) than in 2000 (19%) at seed stations in the secondary stand, which fits the expectation well; but seed consumption was higher in 2000 (22%) than in 2001 (2%)

in the primary stand (Xiao, 2003). For *Q. serrata*, seed consumption was similar in both primary (62% in 2001; 65% in 2002) and secondary (54% in 2001; 50% in 2002) forests (Xiao, 2003). Though it is generally believed that mast seeding or high seed availability reduces the rate of seed removal and seed caching (Kelly, 1994; Jansen, 2003; see also Hoshizaki and Hulme, 2002), our results in the Dujianyan forest were not completely consistent with this hypothesis (Xiao, 2003). There are no simple explanations for this complicated phenomenon in our 3-year study. Similar phenomena also occurred in studies carried out by Hoshizaki and Hulme (2002). Long-term studies are essential for exploring the interactions between community-level seed availability and seed-eating animals, and understanding the role of these interactions in natural regeneration and population dynamics of co-existing nut-bearing tree species.

There may exist differences between geographical regions in seed fate under rodent influence, which is poorly understood at present. There are obvious differences between the subtropical Dujianyan forest and the temperate Donglingshan forest in climate and soil condition. Seed burying by rodents is more important for seedling establishment in the Donglingshan forest with drier weather and more compacted soil than that in Dujiangyan forest with a humid climate and friable soil. Indeed, seeds are often well buried in both the Donglingshan and Dujiangyan forests. However, the proportion of seedlings is higher in the subtropical region, probably because of the difference in precipitation between the two regions. In dry years, very few seeds of *P. armeniaca* become seedlings in Donglingshan forest. Similar results were also obtained by Forget (1997) and Jansen (2003).

The most challenging difficulty in studies of seed fate influenced by rodents lies in the complex interactions between plant seeds and rodents. Though tightly co-evolved plant–vertebrate interactions are rare in nature in seed-dispersal systems (Herrera, 1985), long-term mutualism/ predation relationships between rodents and plant seeds have occurred for a long time since rodents radiated dating from the Palaeocene (Price and Jenkins, 1986; Vander Wall, 2001). The seed–rodent system contains two interactions: mutualism and predation. Rodents eat plant seeds (predation), but rodents also benefit plants through dispersing and burying seeds, which promotes fitness of plants (mutualism). Plants have to trade off between mutualism and predation. On the one hand, plant seeds should have traits attracting rodents to harvest their seeds (competing with other plant seeds). The attracting traits of seeds include increased seed nutrition value or seed size. On the other hand, plants also have traits that serve to reduce over-predation by rodents. The defensive traits include increased tannin content in seeds, hardness of seed hulls, and mast seeding. Similarly, rodents also have to trade off between mutualism and predation. On the one hand, caching behaviour potentially benefits plant seeds. On the other hand, rodents should be effective predators with the capacity to detoxify seeds or open hard seed hulls. As suggested by Zhang (2003), mutualism may promote co-existence in prey–predator systems. However, it is not clear whether mutualistic/predatory relationships between plant seeds and rodents affect the co-evolution of functional traits.

Studies on the interactions between seed-caching rodents and the tree seeds for which they forage have begun to explore the mutualistic/predatory relationships, which may lead to co-evolution/co-adaptation of functional traits in seed–rodent systems (e.g. Smith, 1970, 1975; Vander Wall, 1993, 2002b, 2003; Jansen and Forget, 2001; Hoshizaki and Hulme, 2002; Hulme, 2002; Jansen et al., 2002; Theimer, Chapter 17, this volume). Though many previous studies have provided evidence to understand the relationships between seed-eating rodents and plant seeds, little is known about the evolutionary changes in functional traits for both rodents and tree seeds. Future research should include short-term and long-term systematic investigations on the effects of plant seeds (including seed traits) on rodent behaviour and other functional traits as well

as fitness of rodents, and the rodents' selective force acting on seed traits (e.g. seed size) at different local, regional or global scales. In addition, it is also important to explore the interesting question of how the evolutionary interactions between rodent-dispersed tree species and seed-caching rodents shape the processes of forest regeneration and succession, and influence tree species diversity.

Acknowledgements

We are very grateful to P.-M. Forget, S.B. Vander Wall, W.S. Longland and K. Hoshizaki for their critical comments and valuable suggestions for this manuscript. Funds were provided by the Innovation Programs of the Chinese Academy of Sciences (KSCX2-SW-105 and KSCX2-SW-103), Key Project of CNSF 'Biodiversity Changes of Yangtze River Valley, Sustainable Utilities and Regional Ecological Safety' (G2000046802), and National Science Foundation of China (No. 39893360).

References

Abbott, H.G. and Quink, T.F. (1970) Ecology of eastern white pine seed caches made by small forest mammals. *Ecology* 51, 271–278.

Brewer, S.W. (2001) Predation and dispersal of large and small seeds of a tropical palm. *Oikos* 92, 245–255.

Brewer, S.W. and Rejmánek, M. (1999) Small rodents as significant dispersers of tree seeds in a Neotropical forest. *Journal of Vegetation Science* 10, 165–174.

Forget, P.-M. (1990) Seed-dispersal of *Vouacapoua americana* (Caesalpiniaceae) by caviomorph rodents in French Guiana. *Journal of Tropical Ecology* 6, 459–468.

Forget, P.-M. (1991) Scatterhoarding of *Astrocaryum paramaca* by *Proechimys* in French Guiana; comparison with *Myoprocta exilis*. *Tropical Ecology* 32, 155–167.

Forget, P.-M. (1992) Seed removal and seed fate in *Gustavia superba* (Lecythidaceae). *Biotropica* 24, 408–414.

Forget, P.-M. (1993) Post-dispersal predation and scatterhoarding of *Dipteryx panamensis* (Papilionaceae) seeds by rodents in Panama. *Oecologia* 94, 255–261.

Forget, P.-M. (1997) Effects of microhabitat on seed fate and seedling performance in two rodent-dispersed tree species in rain forest in French Guiana. *Journal of Ecology* 85, 693–703.

Forget, P.-M. and Milleron, T. (1991) Evidence for secondary seed dispersal by rodents in Panama. *Oecologia* 87, 596–599.

Forget, P.-M., Milleron, T. and Feer, F. (1998) Patterns in post-dispersal seed removal by neotropical rodents and seed fate in relation to seed size. In: Newbery, D.M., Prins, H.T. and Brown, N.D. (eds) *Dynamics of Tropical Communities.* Blackwell Science, Oxford, pp. 25–49.

Hadj-Chikh, L.Z., Steele, M.A. and Smallwood, P.D. (1996) Caching decisions by grey squirrels: a test of the handling time and perishability hypotheses. *Animal Behavior* 52, 941–948.

Hallwachs, W. (1994) The clumsy dance between agoutis and plants: scatter-hoarding by Costa Rican dry forest agoutis (*Dasyprocta punctata*: Dasyproctidae: Rodentia). PhD thesis, Cornell University, Ithaca, New York.

Herrera, C.M. (1985) Determinants of plant–animal coevolution: the case of mutualistic dispersal of seeds by vertebrates. *Oikos* 44, 132–141.

Hoshizaki, K. and Hulme, P.E. (2002) Mast seeding and predator-mediated indirect interactions in a forest community: evidence from post-dispersal fate of rodent-generated caches. In: Levey, D., Silva, W.R. and Galetti, M. (eds) *Seed Dispersal and Frugivory: Ecology, Evolution and Conservation.* CAB International, Wallingford, UK, pp. 227–239.

Hulme, P.E. (2002) Seed-eaters: seed dispersal, destruction and demography. In: Levey, D., Silva, W.R. and Galetti, M. (eds) *Seed Dispersal and Frugivory: Ecology, Evolution and Conservation.* CAB International, Wallingford, UK, pp. 257–273.

Hulme, P.E. and Hunt, M.K. (1999) Rodent post-dispersal seed predation in deciduous woodland: predator response to absolute and relative abundance of prey. *Journal of Animal Ecology* 68, 417–428.

Hurly, T.A. and Robertson, R.J. (1987) Scatter-hoarding by territorial red squirrels: a test of the optimal density model. *Canadian Journal of Zoology* 65, 1247–1252.

Iida, S. (1996) Quantitative analysis of acorn transportation by rodents using magnetic locator. *Vegetatio* 124, 39–43.

Ims, R.A. (1990) The ecology and evolution of reproductive synchrony. *Trends in Ecology and Evolution* 5, 135–140.

Jacobs, L.F. (1992) The effect of handling time on the decision to cache by grey squirrels. *Animal Behavior* 43, 522–524.

Jansen, P.A. (2003) Scatter-hoarding and tree regeneration: ecology of nut dispersal in a Neotropical rainforest. PhD thesis, Wageningen University, Wageningen, The Netherlands.

Jansen, P.A. and Forget, P.-M. (2001) Scatter-hoarding rodents and tree regeneration. In: Bongers, F., Charles-Dominique, P., Forget, P.-M. and Théry, M. (eds) *Nouragues: Dynamics and Plant–Animal Interactions in a Neotropical Rainforest.* Kluwer Academic Publisher, Dordrecht, The Netherlands, pp. 275–288.

Jansen, P.A., Bartholomeus, M., Bongers, F., Elzinga, J.A., den Ouden, J. and Van Wieren, S.E. (2002) The role of seed size in dispersal by a scatterhoarding rodent. In: Levey, D., Silva, W.R. and Galetti, M. (eds) *Seed Dispersal and Frugivory: Ecology, Evolution and Conservation.* CAB International, Wallingford, UK, pp. 209–225.

Kelly, D. (1994) The evolutionary ecology of mast seeding. *Trends in Ecology and Evolution* 9, 466–470.

Kollmann, J., Coomes, D.A. and White, S.M. (1998) Consistencies in post-dispersal seed predation of temperate fleshy-fruited species among seasons, years and sites. *Functional Ecology* 12, 683–690.

Levey, D.J. and Sargent, S. (2000) A simple method for tracking vertebrate-dispersed seeds. *Ecology* 81, 267–274.

Li, H.-J. (2002) Effect of small rodents on seed regeneration of forest. PhD thesis, Institute of Zoology, Chinese Academy of Sciences, Beijing, P.R. China. (In Chinese with English summary.)

Li, H.-J. and Zhang, Z.-B. (2003) Effect of rodents on acorn dispersal and survival of the Liaodong oak (*Quercus liaotungensis* Koidz.). *Forest Ecology and Management* 176, 387–396.

Longland, W.S., Jenkins, S.H., Vander Wall, S.B., Veech, J.A. and Pyare, S. (2001) Seedling recruitment in *Oryzopsis hymenoides*: are desert granivores mutualists or predators? *Ecology* 82, 3131–3148.

Meng, Z.-B. and Zhang, Z.-B. (1997) The species and habitats of bird and mammal and the characteristics of rodent community in the mountain area of Beijing. In: Chen, L.-Z. (ed.)

Studies on Structure and Function of Forest Ecosystem in Warm Temperate Zone. Science Press, Beijing, pp. 76–87 (in Chinese).

Meng, Z.-B. and Zhang, Z.-B. (1999) Diversity of small rodents and their role in forest regeneration in Dongling Mountain. In: Ma, K.-P. (ed.) Ecosystem Diversity of Key Areas in China. Zhejiang Press of Sciences and Technology, Huangzhou, pp. 103–108 (in Chinese).

Miyaki, M. and Kikuzawa, K. (1988) Dispersal of Quercus mongolica acorns in a broad-leaved deciduous forest: 2. Scatterhoarding by mice. Forest Ecology and Management 25, 9–16.

Ostfeld, R.S., Jones, G.G. and Wolff, J.O. (1996) Of mice and mast: ecological connections in eastern deciduous forests. BioScience 46, 323–330.

Price, M.V. and Jenkins, S.H. (1986) Rodents as seeds consumers and dispersers. In: Murray, D.R. (ed.) Seed Dispersal. Academic Press, Sydney, pp. 191–235.

Smith, C.C. (1970) The coevolution of pine squirrels (Tamiasciurus) and conifers. Ecological Monographs 38, 31–63.

Smith, C.C. (1975) The coevolution of plants and seed predators. In: Gilbert, L.E and Raven, P.H. (eds) The Coevolution of Animals and Plants. University of Texas Press, Austin, Texas, pp. 53–77.

Sun, S.-C. and Chen, L.-Z. (2000) Seed demography of Quercus liaotungensis in Dongling Mountain Region. Acta Physioecologica Sinica 24, 215–221. (In Chinese with English summary.)

Theimer, T.C. (2001) Seed scatterhoarding by white-tailed rats: consequences for seedling recruitment by an Australian rain forest tree. Journal of Tropical Ecology 17, 177–189.

Vander Wall, S.B. (1990) Food Hoarding in Animals. University of Chicago Press, Chicago, Illinois, 445 pp.

Vander Wall, S.B. (1993) Cache site selection by chipmunks (Tamias spp.) and its influence on the effectiveness of seed dispersal in Jeffrey pine (Pinus jeffreyi). Oecologia 96, 246–252.

Vander Wall, S.B. (1994) Seed fate pathways of antelope bitterbrush: dispersal by seed-caching yellow pine chipmunks. Ecology 75, 1911–1926.

Vander Wall, S.B. (1995a) The effects of seed value on the caching behavior of yellow pine chipmunks. Oikos 74, 533–537.

Vander Wall, S.B. (1995b) Dynamics of yellow pine chipmunk (Tamias amoenus) seed caches: underground traffic in bitterbrush seeds. Écoscience 2, 261–266.

Vander Wall, S.B. (2001) The evolutionary ecology of nut dispersal. Botanical Review 67, 74–117.

Vander Wall, S.B. (2002a) Secondary dispersal of Jeffrey pine seeds by rodent scatter hoarders: the roles of pilfering, recaching, and a variable environment. In: Levey, D., Silva, W.R. and Galetti, M. (eds) Seed Dispersal and Frugivory: Ecology, Evolution and Conservation. CAB International, Wallingford, UK, pp. 193–208.

Vander Wall, S.B. (2002b) Masting in animal-dispersed pines facilitates seed dispersal. Ecology 83, 3508–3516.

Vander Wall, S.B. (2003) Effects of seed size of wind-dispersed pines (Pinus) on secondary seed dispersal and the caching behavior of rodents. Oikos 100, 25–34.

Vander Wall, S.B. and Joyner, J.W. (1998) Recaching of Jeffrey pine (Pinus jeffreyi) seeds by yellow pine chipmunks (Tamias amoenus): potential effects on plant reproductive success. Canadian Journal of Zoology 76, 154–162.

Wang, B.C. and Smith, T.B. (2002) Closing the seed dispersal loop. Trends in Ecology and Evolution 17, 379–385.

Wang, W. and Ma, K.-P. (1999) Predation and dispersal of Quercus liaotungensis acorns by Chinese rock squirrel and Eurasian jay. Acta Botanica Sinica 41, 1142–1144. (In Chinese with English summary.)

Wang, W. and Ma, K.-P. (2001) Predation and dispersal of Quercus liaotungensis Koidz. acorns by animals in Dongling Mountain, Northern. I. Effect of rodents removal on loss of acorns. Acta Ecologica Sinica 21, 204–210. (In Chinese with English summary.)

Wang, W., Ma, K.-P. and Liu, C.-R. (1999) Removal and predation of Quercus liaotungensis acorns by animals. Ecological Research 14, 225–232.

Wang, W., Ma, K.-P. and Liu, C.-R. (2000) Seed shadow of Quercus liaotungensis in a broad-leaved forest in Doling Mountain. Acta Botanica Sinica 42, 195–202.

Wolff, J.O. (1996) Population fluctuations of mast-eating rodents are correlated with production of acorns. Journal of Mammalogy 77, 850–856.

Xiao, Z.-S. (2003) Effects of small mammals on tree seed fates and forest regeneration in Dujiangyan Region, China. PhD thesis, Institute of Zoology, Chinese Academy of

Sciences, Beijing, P.R. China. (In Chinese with English summary.)

Xiao, Z.-S. and Zhang, Z.-B. (2003) How to trace seeds and fruits dispersed by frugivorous animals: a review. *Biodiversity Science* 11, 248–255. (In Chinese with English summary.)

Xiao, Z.-S., Wang, Y.-S. and Zhang, Z.-B. (2001) Seed bank and the factors influencing it for three Fagaceae species in Dujiangyan Region, P. R. China. *Chinese Biodiversity* 9, 373–381. (In Chinese with English summary.)

Xiao, Z.-S., Wang, Y.-S., Zhang, Z.-B. and Ma, Y. (2002) Preliminary studies on the relationships between communities of small mammals and types of habitats in Dujiangyan Region. *Chinese Biodiversity* 10, 163–169. (In Chinese with English summary.)

Xiao, Z.-S., Zhang, Z.-B. and Wang, Y.-S. (2003a) Observations on tree seed selection and caching by Edward's long-tailed rat (*Leopoldamys edwardsi*). *Acta Theriologica Sinica* 23, 208–213. (In Chinese with English summary.)

Xiao, Z.-S., Zhang, Z.-B. and Wang, Y.-S. (2003b) The ability to discriminate weevil-infested nuts by rodents: potential effects on regeneration of nut-bearing plants. *Acta Theriologica Sinica* 23, 312–320.

Xiao, Z.-S., Zhang, Z.-B., Lu, J.-Q. and Chen, J.-R. (2004) Repeated caching of plant seeds by small rodents. *Chinese Journal of Zoology* 39, 94–99. (In Chinese with English summary.)

Yan, W.-J. (2000) Effects of small mammals on seed bank of *Quercus liaotungensis* oak. MSc thesis, Beijing Normal University, Beijing, P.R. China.

Yu, X.-D., Zhou, H.-Z., Luo, T.-H., He, J. and Zhang, Z.-B. (2001) Insect infestation and acorn fate in *Quercus liaotungensis*. *Acta Entomology Sinica* 44, 518–524. (In Chinese with English abstract.)

Zhang, Z.-B. (2003) Mutualism or cooperation among competitors promotes coexistence and competitive ability. *Ecological Modeling* 164, 271–282.

Zhang, Z.-B. and Wang, F.-S. (2001a) Effect of burial on acorn survival and seedling recruitment of Liaodong oak (*Quercus liaotungensis*) under rodent predation. *Acta Theriologica Sinica* 21, 35–43.

Zhang, Z.-B. and Wang, F.-S. (2001b) Effect of rodents on seed dispersal and survival of wild apricot (*Prunus armeniaca*). *Acta Ecologica Sinica* 21, 839–845.

17 Rodent Scatterhoarders as Conditional Mutualists

Tad C. Theimer

*Department of Biological Sciences, Northern Arizona University,
Flagstaff, AZ 86011, USA*

Introduction

Seeds dispersed by vertebrates can face potentially very different seed fate pathways. One of these is via vertebrates that pose little threat to the seed itself, either dropping the seed in the course of fruit handling or passing the seed relatively unharmed through the gut ('benign dispersers' or 'legitimate' dispersers of Schupp (1993)). Alternatively, seeds could be dispersed by vertebrates that potentially act as either seed predators or seed dispersers by consuming and damaging a significant proportion of seeds they handle. One set of animals that act in this way are those that ingest seeds and pass them through the gut (Janzen, 1981; see Beck, Chapter 6, this volume). Another set includes animals that temporarily cache seeds for later consumption, and occasionally fail to recover some of these seeds, thereby acting inadvertently as seed dispersers (Vander Wall, 1990). Because of the large cost in seeds destroyed by the latter two groups, and the difficulty in estimating the relative benefits of dispersal by these animals to plant recruitment, whether they play the role of antagonistic seed predator or mutualistic seed disperser is often equivocal (see Hulme, 2002; Beck, Chapter 6, this volume).

Variation in the outcome of species interactions like that exhibited by animals that can potentially destroy or disperse seeds has been documented in several systems that traditionally have been viewed as mutualisms (e.g. ant–membracid mutualisms; Cushman and Whitham, 1989; Billick and Tonkel, 2003). In these systems, termed 'conditional mutualisms', variation in ecological and life-history factors in either space or time can shift the outcome of a species interaction to multiple points along the continuum from antagonistic to mutualistic (Bronstein, 1994).

In this chapter, I argue that the interaction between plants and the rodents that scatterhoard their seeds (i.e. cache seeds individually or in small groups in numerous cache locations) can be viewed as a conditional mutualism that depends in part on two factors that can vary in time and space:

1. The relative abundance of seeds versus scatterhoarders (the seed:scatterhoarder ratio).
2. The potential recruitment of seedlings that are not handled by scatterhoarders versus recruitment when seeds are handled by scatterhoarders.

In a review of conditional mutualisms, Bronstein (1994) argued that variation in the outcome of species interactions was more likely when: (i) the mutualism was facultative rather than obligate; (ii) a third species was intimately involved in the interaction;

and (iii) the benefit of the interaction was a function of the relative abundance of the partners. All or several of these conditions are met by the scatterhoarder–plant interaction. First, although some scatterhoarder–plant interactions have been argued to be obligate (Hallwachs, 1986; Smythe, 1989; Forget, 1990; Peres et al., 1997; Asquith et al., 1999), many appear to be facultative, with at least some possibility for plant recruitment in the absence of the scatterhoarder. Second, one advantage of dispersal by scatterhoarding in some systems is escape from an independent mortality agent acting on undispersed seeds in the form of other seed predators and pathogens. Third, the number of seeds available relative to scatterhoarder density affects scatterhoarder behaviour, both in the proclivity to cache seeds rather than eat them and in the probability that seeds may survive caching.

In this chapter, I develop a simple conceptual model of how the outcome of the scatterhoarder–plant interaction depends on both the relative abundance of each partner and the challenges to recruitment faced by the plant. I also describe ways in which both factors can be altered by natural and anthropogenic changes that shift the outcome of the interaction, resulting in the type of context-dependent scenarios described by Dennis et al. (Chapter 7, this volume). In developing the model, I have considered primarily the interaction between a single scatterhoarder species and a single plant species, independent of the relative effectiveness of one scatterhoarder species to another or to other dispersers (disperser effectiveness sensu Herrera and Jordano, 1981; Schupp, 1993), or of how the same scatterhoarder can act as a predator for some plant species and a mutualist of others (e.g. differential caching and handling of red and white oaks by squirrels) (Hadj-Chikh et al., 1996; Steele et al., 2001; Steele et al., Chapter 12, this volume).

Conditionality and the Seed to Scatterhoarder Ratio

Large fruit and seed crops have been argued to be important for satiating seed predators (Janzen, 1971a; Silvertown, 1980) and, when followed by periods of food scarcity, for stimulating seed hoarding behaviour (Smythe, 1970; De Steven and Putz, 1984; Schupp, 1990; Vander Wall, 1990; Sork, 1993; Forget et al., 2001). The effect of seed/fruit abundance on scatterhoarder behaviour has been studied at three levels: (i) the amount of the crop handled by scatterhoarders; (ii) the fate of seeds initially handled (either cached or eaten); and (iii) the probability that cached seeds survive to germinate. In some cases, high seed and fruit availability has been shown to increase the proportion of the crop that escapes handling altogether, but even in years of high seed production entire crops can sometimes be harvested by scatterhoarders. Likewise, large crops have been shown to increase the ratio of seeds cached versus those eaten, though this too is not always true. Finally, although the assumption that increased caching would translate into higher probability of seed survival in caches is intuitive, studies linking seed abundance and cache fate are relatively rare. Vander Wall (2002) demonstrated that, although rodent scatterhoarders cached nearly all seeds in years of both high and low pine (Pinus spp.) seed crops, fewer seeds were consumed and more caches survived to time of germination in years of large seed crops. Likewise, caches of seeds of the tropical rainforest tree Beilschmedia bancroftii made by white-tailed rats lasted significantly longer in a year of heavy seed fall than in years when no seeds were produced (Theimer, 2001). Hoshizaki and Hulme (2002) quantified seed production for three tree species (Aesculus turbinata, Fagus crenata and Quercus mongolica), densities of two rodents (Apodemus speciosus and Eothenomys andersoni) and seed fate for Aesculus turbinata over several years that varied in total seed and rodent abundance. They found that the percentage of seeds initially cached (versus eaten) and moved to secondary caches was highest in a year of greatest seed availability. However, they found no detectable difference in cache survival across years, in part because the number of experimental seeds surviving to germination was very small in all years.

I developed a simple graphical model to describe the hypothetical relationship between survival of undispersed seeds and scatterhoarder cached seeds as a function of seed abundance (Fig. 17.1). Because the relative abundance of scatterhoarding rodents often varies in space and time, crop size alone will not predict seed fate as well as a measure of the relative abundance of seed/fruit availability to scatterhoarders. Therefore, I used the ratio of seeds : scatterhoarders in developing the first stage of this model. Figure 17.1 shows the proportion of a seed crop handled by scatterhoarders approximating 100% at low seed : scatterhoarder ratios and eventually declining as the number of seeds relative to scatterhoarders increases. The proportion of seeds that escape handling by scatterhoarders would obviously be the inverse function. The proportion of the crop that potentially survives in caches to germination is assumed to be 0 at low seed : scatterhoarder ratios, where all caches would be retrieved, and increases with increasing seed : scatterhoarder ratios, where the larger number of caches made increases the chance that some caches would escape recovery. At higher seed : scatterhoarder ratios, animals could stop caching entirely in the face of a surfeit of food, but, even if cache rates remained

constant, the proportion of the crop surviving in caches would decrease relative to the total seed crop. This model assumes that all seeds would be handled by scatterhoarders at low seed : scatterhoarder ratios and that, at ratios between the origin and Line X, all seeds handled would eventually be eaten. At seed : scatterhoarder ratios between Lines X and Y, all seeds are handled by scatterhoarders, but the probability that cached seeds survive increases with increasing seed : scatterhoarder ratios. At higher ratios beyond Line Y, some seeds escape detection by scatterhoarders and the number escaping detection increases with increasing seed : scatterhoarder ratio.

This model is consistent with the hypothesis that the role of scatterhoarders is conditional, depending upon the relative abundance of scatterhoarders and seeds. At low seed : scatterhoarder ratios (below Line X) scatterhoarders act as antagonistic predators. Beyond Line X, scatterhoarders have the potential to act as mutualistic seed dispersers, but this depends upon the relative success of recruitment from seeds that escape scatterhoarders compared to that from seeds cached by scatterhoarders. If a plant species was entirely dependent on scatterhoarders for recruitment (scatterhoarders act as obligate mutualists), the number of seedling recruits would be determined entirely by the number of caches surviving (line with solid circles in Fig. 17.1). This could be the case for seeds that require burial for germination or for seedling establishment but which rarely become buried by other means (Hallwachs, 1986; Smythe, 1989; Forget, 1994; Asquith et al., 1999). Seeds escaping scatterhoarder handling at high crop sizes would contribute nothing to recruitment. However, even this case is conditional, as the obligate mutualist acts as a seed predator at low seed : scatterhoarder ratios.

In contrast, in cases where cached seeds never successfully germinated (scatterhoarders act as seed predators), seedling recruitment would be determined by the probability of seed escape (line with open boxes in Fig. 17.1). This could be the case when small seeds are cached too deeply to

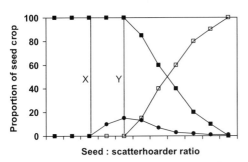

Fig. 17.1. Hypothetical relationship between the proportion of the seed crop handled by scatterhoarders (solid boxes), the proportion escaping handling (open boxes) and the proportion of cached seeds surviving to germination (solid circles) versus the relative abundance of seeds and scatterhoarders (seed : scatterhoarder ratio).

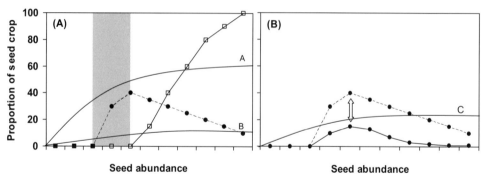

Fig. 17.2. (A) Hypothetical relationship between the proportion of the seed crop that successfully recruits from scatterhoarder caches (solid circles) and the proportion that escapes scatterhoarder handling (open boxes) versus crop size. The potential recruitment of the same plant in the absence of the scatterhoarder could be higher than for scatterhoarder cached seeds (Line A) or lower than that of scatterhoarder cached seeds (Line B). (B) If Line C is the probability of seedling recruitment in the absence of scatterhoarders, factors that shift the probability of cache survival from the solid line to the dashed line will cause a shift in the role of scatterhoarders from antagonists to mutualists.

allow them to germinate (Hulme, 2002) or when scatterhoarders excise embryos from cached seeds to prevent them from germinating (Fox, 1982; Steele *et al.*, 2001). This interaction could be conditional if cached seeds could germinate under a special set of conditions, such as a shift in the behaviour of the scatterhoarder to excise embryos under some circumstances but not others.

Although most studies compare the relative success of seeds dispersed by scatterhoarders to undispersed seeds, the overall benefit of scatterhoarding for a plant should be judged by comparing potential seedling recruitment in the absence of scatterhoarders altogether. For example, in Fig. 17.2, I show two scenarios for recruitment in the absence of scatterhoarders as it relates to crop size. In the absence of the scatterhoarder (Line A), recruitment would be greater than that from scatterhoarder caches and scatterhoarders would be poor mutualists. Note that for this model I have assumed for simplicity that the scatterhoarder population remains constant across varying crop sizes and thus use crop size rather than seed:scatterhoarder ratio as the independent variable. I have also assumed that seedling recruitment in the absence of scatterhoarders will increase with increasing crop size but will reach an asymptote as

density-dependent factors slow recruitment at high crop sizes.

In situations where seedling recruitment in the absence of scatterhoarders would follow Line A, if scatterhoarders remove the entire seed/fruit crop so that some cached seeds survive but no undispersed seeds escape detection (the shaded area of Fig. 17.2A), an 'apparent mutualism' can arise because the only seeds that survive to give rise to seedlings are from rodent caches. In this situation, all seedlings arise from rodent-cached seeds and plant dependence on scatterhoarder dispersal seems apparent. However, if recruitment would be higher in the total absence of scatterhoarders, then scatterhoarders have an overall negative effect on potential plant fitness and they act as antagonists. For example, scatterhoarding white-tailed rats on my Australian study plot removed the entire crop of yellow walnut (*Beilschmedia bancroftii*) seeds well before any seeds reached the time for germination (Theimer, 2001). Although none of the subset of rat-cached seeds I followed survived to germination, all seedlings subsequently arose on the study plot from buried seeds that were placed in microsites consistent with being cached there by rats. Thus, all seedlings I found were apparently from rat-cached seeds (Theimer, 2001). In contrast, a subset of 40 seeds fell naturally into a vertebrate exclosure and were subsequently exposed to all natural processes except those caused by terrestrial birds and mammals. Five years later, only two seedlings survived outside the exclosure out of the roughly 300 seeds that were removed by rats (0.7%) (a very conservative estimate based on one line transect at the height of fruit fall). In contrast, of the 40 seeds protected in the exclosure, five seedlings have arisen and are still alive (12%). This suggests that seedling recruitment would be higher in the complete absence of scatterhoarding rats. As a result, rats are effective dispersers in that seedlings apparently arise from rare caches that are not recovered, but they have a negative effect on the potential plant recruitment that would be realized in their absence.

In contrast, if potential recruitment in the absence of scatterhoarders is low relative to that from cached seeds (Line B in Fig. 17.2), then scatterhoarders would act as mutualists and seedling recruitment would be primarily from cached seeds at low crop sizes and from a combination of caches and seeds escaping scatterhoarders at high crop sizes. The mutualism would be conditional, however, under two scenarios. First, if the probability of cached-seed survival changed so that it fell below the number that could potentially be recruited at a given crop size, then scatterhoarders could act as predators under some conditions and mutualists under others. For example, in arid areas, the ability of rodents to detect caches via olfaction depends on soil moisture, with lower detection rates under very dry conditions (Vander Wall, 1993, 1995, 2000). As a result, cache survival could be significantly lower in a moist year compared to a dry year even though scatterhoarder : seed ratios were similar (Fig. 17.2B), and scatterhoarders could act as antagonists in wet years and as mutualists in dry years. The other scenario under which the mutualism would be conditional is if conditions could cause a shift in the recruitment probability for undispersed seeds, that is, a shift from Line B to A in Fig. 17.2, as discussed in the next section.

Conditionality and Variation in Plant Dependence on Scatterhoarders for Recruitment

For plants with seeds that can germinate whether or not they are dispersed by scatterhoarders, the benefit of dispersal must outweigh the often very high cost of seeds eaten by scatterhoarders. In many cases, this is possible because the probability of recruitment in the absence of the scatterhoarder is very small (Line B in Fig. 17.2), often due to abiotic or biotic conditions that limit recruitment. This can occur when: (i) undispersed seeds and the seedlings arising from them suffer high mortality from seed predators or pathogens, and (ii) scatterhoarders act as directed dispersal agents by moving seeds to specific microsites required for successful plant germination and establishment that seeds would rarely reach by other means (Howe and Smallwood, 1982).

In the first case, the mutualism may be conditional on the presence of the mortality agent acting on undispersed seeds, and the outcome of the interaction can change as the strength of the mortality factor changes. Undispersed seeds of several large-seeded neotropical tree species suffer heavy predation by insects and large vertebrates, and scatterhoarding by rodents may allow escape from these mortality agents (e.g. Janzen, 1971b; Kiltie, 1981; Wright, 1983; Hallwachs, 1986; Smythe, 1989; Forget et al., 1994; Forget, 1996; Wenny, 1999). As a result, if the populations of vertebrate or invertebrate seed predators are altered, the role of scatterhoarding rodents could also change. White-lipped and collared peccaries (Tayassu pecari and Pecari tajacu) have been implicated as important seed and seedling predators in the Neotropics (see Beck, Chapter 6, this volume), but both are subjected to heavy hunting pressure that could potentially change peccary–rodent seed dynamics (Peres, 1996). Silman et al. (2003) documented that, after peccaries had been extirpated from an area, overall seedling recruitment of Astrocaryum murumuru was significantly higher and less reliant on safe sites near objects. Although they interpreted their results as strong evidence for the impact of peccary seed predators on seedling recruitment, their data also suggest that the importance of rodents scatterhoarding seeds near objects can become less important in the absence of the strongly interacting mortality agent. Given that habitat fragmentation and hunting can greatly reduce peccary populations, loss of this mortality agent from the system could shift the relative importance of rodent scatterhoarders for plant recruitment.

Changes in invertebrate seed predation could also shift in time and space. For example, Janzen (1971b) argued that bruchid attack on *Scheelea* palm seeds would be minimal under young trees that had never fruited and in areas where water could carry seeds some distance from the parent tree. As a result, bruchid effects would vary depending upon palm age and location. Likewise, in areas where peccaries have been reduced by hunting, undispersed seeds of *Carapa procera* (Meliaceae), a subcanopy rainforest tree, suffered heavy mortality via both mammal seed predators and infestation by moths, resulting in the majority of seedling recruitment in most years arising from rodent-dispersed seeds (Forget, 1996). However, in the El Niño year of 1997, moth infestation was relatively low and seedlings established from undispersed seeds (P.-M. Forget, personal communication).

For the case of directed dispersal by scatterhoarders, the strength of the mutualism hinges on the dependence of the plant on specific microsites for establishment. Although microsite dependence is often treated as if it is a fixed requirement, a number of environmental factors can shift the relative importance of microsites for plant establishment and thereby shift the role of scatterhoarders from mutualists to antagonists. I will focus on two examples to illustrate this point, one that illustrates how natural variation can potentially shift the importance of microsites and one that illustrates how anthropogenic effects can do so.

In many semi-arid and arid regions, tree establishment may be facilitated by the presence of nurse shrubs, and movement of seeds by scatterhoarders to sites near nurse structures could be interpreted as directed dispersal to these beneficial microsites (Vander Wall, 1997). In some systems, like that of pinyon pines (*Pinus monophylla* and *Pinus edulis*), burial of seeds as well as movement to nurse shrubs is necessary for seedling establishment (Chambers *et al.*, 1999; Chambers, 2001). Scatterhoarding rodents cache a substantial proportion of seeds in these microsites and therefore act as important mutualists in this system (Vander Wall, 1997). However, movement to nurse shrubs may not be beneficial for all species or under all conditions. Callaway *et al.* (1996) showed that congeneric pines growing in the same area responded very differently to the same nurse plant, with one species benefiting from the association while seedlings of two other species suffered because they were inferior competitors for resources compared to the nurse plant. Within a plant species, the importance of nurse plants could also vary with changing conditions. Recent work in northern Patagonia has shown that establishment of the conifer *Austrocedrus chilensis* shows strong climatic variation in nurse dependence, with plant establishment dependent on nurse structures in years with moderate weather and independent of nurse structures in cool, wet years (Kitzberger *et al.*, 2000). Studies of biennial plants (de Jong and Klinkhamer, 1988) and bunchgrasses (Greenlee and Callaway, 1996) likewise showed a shift from facilitation to competition in nurse plant interactions as abiotic conditions changed. Thus, whether movement of seeds to microsites like these by scatterhoarders is a positive or negative interaction for a plant species could be conditional on natural variation in abiotic conditions across space and time.

Anthropogenic changes in natural disturbance regimes could also change plant dependence on microsites and the role of rodent scatterhoarders. In the ponderosa pine forests of the south-western USA, low intensity ground fires were historically relatively common and would have created unique challenges for seedling survival, but livestock grazing and fire suppression over the past century have changed the conditions recruiting seedlings face (Cooper, 1960; Covington and Moore, 1994). Currently, these forests have higher stem densities and lack the widely spaced clumps of trees that apparently characterized historical forests, suggesting that seedling recruitment has increased in the absence of fire. One way this may have occurred is by changing microsite dependence of ponderosa pine seedlings. White (1985) has argued that, in the presence of ground fires, recruitment of ponderosa pine may have required germination in sites that would remain relatively free of fire in successive years. He suggested that, when fire consumed downed logs, it created a heavily burned patch that would act as a relatively fire-free island in successive years. We have recently documented that rodent scatterhoarders cache seeds preferentially near downed logs in unburned forests (a pattern similar to that described by Sullivan (1978) for Douglas fir seeds) and differentially move seeds to areas of charred soil in areas that have been burned (Compton and Theimer, in preparation). Seeds buried near logs by scatterhoarders could potentially survive when fire destroys the overlying downed log and later germinate in the kind of safe site White (1985) envisaged. Preliminary experiments we have undertaken by placing seeds in simulated caches near logs and slash piles suggest some seeds could survive these conditions. Alternatively, seeds could be moved to these charred areas subsequent to fire. Scatterhoarders at one of our study sites showed a significant movement of ponderosa pine seeds from uncharred to charred areas (Compton and Theimer, in preparation). In either case, anthropogenic changes in natural fire regimes could have shifted the role of rodents from a historical role of directed dispersal agents that moved seeds to fire-resistant microsites, to the role of primarily seed predators in the absence of fire when seedlings no longer depend on scatterhoarders for dispersal to safe sites. Although this scenario is currently hypothetical, it illustrates the potential for anthropogenic alteration of natural disturbance regimes to change the role of scatterhoarders due to the conditional nature of the interaction.

Potential Effects of Forest Fragmentation on the Conditional Mutualism

Forest fragmentation could potentially alter both the seed : scatterhoarder ratio and the relative ability of seedlings to recruit in the absence of scatterhoarders through its effect on other strongly interacting agents of seed/seedling mortality. The effect on seed : scatterhoarder ratios will depend on both the response of seeding/fruiting trees to fragmentation and the response of the scatterhoarder. In the simplest scenario, if seed-producing trees of a particular species were uniformly distributed in undisturbed forest, then the number of trees of that species initially remaining in a forest fragment should be linearly related to fragment size (Line B in Fig. 17.3A). For trees with non-uniform adult distribution, effects of fragmentation would be more idiosyncratic. In species with highly aggregated distributions, for example, relative tree abundance (and therefore seed abundance) could vary unpredictably among fragments of similar size simply because of variation in whether one or more aggregations were included in the fragment. Tree density in fragments could also change through time. Large trees have been shown to experience higher mortality in small fragments (Laurance et al., 2000) and the shift in community composition towards a greater proportion of redural species is also more rapid due to larger edge effects (Laurance et al., 1998; Tabarelli et al., 1999). Finally, the relationship between tree abundance and seed production could be fundamentally altered in fragments if productivity of trees was changed in fragmented habitats (e.g. Ganzhorn, 1995).

Animals also vary in their response to fragmentation (Laurance, 1994; Harrington et al., 2001; see Dennis et al., Chapter 7, this volume), and Fig. 17.3A illustrates four hypothetical responses. Some species are highly sensitive to fragmentation and their

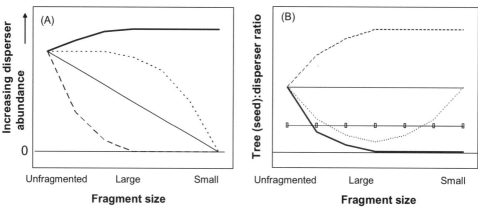

Fig. 17.3. (A) Four of several ways in which scatterhoarder abundance could be altered by fragmentation. Note that degree of fragmentation increases to the right, so fragment size decreases. The light solid line represents a species with abundance that decreases linearly with decreasing fragment size, the heavy solid line represents a species that increases with decreasing fragment size, the dotted line represents a species with populations that remain stable until a threshold fragment size is reached and the dashed line represents a species with populations that decline rapidly even in relatively large fragments. (B) The hypothetical seed : scatterhoarder ratio for the scatterhoarders described in A if the tree population (and therefore the seed abundance) declines linearly with decreasing fragment size. The box line represents the hypothetical minimum seed : disperser ratio necessary for seedling recruitment in the presence of scatterhoarders. For scatterhoarder species that either increase with fragmentation (heavy solid line) or remain stable until a threshold is reached (dotted line), seed : scatterhoarder ratios can fall below the level that allows seedling recruitment, and scatterhoarders act as antagonists.

populations decline to low levels in all but the largest fragments (dashed line in Fig. 17.3A), some show populations that decline relatively linearly with decreasing fragment size (solid line in Fig. 17.3A), and some remain stable in large and intermediate fragments and decline only in small fragments (dotted line in Fig. 17.3A). Finally, some species could increase as fragment size decreases, perhaps due to an ability to exploit edge habitats or the surrounding matrix (heavy, solid line in Fig. 17.3A).

For any one scatterhoarding rodent, the effect of fragmentation on its abundance relative to that of seeds could shift its role from seed disperser to seed predator. Figure 17.3B illustrates how the relative abundance of seeds and scatterhoarders could change with fragmentation for a scatterhoarding species exhibiting each of the four responses to fragmentation illustrated in Fig. 17.3A. In this example, for simplicity I have assumed that tree and, therefore, seed abundance decline linearly with decreasing fragment size, but, as pointed out above, this function

could vary greatly, depending upon spatial distribution of seed-producing trees and the tree response to fragmentation. I have further assumed that there is a ratio of seed : scatterhoarder above which some seeds will survive in caches to germination but below which probability of cache survival approaches zero (e.g. beyond Line X on Fig. 17.1). This model suggests that, if the seed : scatterhoarder ratio in continuous forest were above the threshold for seed survival, so that scatterhoarders were effective seed dispersers in these areas, decreasing fragment size would lead to a shift from disperser to predator in two cases. In the first, scatterhoarder species with populations that either remained stable or increased with increasing fragmentation could become so numerous that probability of seed survival falls to 0 and scatterhoarders would act as seed predators. In the second, less intuitive case, scatterhoarder populations that remained stable in large and intermediate fragments and decreased in small fragments could act as predators in the intermediate fragments but return to levels for effective dispersal in smaller fragments.

Several studies suggest that fragmentation could potentially alter the relative abundance of rodents and shift the scatterhoarder : seed ratio (Rosenblatt *et al.*, 1999, Nupp and Swihart, 2000; Donoso *et al.* 2004). For example, biomass of granivorous rodents in the Midwestern USA showed a curvilinear relationship with log fragment area, with lowest biomasses occurring in fragments of intermediate size (Nupp and Swihart, 2000). Although scatterhoarding behaviour was not studied, several species known to cache seeds showed the same curvilinear relationship (white-footed mice, *Peromyscus leucopus*, and eastern chipmunks, *Tamias striatus*), while others showed a negative response to decreasing fragment size (grey squirrels, *Sciurus carolinensis*) or no response (fox squirrels, *Sciurus niger*). The model presented here suggests that white-footed mice and chipmunks would act more as seed predators in small fragments and as seed dispersers in intermediate fragments, while fox squirrels would act increasingly as seed predators as fragment sizes decreased. In a similar example where seed fate was monitored, abundance of wood mice (*Apodemus sylvaticus*) in small fragments of Spanish juniper was nearly nine times that in contiguous forest (Santos and Telleria, 1994) and the significantly lower abundance of seedlings in fragments was hypothesized to be the result of increased seed predation by these mice. Studies demonstrating idiosyncratic effects of fragmentation on the seed–scatterhoarder interaction could be due in part to natural variation in seed : scatterhoarder ratios among and within fragments and intact forest, or to variation in the effects on other strongly interacting agents.

Because the seed–scatterhoarder mutualism is often dependent on strong interaction from other biotic agents like seed predators and pathogens, the response of these interactors to fragmentation can alter the relative dependence on scatterhoarding for tree recruitment. If the seed predators or pathogens that lower recruitment in intact forest are absent in fragments, potential tree recruitment in the absence of scatterhoarders could increase (e.g. shift from Line B to A in Fig. 17.2) and scatterhoarders would no longer act as mutualists. For example, Pizo (1997) showed that rodent densities and insect seed predators varied markedly between a large (49,000 ha) reserve and a small (250 ha) fragment.

The outcome of seed–scatterhoarder interactions in both fragmented and continuous forest will be complicated by community interactions (e.g. Dennis *et al.*, Chapter 7, this volume). From the plant perspective, the relative value of seeds of any one tree species compared to others in the fragment or the surrounding matrix could result in frequency-dependent foraging behaviour by the scatterhoarder (Hulme, 1996; Hulme and Hunt, 1999; Hoshizaki and Hulme, 2002). For example, if the seeds of two plant species were equally abundant in an area, and both were known to be cached by a scatterhoarder, the seed : scatterhoarder ratio could be approximated by combining the abundances of both plant species. However, if the seeds of one species were strongly preferred over the other, the effective seed : scatterhoarder ratio experienced by the preferred species could be considerably lower than for the less preferred species, altering the probability of recruitment and the outcome of the seed–scatterhoarder interaction. Likewise, the total abundance of all species competing for seeds may be a better estimator of the probability of cache survival in some cases, as scatterhoarders may experience much smaller realized crop sizes in the presence of competing species than in their absence. Finally, overall cache survival could be lowered in areas with species that can pilfer caches but do not create them.

Short-term versus Long-term Conditionality in Mutualism: When Does the Paradigm Matter?

Considering the seed–scatterhoarder system as a conditional mutualism because rodents act as seed predators at low seed : scatterhoarder ratios and as dispersers at high ratios could be considered as trivial as considering trees mutualists of rodents when they produce seeds and antagonists when they do not. However, it is exactly within such temporally and spatially limited contexts that many studies are forced to attempt to define the relationship. As a result, recognizing the short-term conditionality of the relationship could help explain the significant variation in the outcomes of the interaction over small spatial scales and the inconsistent response when comparing fragmented and continuous landscapes (Hallwachs, 1986; Asquith *et al.*, 1999; Wright *et al.*, 2000; Chauvet, 2001; Feer and Forget, 2002).

The evolutionary role that scatterhoarders play will be determined by the outcome of the interaction averaged over broad spatial scales and long time periods. As a result, the conditional mutualism paradigm would be most important when it suggests how anthropogenic effects may alter long-established interactions (Fig. 17.4). Three examples described in this chapter are relevant in this context:

1. The potential for human-induced climate change to alter the importance of directed dispersal to nurse plants.
2. The potential for disruption of natural fire cycles in fire-adapted communities

to change the conditions for successful recruitment and thereby the importance of scatterhoarders as mutualists.
3. The disruption of long-standing seed–seed-predator–scatterhoarder interactions through hunting and habitat fragmentation.

Ultimately, the value of a conceptual model like that of conditional mutualism depends on its ability to define new avenues of research and alter our view of system interactions. The utility of this approach applied to scatterhoarding rodents is that it focuses attention on those general conditions that may affect their role and highlights how both natural variation and anthropogenic changes could interact to change those conditions. Finally, it underscores the importance of verifying system and site-specific interactions (e.g. Hoshizaki *et al.*, 1999) and cautions against generalizations about the role of a scatterhoarder species based on studies carried out under different abiotic or biotic conditions.

Acknowledgements

I thank C.A. Gehring for critical input at several stages during the development of these ideas and Lee Ann Compton and

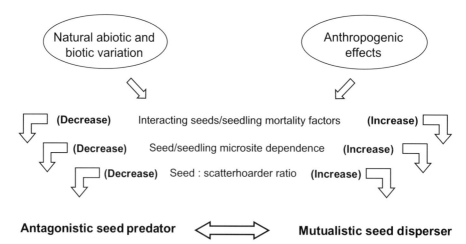

Fig. 17.4. The conditional nature of the role of scatterhoarding rodents along the continuum from antagonistic seed predators to mutualistic seed dispersers depends on whether anthropogenic and natural variation increase or decrease seed and scatterhoarder abundances, interacting mortality agents of seeds and seedlings and microsite dependence.

Pierre-Michel Forget for allowing use of unpublished data. S. Chauvet, A. Dennis, P.-M. Forget and T.G. Whitham provided valuable feedback. Funding in part from ERI 35AK-09 and NSF DEB95-03217.

References

Asquith, N.M., Terborgh, J., Arnold, A.E. and Riveros, C.M. (1999) The fruits the agouti ate: *Hymenaea courbaril* seed fate when its disperser is absent. *Journal of Tropical Ecology* 15, 229–235.

Billick, I. and Tonkel, K. (2003) The relative importance of spatial vs. temporal variability in generating a conditional mutualism. *Ecology* 84, 289–295.

Bronstein, J.L. (1994) Conditional outcomes in mutualistic interactions. *Trends in Ecology and Evolution* 9, 214–217.

Callaway, R.M., Delucia, E.H., Moore, D., Nowak, R. and Schlesinger, W.H. (1996) Competition and facilitation: contrasting effects of *Artemesia tridentata* on desert vs. montane pines. *Ecology* 77, 2130–2141.

Chambers, J.C. (2001) *Pinus monophylla* establishment in an expanding *Pinus–Juniperus* woodland: environmental conditions, facilitation and interacting factors. *Journal of Vegetation Science* 12, 27–40.

Chambers, J.C., Vander Wall, S.B. and Schupp, E.W. (1999) Seed and seedling ecology of pinon and juniper species in the pygmy woodlands of western North America. *The Botanical Review* 65, 1–36.

Chauvet, S. (2001) Effects of forest fragmentation on plant–animal interactions: consequences for plant regeneration. PhD thesis, University of Paris VI, Paris, France.

Cooper, C.F. (1960) Changes in vegetation, structure, and growth of southwestern pine forests since white settlement. *Ecological Monographs* 30, 129–164.

Covington, W.W. and Moore, M.M. (1994) Southwestern ponderosa pine forest structure: changes since Euro-American settlement. *Journal of Forestry* 92, 39–47.

Cushman, J.H. and Whitham, T.G. (1989) Conditional mutualism in a membracid–ant association: temporal, age-specific, and density-dependent effects. *Ecology* 70, 1040–1047.

de Jong, T.J. and Klinkhamer, P.G. (1988) Seedling establishment of biennials *Cirsium vulgare* and *Cynoglossum officinale* in a sand-dune area: the importance of water for differential survival and growth. *Journal of Ecology* 76, 393–402.

De Steven, D. and Putz, F.E. (1984) Impact of mammals on early recruitment of a tropical tree. *Oikos* 43, 207–216.

Donoso, D.S., Grez, A.A. and Simonetti, J.A. (2004) Effects of forest fragmentation on the granivory of differently sized seeds. *Biological Conservation* 115, 63–70.

Feer, F. and Forget, P.-M. (2002) Spatio-temporal variations in post-dispersal seed fate. *Biotropica* 34, 555–566.

Forget, P.-M. (1990) Seed dispersal of *Vouacapoua americana* (Caesalpiniaceae) by caviomorph rodents in French Guiana. *Journal of Tropical Ecology* 6, 459–468.

Forget, P.-M. (1994) Recruitment pattern of *Voucapa americana* (Caesalpinaceae), a rodent-dispersed tree species in French Guiana. *Biotropica* 26, 408–419.

Forget, P.-M. (1996) Removal of seeds of *Carapa procera* (Meliceae) by rodents and their fate in rainforest in French Guiana. *Journal of Tropical Ecology* 12, 751–761.

Forget, P.-M., Munoz, E. and Leigh, E.G. Jr (1994) Predation by rodents and bruchid beetles on seeds of *Scheelea* palms on Barro Colorado Island, Panama. *Biotropica* 26, 420–426.

Forget, P.-M., Hammond, D., Milleron, T. and Thomas, R. (2001) Seasonality of fruiting and food hoarding by rodents in Neotropical forests: consequences for seed dispersal and seedling recruitment. In: Levey, D.J., Silva, W.R. and Galetti, M. (eds) *Seed Dispersal and Frugivory: Ecology, Evolution and Conservation*. CAB International, Wallingford, UK, pp. 241–246.

Fox, J.F. (1982) Adaptation of gray squirrel behavior to autumn germination by white oak acorns. *Evolution* 36, 800–809.

Ganzhorn, J.U. (1995) Low-level forest disturbance effects on primary production, leaf chemistry and fruiting. *Ecology* 76, 2084–2096.

Greenlee, J. and Callaway, R.M. (1996) Effects of abiotic stress on the relative importance of interference and facilitation in montane bunchgrass communities in western Montana. *American Naturalist* 148, 386–396.

Hadj-Chikh, L.Z., Steele, M.A. and Smallwood, P.D. (1996) Caching decisions by grey squirrels: a test of the handling-time and perishability hypotheses. *Animal Behaviour* 52, 941–948.

Hallwachs, W. (1986) Agoutis (*Dasyprocta punctata*) the inheritors of guapinol (*Hymenaea courbaril*: Leguminosae). In: Estrada, A. and Fleming, T.H. (eds) *Frugivores and Seed Dispersal*. Dr W. Junk Publishers, Dordrecht, The Netherlands, pp. 285–304.

Harrington, G.N., Freeman, A.N.D. and Crome, F.H.J. (2001) The effects of fragmentation of an Australian tropical rain forest on populations and assemblages of small mammals. *Journal of Tropical Ecology* 17, 225–240.

Herrera, C.M. and Jordano, P. (1981) *Prunus mahaleb* and birds: the high efficiency seed dispersal system of a temperate fruiting tree. *Ecological Monographs* 51, 203–218.

Hoshizaki, K. and Hulme, P.E. (2002) Mast seeding and predator-mediated indirect interactions in a forest community: evidence from post-dispersal fate of rodent-generated caches. In: Levey, D.J., Silva, W.R. and Galetti, M. (eds) *Seed Dispersal and Frugivory: Ecology, Evolution and Conservation*. CAB International, Wallingford, UK, pp. 227–239.

Hoshizaki, K., Suzuki, W. and Nakashizuka, T. (1999) Evaluation of secondary dispersal in a large-seeded tree *Aesculus turbinata*: a test of directed dispersal. *Plant Ecology* 144, 167–176.

Howe, H.F. and Smallwood, P.D. (1982) Ecology of seed dispersal. *Annual Review of Ecology and Systematics* 13, 201–238.

Hulme, P.E. (1996) Natural regeneration of yew (*Taxus baccata* L.): microsite, seed or herbivore limitation? *Journal of Ecology* 84, 853–861.

Hulme, P.E. (2002) Seed-eaters: seed dispersal, destruction and demography. In: Levey, D.J., Silva, W.R. and Galetti, M. (eds) *Seed Dispersal and Frugivory: Ecology, Evolution and Conservation*. CAB International, Wallingford, UK, pp. 257–271.

Hulme, P.E. and Hunt, M.K. (1999) Rodent post-dispersal seed predation in deciduous woodland: predator response to absolute and relative abundance of prey. *Journal of Animal Ecology* 68, 417–428.

Janzen, D.H. (1971a) Seed predation by animals. *Annual Review of Ecology and Systematics* 2, 465–492.

Janzen, D.H. (1971b) The fate of *Scheelea rostrata* fruits beneath the parent tree: predispersal attack by bruchids. *Principes* 15, 89–101.

Janzen, D.H. (1981) *Enterolobium cyclocarpum* seed passage rate and survival in horses, Costa Rican Pleistocene seed dispersal agents. *Ecology* 62, 593–601.

Kiltie, R.A. (1981) Distribution of palm fruits on a rain forest floor: why white-lipped peccaries forage near objects. *Biotropica* 13, 141–145.

Kitzberger, T., Steinaker, D.F. and Veblen, T.T. (2000) Effects of climatic variability on facilitation of tree establishment in northern Patagonia. *Ecology* 81, 1914–1924.

Laurance, W.F. (1994) Rainforest fragmentation and the structure of small mammal communities in tropical Queensland. *Biological Conservation* 69, 23–32.

Laurance, W.F., Ferreira, L.V., Rankin-De Merona, J.M., Laurance, S.G., Hutchings, R.W. and Lovejoy, T.E. (1998) Effects of forest fragmentation on recruitment patterns in Amazonian tree communities. *Conservation Biology* 12, 460–464.

Laurance, W.F., Delamonica, P., Laurance, S.G., Vascocelos, H.L. and Lovejoy, T.E. (2000) Rainforest fragmentation kills big trees. *Nature* 404, 836.

Nupp, T.E. and Swihart, R.K. (2000) Landscape-level correlates of small-mammal assemblages in forest fragments of farmland. *Journal of Mammalogy* 81, 512–526.

Peres, C.A. (1996) Population status of white-lipped and collared peccaries in hunted and unhunted Amazonian forests. *Biological Conservation* 77, 115–123.

Peres, C.A., Schiesari, L.C. and Dias-Leme, C.L. (1997) Vertebrate predation of Brazil-nuts (*Bertholletia excelsa*, Lecythidaceae), an agouti-dispersed Amazonian seed crop: a test of the escape hypothesis. *Journal of Tropical Ecology* 13, 69–79.

Pizo, M.A. (1997) Seed dispersal and predation in two populations of *Cabralea canjerana* (Meliaceae) in the Atlantic forest of southeastern Brazil. *Journal of Tropical Ecology* 13, 559–578.

Rosenblatt, D.L., Heske, E.J., Nelson, S.L., Barber, D.M., Miller, M.A. and MacAllister, B. (1999) Forest fragments in east-central Illinois: islands or habitat patches for mammals? *American Midland Naturalist* 141, 115–123.

Santos, T. and Telleria, J.L. (1994) Influence of forest fragmentation on seed consumption and dispersal of Spanish juniper *Juniperus thurifera*. *Biological Conservation* 70, 129–134.

Schupp, E.W. (1990) Annual variation in seedfall, postdispersal predation, and recruitment of a neotropical tree. *Ecology* 71, 504–514.

Schupp, E.W. (1993) Quantity, quality and the effectiveness of seed dispersal by animals. *Vegetatio* 107/108, 15–29.

Silman, M.R., Terborgh, J.W. and Kiltie, R.A. (2003) Population regulation of a dominant rain forest tree by a major seed predator. *Ecology* 84, 431–438.

Silvertown, J.W. (1980) The evolutionary ecology of mast seeding in trees. *Biological Journal of the Linnean Society* 14, 235–250.

Smythe, N. (1970) Relationships between fruiting seasons and seed dispersal methods in a neotropical forest. *American Naturalist* 104, 25–35.

Smythe, N. (1989) Seed survival in the palm *Astrocaryum standleyanum*: evidence for dependence upon its seed dispersers. *Biotropica* 21, 50–56.

Sork, V.L. (1993) Evolutionary ecology of mast-seeding in temperate and tropical oaks (*Quercus* spp.). *Vegetatio* 107/108, 133–147.

Steele, M.A., Turner, G., Smallwood, P.D., Wolff, J.O. and Radillo, J. (2001) Cache management by small mammals: experimental evidence for the significance of acorn-embryo excision. *Journal of Mammalogy* 82, 35–42.

Sullivan, T.P. (1978) Lack of caching of direct-seeded Douglas-fir seeds by mice. *Canadian Journal of Zoology* 56, 1214–1216.

Tabarelli, M., Montovani, W. and Peres, C. (1999) Effects of habitat fragmentation on plant guild structure in the montane Atlantic forest of southeastern Brazil. *Biological Conservation* 10, 119–127.

Theimer, T.C. (2001) Seed scatterhoarding by white-tailed rats: consequences for seedling recruitment by an Australian rain forest tree. *Journal of Tropical Ecology* 17, 177–189.

Vander Wall, S.B. (1990) *Food Hoarding in Animals*. The University of Chicago Press, Chicago, Illinois, 445 pp.

Vander Wall, S.B. (1993) Cache site selection by chipmunks (*Tamias* spp.) and its influence on the effectiveness of seed dispersal in Jeffrey pine (*Pinus jeffreyi*). *Oecologia* 96, 246–252.

Vander Wall, S.B. (1995) Influence of substrate water on the ability of rodents to find buried seeds. *Journal of Mammalogy* 76, 851–856.

Vander Wall, S.B. (1997) Dispersal of singleleaf pinon pine (*Pinus monophylla*) by seed-caching rodents. *Journal of Mammalogy* 78, 181–191.

Vander Wall, S.B. (2000) The influence of environmental conditions on cache recovery and cache pilferage by yellow pine chipmunks (*Tamias amoenus*) and deer mice (*Peromyscus maniculatus*). *Behavioral Ecology* 11, 544–549.

Vander Wall, S.B. (2002) Masting in animal-dispersed pines facilitates seed dispersal. *Ecology* 83, 3508–3516.

Wenny, D.G. (1999) Two stage dispersal of *Guarea glabra* and *G. kunthiana* (Meliaceae) in Monteverde, Costa Rica. *Journal of Tropical Ecology* 5, 481–496.

White, A.S. (1985) Presettlement regeneration patterns in a southwestern ponderosa pine stand. *Ecology* 66, 589–594.

Wright, S.J. (1983) The dispersion of eggs by a bruchid beetle among *Scheelea* palm seeds and the effect of distance to the parent palm. *Ecology* 65, 1016–1021.

Wright, S.J., Zeballos, H., Dominguez, I., Gallardo, M.M., Moreno, M.C. and Ibanez, R. (2000) Poachers alter mammal abundance, seed dispersal, and seed predation in a neotropical forest. *Conservation Biology* 14, 227–239.

18 Diplochory and the Evolution of Seed Dispersal

Stephen B. Vander Wall[1] and William S. Longland[2]

[1]Department of Biology-314, [2]USDA Agricultural Research Service, 920 Valley Road, and [1,2]The Program in Ecology, Evolution, and Conservation Biology, University of Nevada, Reno, NV 89557, USA

Introduction

Dispersal of seeds and other plant propagules is often a complex, multi-step process. Each step is a discrete movement of a seed across a landscape. Studies of seed fates help to reveal the ecological processes influencing seed movements and seedling establishment, clarify the benefits of seed dispersal, and give insight into how different modes of seed dispersal might have evolved.

We define two classes of multi-step seed movements. We feel that drawing these distinctions will help to reveal some of the ecological and evolutionary processes that influence seed fates. First, there are movements that consist of a sequence of steps all involving the same basic mechanism. For example, there is wind dispersal, in which a seed initially moves from a plant to the ground, and then moves across the surface, sometimes over long distances, before arriving at its final destination (Matlack, 1989; Johnson and Fryer, 1992; Greene and Johnson, 1997). Similarly, a squirrel or jay may move an acorn from a tree to a cache site and then the same or a different animal retrieves and recaches the nut elsewhere (Vander Wall and Joyner, 1998; Soné and Kohno, 1999). In both of these examples, the

initial movement to the ground beyond the canopy of the parent plant is commonly called primary dispersal, and any significant subsequent horizontal movements have been termed secondary dispersal (Chambers and MacMahon, 1994).

Second, there are movements that consist of a sequence of steps in which two steps (usually the first and second) are by distinctly different mechanisms. This is known as diplochory or 'two-phase' dispersal (van der Pijl, 1969). Phase one is the initial mode of seed movement away from the influence of the parent plant, and phase two refers to subsequent horizontal movements by any other mechanism. We have identified five seed dispersal syndromes that demonstrate diplochory: (i) wind assisted by gravity followed by scatterhoarding by rodents or corvids; (ii) ballistic dispersal followed by myrmecochory; (iii) endozoochory followed by burial of faeces by dung beetles; (iv) endozoochory followed by scavenging of seeds from faeces or regurgitate by scatterhoarding rodents; and (v) endozoochory followed by myrmecochory. Other examples exist but are not well documented. Three-phase dispersal systems appear to be rare (Clifford and Monteith, 1989; Aronne and Wilcock, 1994). For example, seeds of quinine bush (*Petalostigma pubescens*) are initially dispersed by frugivores followed by ballistic dispersal from faeces followed by myr-mecochory (Clifford and Monteith, 1989).

Two-phase seed dispersal systems are simply special cases of secondary dispersal. But seed dispersal involving two distinct mechanisms of seed movement deserves special attention because two-phase dispersal can offer more benefits to plants (see below; Vander Wall and Longland, 2004) and the most effective mechanism can shift temporally or spatially from one mode of dispersal (i.e. phase one) to another (phase two).

We have three objectives. First, we describe the five two-phase seed dispersal systems mentioned above. For each type, we describe how seeds, fruits and, where relevant, animals are adapted to the two phases of dispersal and describe the advantages and disadvantages of each phase of seed dispersal to plants. Second, we summarize how two different modes of seed dispersal can work together to increase the effectiveness of dispersal (*sensu* Schupp, 1993). Third, we discuss how two-phase seed dispersal systems may have contributed to evolutionary change in dispersal mode. We will show that two-phase dispersal has played an important role in the evolution of seed dispersal mechanisms.

Two-phase Seed Dispersal Systems

Wind and scatterhoarding animals

The clearest examples of plants being dispersed by a combination of wind and scatterhoarding animals are pines (*Pinus*). In general, the cones of these pines open, wind scatters the winged seeds, and ani-mals gather the seeds from the ground and transport most to cache sites, where a small fraction of the seeds survive and germinate the following spring. Most of the half-dozen species of pines that are known to be dispersed in this way have relatively large cones that open at maturity (e.g. Jeffrey pine, *P. jeffreyi*, and sugar pine, *P. lambertiana*) or after fire (i.e. Coulter pine,

P. coulteri) (Vander Wall, 1992, 2002; Borchert et al., 2003). Eastern white pine (*P. strobus*) and ponderosa pine (*P. ponderosa*) have smaller cones and smaller seeds (Abbott and Quink, 1970; Vander Wall, 2002). All species have seeds with a well-developed wing that autorotates as the seed falls. This rotation acts to decrease the seed's descent velocity and increase the opportunity for horizontal air movements or up draughts to move the seed away from the parent plant (Greene and Johnson, 1989; Nathan et al., 2002). Those pine species most attractive to scatterhoarding animals, however, have large seeds with heavy wing loadings relative to pine seeds dispersed solely by wind (Johnson et al., 2003), which makes wind a relatively ineffective means of primary dispersal. Most seeds are likely to land within one tree height of the parent tree.

Chipmunks, mice, ground squirrels and, to a lesser extent, jays gather these scattered seeds very rapidly. Removal rates of simulated wind-dispersed pine seeds are rapid, generally about 10–90%/day (Vander Wall, 1994, 2002; Thayer and Vander Wall, 2005). The variation is caused by both pine species and setting. Relatively large, winged seeds like those of sugar pine and Jeffrey pine placed in shrubby habitats generally disappear rapidly, while small, winged seeds like those of lodgepole pine (*P. contorta*), which are dispersed solely by the wind, disappear much more slowly (Vander Wall, 1994). For the large seeds favoured by rodents and corvids, removal rates are sufficiently great that animals har-vest nearly all of the available seeds between seed fall and the onset of winter (~2 months).

Many wind-dispersed seeds are eaten by animals that do not serve as seed dis-persers, such as black bears, quail, mule deer and nuthatches (Kuhn and Vander Wall, unpublished results). Further, a small por-tion (usually < 5%) of the seeds gathered by chipmunks and jays are eaten immediately. But most of the seeds gathered by chip-munks and jays are transported to cache sites. Yellow pine chipmunks (*Tamias amoenus*), for example, typically transport seeds up to 60–70 m from the source tree (max = 82 m; Vander Wall, 1992, 1993, 2003) whereas Steller's jays (*Cyanocitta stelleri*) carry seeds 50–400 m (Thayer and Vander Wall, 2005). Nutcrackers (*Nucifraga* spp.) routinely carry seeds 5–10 km to cache sites (Vander Wall and Balda, 1977; Tomback, 1978). Corvids and rodents usually cache seeds in the soil or plant litter 5–30 mm deep. Rodents usually prefer to cache in open areas under or around shrubs, whereas corvids cache in a wide variety of habitats from closed-canopy forests to subalpine tundra (Tomback, 1978; Vander Wall, 1993; Thayer and Vander Wall, 2005).

Caches of chipmunks and jays do not last long. Most cached seeds are removed within days or weeks (Vander Wall and Joyner, 1998; Vander Wall, 2002). Much of this cache removal probably represents cache retrieval by the hoarder (Vander Wall, 2000), but studies using artificial caches demonstrate that many of the caches are pilfered by other animals (Vander Wall, 1998). Many of these recovered seeds are eaten but most are recached elsewhere, usually within 50 m of the original cache (Vander Wall and Joyner, 1998; Vander Wall, 2002). The number of cached seeds available gradually declines as winter approaches due to consumption by animals (Vander Wall, 2002; Thayer and Vander Wall, 2005) and because rodents transfer some of the seeds to their winter larders. By spring, from 1 to 14% of the seeds originally cached by yellow pine chipmunks still remain buried and viable (Vander Wall and Joyner, 1998; Vander Wall, 2002). As the soil warms, these seeds germinate (Munger, 1917; West, 1968; Saigo, 1969; Tomback, 1978). The number of germinants is disproportionately greater in mast years than non-mast years (Vander Wall, 2002). It is probably uncommon for seedlings to arise from seeds dispersed solely by the wind for large-seeded, winged pines like sugar and Jeffrey pines because animals eat or cache nearly all the seeds.

Two-phase seed dispersal in these pines confers many advantages. First, wind dispersal (phase-one dispersal) serves to scatter the seeds and appears to be an effective way of avoiding heavy seed predation by animals that feed on aggregated seeds (e.g. birds, bears, some rodents) or directly on cones (e.g. Douglas squirrels and red squirrels, *Tamiasciurus douglasi* and *T. hudsonicus*, respectively) and that do not act as seed dispersal agents. Second, although it is probably uncommon, wind dispersal has the potential to move some propagules great distances and thus promote colonization of new habitat patches. Third, the rapid removal of seeds from the forest floor (phase-two dispersal) reduces seed predation by non-seed-caching species. Fourth, caching by rodents and corvids places most seeds in a substrate and at a depth that is suitable for seedling establishment. Scatterhoarding animals represent an effective means of seed burial, which can be difficult for large seeds. Fifth, transport by animals, especially corvids, also has the potential to colonize new patches of suitable habitat. And, finally, cache site selection by rodents (and to a lesser extent corvids) places seeds in environments that favour seedling establishment. Compared to wind, which is random with regard to suitable establishment sites, the behaviour of chipmunks is directed (coincidentally) towards high quality establishment sites in open, shrubby areas. Although the evolution of large seeds has reduced the median and maximum distances to which these pine seeds are likely to be carried by wind, advantages gained by the second phase of seed dispersal appear to more than compensate for the loss in dispersal distance in the first phase.

Ballistic dispersal and myrmecochory

Many plants enhance dispersal distance over that possible by passive release using explosive ejection or 'ballistic dispersal' of seeds from pods, capsules or other structures. A wide variety of plant taxa, from herbaceous annuals in mesic deciduous forests to sclerophyllous shrubs and trees in arid Mediterranean chaparral, and an equally impressive diversity of mechanisms that have evolved for ejecting seeds (Stamp and Lucas, 1983) provide testimony to potential fitness advantages offered by ballistic ejection. Myrmecochory is also a common means of seed dispersal in these communities. Many plant species have combined primary seed dispersal by ballistic ejection with subsequent secondary dispersal by ants (Berg, 1966; Beattie and Lyons, 1975; Bulow-Olsen, 1984; Ohkawara and Higashi, 1994; Auld, 1996, 2001).

Plants that have adopted ant dispersal, myrmecochores, often have seeds with elaiosomes, appendages on seeds rich in lipids or carbohydrates that serve as a food reward for ants (Berg, 1966; Nakanishi, 1994). Elaiosomes typically comprise ~10–50% of total diaspore mass, but those seeds that are dispersed ballistically usually have relatively small elaiosomes (Nakanishi, 1994). Ants usually transport elaiosome-bearing seeds to their nests, consume the elaiosome, and then either leave viable seeds buried in a nest chamber or discard seeds in refuse piles outside the nest.

Even when these two modes of seed dispersal are combined, the total dispersal distance is short relative to other means of dispersal. Ballistic dispersal typically scatters seeds 0–5 m from the parent plant (Berg, 1966; Beattie and Lyons, 1975; Bulow-Olsen, 1984; Nakanishi, 1994; Auld, 1996; Lisci and Pacini, 1997). Subsequent transport of seeds by ants usually moves seeds only 1–2 m (Culver and Beattie, 1978; Ohkawara and Higashi, 1994; Passos and Ferreira, 1996), although reports of seeds being hauled >10 m exist (Andersen, 1988; Lisci and Pacini, 1997). Total seed dispersal distance can be the sum of these two distances, but overall dispersal distances can also be reduced when ants carry seeds back towards the parent plant (Beattie and Culver, 1981; Westoby and Rice, 1981). Myrmecochory can diminish ballistic dispersal in another way. Plant species with seeds bearing elaiosomes for secondary dispersal by ants have shorter ballistic dispersal distances than those that use only ballistic dispersal and lack elaiosomes (Stamp and Lucas, 1983). Long-range dispersal of myrmecochores can occur by other means (e.g. Vellend *et al.*, 2003).

Seed dispersal by ants is very common in mediterranean environments of Australia, which has approximately 1500 myrmecochorous plant species (Berg, 1975; Westoby and Rice, 1981), and South Africa, which has approximately 1300 such species (Bond and Slingsby, 1983). Because ballistic seed dispersal is also common in Australia, the combination of these dispersal agents occurs relatively frequently there. The same may be true in South Africa, but documentation appears to be lacking. Ballistic/ant dispersal has been found in a scattering of plants from other mediterranean regions, for example coastal California (Berg, 1966; Bullock, 1989) and southern Europe (Espadaler and Gomez, 1996; Lisci and Pacini, 1997). This two-phase dispersal system has also been documented in plant species from more temperate environments such as the eastern USA, central and northern Europe, and Japan (Culver and Beattie, 1978; Bulow-Olsen, 1984; Nakanishi, 1994; Ohkawara and Higashi, 1994).

Ant species differ in their effectiveness as secondary dispersers following ballistic dispersal (Bullock, 1989; Stamp and Lucas, 1990). Interactions between ants and myrmecochores are not highly species specific (Davidson and Morton, 1984; Passos and Ferreira, 1996). In general, omnivorous ants that remove elaiosomes before discarding seeds are often involved in mutualistic interactions with plants. By contrast, granivorous harvester ants (e.g. *Pogonomyrmex* and *Veromessor*) collect seeds lacking elaiosomes and typically cache them in underground nest chambers, where the seeds may be consumed directly by the ants or by microbes within the nest (Berg, 1966; Beattie and Lyons, 1975). For seeds taken by these ants, seedling establishment is rare. Even among those ant taxa that consume elaiosomes, different ant species may differ considerably in their effectiveness as seed dispersers. Some of these ant species consume seeds while others consume only the elaiosomes (Espadaler and Gomez, 1996; Passos and Ferreira, 1996).

There are numerous apparent advantages of this two-phase seed dispersal system. The explosive ejection of seeds scatters them, thereby reducing losses to a variety of seed predators, including ground beetles, insect larvae, slugs, birds and rodents that specialize on aggregated seeds (Beattie and Lyons, 1975; Culver and Beattie, 1978; Heithaus, 1981; Ohara and Higashi, 1987; Ohkawara and Higashi, 1994). It might also reduce seedling competition (Beattie, 1985). Secondary dispersal by ants can enhance overall dispersal distance of seeds over that achieved by ballistic ejection alone (Westoby and Rice, 1981; Andersen, 1988; Stamp and Lucas, 1990; Ohkawara and Higashi, 1994; Passos and Ferreira, 1996; Lisci and Pacini, 1997), but these gains are so small that it seems unlikely that the benefits of secondary dispersal by ants lies in increased dispersal distance (Culver and Beattie, 1978). Burial of seeds by ants that discard them underground reduces losses to above-ground seed predators (Beattie and Lyons, 1975; Culver and Beattie, 1978; Heithaus, 1981; Ohkawara and Higashi, 1994). In fire-prone environments, burial can provide a safe site from intense heat during fires (Beattie, 1985; Hughes and Westoby, 1992). However, ants can bury seeds too deeply within their nests to permit seedlings to emerge at the soil surface (Andersen, 1988; Hughes and Westoby, 1992).

Probably the most important benefit of phase-two dispersal is that ants often discard seeds in nests or refuse piles that can provide ideal microsites for germination and seedling establishment. Ant nests and refuse piles offer nutrient-rich substrates (high levels of nitrogen and phosphorus) for seedlings (Culver and Beattie, 1978, 1980; Beattie and Culver, 1981, 1983; Davidson and Morton, 1981, 1984; Stamp and Lucas, 1990), which can be essential to survival because many myrmecochores occur on nutrient-poor soils (Beattie and Culver, 1981; Davidson and Morton, 1981). These sites also usually have more aerated soils with greater moisture-holding capacity. However, Rice and Westoby (1986) and Higashi et al. (1989) found no support for the nutrient hypothesis at their sites in Australia and Japan; nitrogen and phosphorous concentrations around myrmecochorous seedlings did not differ from local background concentrations. Bond and Stock (1989) also

report a case of ants dispersing seeds to low nutrient microsites relative to seeds not dispersed by ants. Regardless of the underlying mechanism, increased rates of seedling establishment have been demonstrated for seeds of ballistically dispersed plant species that are deposited in ant refuse piles (Culver and Beattie, 1980; Davidson and Morton, 1984; Hanzawa et al., 1988; Passos and Ferreira, 1996).

Endozoochory and dung beetles

Herbivores and frugivores consume plant material often rich in seeds. Most fruit consumed by frugivores is adapted for this means of dispersal (e.g. Howe and Smallwood, 1982). The foliage ingested by herbivores can also contain seeds, either as contaminants or as intentionally eaten items. Janzen's (1984) 'foliage is the fruit' hypothesis suggests that some plants have evolved to use herbivores as vehicles of seed dispersal. Some studies have found support for Janzen's predictions (Quinn et al., 1994; Ortmann et al., 1998) and some have not (Collins and Uno, 1985; Dinerstein, 1989). In either case, the faeces of many animals contain seeds of a variety of plants varying in size and adaptation to this means of dispersal. Residence time in the gut of animals ranges from minutes to weeks, and seeds can be dispersed from metres to kilometres. However, seeds dispersed in faeces can be destroyed by a variety of biotic and abiotic factors including insects, fungi, rodents and desiccation (Estrada and Coates-Estrada, 1991).

Dung beetles (subfamily Scarabaeinae) are important components of ecological communities in warm temperate and tropical regions of the world (Davis et al., 2002) that play an important role in the removal of animal dung. There is a great diversity of dung beetles with a range of dung acquisition and burying behaviours from burrowers that bury dung directly below the dung deposit to ball rollers that form dung balls that they roll away to burial sites (Halffter and Edmonds, 1982). The means of dung burial is important because it influences how quickly the dung gets buried, the size of buried seeds, the depth of dung and seed burial, and the degree to which dung beetles actively sort out seeds as they provision their nests (Andresen, 1999; Andresen and Feer, Chapter 20, this volume). Serious study of this means of seed dispersal began only in the early 1990s (Estrada and Coates-Estrada, 1991), but there is great potential for dung beetles to serve as important agents of secondary seed dispersal.

Rodents play an important role in dung beetle foraging ecology. Rodents (e.g. *Lyomys*, *Peromyscus* and *Proechimys*) remove seed from dung on the ground surface and from dung buried by dung beetles (Janzen, 1982a,b; Estrada and Coates-Estrada, 1991; Shepherd and Chapman, 1998; Feer and Forget, 2002). The odour of dung helps rodents to detect seeds (Janzen, 1982a; Estrada and Coates-Estrada, 1991; Andresen, 1999, 2002b). Dung beetles do not consume seeds so they do not compete directly with rodents, but rodents can interfere with the nesting behaviour of the beetles by excavating their nests and removing seeds from the dung provisions. Nests and feeding chambers within 2–3 cm of the ground surface are vulnerable to rodents. The beetles can reduce this interference by burying dung deeply. Movement and burial of dung by dung beetles decreases the probability that seeds will be removed by rodents (Estrada and Coates-Estrada, 1991; Shepherd and Chapman, 1998; Andresen, 1999, 2002a; Vulinec, 2000, 2002). Dung beetles bury dung and its seed contaminants from 1 to 25 cm deep, with most dung buried 1–5 cm deep (Estrada and Coates-Estrada, 1991; Andresen, 1999, 2001, 2002b). This is a mixed blessing for the seeds contained in the dung. Shallow burial of dung results in relatively high probability that seeds will be removed from provisioned nests by rodents (Estrada and Coates-Estrada, 1991; Shepherd and Chapman, 1998; Andresen, 1999), but deep burial often leads to failed seedling emergence (Shepherd and Chapman, 1998; Andresen, 2001, 2002a). Seedling establishment is likely to occur in a narrow range of depths where seed removal by rodents

is low but where seedlings can still emerge (Estrada and Coates-Estrada, 1991; Shepherd and Chapman, 1998).

Another way for dung beetles to avoid interference with rodents is to sort seeds out of dung provisions while preparing nests. Janzen (1982a) has shown that, if dung does not contain seeds, rodents will learn to ignore it. Eliminating large seeds also probably increases the beetles' provisioning efficiency. Small seeds are more likely to be buried than large seeds and are buried over a broader range of depths (Estrada and Coates-Estrada, 1991; Shepherd and Chapman, 1998; Feer, 1999; Andresen, 2002a). Proportion buried was inversely correlated with seed length in both burrowers and ball rollers. Andresen (1999) used beads embedded in faeces to demonstrate an inverse correlation between bead (seed) size and proportion of beads buried. Ball rollers bury fewer seeds than burrowers (Estrada and Coates-Estrada, 1991; Feer, 1999; Vulinec, 2000). Small dung beetles bury fewer seeds because they bury less dung (Vulinec, 2000).

Primary dispersal by herbivores and frugivores often serves to lower density-dependent seed mortality near the parent plant (Janzen, 1970; Connell, 1971; Estrada and Coates-Estrada, 1991; Shepherd and Chapman, 1998). The potential benefits of secondary seed dispersal by dung beetles, in addition to the obvious advantage of seed burial (e.g. Andresen, 2003), include the deposition of seeds amongst quantities of organic fertilizer, which increases seedling growth rates (Bornemissza and Williams, 1970; MacQueen and Beirne, 1975). Rapid burial of dung prevents the dung and seeds from drying out (Halffter and Edmonds, 1982; Wicklow *et al.*, 1984), which could interfere with seed germination and seedling establishment. The scattering of seeds from a dung pile by dung beetles into numerous buried dung deposits probably serves to reduce competition among seedlings in a dung deposit (Howe, 1989; Estrada and Coates-Estrada, 1991; Andresen, 1999). Beetles compete with flies and fungi, as well as rodents, for dung resources. The effects of fly larvae and fungi on seeds and seedlings appear to be unknown, but, if larvae and fungi interfere with seedling establishment, then rapid burial of dung by dung beetles may have positive indirect effects on plants by reducing seed mortality caused by flies and fungi. Finally, since dung beetles do not eat seeds they do not impose a cost of dispersal to the plants in embryos as do some other agents of seed dispersal, such as scatterhoarding rodents.

Endozoochory and scatterhoarding rodents

As described in the previous section, animal faeces are often rich in seeds. Further, many frugivores spit or regurgitate large seeds after carrying them away from the parent plant. Rodents actively gather many of these seeds. For example, when spiny pocket mice (*Liomys salvini*) were offered horse dung containing guanacaste (*Enterolobium cyclocarpum*) seeds in field studies, they removed nearly all of the seeds within 1 night (Janzen, 1982a,b). Removal of seeds from faeces or regurgitate by mice (*Peromyscus* spp.), agoutis (*Dasyprocta punctata*) and spiny rats (*Proechimys* spp.) is also rapid (Estrada and Coates-Estrada, 1991; Forget and Milleron, 1991; Andresen, 1999; Wenny, 1999, 2001; Feer and Forget, 2002; Jansen and den Ouden, Chapter 22, this volume). It has generally been assumed, with little supporting evidence, that most of these seeds are eaten by rodents, either immediately or later in their burrows. Although this may be true for many seeds, the ultimate fate of seeds removed by rodents from faeces and regurgitate has, until recently, received little attention. Because many of the rodents that remove these seeds are also avid scatterhoarders, there is great potential for removed seeds to be secondarily dispersed by these rodent scavengers. There is no reason to suspect that these rodents are any less effective at dispersing seeds taken from faeces than they are for seeds that they find in other situations.

Tropical forest rodents are attracted to both the odour of faeces and the sound of faeces falling from tree canopies (Janzen, 1982a; Estrada and Coates-Estrada, 1991; Andresen, 1999). Feer and Forget (2002) found that the presence of dung had no effect on the rate of seed removal by spiny rats, indicating that faeces did not inhibit seed gathering. Using a string and tag marking system, Wenny (1999) found that 27–46% of *Guarea glabra* (Meliaceae) seeds removed from regurgitate were scatterhoarded by agoutis. Using similar methods, Feer and Forget (2002) found that 16% of *Chrysophyllum lucentifolium* (Sapotaceae) seeds were cached by spiny rats. Secondary dispersal distances were typically < 5–15 m, and cached seeds were slightly farther from source trees than seeds dispersed by frugivores (phase-one dispersers) (Forget and Milleron, 1991; Wenny, 1999; Feer and Forget, 2002). Cache sites were usually 1–3 cm deep in soil (Wenny, 1999) or under plant litter (Feer and Forget, 2002). Over one-third of the seeds cached (34% of *Guarea glabra* seeds and 38% of *Chrysophyllum lucentifolium* seeds) remained in place until the time of seed germination.

Secondary dispersal by rodents represents a significant cost to plants because the rodents eat a large proportion of the seeds that they handle. However, this may be a case of compensatory mortality because many of these seeds probably would have died anyway. Seeds in faeces and regurgitate are susceptible to a variety of seed predators including peccaries, non-seed-caching rodents and weevils (Forget and Milleron, 1991). Seeds in faeces are often attacked by fungal pathogens and, if seeds manage to avoid predation by animals and microbes, they are susceptible to desiccation on the soil surface (Wenny, 1999). When seeds are buried by rodents, they escape many of these fates. Buried seeds are in more favourable environments to maintain seed viability. Furthermore, dung deposits often contain many seeds and, if these seeds were to germinate in the dung, seedlings would probably experience intense competition (Howe, 1989). Rodents that harvest seeds from dung often distribute them to many different cache sites where sibling competition is less (Wenny, 1999; Feer and Forget, 2002;

see Wenny, Chapter 21, this volume) and the risk of eventual consumption by seed eaters is reduced. In addition, secondary dispersal by rodents can move seeds into new, more favourable microhabitats. Wenny (1999), for example, found that agoutis moved seeds into more open microhabitats where the probability of seedling establishment was greater. Finally, and perhaps most importantly, the burial sites provided by rodents are often ideal for seed germination and seedling establishment. Two-phase dispersal might help account for the clumped dispersion of some plants. Fragoso (1997) suggested that the clumped pattern of palm (*Maximiliana maripa*) establishment in some Amazon forests may be the result of interaction between long range dispersal of palm nuts by tapirs (*Tapirus terrestris*) and secondary dispersal in the vicinity of latrines by nut-caching rodents.

Endozoochory and myrmecochory

Two-phase seed dispersal involving frugivory followed by myrmecochory has been documented relatively frequently in diverse tropical forest systems (Roberts and Heithaus, 1986; Davidson, 1988; Kaufmann et al., 1991; Böhning-Gaese et al., 1999; Passos and Oliveira, 2002; Pizo et al., Chapter 19, this volume). For example, *Miconia* fruits (Melastomataceae) are consumed by avian and mammalian frugivores, and the seeds are subsequently harvested from their faeces by at least two species of *Pheidole* ants (Levey and Byrne, 1993). These ants cached some seeds in twig nests and deposited others in refuse piles among leaf litter outside colonies. While both cached seeds and discarded seeds sometimes established seedlings, significantly more seedlings resulted from the former. Twig nests decay quickly and the turnover time of ant colonies is short relative to the time that seeds remain viable (Levey and Byrne, 1993).

Sequential seed dispersal by frugivores and ants may be substantially more common than the existing literature implies. For example, dispersal by frugivores followed by ants has been documented in a few epiphytic plants, but fruit and seed morphological features suggesting such sequential, two-phase dispersal are found in at least four families of tropical epiphytes (Davidson, 1988; Davidson and Epstein, 1989; Kaufmann et al., 1991). Other authors have noted ants harvesting and transporting seeds from faeces or from fruits dropped by frugivores (Nakanishi, 1988; Traveset, 1994; Compton et al., 1996; Passos and Oliveira, 2002; see Pizo et al., Chapter 19, this volume). While such a dispersal syndrome could conceivably involve either frugivores or foliage-eating herbivores, documented examples appear to involve only frugivores. However, Davidson and Morton (1984) speculated that some African and New World *Acacia* species may be secondarily dispersed by ants after the seed pods are eaten and the seeds within are defecated by ruminant ungulates.

Diaspores adapted for these two modes of dispersal generally offer two distinct food rewards: fruit pulp attractive to vertebrates frugivores and a lipid-rich elaiosome (see Mayer et al., Chapter 10, this volume) or other edible appendage that passes more or less intact through a vertebrate's digestive tract and attracts ants (Davidson and Epstein, 1989; Kaufmann et al., 1991; Aronne and Wilcock, 1994; Böhning-Gaese et al., 1999). However, the presence of an elaiosome or other appendage is not necessary to make seeds attractive to some ants that act as secondary dispersers of seeds deposited by frugivores (Roberts and Heithaus, 1986; Levey and Byrne, 1993; Leal and Olivera, 1998; Passos and Oliveira, 2003; Pizo et al., Chapter 19, this volume). Apparently, for such seeds, the cotyledons and embryo serve as the reward (as in seed dispersal by scatterhoarding vertebrates) and dispersal is probably only effective because stored seeds get 'lost' in nest galleries and many ant species move their nest sites frequently, thus providing opportunities for neglected seeds to germinate (Levey and Byrne, 1993). As in other frugivore-dispersed and ant-dispersed seeds discussed earlier, frugivores can transport seeds from metres to many kilometres whereas ants seldom carry seeds more than a few metres.

Dispersal systems utilizing endozoo-chory followed by myrmecochory most often involve fairly generalized interactions. Multiple frugivore species tend to interact with a given plant species, and may often differ in effectiveness as dispersal agents (Traveset, 1994; Compton et al., 1996), as is also often the case with the ant taxa that serve as secondary dispersers (Roberts and Heithaus, 1986; Kaspari, 1993; Levey and Byrne, 1993; Traveset, 1994). A striking exception to this pattern is the tree *Commiphora guillaumini*, in which the Lesser Vasa parrot (*Coracopsis nigra*) and a single ant species (*Aphaenogaster swammerdami*) are considerably more important than alternative primary and secondary dispersers, but even in this example, other dispersers operate occasionally (Böhning-Gaese et al., 1999). In cases where this type of two-phase dispersal system could operate in temperate environments, seed harvester ants may be involved (Davidson and Morton, 1984; Gonzalez-Espinosa and Quintana-Ascencio, 1986; Aronne and Wilcock, 1994).

The main benefit of phase-one dispersal in this two-phase dispersal system is probably the avoidance of density-dependent seed mortality near the parent tree and relatively long-range dispersal. Additionally, seed passage through vertebrate digestive tracts can stimulate germination in some cases (de Figueiredo, 1993; Compton et al., 1996), although in other cases it can inhibit it (Roberts and Heithaus, 1986) or have no effect on germination (Kaufmann et al., 1991). Deposition of seeds in ant nests or refuse piles can reduce seed predation by granivorous insects and rodents (Janzen, 1971; Perry and Fleming, 1980; Pizo and Oliveira, 1998). Ants may reduce fungal attack on seeds by cleaning them of perishable fruit pulp or faecal material (Leal and Oliveira, 1998), and some ants appear to produce secretions with fungistatic characteristics that they spread on seeds (Levey and Byrne, 1993). The dissemination of seeds by ants from frugivore faeces or discarded fruits could serve to reduce seedling competition (Roberts and Heithaus, 1986; Kaufmann et al., 1991; Byrne and Levey, 1993; Kaspari, 1993), although seeds in ant nests can also be

clumped (Levey and Byrne, 1993; Böhning-Gaese et al., 1999). For some plants, removal of a seed's aril by ants can facilitate germination (Pizo and Oliveira, 1998, 2001; Passos and Oliveira, 2002). But one of the most important benefits of ant dispersal is that the nest is a highly favourable site in which seeds can germinate and establish seedlings (Passos and Oliveira, 2002, 2003).

Evolutionary Transition in the Mode of Seed Dispersal

It is not uncommon to find two or more distinctly different modes of seed dispersal within a taxonomic group. For example, Ridley (1930) states that members of the Leguminosae are adapted for dispersal by wind, water, frugivorous animals, ants, adhesion or ballistics. Even within a genus of legumes (e.g. *Acacia*) there is evidence for dispersal by mammals, birds, ants, water or wind (Davidson and Morton, 1984). Other examples of a diversity of seed dispersal mechanisms within a taxon can be found in van der Pijl (1969), Kaufmann et al. (1991), Aronne and Wilcock (1994), and Auld (2001). Distinctly different means of seed dispersal within a taxon suggest that new means of seed dispersal can evolve from an ancestral form that often bears little resemblance to the new, derived form of dispersal. How do these major changes in the mechanisms of seed dispersal come about, and what conditions foster these changes?

We propose two general models for the origin of new modes of seed dispersal. First, a change in the mode of seed dispersal could originate when a second means of primary dispersal is acquired, resulting in two simultaneous, competing means of primary seed dispersal. Van der Pijl (1969) states that this situation is common among plants. For example, many cases of seed dispersal by frugivores (endozoochory) may have evolved from fruits that initially had some sort of abiotic dispersal (e.g. wind) (Regal, 1977; Tiffney, 1984; Herrera, 1989). The new mode of dispersal may initially have been 'incidental' and was probably based on an antagonistic interaction (Thompson, 1982) such as seed predation resulting from the consumption of fruits by a vertebrate before dispersal could occur. But some seeds might have survived passage through the animal gut, resulting in a second alternative means of seed dispersal parallel, in a sense, to that of the original abiotic dispersal. Over evolutionary time, seeds might have evolved traits that increased their probability of survival in vertebrate guts, increasing the effectiveness of the new means of dispersal. This increase in dispersal effectiveness would favour the evolution of traits that attract or reward the animal dispersers. A complete transition from the original mode to the new form of dispersal might occur if the newly acquired mode of dispersal increased the overall effectiveness of seed dispersal. This model is illustrated in Fig. 18.1A.

An example of this model of evolutionary transition in dispersal mode can be seen in *Prunus*. The ancestral form of dispersal in *Prunus* is by frugivores and most modern species are dispersed this way, but numerous species in Asia and several species in south-western North America are dispersed by seed-caching rodents (Beck and Vander Wall, unpublished results). Almonds (*Prunus amygdalus*) are the most familiar example of the latter mode of dispersal (Kester *et al.*, 1991; Vander Wall, 2001). Unlike most species of *Prunus*, almonds are encased in a dry, inedible husk. The likely evolutionary transition between almonds and their fleshy-fruited relatives probably began when rodents gathered fruits that were morphologically adapted to dispersal by frugivores directly from plants and stripped away and ate or discarded the pulp and cached the nut. For a time, this ancestor of the modern almond may have been dispersed by two simultaneous mechanisms, frugivores and seed cachers, but over evolutionary time the fruits of this plant changed

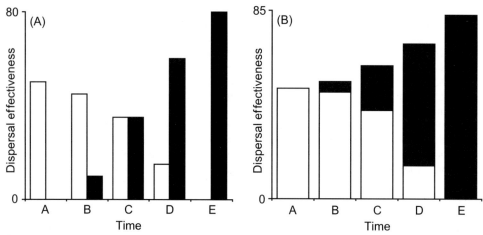

Fig. 18.1. Two models of evolutionary change in dispersal mode. Open bars represent an original form of dispersal, and closed bars represent a new mode of dispersal that replaces the original form over five stages (A–E) of evolutionary time. (A) A plant originally has only one means of dispersal, but by stage B it acquires a new form of primary dispersal, resulting in two simultaneous and 'competing' modes of primary dispersal. In each subsequent stage, the overall effectiveness of seed dispersal increases. As the plant adapts to the new means of dispersal, the effectiveness of the new dispersal agent increases while the original mode of dispersal becomes less effective. By stage E, the new form of dispersal has completely replaced the original form. (B) A plant originally has only one mode of dispersal, but by stage B a new means of secondary dispersal (phase two) is acquired that begins after primary dispersal. In this case, the two dispersal modes are sequential. In each subsequent stage, the overall effectiveness of seed dispersal increases. As the plant adapts to phase-two dispersal, the effectiveness of the new means of dispersal increases while phase-one dispersal becomes less effective. By stage E, phase-two dispersal has completely replaced phase-one dispersal.

to favour seed-caching rodents. In this case, the fleshy fruit that was the reward for dispersal by frugivores was lost when equally or more effective granivores took over as the new dispersers. The ecological and evolutionary context for this evolutionary change is described by Beck and Vander Wall (unpublished results). Davidson and Morton (1984) describe a similar scenario for the evolutionary transition from avian frugivory to myrmecochory in Australian *Acacia*.

This type of evolutionary transition in dispersal mode might also be associated with the extinction of the original disperser. For example, the extinction of large seed-dispersing herbivores and frugivores at the end of the Pleistocene may have created the conditions for a rapid shift in dispersal mode (e.g. Janzen, 1985; Hallwachs, 1986). At the time of an extinction, the means of seed dispersal might have been at stage B of Fig. 18.1A, and the extinction may have set the stage for a quick transition to stage E.

A second way that a new form of seed dispersal might originate would be to acquire a second means of dispersal (phase-two dispersal) beginning at the end of original primary dispersal. In this model, the two forms of dispersal are sequential rather than simultaneous, such as scatterhoarding of initially wind-dispersed seeds. As with the simultaneous model of evolution of new forms of seed dispersal, the initial steps in phase-two dispersal were probably antagonistic, with foragers acting as important seed predators of wind-dispersed seeds. But the storage of a few of the seeds may have set the stage for an evolutionary transition to a new mode of seed dispersal. If the second phase of dispersal was more effective in getting seeds to favourable establishment sites than primary dispersal, as was generally the case in the five types of diplochory described here, then plants might evolve fruits and seeds that attract phase-two dispersers. The first phase of seed dispersal eventually might be reduced, lost or modified to the extent that phase-one dispersal simply becomes a means of presenting seeds to phase two dispersal agents (e.g. nut fall in plants dispersed by scatterhoarding animals; Vander Wall, 2001). A graphical illustration of this process is portrayed in Fig. 18.1B. In this model, if the original (phase one) disperser continues to offer some advantage to the plant that is not provided by the new phase-two disperser (as we argue below for the examples of diplochory we have discussed here), then the evolutionary transition may be maintained at a stable form of diplochory (e.g. at stage C or D). If, however, the new phase-two disperser can offer the same or enhanced benefits that were originally facilitated by the phase-one disperser, diplochory may be a transitional condition during a shift from the old to the new dispersal agent.

There are many examples of the evolution of new modes of dispersal from sequential, two-phase dispersal systems. Walnuts (*Juglans*) and hickory nuts (*Carya*), which are dispersed by nut-caching squirrels, have ancestors that were wind dispersed. The oldest known members of the group (Juglandaceae) are *Caryanthus* and *Casholdia* from the Palaeocene of Europe that had small, winged nutlets (Crane and Manchester, 1982; Manchester, 1987). Modern members of the family include *Pterocarya* and *Platycarya*, which also have small nuts attached to papery wings. It seems likely that the ancestors of modern walnuts and hickory nuts passed through an intermediate stage of initial dispersal by the wind followed by animal scatterhoarding. Eventually, the original form of seed dispersal was lost because animal caching is a much more effective means of dispersal (*sensu* Schupp, 1993) than wind. Likewise, fossil evidence indicates that ancestors of the Fagaceae were wind dispersed (Crepet, 1989; Crepet and Nixon, 1989), but today all members of the family in north temperate and tropical regions of the world (i.e. *Fagus*, *Quercus*, *Lithocarpus*, *Castanea* and *Castanopsis*) are dispersed by scatterhoarding rodents and corvids (Johnson and Adkisson, 1985; Nilsson, 1985; Vander Wall, 2001; Roth and Vander Wall, 2005). Hazelnuts (*Corylus*) are the only animal-dispersed members of the family Betulaceae, with all other members of the group being wind dispersed. The oldest known member of the lineage is *Paleocarpinus* from the Palaeocene of

Montana that had small nutlets attached to involucral bracts and appears to have been wind dispersed (Crane, 1989). In all of these genera, the initial form of abiotic dispersal has been abandoned in favour of more effective animal-mediated dispersal.

The situation in pines is more complicated. Pines exhibit a range of seed morphologies from relatively small, winged seeds that are wind dispersed (e.g. eastern white pine and lodgepole pine, *Pinus contorta*) to large, wingless seeds dispersed by animals (e.g. piñon pines (subsection *Cembroides*) and stone pines (subsection *Cembrae*)). It seems clear that the wingless seeded pines evolved from wind-dispersed ancestors (Lanner, 1982). Some pines have intermediate morphologies with relatively large seeds with large functional wings. The seeds of this latter group, including Jeffrey pine, Coulter pine and sugar pine, are dispersed by a sequential combination of wind and scatterhoarding rodents. Seed dispersal of these latter species fits that of model two and appears to be at stage D of Fig. 18.1B. For the wingless-seeded stone pines that are dispersed primarily by nutcrackers (*Nucifraga*), which forage directly on cones, model one (Fig. 18.1A) seems to be the most likely scenario for the evolution of animal-mediated seed dispersal. For piñon pines, which are dispersed by a combination of jays that feed on the cones and rodents that gather seeds from the ground (Vander Wall and Balda, 1977; Vander Wall, 1988, 1997; Hollander and Vander Wall, 2004), a combination of the two models might best account for the acquisition of animal-dispersed seeds.

Advantages of Two-phase Seed Dispersal

Two-phase seed dispersal systems do not always represent forms of seed dispersal in evolutionary transition. In fact, most forms of diplochory appear to be relatively stable with regard to the combined modes of dispersal. For example, dispersal by a combination of frugivores and dung beetles will never shift to dispersal solely by dung

beetles (i.e. dung beetles taking seeds directly from the plant), simply because dung beetles are not attracted to seeds, but to the dung in which seeds are embedded. Similarly, it seems unlikely that plants that produce seeds which ants and rodents take from dung could evolve a means of seed dispersal that did not first include phase-one dispersal by frugivores or herbivores. Why do these and other forms of diplochory persist? Vander Wall and Longland (2004) argue that two-phase seed dispersal systems have evolved and persist in some plants because diplochory is a means by which those plants can realize more of the benefits of seed dispersal.

Potential benefits of seed dispersal fall into three categories: *escape* from density-dependent seed predators and pathogens near the parent plant (Connell, 1971; Janzen, 1971), *colonization* of suitable habitat at some relatively great distance from the parent plant (Clark *et al.*, 1998; Cain *et al.*, 2000; Nathan *et al.*, 2002), and *directed dispersal* to specific sites that offer a disproportionately high probability of seedling establishment (Howe and Smallwood, 1982; Wenny, 2001). The ideal dispersal system would maximize all three benefits, but few, if any, single means of dispersal seem capable of doing so. Phase-one dispersal is generally effective at escaping predators and intense seedling competition near the parent and sometimes colonizing new areas of suitable habitat. But seeds are deposited on the ground surface, often in dense concentrations where they are subjected to post-dispersal seed predators and, if they germinate, to intense seedling competition. Further, phase-one means of dispersal (i.e. wind, ballistic mechanisms, frugivores, herbivores) move seeds more or less at random with regard to favourable establishment sites. Phase-two dispersers (i.e. seed-caching rodents, ants, dung beetles), on the other hand, usually move seeds below ground where they have a much higher chance of surviving and germinating. Phase-two dispersal is generally not important in colonizing new habitats (i.e. crossing barriers) because dispersal distances are short, but it provides escape from post-dispersal seed predators and it is often non-random, directed towards specific, high-quality sites that favour seedling establishment. For example, herbivores provide a vehicle by which seeds can move long distances from a parent plant (providing escape and ability to colonize), but the dung deposit is seldom a safe or favourable site for seedling establishment. Dung beetles move seeds to much more favourable sites below ground (Andresen, 2001). Likewise, wind scatters pine seeds, avoiding cone-foraging squirrels that act solely as seed predators, and occasionally colonizing suitable habitat far from the parent plant, but seeds fall on the ground surface at random locations. Chipmunks and other rodents gather these seeds, move them to forest openings, and bury them, greatly improving the probability of seedling establishment (Vander Wall, 1993). Of the five types of diplochory described here, only ballistic dispersal combined with myrmecochory fails to realize all three of the benefits of dispersal, but this seed dispersal system effectively combines escape of predators with the targeting of high quality establishment sites.

The two steps of two-phase seed-dispersal systems have different effects on seeds (Culver and Beattie, 1978; Davidson, 1988; Wenny, 1999). Plants whose seeds are dispersed in two phases often benefit by combining the advantages of two different modes of dispersal while minimizing the costs in the form of seeds destroyed (Vander Wall and Longland, 2004). It appears that multi-step seed-dispersal systems are not random combinations of two or more modes of seed dispersal, but combinations that have evolved and persist because they are effective at realizing more of the benefits of seed dispersal.

Acknowledgements

We thank Michael Ashley, Maurie Beck, Gene Schupp, Dan Wenny and Jim Young for their comments on the manuscript.

References

Abbott, H.G. and Quink, T.F. (1970) Ecology of eastern white pine seed caches made by small forest mammals. *Ecology* 51, 271–278.

Andersen, A.N. (1988) Dispersal distance as a benefit of myrmecochory. *Oecologia* 75, 507–511.

Andresen, E. (1999) Seed dispersal by monkeys and the fate of dispersed seeds in a Peruvian rain forest. *Biotropica* 31, 145–158.

Andresen, E. (2001) Effects of dung presence, dung amount and secondary dispersal by dung beetles on the fate of *Micropholis guyanensis* (Sapotaceae) seeds in Central Amazonia. *Journal of Tropical Ecology* 17, 61–78.

Andresen, E. (2002a) Dung beetles in a Central Amazonian rainforest and their ecological role as secondary seed dispersers. *Ecological Entomology* 27, 257–270.

Andresen, E. (2002b) Primary seed dispersal by red howler monkeys and the effect of defecation patterns on the fate of dispersed seeds. *Biotropica* 34, 261–272.

Andresen, E. (2003) Effect of forest fragmentation on dung beetle communities and functional consequences for plant regeneration. *Ecography* 26, 87–97.

Aronne, G. and Wilcock, C.C. (1994) First evidence of myrmecochory in fleshy-fruited shrubs of the Mediterranean region. *New Phytologist* 127, 781–788.

Auld, T.D. (1996) Ecology of the Fabaceae in the Sydney region: fire, ants and the soil seedbank. *Cunninghamia* 4, 531–551.

Auld, T.D. (2001) The ecology of the Rutaceae in the Sydney region of south-eastern Australia: poorly known ecology of a neglected family. *Cunninghamia* 7, 213–239.

Beattie, A.J. (1985) *The Evolutionary Ecology of Ant–Plant Mutualisms.* Cambridge University Press, New York, 182 pp.

Beattie, A.J. and Culver, D.C. (1981) The guild of myrmecochores in the herbaceous flora of West Virginia forests. *Ecology* 62, 107–115.

Beattie, A.J. and Culver, D.C. (1983) The nest chemistry of two seed-dispersing ant species. *Oecologia* 56, 99–103.

Beattie, A.J. and Lyons, N. (1975) Seed dispersal in *Viola* (Violaceae): adaptations and strategies. *American Journal of Botany* 62, 714–722.

Berg, R.Y. (1966) Seed dispersal of *Dendromecon*: its ecologic, evolutionary, and taxonomic significance. *American Journal of Botany* 53, 61–73.

Berg, R.Y. (1975) Myrmecochorous plants in Australia and their dispersal by ants. *Australian Journal of Botany* 23, 475–508.

Böhning-Gaese, K., Gaese, B.H. and Rabemanantsoa, S.B. (1999) Importance of primary and secondary seed dispersal in the Malagasy tree *Commiphora guillaumini*. *Ecology* 80, 821–832.

Bond, W.J. and Slingsby, P.J. (1983) Seed dispersal by ants in shrublands of the Cape Province and its evolutionary implications. *South African Journal of Science* 79, 231–233.

Bond, W.J. and Stock, W.D. (1989) The costs of leaving home: ants disperse myrmecochorous seeds to low nutrient sites. *Oecologia* 81, 412–417.

Borchert, M., Johnson, J.M., Schreiner, D. and Vander Wall, S.B. (2003) Early postfire seed dispersal, seedling establishment and seedling mortality of Coulter pine (*Pinus coulteri* D. Don) in central coastal California, USA. *Plant Ecology* 168, 207–220.

Bornemissza, G.F. and Williams, C.H. (1970) An effect of dung beetle activity on plant yield. *Pedobiologia* 10, 1–7.

Bullock, S.H. (1989) Life history and seed dispersal of the short-lived chaparral shrub *Dendromecon rigida* (Papaveraceae). *American Journal of Botany* 76, 1506–1517.

Bulow-Olsen, A. (1984) Diplochory in *Viola*: a possible relation between seed dispersal and soil seed bank. *American Midland Naturalist* 112, 251–260.

Byrne, M.M. and Levey, D.J. (1993) Removal of seeds from tropical frugivore defecations by ants in a Costa Rican rain forest. *Vegetatio* 107/108, 363–374.

Cain, M.L., Milligan, B.G. and Strand, A.E. (2000) Long-distance seed dispersal in plant populations. *American Journal of Botany* 87, 1217–1227.

Chambers, J.C. and MacMahon, J.A. (1994) A day in the life of a seed: movements and fates of seeds and their implications for natural and managed systems. *Annual Review of Ecology and Systematics* 25, 263–292.

Clark, J.S., Fastie, C., Hurtt, G., Jackson, S.T., Johnson, C., King, G.A., Lewis, M., Lynch, J., Pacala, S., Prentice, C., Schupp, E.W., Webb, T.I. and Wyckoff, P. (1998) Reid's paradox of rapid plant migration. *BioScience* 48, 13–24.

Clifford, H.T. and Monteith, G.B. (1989) A three phase seed dispersal mechanism in Australian quinine bush (*Petalosigma pubescens* Domin). *Biotropica* 21, 284–286.

Collins, S.L. and Uno, G.E. (1985) Seed predation, seed dispersal, and disturbance in grasslands: a comment. *American Naturalist* 125, 866–872.

Compton, S.G., Craig, A.J.F.K. and Waters, I.W.R. (1996) Seed dispersal in an African fig tree: birds as high quantity, low quality dispersers? *Journal of Biogeography* 23, 553–563.

Connell, J.H. (1971) On the role of natural enemies in preventing competitive exclusion in some marine animals and rain forest trees. In: Den Boer, P.J. and Gradwell, G. (eds) *Dynamics of Populations*. Centre for Agricultural Publishing and Documentation, Wageningen, The Netherlands, pp. 298–312.

Crane, P.R. (1989) Early fossil history and evolution of the Betulaceae. In: Crane, P.R. and Blackmore, S. (eds) *Evolution, Systematics, and Fossil History of the Hamamelidae. Vol. 2. 'Higher' Hamamelidae*. Clarendon Press, Oxford, pp. 87–116.

Crane, P.R. and Manchester, S.R. (1982) An extinct juglandaceous fruit from the Upper Palaeocene of southern England. *Botanical Journal of the Linnean Society* 85, 89–101.

Crepet, W.L. (1989) History and implications of the early North American fossil record of Fagaceae. In: Crane, P.R. and Blackmore, S. (eds) *Evolution, Systematics, and Fossil History of the Hamamelida. Vol. 2. 'Higher' Hamamelidae*. Clarendon Press, Oxford, pp. 45–66.

Crepet, W.L. and Nixon, K.C. (1989) Earliest megafossil evidence of Fagaceae: phylogenetic and biogeographic implications. *American Journal of Botany* 76, 842–855.

Culver, D.C. and Beattie, A.J. (1978) Myrmecochory in *Viola*: dynamics of seed–ant interactions in some West Virginia species. *Journal of Ecology* 66, 53–72.

Culver, D.C. and Beattie, A.J. (1980) The fate of *Viola* seeds dispersed by ants. *American Journal of Botany* 67, 710–714.

Davidson, D.W. (1988) Ecological studies of Neotropical ant gardens. *Ecology* 69, 1138–1152.

Davidson, D.W. and Epstein, W.W. (1989) Epiphytic associations with ants. In: Luttge, U. (ed.) *Phylogeny and Ecophysiology of Epiphytes*. Springer-Verlag, Berlin, pp. 200–233.

Davidson, D.W. and Morton, S.R. (1981) Myrmecochory in some plants (F. Chenopodiaceae) of the Australian arid zone. *Oecologia* 50, 357–366.

Davidson, D.W. and Morton, S.R. (1984) Dispersal adaptations of some *Acacia* species in the Australian arid zone. *Ecology* 65, 1038–1051.

Davis, A.L., Scholtz, C.H. and Philips, T.K. (2002) Historical biogeography of scarabaeine dung beetles. *Journal of Biogeography* 29, 1217–1256.

de Figueiredo, R.A. (1993) Ingestion of *Ficus enormis* seeds by howler monkeys (*Alouatta fusca*) in Brazil: effects on seed germination. *Journal of Tropical Ecology* 9, 541–543.

Dinerstein, E. (1989) The foliage-as-fruit hypothesis and the feeding behavior of South Asian ungulates. *Biotropica* 21, 214–218.

Espadaler, X. and Gomez, C. (1996) Seed production, predation and dispersal in the Mediterranean myrmecochore *Euphorbia characias* (Euphorbiaceae). *Ecography* 19, 7–15.

Estrada, A. and Coates-Estrada, R. (1991) Howler monkey (*Alouatta palliata*), dung beetles (Scarabaeidae) and seed dispersal: ecological interactions in the tropical rain forest of Los Tuxtlas, Mexico. *Journal of Tropical Ecology* 7, 459–474.

Feer, F. (1999) Effects of dung beetles (Scarabaeidae) on seeds dispersed by howler monkeys (*Alouatta seniculus*) in the French Guianan rain forest. *Journal of Tropical Ecology* 15, 129–142.

Feer, F. and Forget, P.-M. (2002) Spatio-temporal variation in post-dispersal seed fate. *Biotropica* 34, 555–566.

Forget, P.-M. and Milleron, T. (1991) Evidence for secondary seed dispersal by rodents in Panama. *Oecologia* 87, 596–599.

Fragoso, J.M.V. (1997) Tapir-generated seed shadows: scale-dependent patchiness in the Amazon rain forest. *Journal of Ecology* 85, 519–529.

Gonzales-Espinosa, M. and Quintana-Ascencio, P.F. (1986) Seed predation and dispersal in a dominant desert plant: *Opuntia*, ants, birds, and mammals. In: Estrada, A. and Fleming, T.H. (eds) *Frugivores and Seed Dispersal*. Dr W. Junk Publishers, Dordrecht, The Netherlands, pp. 273–284.

Greene, D.F. and Johnson, E.A. (1989) A model of wind dispersal of winged or plumed seeds. *Ecology* 70, 339–347.

Greene, D.F. and Johnson, E.A. (1997) Secondary dispersal of tree seeds on snow. *Journal of Ecology* 85, 329–340.

Halffter, G. and Edmonds, W.D. (1982) *The Nesting Behavior of Dung Beetles (Scarabaeinae). An Ecological and Evolutive Approach*. Instituto de Ecologia, Mexico DF, 176 pp.

Hallwachs, W. (1986) Agoutis (*Dasyprocta punctata*): the inheritors of guapinol (*Hymenaea courbaril*: Leguminosae). In: Estrada, A. and Fleming, T.H. (eds) *Frugivores and Seed Dispersal*. Dr W. Junk Publishers, Dordrecht, The Netherlands, pp. 285–304.

Hanzawa, F.M., Beattie, A.J. and Culver, D.C. (1988) Directed dispersal: demographic analysis of an ant–seed mutualism. *American Naturalist* 131, 1–13.

Heithaus, E.R. (1981) Seed predation by rodents on three ant-dispersed plants. *Ecology* 62, 136–145.

Herrera, C.M. (1989) Seed dispersal by animals: a role in angiosperm diversification? *American Naturalist* 133, 309–322.

Higashi, S., Tsuyuzaki, S., Ohara, M. and Ito, F. (1989) Adaptive advantages of ant dispersed seeds in the myrmecochorous plant, *Trillium tschonoskii* (Lilliaceae). *Oikos* 54, 389–394.

Hollander, J.L. and Vander Wall, S.B. (2004) Effectiveness of six species of rodents as dispersers of singleleaf piñon pine (*Pinus monophylla*). *Oecologia* 138, 57–65.

Howe, H.F. (1989) Scatter- and clump-dispersal and seedling demography: hypothesis and implications. *Oecologia* 79, 417–426.

Howe, H.F. and Smallwood, P.D. (1982) Ecology of seed dispersal. *Annual Review of Ecology and Systematics* 13, 201–228.

Hughes, L. and Westoby, M. (1992) Fate of seeds adapted for dispersal by ants in Australian sclerophyll vegetation. *Ecology* 73, 1285–1299.

Janzen, D.H. (1970) Herbivores and the number of tree species in tropical forests. *American Naturalist* 104, 501–528.

Janzen, D.H. (1971) Seed predation by animals. *Annual Review of Ecology and Systematics* 2, 465–492.

Janzen, D.H. (1982a) Attraction of *Liomys* mice to horse dung and the extinction of this response. *Animal Behavior* 30, 483–489.

Janzen, D.H. (1982b) Removal of seeds from horse dung by tropical rodents: influence of habitat and amount of dung. *Ecology* 63, 1887–1900.

Janzen, D.H. (1984) Dispersal of small seeds by big herbivores: foliage is the fruit. *American Naturalist* 123, 338–353.

Janzen, D.H. (1985) *Spondias mombin* is culturally deprived in megafauna-free forest. *Journal of Tropical Ecology* 1, 131–155.

Johnson, E.A. and Fryer, G.I. (1992) Physical characterization of seed microsites – movement on the ground. *Journal of Ecology* 80, 823–836.

Johnson, J.M., Vander Wall, S.B. and Borchert, M. (2003) A comparative analysis of seed and cone characteristics and seed-dispersal strategies of three pines in the subsection *Sabinianae*. *Plant Ecology* 168, 69–84.

Johnson, W.C. and Adkisson, C.S. (1985) Dispersal of beech nuts by blue jays in fragmented landscapes. *American Midland Naturalist* 113, 319–324.

Kaspari, M. (1993) Removal of seeds from Neotropical frugivore droppings: ant responses to seed number. *Oecologia* 95, 81–88.

Kaufmann, S., McKey, D.B., Hossaert-McKey, M. and Horvitz, C.C. (1991) Adaptations for a two-phase seed dispersal system involving vertebrates and ants in a hemiepiphytic fig (*Ficus microcarpa*: Moraceae). *American Journal of Botany* 78, 971–977.

Kester, D.E., Gradziel, T.M. and Grasselly, C. (1991) Almonds (*Prunus*). *Acta Horticulturae* 290, 701–758.

Lanner, R.M. (1982) Avian seed dispersal as a factor in the ecology and evolution of limber and whitebark pines. In: Dancik, B.P. (ed.) *Proceedings 6th North American Forest Biology Workshop*, Edmonton, Alberta, pp. 15–48.

Leal, I.R. and Oliveira, P.S. (1998) Interactions between fungus-growing ants (*Attini*), fruits and seeds in cerrado vegetation in southeast Brazil. *Biotropica* 30, 170–178.

Levey, D.J. and Byrne, M.M. (1993) Complex ant–plant interactions: rain forest ants as secondary dispersers and post-dispersal seed predators. *Ecology* 74, 1802–1812.

Lisci, M. and Pacini, E. (1997) Fruit and seed structural characteristics and seed dispersal in *Mercurialis annua* L. (Euphorbiaceae). *Acta Societatis Botanicorum Poloniae* 66, 379–386.

MacQueen, A. and Beirne, B.P. (1975) Effects of cattle dung and dung beetle activity on growth of beardless wheatgrass in British Columbia. *Canadian Journal of Plant Science* 55, 961–967.

Manchester, S.R. (1987) The fossil history of the Juglandaceae. *Monographs in Systematic Botany* 21, 1–137.

Matlack, G.R. (1989) Secondary dispersal of seed across snow in *Betula lenta*, a gap-colonizing tree species. *Journal of Ecology* 77, 853–869.

Munger, T.T. (1917) Western yellow pine in Oregon. *US Department of Agriculture, Bulletin* 418, 1–48.

Nakanishi, H. (1988) Myrmecochores in warm-temperate zone of Japan. *Japanese Journal of Ecology Tokyo* 38, 169–176.

Nakanishi, H. (1994) Myrmecochorous adaptations of *Corydalis* species (Papaveraceae) in southern Japan. *Ecological Research* 9, 1–8.

Nathan, R., Katul, G.G., Horn, H.S., Thomas, S.M., Oren, R., Avissar, R., Pacala, S.W. and Levin, S.A. (2002) Mechanisms of long-distance dispersal of seeds by wind. *Nature* 418, 409–413.

Nilsson, S.G. (1985) Ecological and evolutionary interactions between reproduction of beech *Fagus sylvatica* and seed eating animals. *Oikos* 44, 157–164.

Ohara, M. and Higashi, S. (1987) Interference by ground beetles with the dispersal by ants of seeds of *Trillium* species (Liliaceae). *Journal of Ecology* 75, 1091–1098.

Ohkawara, K. and Higashi, S. (1994) Relative importance of ballistic and ant dispersal in two diplochorous *Viola* species (Violaceae). *Oecologia (Berlin)* 100, 135–140.

Ortmann, J., Schacht, W.H., Stubbendieck, J. and Brink, D.R. (1998) The 'foliage is the fruit' hypothesis: complex adaptations in buffalograss (*Buchloe dactyloides*). *American Midland Naturalist* 140, 252–263.

Passos, L. and Ferreira, S.O. (1996) Ant dispersal of *Croton priscus* (Euphorbiaceae) seeds in a tropical semideciduous forest in Southeastern Brazil. *Biotropica* 28, 697–700.

Passos, L. and Oliveira, P.S. (2002) Ants affect the distribution and performance of seedlings of *Clusia criuva*, a primarily bird-dispersed rain forest tree. *Journal of Ecology* 90, 517–528.

Passos, L. and Oliveira, P.S. (2003) Interactions between ants, fruits and seeds in a restinga forest in south-eastern Brazil. *Journal of Tropical Ecology* 19, 261–270.

Perry, A.E. and Fleming, T.H. (1980) Ant and rodent predation on small, animal-dispersed seeds in a dry tropical forest. *Brenesia* 17, 11–22.

Pizo, M.A. and Oliveira, P.S. (1998) Interaction between ants and seeds of a nonmyrmecochorous neotropical tree, *Cabralea canjerana* (Meliaceae), in the Atlantic forest of southeast Brazil. *American Journal of Botany* 85, 669–674.

Pizo, M.A. and Oliveira, P.S. (2001) Size and lipid content of nonmyrmecochorous diaspores: effects on the interaction with litter-foraging ants in the Atlantic rain forest of Brazil. *Plant Ecology* 157, 37–52.

Quinn, J.A., Mowrey, D.P., Emanuele, S.M. and Whalley, R.D.B. (1994) The 'foliage is the fruit' hypothesis: *Buchloe dactyloides* (Poaceae) and the shortgrass prairie of North America. *American Journal of Botany* 81, 1545–1554.

Regal, P.J. (1977) Ecology and evolution of flowering plant dominance. *Science* 196, 622–629.

Rice, B. and Westoby, M. (1986) Evidence against the hypothesis that ant-dispersed seeds reach nutrient-enriched microsites. *Ecology* 67, 1270–1274.

Ridley, H.N. (1930) *The Dispersal of Plants Throughout the World*. L. Reeve and Company, Limited, Ashford, UK, 745 pp.

Roberts, J.T. and Heithaus, E.R. (1986) Ants rearrange the vertebrate-generated seed shadow of a Neotropical fig tree. *Ecology* 67, 1046–1051.

Roth, J.K. and Vander Wall, S.B. (2005) Importance of primary and secondary seed dispersal of Sierra bush chinquapin (Fagaceae) by scatter-hoarding rodents. *Ecology* (in press).

Saigo, B.W. (1969) The relationship of non-recovered rodent caches to the natural regeneration of ponderosa pine. MSc thesis, Oregon State University, Corvallis, Oregon.

Schupp, E.W. (1993) Quantity, quality, and the effectiveness of seed dispersal by animals. *Vegetatio* 107/108, 15–29.

Shepherd, V.E. and Chapman, C.A. (1998) Dung beetles as secondary seed dispersers: impact on seed predation and germination. *Journal of Tropical Ecology* 14, 199–215.

Soné, K. and Kohno, A. (1999) Acorn hoarding by the field mouse, *Apodemus speciosus* Temminck (Rodentia: Muridae). *Journal of Forest Research* 4, 167–175.

Stamp, N.E. and Lucas, J.R. (1983) Ecological correlates of explosive seed dispersal. *Oecologia* 59, 272–278.

Stamp, N.E. and Lucas, J.R. (1990) Spatial patterns and dispersal distances of explosively dispersing plants in Florida sandhill vegetation. *Journal of Ecology* 78, 589–600.

Thayer, T.C. and Vander Wall, S.B. (2005) The interactions between sugar pines (*Pinus lambertiana*), Steller's jays (*Cyanocitta stelleri*) and yellow pine chipmunks (*Tamias amoenus*). *Journal of Animal Ecology* (in press).

Thompson, J.N. (1982) *Interaction and Coevolution*. John Wiley & Sons, New York, 179 pp.

Tiffney, B.H. (1984) Seed size, dispersal syndromes, and the rise of the angiosperms: evidence and hypothesis. *Annals of the Missouri Botanical Garden* 71, 551–576.

Tomback, D.F. (1978) Foraging strategies of Clark's nutcrackers. *Living Bird* 16, 123–161.

Traveset, A. (1994) Cumulative effects on the reproductive output of *Pistacia terebinthus* (Anacardiaceae). *Oikos* 71, 152–162.

van der Pijl, L. (1969) *Principles of Dispersal in Higher Plants*. Springer-Verlag, New York, 153 pp.

Vander Wall, S.B. (1988) Foraging of Clark's nutcrackers on rapidly changing pine seed resources. *Condor* 90, 621–631.

Vander Wall, S.B. (1992) The role of animals in dispersing a 'wind-dispersed' pine. *Ecology* 73, 614–621.

Vander Wall, S.B. (1993) Cache site selection by chipmunks (*Tamias* spp.) and its influence on the effectiveness of seed dispersal in Jeffrey pine (*Pinus jeffreyi*). *Oecologia* 96, 246–252.

Vander Wall, S.B. (1994) Removal of wind-dispersed pine seeds by ground-foraging vertebrates. *Oikos* 69, 125–132.

Vander Wall, S.B. (1997) Dispersal of singleleaf piñon (*Pinus monophylla*) by seed-caching rodents. *Journal of Mammalogy* 78, 181–191.

Vander Wall, S.B. (1998) Foraging success of granivorous rodents: effects of varation in seed and soil water on olfaction. *Ecology* 79, 233–241.

Vander Wall, S.B. (2000) The influence of environmental conditions on cache recovery and cache pilferage by yellow pine chipmunks (*Tamias amoenus*) and deer mice (*Peromyscus maniculatus*). *Behavioral Ecology* 11, 544–549.

Vander Wall, S.B. (2001) The evolutionary ecology of nut dispersal. *Botanical Review* 67, 74–117.

Vander Wall, S.B. (2002) Secondary dispersal of Jeffrey pine seeds by rodent scatter hoarders: the roles of pilfering, recaching, and a variable environment. In: Levey, D., Silva, W.R. and Galetti, M. (eds) *Seed Dispersal and Frugivory: Ecology, Evolution and Conservation*. CAB International, Wallingford, UK, pp. 193–208.

Vander Wall, S.B. (2003) Effects of seed size of wind-dispersed pines (*Pinus*) on secondary seed dispersal and the caching behavior of rodents. *Oikos* 100, 25–34.

Vander Wall, S.B. and Balda, R.P. (1977) Coadaptations of the Clark's nutcracker and the piñon pine for efficient seed harvest and dispersal. *Ecological Monographs* 47, 89–111.

Vander Wall, S.B. and Joyner, J.W. (1998) Recaching of Jeffrey pine (*Pinus jeffreyi*) seeds by yellow pine chipmunks (*Tamias amoenus*): potential effects on plant reproductive success. *Canadian Journal of Zoology* 76, 154–162.

Vander Wall, S.B. and Longland, W.S. (2004) Diplochory: are two seed dispersers better than one? *Trends in Ecology and Evolution* 19, 155–161.

Vellend, M., Myers, J.A., Gardescu, S. and Marks, P.L. (2003) Dispersal of *Trillium* seeds by deer: implications for long-distance migration of forest herbs. *Ecology* 84, 1067–1072.

Vulinec, K. (2000) Dung beetles (Coleoptera: Scarabaeidae), monkeys, and conservation in Amazonia. *Florida Entomologist* 83, 229–241.

Vulinec, K. (2002) Dung beetle communities and seed dispersal in primary forest and disturbed land in Amazonia. *Biotropica* 34, 297–309.

Wenny, D.G. (1999) Two-stage dispersal of *Guarea glabra* and *G. kunthiana* (Meliaceae) in Monteverde, Costa Rica. *Journal of Tropical Ecology* 15, 481–496.

Wenny, D.G. (2001) Advantages of seed dispersal: a re-evaluation of directed dispersal. *Evolutionary Ecology Research* 3, 51–74.

West, N.E. (1968) Rodent-influenced establishment of ponderosa pine and bitterbrush seedlings in central Oregon. *Ecology* 49, 1009–1011.

Westoby, M. and Rice, B. (1981) A note on combining two methods of dispersal-for-distance. *Australian Journal of Ecology* 6, 189–192.

Wicklow, D.T., Kumar, R. and Lloyd, J.E. (1984) Germination of blue grama seeds buried by dung beetles (Coleoptera: Scarabaeidae). *Environmental Entomology* 13, 878–881.

19 Ants as Seed Dispersers of Fleshy Diaspores in Brazilian Atlantic Forests

Marco A. Pizo,[1] Luciana Passos[2] and Paulo S. Oliveira[3]

[1]Departamento de Botânica, C.P. 199, Universidade Estadual Paulista, 13506-900 Rio Claro SP, Brazil; [2]Departamento de Botânica, C.P. 6109, Universidade Estadual de Campinas, 13083-970 Campinas SP, Brazil; [3]Departamento de Zoologia, C.P. 6109, Universidade Estadual de Campinas, 13083-970 Campinas SP, Brazil

Introduction

Most studies on seed dispersal of tropical species have hitherto focused mostly on fruit consumption and seed deposition patterns generated by primary seed dispersers (Estrada and Fleming, 1986; Fleming and Estrada, 1993; Levey et al., 2002), but recent studies have emphasized the importance of post-dispersal events for seed fate and demography of plant species (see Chambers and MacMahon, 1994; Andresen, 1999). For instance, seed removal by ants from frugivore defecations has been shown to affect seed distribution in tropical forests (Roberts and Heithaus, 1986; Kaspari, 1993; Pizo and Oliveira, 1999), and this in turn may have a marked influence on seedling growth and survival (Levey and Byrne, 1993; Böhning-Gaese et al., 1999; Passos and Oliveira, 2002). Ants can transport fruits that have fallen from parent plants, acting as primary seed dispersers, or fruits and seeds dropped by vertebrate frugivores, serving as secondary seed dispersers. In either case, ant-mediated seed dispersal can affect plant recruitment (Böhning-Gaese et al., 1999; Passos and Oliveira, 2002). Fallen fruits can weigh up to 400 kg/ha/year in humid forests of south-east Brazil (Morellato, 1992).

Such considerable fruitfall applies not only on a community-wide basis, but also for individual plants. For instance, over 50% of the seed crop produced by Ficus trees in a Bornean rainforest falls beneath parent plants (Laman, 1996), and in the Brazilian Atlantic forest ≈ 30% of the diaspores taken by birds from Cabralea canjerana (Meliaceae) trees drop to the ground below the parent tree, which, for some especially fecund trees, may represent nearly 8000 diaspores over the entire fruiting season (Pizo, 1997).

The abundance and diversity of ants in the tropics are remarkable, and ground-dwelling ants are perhaps the most likely organisms to encounter diaspores on the floor of tropical forests, where ant density may exceed 800 workers/m^2 (Hölldobler and Wilson, 1990). For instance, in the 1500 ha of rainforest at La Selva Biological Station (Costa Rica), densities of ant colonies exceed 4 nests/m^2 (Kaspari, 1993), and at least 437 ant species can be encountered (Longino et al., 2002). Given the heavy fruitfall and the diversity and density of tropical ground-dwelling ants (Byrne, 1994), a wide range of interactions between ants and fleshy fruits (hereafter called diaspores, sensu van der Pijl, 1982) is expected (Kaspari, 1993; Leal and Oliveira, 1998; Pizo and Oliveira, 2000, 2001; Passos and Oliveira, 2003). Indeed, in the past decade, subtle relations involving ants and fleshy diaspores have been documented in tropical environments (Beattie and Hughes, 2002). Even leaf-cutter ants (tribe Attine), traditionally considered pests, have been shown to positively influence the biology of seeds (Farji-Brener and Silva, 1996; Wirth *et al.*, 2002).

We have been investigating the interactions between ground-dwelling ants and vertebrate-dispersed diaspores in the Brazilian Atlantic forest. Results from a series of field studies have both improved our understanding of these interactions and raised relevant questions for future research (Oliveira *et al.*, 1995; Pizo and Oliveira, 1998, 1999, 2000, 2001; Passos and Oliveira, 2002, 2003). In this chapter we first provide an overview of our recent findings by describing:

1. The ant and plant species involved in these interactions.
2. The attributes of ants and diaspores that mediate the interaction.
3. The possible consequences of the interaction for plants.

We then examine how these interactions vary spatially by comparing the patterns of selected ant–diaspore interactions occurring at our two main study areas in the Brazilian Atlantic forest. Possible causes underlying such patterns are examined and discussed, and avenues of future research are suggested. Although ants may climb on to plants to gather diaspores (Lu and Mesler, 1981; Dalling and Wirth, 1998) and very often remove seeds from vertebrate defecations (Kaspari, 1993; Levey and Byrne, 1993; Pizo and Oliveira, 1999), our focus here is on interactions involving ants and fallen (either directly from plants or dropped by primary seed dispersers) diaspores on the forest floor.

Study Areas

Our studies were carried out at two sites in extensive, well-preserved Atlantic forest reserves in São Paulo State, south-east

Brazil: the lowland forest of Parque Estadual Intervales (49,000 ha) at Saibadela Research Station (hereafter PEI; 24°14'S, 48°04'W; 70 m above sea level), and the sandy plain forest (locally called 'restinga' forest) that grows on the lowest portion of Parque Estadual da Ilha do Cardoso (hereafter PEIC; 25°03'S, 47°53'W; 2–3 m above sea level), a 22,500 ha island. Separated 94 km from each other, PEI and PEIC sites differ in several aspects, including soil, annual rainfall, and fruit production (Table 19.1). Old-growth forest (*sensu* Clark, 1996) predominates at PEI; the understorey is open and the canopy is 25–30 m tall (Almeida-Scabbia, 1996). The vegetation at PEIC is characterized by 5–15 m tall trees forming an open canopy, with abundant bromeliads on the ground layer (Barros *et al.*, 1991). At both sites, there is a cool and less humid season between April and August, and a warm, wet season from September to May. No well-marked dry season occurs at PEI because no month receives less than 100 mm of rainfall.

Ant Attendance to Diaspores

A great variety of ant and diaspore species interact on the floor of the Atlantic forest. At PEI, 36 ant species (17 genera, four subfamilies) and 56 species of diaspores (40 genera, 28 families) are potentially involved in these interactions. At PEIC, potential ant–diaspore interactions include 48 ant species (19 genera, four subfamilies) and 44 plant species (40 genera, 26 families). Myrmicinae was the most frequently recorded ant subfamily at both sites (25 and 36 species at PEI and PEIC, respectively), followed by Ponerinae (seven and five species, respectively). The subfamilies Formicinae, Dorylinae and Dolichoderinae were also recorded, but much less frequently. Plant species whose diaspores were exploited by ants included trees, shrubs, herbs, lianas, epiphytes, hemiepiphytes and parasites. Diaspores ranged in weight from 0.02 g to 29.5 g. Complete lists of ant and plant species names for PEI and

Table 19.1. Summary of abiotic and biotic features of Parque Estadual Intervales (PEI) and Parque Estadual da Ilha do Cardoso (PEIC) related to the interactions between ants and fleshy, vertebrate-dispersed diaspores on the forest floor.

	PEI	PEIC
Soil	Alluvial, rich	Sandy, poor
Annual precipitation (mm)	4000	2200
Fruit production	High	Low
Litter depth	Thick	Thin
(mean no. of leaves ± SD)	(3.4 ± 1.1)	(1.6 ± 1.3)
Biomass of litter arthropods (g/trap)	High	Low
(mean ± SD)[a]	(0.14 ± 0.15)	(0.06 ± 0.05)
No. of ant species that exploit diaspores	36	48
No. of diaspore species exploited by ants	56	44
Frequency of interaction (interactions/km)[b]	7.4	2.2
Abundance of ponerines (%)[c]		
Pachycondyla striata	15	31
Odontomachus chelifer	6	23
Diaspores exploited by ponerines (%)		
Pachycondyla striata	38	80
Odontomachus chelifer	23	45
Interactions with ponerines (%)[d]	1	35

[a]Arthropods sampled with 50 and 41 pitfall traps (plastic cups 6.5 cm wide × 8.0 cm height) run for 24 h at PEI and PEIC, respectively; see text for details.
[b]5-km and 1.4-km trails sampled monthly during 2 consecutive years at PEI and PEIC, respectively.
[c]Percentage of tuna baits (n = 100) with each ant species.
[d]*Pachycondyla striata* and *Odontomachus chelifer* pooled.

PEIC are given in Pizo and Oliveira (2000) and Passos and Oliveira (2003), respectively.

Ants treated the diaspores in different ways. Ants were observed removing whole diaspores, tearing pieces off diaspores, or collecting liquids from them (the last two behaviours were more common). The behaviour of ants towards diaspores depended in part on the size of the diaspore relative to the size of the ant: (i) large ponerine ants (total length 1–1.5 cm, mainly *Pachycondyla striata* and *Odontomachus chelifer*) individually removed diaspores of up to 1 g to their nests (Fig. 19.1A, 19.1B); (ii) small and medium-sized ants (e.g. *Pheidole*, *Crematogaster*) recruited 1–110 workers and fed on the diaspore on the spot (Fig. 19.1C), but diaspores ≤ 0.05 g were occasionally transported; and (iii) some *Solenopsis* species covered the diaspore with soil before collecting liquid and solid food from it (Fig. 19.1D). Although large diaspores (> 1 g) were generally consumed on the spot (Figs 19.1E, 19.1F, 19.2A), heavy fruits containing small

seeds (e.g. *Psidium* spp.; > 10 g fruit weight) often had the seeds removed with bits of pulp attached by *Pachycondyla* and *Odontomachus*, or by large attines such as *Acromyrmex* (Pizo and Oliveira, 2000; Passos and Oliveira, 2003).

The distance of diaspore displacement varied greatly, depending mostly on the relative sizes of the diaspore and the ant. Small diaspores (< 0.10 g) can be displaced up to 100 m by leaf-cutter ants (tribe Attini) (Fig. 19.2A–D; Dalling and Wirth, 1998; Leal and Oliveira, 1998), whereas medium- to large-sized diaspores were moved, if ever, for only a few metres. Large ponerines can move diaspores for 10 m or more (Fig. 19.1A; Horvitz, 1981), while small ants (< 0.5 cm) usually do not carry diaspores beyond 2 m (Pizo and Oliveira, 1999). Inside the nests of ponerines, fleshy portions of diaspores (either pulp or aril) serve as food for larvae and adults (Horvitz, 1981; Pizo and Oliveira, 2001). Residence time of seeds inside captive colonies of the ponerine ant *Pachycondyla striata* ranged from 2 to 9 days, after

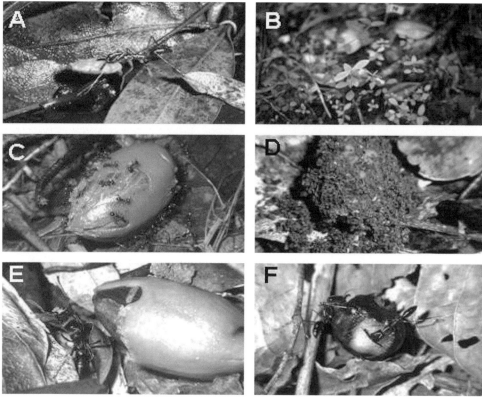

Fig. 19.1. Ponerinae and Myrmicinae ants and fleshy, vertebrate-dispersed diaspores in the Brazilian Atlantic forest. (A) *Odontomachus chelifer* carrying a fallen diaspore of *Clusia criuva* (Clusiaceae) to its nest; (B) seedlings of *Clusia criuva* (Clusiaceae), *Myrcia rostrata* and *M. bicarinata* (Myrtaceae) growing on a nest of *O. chelifer*; (C) *Pheidole* sp. removing bits of the aril of *Virola bicuhyba* (Myristicaceae); (D) a diaspore of *V. bicuhyba* covered with soil by *Solenopsis* sp.; (E) *Pachycondyla striata* removing the aril of *V. bicuhyba*; (F) *Odontomachus chelifer* exploiting a fruit of *Eugenia* sp. (Myrtaceae).

which the intact, cleaned seeds (i.e. without the fleshy portion) are deposited on refuse piles outside the nest. Attine ants also discard intact seeds in refuse piles outside their nests (Fig. 19.2D; Dalling and Wirth, 1998; Leal and Oliveira, 1998). Other myrmicines (e.g. *Pheidole*) prey upon some of the seeds they collect, but also cache intact seeds inside their nests where germination and seedling establishment occasionally occur (Levey and Byrne, 1993; M.A. Pizo, unpublished data). For diaspores exploited by ants on the spot (Figs 19.1E, 19.1F, 19.2A), the piecemeal removal of pulp or aril usually lasts less than 24 h (Pizo and Oliveira, 2001). Further details on ant behaviour towards fleshy diaspores of a variety of Atlantic forest plant species are given by Pizo and

Oliveira (2000) and Passos and Oliveira (2003).

Ant and Diaspore Attributes Mediating the Interaction

Field observations indicate that the sizes of ants and diaspores are key factors in the way that they interact. Paralleling what happens with the interaction between ants and myrmecochorous diaspores (*sensu* van der Pijl, 1982), which is influenced not only by the size of diaspores, but also by the presence of a lipid-rich appendage called elaiosome (Hughes and Westoby, 1992; Gorb and Gorb, 1995; Mark and Olesen,

Fig. 19.2. Attini ants and fleshy, vertebrate-dispersed diaspores. (A) *Atta sexdens* removing the aril of *Copaifera langsdorffii* (Caesalpiniaceae); (B) *Atta sexdens* carrying a seed of *C. langsdorffii* to its nest; (C) fungi infestation on a diaspore of *C. langsdorffii* not exploited by attines; (D) seeds and seedlings of *Prunus sellowii* (Rosaceae) growing on a nest of *Acromyrmex* sp.

1996), the chemical composition of verte-brate-dispersed diaspores, particularly the lipid content, also plays a role in the inter-action with ants. The importance of lipids, particularly fatty acids, as possible media-tors of ant–diaspore interactions has been stressed by several authors (e.g. Marshall *et al.*, 1979; Skidmore and Heithaus, 1988; Brew *et al.*, 1989), and vertebrate-dispersed diaspores have a fatty acid composition remarkably similar to the elaiosomes of myrmecochorous diaspores (Hughes *et al.*, 1994; Pizo and Oliveira, 2001; see Mayer *et al.*, Chapter 10, this volume). The verte-brate-dispersed diaspores with which ants interact in Atlantic forests have a broad range of sizes and lipid content in the fleshy portion exploited by the ants (Pizo and Oliveira, 2000). How ants respond to variation in these features and how such responses affect the biology of vertebrate-dispersed diaspores were thoroughly inves-tigated at PEI (Pizo and Oliveira, 2001).

We studied the interactions between ground-dwelling ants and six selected orni-thochorous diaspore species at PEI: *Virola bicuhyba* (Myristicaceae); *Eugenia sticto-sepala* (Myrtaceae); *Cabralea canjerana* (Meliaceae); *Citharexylum myrianthum* (Verbenaceae); *Alchornea glandulosa* and *Hyeronima alchorneoides* (Euphorbiaceae) (hereafter referred to by their generic names). These diaspores were chosen because they represent three discrete size classes that encompass the size range of most of the fleshy diaspores produced in the study site (M.A. Pizo, M. Galetti and L.P.C. Morellato, unpublished data). *Alchornea* and *Hyeronima* have small diaspores (< 0.1 g), *Cabralea* and *Citharexylum* have medium-sized diaspores (both 0.9 g), whereas the diaspores of *Virola* and *Eugenia* are much larger (3.5 and 5.8 g, respec-tively). Moreover, the selected diaspores also represent two extremes relative to the lipid content of their fleshy portions; the arils of *Virola*, *Cabralea* and *Alchornea* are lipid-rich (> 60% of dry mass), while the pulp of *Eugenia*, *Citharexylum* and *Hyeronima* is lipid-poor (< 8% of dry mass; lipid analysis follows Bligh and Dyer, 1959). With this set of diaspore species, we investigated the time to discovery, recruit-ment behaviour, attendance, diaspore clean-ing, removal and displacement distance of diaspores by ants.

Results from this series of investigations are summarized in Fig. 19.3. Ants generally

Diaspore size

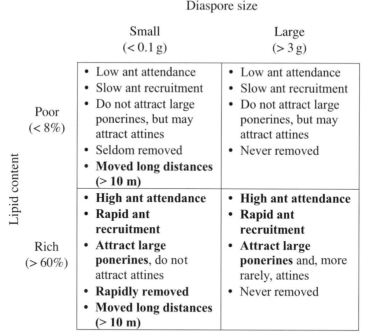

Fig. 19.3. Summary of ant responses to fleshy, vertebrate-dispersed diaspores found by ants on the floor of the Atlantic forest at the Parque Estadual Intervales (PEI). Ant responses are categorized according to the size and lipid content (on a dry mass basis) of diaspores. Responses that are most likely to benefit diaspores are in bold.

discovered diaspores on the forest floor rapidly (on average < 8 min). Time to discover a diaspore was not influenced either by diaspore size (ANOVA: $F = 1.27$, df = 2, $P = 0.28$) or lipid content ($F = 0.227$, df = 1, $P = 0.64$). Lipid content influenced positively the recruitment rate of ants to large and medium-sized diaspores (within size class comparisons with Kolmogorov–Smirnov tests: both $P < 0.001$), and ant attendance on a daily basis. Lipid-rich diaspores were attended day and night by a greater number of ants than lipid-poor ones. Diaspore size influenced negatively both removal rate (Spearman rank correlation: $r_s = -0.93$, $n = 6$, $P < 0.05$) and displacement distance ($r_s = -0.94$, $n = 6$, $P = 0.02$). Based on these results, Pizo and Oliveira (2001) predicted that small, lipid-rich diaspores would be more likely to benefit from interactions with ants at PEI.

The attachment of the fleshy portion to the seed influences how quickly ants clean seeds. For instance, other things being equal,

a seed with a loosely attached pulp or aril is cleaned more rapidly than a seed wrapped by a firmly attached pulp. Moreover, large ants with powerful mandibles (Figs 19.1E, 19.1F, 19.2A) tend to clean seeds more rapidly than small ants (ranges 2–24 h and 8–42 h, respectively; Fig. 19.1C, 19.1D). Because most of the diaspore-exploiting ants in the Atlantic forest are attracted to the pulp or aril rather than to the seed itself, the velocity of seed cleaning affects the probability of the exploiting ant being displaced by another, competitively superior ant species.

The diet of ground-dwelling ants is also important. Carnivorous ponerines foraged more frequently on lipid-rich diaspores, whereas fungus-growing attines were more frequently recorded on lipid-poor diaspores (Fig. 19.4). As mentioned previously, the fatty acid composition of vertebrate-dispersed lipid-rich diaspores is similar to that found in the elaiosomes of myrmecochorous diaspores, which have been hypothesized to mimic insect prey (Hughes *et al.*,

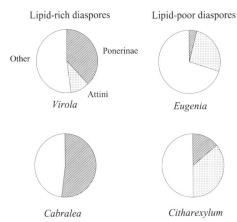

Lipid-rich diaspores   Lipid-poor diaspores

Other   Ponerinae

Attini

Virola   *Eugenia*

Cabralea   *Citharexylum*

Fig. 19.4. Attendance by Ponerinae, Attini and other ants on lipid-rich (> 60% of lipids in aril dry mass; *Virola bicuhyba* and *Cabralea canjerana*) and lipid-poor (< 8% of lipids in pulp dry mass; *Eugenia stictosepala* and *Citharexylum myrianthum*) diaspores placed on the forest floor of Parque Estadual Intervales (PEI). Fifty diaspores were used for each species.

1994; Beattie and Hughes, 2002). Thus it is not surprising that carnivorous ponerines would prefer lipid-rich diaspores. Why attine ants, in contrast, exploit lipid-poor diaspores more often than lipid-rich ones should be investigated in greater detail (see Beattie, 1991).

Effects of Ants on Seeds and Seedlings

We have shown that ground-dwelling ants frequently interact with a plethora of vertebrate-dispersed diaspores in the Brazilian Atlantic forest. A key issue for this interaction is whether ants have any significant impact on population recruitment of these plants (see Horvitz and Schemske, 1986). Potential benefits to vertebrate-dispersed diaspores secondarily dispersed by ants are similar to benefits to myrmecochorous diaspores primarily dispersed by ants (Beattie and Hughes, 2002), and include enhancement of germination success, escape from predation and directed dispersal (*sensu* Howe and Smallwood, 1982), i.e. the placement of seeds in sites where

seedlings find better conditions for establishment and development than the surrounding environment.

By removing the pulp or aril from fleshy diaspores, ants reduce fungi infestation of seeds (Fig. 19.2C) and can increase germination success by 19 to 63% (Oliveira *et al.*, 1995; Leal and Oliveira, 1998; Pizo and Oliveira, 1998, 2001; Passos and Oliveira, 2002, 2003). When ants remove the seeds from beneath parent plants, they help the seeds escape from density- or distance-oriented seed predators (Janzen, 1970; Pizo and Oliveira, 1998). Using exclosure experiments to compare the removal rates of caged (no access to vertebrates) and uncaged diaspores (free access to vertebrates and ants), we have shown that: (i) ants removed from 0 to 91% of the diaspores in 24 h (small, lipid-rich diaspores experienced the highest removal rates), and (ii) removal attributed to vertebrate seed predators (estimated by the difference in diaspore removal between caged and uncaged treatments) increased with the size of the diaspore (up to the threshold represented by the size of *Eugenia* (5.8 g), which is too big to be exploited by most of the rodents at PEI; Vieira *et al.*, 2003) because large diaspores are not transported by ants, thus becoming available to rodents (Fig. 19.5; Pizo and Oliveira, 2001). Therefore, if escape from seed predators is a benefit accrued by vertebrate-dispersed diaspores as a result of their interaction with ants, this benefit is greatest for small, lipid-rich diaspores, which are rapidly removed by ants (especially primarily carnivorous, ponerine ants).

Directed dispersal of *Clusia* and *Guapira* seeds to ponerine ant nests

The lipid-rich, arillate diaspores of *Clusia criuva* (Clusiaceae; diaspore mass = 0.10 ± 0.05 g, $n = 150$; 83.4% lipids on a dry mass basis; Passos and Oliveira, 2002), and the protein-rich fruits of *Guapira opposita* (Nyctaginaceae; 0.25 ± 0.04 g, $n = 30$; 28.4% protein; Passos, 2001) are dispersed by many bird species in the Atlantic forest

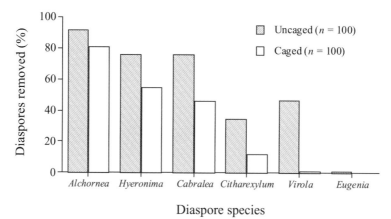

Fig. 19.5. Percentage of caged (access to ants only) and uncaged (access to ants and vertebrates) diaspores removed in a 24-h period from the forest floor of Parque Estadual Intervales (PEI). Diaspores are arranged from left to right according to their size: small (*Alchornea glandulosa* and *Hyeronima alchorneoides*; < 0.1 g), medium (*Cabralea canjerana* and *Citharexylum myrianthum*; 0.9 g), and large (*Virola bicuhyba* and *Eugenia stictosepala*; > 3.5 g).

(M.A. Pizo, unpublished data). However, once on the ground, either dropped directly from the parent plant or dispersed by birds, *Clusia* and *Guapira* diaspores are exploited by a diverse assemblage of ground-dwelling ants, but most especially by two primarily carnivorous ponerines – *Pachycondyla striata* and *Odontomachus chelifer* (Fig. 19.1A; Passos, 2001; Passos and Oliveira, 2002). We investigated the effects of these two ponerines on seedlings of *Clusia* and *Guapira* at PEIC in 1998 and 2000, respectively. The number of *Clusia* and *Guapira* seedlings growing on nests of these ant species was compared to control areas (without nests) by establishing paired experimental plots (0.5 m × 0.5 m). *Clusia* seedlings within plots were marked and monitored every 2 months for a year. Additionally, soil samples from ponerine nests and from control plots were analysed for nutrients and physical properties. Further methodological details are given in Passos and Oliveira (2002) and Passos (2001).

Effects of ants on seeds and seedlings of *Clusia* and *Guapira* are summarized in Table 19.2. Overall, ants behaved similarly towards seeds and seedlings of both species, but the plants were not equally affected by the two ponerines. Seedlings of *C. criuva* were more abundant in the vicinity of nests

of *P. striata* (Wilcoxon paired-sample test, $Z = -3.869$, $P < 0.0001$, $n = 21$) and *O. chelifer* ($Z = -2.964$, $P = 0.003$, $n = 20$) than in areas without nests (Fig. 19.1B). Early survival (1 year) of *Clusia* seedlings was greater in nests of *P. striata* than in control areas ($\chi^2 = 14.18$, $P = 0.0002$), but was not affected by *O. chelifer* ($\chi^2 = 1.31$, $P = 0.2526$). *Guapira* seedlings, on the other hand, are more frequent close to *O. chelifer* nests than in sites without such nests ($Z = -4.947$, $P < 0.0001$, $n = 40$; Passos, 2001). This interspecific variation may have been influenced by differences in soil properties around ant nests, in nutrient requirements of the seedlings of both plant species, or by unknown temporal factors. The results with *Clusia* and *Guapira* indicate that, as reported for myrmecochorous diaspores in xeric environments (Davidson and Morton, 1981; Culver and Beattie, 1983; Beattie, 1985), the directed dispersal provided by ponerine ants at the sandy plain forest of PEIC involves the deposition of vertebrate-dispersed diaspores in nutrient-enriched soil close to their nests. Soil enrichment near the nests probably results from the deposition of organic material on adjacent refuse piles (Beattie, 1985). Similar effects of seed dispersal by *Odontomachus* have been shown for seedlings of other vertebrate-dispersed

Table 19.2. Summary of the effects of two ponerine ant species, *Pachycondyla striata* and *Odontomachus chelifer*, on seeds and seedlings of *Clusia criuva* and *Guapira opposita* in the sandy forest of Cardoso Island (PEIC), south-east Brazil. Data with *C. criuva* and *G. opposita* were obtained in 1998 and 2000, respectively. A dash indicates that the ant activity or ant effects were not tested with that particular plant species. Further details in Passos and Oliveira (2002) and Passos (2001).

Ant activity and ant effects on plants	C. criuva		G. opposita	
	P. striata	O. chelifer	P. striata	O. chelifer
Remove fallen diaspores	Yes	Yes	Yes	Yes
Remove diaspores from bird faeces	Yes	Yes	–	?
Discard intact seeds outside nest	Yes	Yes	Yes	Yes
Removal of fleshy portion increases germination[a]	Yes	Yes	–	–
Increased seedling recruitment near nest[b]	Yes	Yes	–	Yes
Increased seedling survival (1 year) near nest[b]	Yes	No	–	–
Increased soil nutrients near nest[b]	Yes	Yes	–	Yes
Higher soil penetrability near nest[b]	–	–	–	Yes
Potential herbivore deterrence near nest[b,c]	–	–	–	Yes

[a]Although the effects of cleaning activity were not assessed for *Guapira*, pulp removal by the authors increased germination in this species.
[b]Compared to random control plots without nests.
[c]Evaluated by recording attack rates by ants on dipteran larvae placed on seedlings growing near the nests and in control plots.

plants such as *Anthurium* sp. (Araceae), *Myrcia rostrata* and *Psidium catleyanum* (Myrtaceae) in the same area (Fig. 19.1B; Passos and Oliveira, 2003).

Spatial Variation in Ant–Diaspore Interactions

Patterns

Although the interactions between ants and vertebrate-dispersed diaspores occur at PEI and PEIC and involve a variety of ant and plant species at both places, there are important differences between the sites (Table 19.1). One difference is the frequency of ant–diaspore interactions, as revealed by the monthly surveys carried out during 2 consecutive years along 5-km and 1.4-km trails at PEI and PEIC, respectively. Any instance of one or more ants in contact with a given diaspore either removing it or collecting material from it was recorded as an interaction. A total of 886 interactions (7.4 interactions/km) were recorded at PEI (Pizo and Oliveira, 2000), whereas only 75 interactions (2.2 interactions/km) were

recorded at PEIC (Passos and Oliveira, 2003).

The between-site difference in the participation of the two most common ponerine species at both sites (*Pachycondyla striata* and *Odontomachus chelifer*) was marked. While these species were responsible for only 1% of the interactions recorded at PEI, they accounted for 35% of the interactions at PEIC. This difference may reflect the abundances of these ants at each site. A survey with 100 tuna fish baits used to evaluate the abundance of ant species at PEI detected *P. striata* and *O. chelifer* in 15% and 6% of the baits, respectively. The same procedure employed at PEIC revealed that *P. striata* is twice as abundant (31% of the baits), while the abundance of *O. chelifer* is almost four times greater (23% of the baits). However, abundance alone is not sufficient to explain the difference found in the number of interactions with fallen diaspores because *O. chelifer* and *P. striata* were apparently more selective towards diaspores at PEI than at PEIC. In the former site, these two ant species exploited 13 and 21 diaspore species, representing respectively 23% and 38% of all plant species used by ants (Pizo and Oliveira, 2000). In contrast, at PEIC

O. chelifer and *P. striata* utilized 20 and 35 diaspore species, which comprise respectively 45% and 80% of all plant species used by ants at this site (Passos and Oliveira, 2003). When we compare the number of interactions involving the two ponerines and selected diaspores shared by both sites, the between-site difference holds not only for lipid-rich, but also for lipid-poor fleshy diaspores (Table 19.3). Therefore, *O. chelifer* and *P. striata* use a greater variety of diaspores at PEIC than at PEI.

Processes

The overall higher frequency of ant–diaspore interactions at PEI compared with PEIC may be related to inter-site differences in the abundance and composition of the ant fauna and/or in the availability of fruits. The abundance of ants (all species pooled), as revealed by the number of ants captured in pitfall traps (see below), did not differ between PEI and PEIC (t-test on $\log(n + 1)$-transformed data: $t = -0.91$, df $= 89$, $P = 0.36$). It is possible, however, that the high abundance of ponerines at PEIC (see above) may lead to the rapid disappearance of fallen diaspores, thus precluding diaspore use by other ants. Although comparative data on fruit availability at the two sites are lacking, the sandy plain forest at PEIC grows on poor-quality soils probably sustaining a lower annual fruit production compared to the forest at PEI that grows on rich alluvial soils. Therefore, it is possible that the higher frequency of ant–diaspore interactions at PEI may result from the greater fruit abundance at this area than at PEIC. Moreover, the rapid removal of fallen fleshy fruits by common ponerines at PEIC can make this resource unavailable to other small, slow-moving ants.

We developed two non-exclusive hypotheses to explain the broader use of diaspores by *P. striata* and *O. chelifer* at PEIC compared to PEI: (i) ant populations from each site may differ in food preferences ('food preference hypothesis'), and (ii) availability of arthropod prey may differ between the two sites ('arthropod availability hypothesis'). If the food preference hypothesis is to explain the observed difference in diaspore use, we predicted that, given the opportunity to choose between fruit and arthropod food items, the ants would choose fruits more frequently at PEIC than at PEI. On the other hand, if the arthropod availability hypothesis is valid, we predicted that leaf litter arthropod biomass would be higher at PEI than at PEIC, thus leading ground-dwelling *P. striata* and *O. chelifer* to rely more frequently upon fleshy diaspores at PEIC to complement their predominantly carnivorous diets.

To test the food preference hypothesis, we performed choice experiments using guava fruits (*Psidium guajava*, Myrtaceae) and cockroaches. Member of a dominant plant family in the Atlantic forest (Mori

Table 19.3. Number of interactions between ants and selected diaspore species recorded at Parque Estadual Intervales (PEI) and Parque Estadual da Ilha do Cardoso (PEIC). Lipid content of the diaspores' fleshy portion (either pulp or aril), total number of interactions with ants and number of interactions involving *Pachycondyla striata* (Ps) and *Odontomachus chelifer* (Oc) are presented.

Diaspore species (family)	Lipid content[a] (dry mass (%))	Interactions at PEI			Interactions at PEIC		
		Total	Ps	Oc	Total	Ps	Oc
Aechmea nudicaulis (Bromeliaceae)	–	25	0	0	43	8	4
Alchornea triplinervia (Euphorbiaceae)	68	39	2	0	15	5	2
Clusia criuva (Clusiaceae)	83	8	0	0	79	20	7
Euterpe edulis (Arecaceae)	20	191	0	0	9	4	0
Maytenus robusta (Celastraceae)	1	45	0	0	9	3	0

[a]Unpublished data for *A. triplinervia*, *E. edulis* and *M. robusta*; data for *C. criuva* from Passos and Oliveira (2002).

et al., 1983; Oliveira-Filho and Fontes, 2000), the guava fruit is a typical lipid-poor fruit like many others with which ants interact (see Pizo and Oliveira, 2000). We used tuna baits to locate five nests of *P. striata* and four of *O. chelifer* at each site. Around the nest entrances, we offered the ants a two-choice food source composed of equivalent pieces (< 1 cm in length) of guava and cockroach placed side by side on a white filter paper (4 cm × 4 cm). The two types of food items were < 1 cm apart. A trial was initiated after the location of the food items by a foraging ant, and terminated after the removal of one item by the ant. A choice was recorded only if the ant antennated both food items prior to selecting one of them. To ensure independence of trials, every tested ant was collected. The filter paper had no apparent effect on ant behaviour (Levey and Byrne, 1993; Pizo and Oliveira, 2000; Passos and Oliveira, 2002).

A frequency test performed using the procedure PROC CATMOD of the SAS Statistical package (SAS Institute, 1987) revealed that ants differed in their food choices ($\chi^2 = 21.54$, df = 1, $P < 0.0001$), although choices were consistent across areas ($\chi^2 = 0.09$, df = 1, $P = 0.76$). *Pachycondyla striata* consistently selected cockroaches in both areas (guava was never chosen first), whereas the results for *O. chelifer* were less clear and revealed intercolony variation (seven colonies chose cockroaches more frequently – 69–87% of choices – while one colony at PEI preferred guava fruits – 56% of choices). Despite this variation, a *t*-test applied to *O. chelifer* pooled data (PROC TTEST, SAS Institute, 1987) showed that cockroaches were selected more frequently than fruit ($t = 4.71$, df = 8.34, $P = 0.0014$). Therefore, we rejected the food preference hypothesis as an explanation for the differences in diaspore use observed between PEI and PEIC.

To test the arthropod availability hypothesis, we compared the biomass of litter arthropods in January–February 2002 at PEI and PEIC by setting 50 and 41 pitfall traps in each area, respectively. Traps consisted of plastic cups (6.5 cm wide × 8.0 cm height) half-filled with alcohol 70%, placed ≥ 10 m apart from each other along one trail in each site. Traps were set over a period of 24 h, after which the arthropods were collected and dried to constant weight in an oven set at 55°C. Total arthropod biomass collected at each trap was then weighed to the nearest 0.01 g.

Results showed that biomass of ground-dwelling arthropods is twice as great at PEI than at PEIC (0.14 ± 0.15 and 0.06 ± 0.05 g per trap, respectively; *t*-test on log-transformed data: $t = 2.15$, df = 88, $P = 0.03$). Between-site difference in leaf litter thickness possibly accounts for the observed difference. By counting the number of leaves intersected by a wood stick inserted into the leaf litter at 63 random points at each site, we found that the litter is twice as thick at PEI than at PEIC (3.4 ± 1.1 and 1.6 ± 1.3 leaves, respectively; *t*-test on $\log(n + 1)$-transformed data: $t = 8.27$, df = 124, $P < 0.001$). Therefore, the arthropod availability hypothesis cannot be rejected. Ponerine ants appear to increase the consumption of fleshy diaspores when faced with a low supply of preferred arthropod prey, producing the between-site difference in diaspore use here reported for PEI and PEIC.

Consequences

Biotic interactions are typically variable in space, and the interactions between ants and vertebrate-dispersed diaspores are not exceptions (Thompson, 1994; Garrido *et al.*, 2002). For instance, although ants commonly consume fallen diaspores and increase seedling establishment of ornithochorous *Commiphora* trees (Burseraceae) in Madagascar, this interaction is absent in South Africa (Böhning-Gaese *et al.*, 1999; Bleher and Böhning-Gaese, 2001). Several factors may account for spatial variation in ant–diaspore interactions. In the *Commiphora* study, the difference in ant communities between Madagascar and South Africa is an underlying reason for the between-site differences observed (Voigt *et al.*, 2002). In the Brazilian Atlantic forest, we suggested that low fruit availability rather than low ant abundance was responsible for the overall lower frequency of interactions recorded at PEIC compared with PEI, and decreased availability of litter arthropods at PEIC might have led predominantly carnivorous ants to use a greater variety of fleshy diaspores at this site than at PEI. As a consequence, diaspore features hypothesized as important for determining the outcome of the interactions with ants at PEI (i.e. small size and high lipid content; Fig. 19.3; see Pizo and Oliveira, 2001) may not necessarily hold for PEIC. For instance, while *Clusia* diaspores fit into these characteristics (mass = 0.10 g, 83.4% lipids; Passos and Oliveira, 2002), *Guapira* fruits are rich in protein, but poor in lipids (28.4% and 0.5%, respectively; Passos, 2001). The same holds true for *Myrcia rostrata* and *Psidium cattleyanum* (10.3% and 1.7% lipids, respectively; M.A. Pizo, unpublished data), whose seedlings are also clustered around *O. chelifer* nests at PEIC (Fig. 19.1B; Passos and Oliveira, 2003).

Therefore the relevance of ponerine ants as seed dispersers of vertebrate-dispersed plants at PEIC can be attributed to a combination of factors: (i) high abundance of *P. striata* and *O. chelifer* ants; (ii) low availability of litter arthropod prey; (iii) great longevity of ponerine nests (one nest of *O. chelifer* tagged in 1995 was still active in 2002; M.A. Pizo, personal observation); and (iv) nutrient-poor sandy soil. While the great longevity of ponerine nests contributes to the establishment of nutrient-enriched sites around ant nests, the nutrient-poor soil increases their relative importance as plant recruitment foci (see Hughes, 1990; Passos and Oliveira, 2002).

Concluding Remarks and Perspectives

The impact of ants on the reproductive output of vertebrate-dispersed tropical plants may be very important. A large percentage of the diaspores produced by such plants is dropped by vertebrate seed dispersers (Böhning-Gaese *et al.*, 1999; Passos and

Oliveira, 2002). We demonstrate that in the Brazilian Atlantic forest ground-dwelling ants interact frequently with fallen vertebrate-dispersed diaspores and that this may render recruitment benefits for some plant species. We are just beginning to understand the complexity of such interactions, and more natural history studies are needed before one can assess the intricacy of plant dispersal systems in tropical forests. It is crucial to weigh the importance of secondary seed dispersal by ants against primary seed dispersal by vertebrates in what concerns seedling establishment and, ultimately, population recruitment (see Böhning-Gaese *et al.*, 1999). In this context, it is important to consider that the spatial scale of seed dispersal provided by ants is usually smaller than that provided by vertebrate seed dispersers, which has important consequences for population recruitment and spatial distribution (Horvitz and Le Corff, 1993; Horvitz *et al.*, 2002). Short-distance dispersal provided by ants may be sufficient if appropriate recruitment sites are located close to the parent plant, which seems not to be rare in tropical forests given the frequency of adult clumped distribution among tree plant species (Hubbell, 1979).

Available evidence indicates that the outcome of the interaction for the plants varies spatially, and therefore we should make progress by studying the interaction in a variety of habitats that differ both in abiotic (e.g. soil nutrient content) and biotic features (e.g. diaspore and animal prey availability, floristic composition, ant fauna). Recently, Garrido *et al.* (2002) stressed the importance of studying different localities for a better understanding of the intricacies involved in the interactions between ants and diaspores, and their possible evolutionary pathways. Data on spatial and intercolony variation in the dietary requirements of ants are also needed. Can predominantly carnivorous ants sustain their colonies on a diet comprised mostly of fleshy fruits? What kinds of benefits, if any, do ants gain by exploiting vertebrate-dispersed diaspores? Working with myrmecochorous diaspores, Morales and Heithaus (1998) showed that elaiosomes positively affected the reproductive potential of ant colonies. Does the same hold true for ants feeding on lipid-rich tissues derived from vertebrate-dispersed diaspores?

Finally, the interaction between ants and vertebrate-dispersed diaspores is not free from human-caused disturbance (Carvalho and Vasconcelos, 1999). Guimarães and Cogni (2002) have recently demonstrated that reduced ant activity at fallen ornithochorous diaspores of *Cupania vernalis* (Sapindaceae) in the edge of a Brazilian forest fragment results in increased seed predation. The study of ant–diaspore interactions in disturbed habitats can add insights to the growing body of knowledge on the effects of habitat disturbance on biotic interactions.

Acknowledgements

We are grateful to the Fundação Florestal and Instituto Florestal do Estado de São Paulo for allowing us to work at Parque Intervales and Ilha do Cardoso, respectively. Special thanks to all the friends that helped us during the fieldwork. We also thank Aricio X. Linhares for help with the statistical analyses, and Steve Vander Wall, Pierre-Michel Forget, and Andrew Beattie for suggestions to the manuscript. Financial support was provided by a doctoral fellowship from FAPESP and CNPq to M.A.P. and L.P., respectively, and by research grants from FAEP/UNICAMP and CNPq to P.S.O.

References

Almeida-Scabbia, R. (1996) Fitossociologia de um trecho de Mata Atlântica no sudeste do Brasil. Masters thesis, Universidade Estadual Paulista, Rio Claro, Brazil.

Andresen, E. (1999) Seed dispersal by monkeys and the fate of dispersed seeds in a Peruvian rain forest. *Biotropica* 31, 145–158.

Barros, F., Melo, M.M.R.F., Chiea, S.A.C., Kirizawa, M., Wanderley, M.G.L. and Jung-Mendaçolli, S.L. (1991) *Flora fanerogâmica da Ilha do Cardoso*. Instituto de Botânica, São Paulo.

Beattie, A.J. (1985) *The Evolutionary Ecology of Ant–Plant Mutualisms*. Cambridge University Press, Cambridge, 182 pp.

Beattie, A.J. (1991) Problems outstanding in ant–plant interactions. In: Huxley, C.R. and Cutler, D.F. (eds) *Ant–Plant Interactions*. Oxford Science Publications, Oxford, pp. 559–576.

Beattie, A.J. and Hughes, L. (2002) Ant–plant interactions. In: Herrera, C.M. and Pellmyr, O. (eds) *Plant–Animal Interactions: an Evolutionary Approach*. Blackwell Publishing, Oxford, pp. 211–235.

Bleher, B. and Böhning-Gaese, K. (2001) Consequences of frugivore diversity for seed dispersal, seedling establishment and the spatial pattern of seedlings and trees. *Oecologia* 129, 385–394.

Bligh, E.G. and Dyer, W.J. (1959) A rapid method of total lipid extraction and purification. *Canadian Journal of Biochemistry and Physiology* 37, 911–917.

Böhning-Gaese, K., Gaese, B.H. and Rabemanantsoa, S.B. (1999) Importance of primary and secondary seed dispersal in the Malagasy tree *Commiphora guillaumini*. *Ecology* 80, 821–832.

Brew, C.R., O'Dowd, D.J. and Rae, I.D. (1989) Seed dispersal by ants: behaviour-releasing compounds in elaiosomes. *Oecologia* 80, 490–497.

Byrne, M.M. (1994) Ecology of twig-dwelling ants in a wet lowland tropical forest. *Biotropica* 26, 61–72.

Carvalho, K.S. and Vasconcelos, H.L. (1999) Forest fragmentation in Central Amazonia and its effects on litter dwelling ants. *Biological Conservation* 91, 151–158.

Chambers, J.C. and MacMahon, J.A. (1994) A day in the life of a seed: movements and fates of seeds and their implications for natural and managed systems. *Annual Review of Ecology and Systematics* 25, 263–292.

Clark, D.B. (1996) Abolishing virginity. *Journal of Tropical Ecology* 12, 735–739.

Culver, D.C. and Beattie, A.J. (1983) Effects of ant mounds on soil chemistry and vegetation patterns in a Colorado montane meadow. *Ecology* 64, 485–492.

Dalling, J.W. and Wirth, R. (1998) Dispersal of *Miconia argentea* seeds by the leaf-cutting ant *Atta colombica*. *Journal of Tropical Ecology* 14, 705–710.

Davidson, D.W. and Morton, S.R. (1981) Myrmecochory in some plants (F. Chenopodiaceae) of the Australian arid zone. *Oecologia* 50, 357–366.

Estrada, A. and Fleming, T.H. (1986) *Frugivores and Seed Dispersal*. Dr W. Junk Publishers, Dordrecht, The Netherlands, 392 pp.

Farji-Brener, A.G. and Silva, J.F. (1996) Leaf cutter ants' (*Atta laevigata*) aid to the establishment success of *Tapirira velutinifolia* (Anacardiaceae) seedlings in a parkland savanna. *Journal of Tropical Ecology* 12, 163–168.

Fleming, T.H. and Estrada, A. (1993) *Frugivory and Seed Dispersal: Ecological and Evolutionary Aspects*. Kluwer Academic Publishers, Dordrecht, The Netherlands, 392 pp.

Garrido, J.L., Rey, P., Cerda, X. and Herrera, C.M. (2002) Geographical variation in diaspore traits of an ant-dispersed plant (*Helleborus foetidus*): are ant community composition and diaspore traits correlated? *Journal of Ecology* 90, 446–455.

Gorb, S.N. and Gorb, E.V. (1995) Removal rates of seeds of five myrmecochorous plants by the ant *Formica polyctena* (Hymenoptera: Formicidae). *Oikos* 73, 367–374.

Guimarães, P.R. and Cogni, R. (2002) Seed cleaning of *Cupania vernalis* (Sapindaceae) by ants: edge effects in a highland forest in southeast Brazil. *Journal of Tropical Ecology* 18, 303–307.

Hölldobler, B. and Wilson, E.O. (1990) *The Ants*. Belknap Press, Cambridge, 732 pp.

Horvitz, C.C. (1981) Analysis of how ant behaviours affect germination in a tropical myrmecochore *Calathea microcephala* (P. & E.) Koernicke (Marantaceae): microsite selection and aril removal by neotropical ants, *Odontomachus, Pachycondyla*, and *Solenopsis* (Formicidae). *Oecologia* 51, 47–52.

Horvitz, C.C. and LeCorff, J. (1993) Spatial scale and dispersion patterns of ant- and bird-dispersed herbs in two tropical lowland forests. *Vegetatio* 107/108, 351–362.

Horvitz, C.C. and Schemske, D.W. (1986) Ant-nest soil and seedling growth in a neotropical ant-dispersed herb. *Oecologia* 70, 318–320.

Horvitz, C.C., Pizo, M.A., Bello y Bello, B., LeCorff, J. and Dirzo, R. (2002) Are plant species that need gaps for recruitment more attractive to seed-dispersing birds and ants than other species? In: Levey, D.J., Silva, W.R. and Galetti, M. (eds) *Seed Dispersal and Frugivory: Ecology, Evolution and Conservation*. CAB International, Wallingford, UK, pp. 145–159.

Howe, H.F. and Smallwood, J. (1982) Ecology of seed dispersal. *Annual Review of Ecology and Systematics* 13, 201–228.

Hubbell, S.P. (1979) Tree dispersion, abundance, and diversity in a tropical dry forest. *Science* 203, 1299–1309.

Hughes, L. (1990) The relocation of ant nest entrances: potential consequences for ant-dispersed seeds. *Australian Journal of Ecology* 16, 207–214.

Hughes, L. and Westoby, M. (1992) Effect of diaspore characteristics on removal of seeds adapted for dispersal by ants. *Ecology* 73, 1300–1312.

Hughes, L., Westoby, M. and Jurado, E. (1994) Convergence of elaiosomes and insect prey: evidence from ant foraging behaviour and fatty acid composition. *Functional Ecology* 8, 358–365.

Janzen, D.H. (1970) Herbivores and the number of tree species in tropical forests. *American Naturalist* 104, 501–529.

Kaspari, M. (1993) Removal of seeds from neo-tropical frugivore feces: ants responses to seed number. *Oecologia* 95, 81–88.

Laman, L.G. (1996) *Ficus* seed shadow in a Bornean rainforest. *Oecologia* 107, 347–355.

Leal, I.R. and Oliveira, P.S. (1998) Interactions between fungus-growing ants (Attini), fruits and seeds in cerrado vegetation in southeast Brazil. *Biotropica* 30, 170–178.

Levey, D.J. and Byrne, M.M. (1993) Complex ant–plant interactions: rain forest ants as secondary dispersers and post-dispersal seed predators. *Ecology* 74, 1802–1812.

Levey, D.J., Silva, W.R. and Galetti, M. (2002) *Seed Dispersal and Frugivory: Ecology, Evolution and Conservation*. CAB International, Wallingford, UK, 511 pp.

Longino, J.T., Coddington, J. and Colwell, R.K. (2002) The ant fauna of a tropical rain forest: estimating species richness three different ways. *Ecology* 83, 689–702.

Lu, K.L. and Mesler, M.R. (1981) Ant dispersal of a neotropical forest floor gesneriad. *Biotropica* 13, 159–160.

Mark, S. and Olesen, J.M. (1996) Importance of elaiosome size to removal of ant-dispersed seeds. *Oecologia* 107, 95–101.

Marshall, D.L., Beattie, A.J. and Bollenbacher, W.E. (1979) Evidence for diglycerides as attractants in an ant–seed interaction. *Journal of Chemical Ecology* 5, 335–343.

Morales, M.A. and Heithaus, E.R. (1998) Food from seed dispersal mutualism shifts sex ratios in colonies of the ant *Aphaenogaster rudis. Ecology* 79, 734–739.

Morellato, L.P.C. (1992) Sazonalidade e dinâmica de ecossistemas florestais na Serra do Japi. In: Morellato, L.P.C. (ed.) *Historia Natural da*

Serra do Japi: Ecologia e Preservação de Uma Area Florestal no Sudeste do Brasil. Editora da UNICAMP, Campinas, Brazil, pp. 98–110.

Mori, S.A., Boom, B.M., Carvalino, A.M. and Santos, T.S. (1983) Ecological importance of Myrtaceae in an eastern Brazilian wet forest. *Biotropica* 15, 68–70.

Oliveira, P.S., Galetti, M., Pedroni, F. and Morellato, L.P.C. (1995) Seed cleaning by *Mycocepurus goeldii* ants (Attini) facilitates germination in *Hymenaea courbaril* (Caesalpiniaceae). *Biotropica* 27, 518–522.

Oliveira-Filho, A.T. and Fontes, M.A.L. (2000) Patterns of floristic differentiation among Atlantic forests in southeastern Brazil and the influence of climate. *Biotropica* 32, 793–810.

Passos, L. (2001) Ecologia da interação entre formigas, frutos e sementes em solo de mata de restinga. PhD thesis, Universidade Estadual de Campinas, Campinas, Brazil.

Passos, L. and Oliveira, P.S. (2002) Ants affect the distribution and performance of *Clusia criuva* seedlings, a primarily bird-dispersed rainforest tree. *Journal of Ecology* 90, 517–528.

Passos, L. and Oliveira, P.S. (2003) Interactions between ants, fruits and seeds in a restinga forest in south-eastern Brazil. *Journal of Tropical Ecology* 19, 261–270.

Pizo, M.A. (1997) Seed dispersal and predation in two populations of *Cabralea canjerana* (Meliaceae) in the Atlantic forest of south-eastern Brazil. *Journal of Tropical Ecology* 13, 559–578.

Pizo, M.A. and Oliveira, P.S. (1998) Interactions between ants and seeds of a nonmyrmecochorous neotropical tree, *Cabralea canjerana* (Meliaceae), in the Atlantic forest of southeast Brazil. *American Journal Botany* 85, 669–674.

Pizo, M.A. and Oliveira, P.S. (1999) Removal of seeds from vertebrate faeces by ants: effects of seed species and deposition site. *Canadian Journal of Zoology* 77, 1595–1602.

Pizo, M.A. and Oliveira, P.S. (2000) The use of fruits and seeds by ants in the Atlantic forest of southeast Brazil. *Biotropica* 32, 851–861.

Pizo, M.A. and Oliveira, P.S. (2001) Size and lipid content of nonmyrmecochorous diaspores: effects on the interaction with litter-foraging ants in the Atlantic rain forest of Brazil. *Plant Ecology* 157, 37–52.

Roberts, J.T. and Heithaus, E.R. (1986) Ants rearrange the vertebrate-generated seed shadow of a neotropical fig tree. *Ecology* 67, 1046–1051.

SAS Institute Inc. (1987) *SAS User's Guide. Statistics, 6th Version.* SAS Institute Inc., Cary, North Carolina.

Skidmore, B.A. and Heithaus, E.R. (1988) Lipid cues for seed carrying by ants in *Hepatica americana*. *Journal of Chemical Ecology* 14, 2185–2196.

Thompson, J.N. (1994) *The Coevolutionary Process.* Chicago University Press, Chicago, Illinois, 340 pp.

van der Pijl, L. (1982) *Principles of Seed Dispersal in Higher Plants.* Springer-Verlag, Berlin, 199 pp.

Vieira, E.M., Pizo, M.A. and Izar, P. (2003) Fruit and seed exploitation by small rodents of the Brazilian Atlantic Forest. *Mammalia* 67, 533–539.

Voigt, F.A., Burkhardt, J.F., Verhaagh, M. and Böhning-Gaese, K. (2002) Regional differences in ant community structure and consequences for secondary dispersal of *Commiphora* seeds. *Ecotropica* 8, 59–66.

Wirth, R., Herz, H., Ryel, R.J., Beyschlag, W. and Hölldobler, B. (2002) *Herbivory of Leaf-cutting Ants: a Case Study of* Atta colombica *in the Tropical Rainforest of Panama.* Springer-Verlag, Berlin, 230 pp.

20 The Role of Dung Beetles as Secondary Seed Dispersers and their Effect on Plant Regeneration in Tropical Rainforests

Ellen Andresen[1] and François Feer[2]

[1]Centro de Investigaciones en Ecosistemas, Universidad Nacional Autónoma de México, AP 27-3, Morelia, CP 58089, Michoacán, México; [2]Muséum National d'Histoire Naturelle, Département Ecologie et Gestion de la Biodiversité, UMR 5176 CNRS-MNHN, 4 avenue du Petit Château, F-91800 Brunoy, France

Introduction

The main biotic processes that affect the fate of seeds during the time between primary dispersal and germination are predation, pathogen attack, and secondary dispersal. The former two generally kill seeds, thus affecting plant regeneration negatively (Fenner, 2000). Secondary dispersal, although much less studied than primary dispersal, may have a greater impact on the patterning of plant communities than primary dispersal (Chambers and MacMahon, 1994) and has often been shown to have a net positive effect on the fitness of plants (Forget and Milleron, 1991; Levey and Byrne, 1993; Böhning-Gaese et al., 1999; Andresen, 2001). When primary dispersal occurs through defecation of seeds by vertebrates, the faecal material that accompanies seeds has potential to affect the fate of dispersed seeds (Janzen, 1986; Andresen, 2001). Faecal material can promote pathogen attack (Jones, 1994), attract animals such as ants (Byrne and Levey, 1993; Engel, 2000) and rodents (Janzen, 1982), which can behave either as seed predators or as secondary dispersers, or attract dung beetles that act as secondary dispersers (Feer, 1999; Andresen, 2002a).

Both adult dung beetles and their larvae feed on vertebrate faeces. After locating a pile of dung, the adults of many species process a portion of it. Dung processing by a beetle generally culminates in the burial under the soil surface in the form of one or several brood or feeding dung balls (Halffter and Edmonds, 1982; Hanski and Cambefort, 1991a). In many terrestrial habitats, large numbers of dung beetles are attracted to dung and, since competition is intense, all dung quickly disappears from the soil surface (Hanski and Cambefort, 1991a). Ecological benefits of this behaviour include soil fertilization and aeration (Mittal, 1993), increased rates and efficiency of nutrient cycling (Nealis, 1977; Miranda et al., 1998), increased plant nutrient uptake and yield (Miranda et al., 1998), prevention of pasture wastage (McKinney and Morley, 1975), control of pest flies and enteric parasites of vertebrates (Bergstrom et al., 1976), and secondary seed dispersal. The latter role, which occurs due to the accidental movement of seeds embedded in dung, has only recently been investigated in tropical rainforest ecosystems (Estrada and Coates-Estrada, 1991; Shepherd and Chapman, 1998; Andresen, 1999, 2001, 2002a; Feer, 1999; Vulinec, 1999, 2002).

Secondary seed dispersal by dung beetles includes both horizontal and/or vertical movement of seeds, and it can affect the regeneration of plants in several ways. Most importantly, seed burial greatly enhances the probability that a seed will escape predation compared to seeds on the surface (Crawley, 2000). Because post-dispersal seed predation is high for most plant species (Janzen, 1971), even small changes in predation pressure can have a large effect on plant demography (Crawley, 2000). Seed burial also affects the microclimate a seed experiences, most importantly less light and more humidity. These changes can affect seed survival, germination and seedling establishment (positively or negatively, depending on the particular requirements of each seed species) (Price and Jenkins, 1986; Chambers and MacMahon, 1994; Fenner, 2000). Deep burial can also have a negative effect on seedling recruitment by preventing elongating seedlings from reaching the soil surface (Fenner, 1987). Finally, by reducing seed clumping in a dung deposit, dung beetles may increase plant regeneration by diminishing the negative effects of seedling competition (Howe, 1989).

Dung beetles of the family Scarabaeidae (taxonomy follows Hanski and Cambefort, 1991a) are speciose and abundant insects in tropical forests worldwide (Nummelin and Hanski, 1989; Gill, 1991; Davis et al., 2001). Because dung beetles generally prefer the dung of large herbivorous (sensu lato) mammals (Hanski, 1989), and because in tropical forests most herbivorous mammals include fruit in their diet (Terborgh, 1986), it is likely that much of the dung exploited by dung beetles in these ecosystems will contain seeds. Thus the potential for secondary dispersal by dung beetles is great, and the consequences for plant regeneration deserve attention.

The outcome of the seed–beetle interaction (Fig. 20.1), i.e. whether a seed is secondarily dispersed to a site favourable for establishment or dies, depends on many factors.

Some factors are intrinsic, those that are related to the participants in the interaction and have a direct effect on it. Intrinsic factors include dung beetle characteristics such as body mass and dung-processing behaviour (rollers versus tunnellers), and seed/seedling characteristics such as seed size and seedling functional morphology. Other factors are extrinsic, those that have an indirect effect on the outcome of the interaction by affecting the composition of the local dung beetle assemblages. These include characteristics associated with the primary disperser such as defecation pattern, or environmental characteristics such as time of day, season and habitat disturbance.

In this chapter we review the current knowledge on secondary seed dispersal by dung beetles and its effect on plant regeneration in tropical rainforests. In the first section, we focus on the intrinsic factors affecting the outcome of the seed–beetle interaction, emphasizing both the beetle's and the plant's perspective. In the second section, we deal with some of the extrinsic factors that indirectly affect the outcome of the interaction. Both sections emphasize little-known aspects of the effects of dung beetles on plant regeneration, pointing out new avenues for future research.

Intrinsic Factors Affecting the Beetle–Seed Interaction

What will ultimately determine the outcome of the beetle–seed interaction, i.e. whether a seed is secondarily dispersed by dung beetles, and whether such a seed is able to establish as a seedling, is the species of seed and the species of beetle that handles it. However, there are two versions of the 'interaction tale': one told by the beetle, and one told by the seed.

The beetle's perspective: amount of seed burial and depth of burial

Dung beetles do not eat seeds, they eat dung. Consequently, seeds present in dung are contaminants from the beetles' perspective. However, competition for dung is intense and it is advantageous for beetles to move a portion of a dung deposit as quickly as possible (Hanski and Cambefort, 1991b). Thus, a conflict occurs and a beetle must

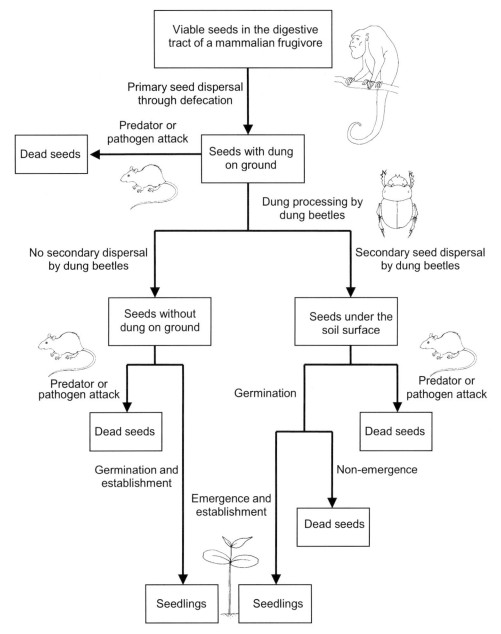

Fig. 20.1. Fate paths for seeds swallowed intact by a frugivorous mammal. Boxes represent states in the life of seeds, while arrows represent processes leading to such states. Factors intrinsic to the dung beetle–seed interaction (e.g. seed size, seed functional morphology, beetle size and beetle dung-processing behaviour), as well as factors extrinsic to it (e.g. frugivore defecation pattern, time of day, season and habitat type) play a role in determining which particular path is followed and whether seeds die or establish as seedlings.

choose whether to invest time in removing seeds from dung. If a seed is not removed, it will remain embedded in the dung that beetles use to provision their underground feeding and nesting chambers. On the other hand, if a beetle removes a seed, it may do so either before or during dung burial. In the first case, the seed will most probably remain on the surface, while, in the second case, the seed will end up buried, not embedded in dung, at some intermediate depth between the soil surface and the beetle's feeding or nesting chamber. A seed's final location will depend on factors like the size of the seed, the size of the beetle and the beetle's dung-processing method.

Seed size

The results of experiments conducted in Mexico (Estrada and Coates-Estrada, 1991), Peru (Andresen, 1999), Brazil (Andresen, 2002a; Andresen and Levey, 2004), French Guiana (Feer, 1999) and Uganda (Shepherd and Chapman, 1998), in which beetles were allowed to bury either seeds or plastic beads (used as seed mimics) on the forest floor, show a clear negative relationship between seed or bead size and the

percentage of these buried by dung beetles (Fig. 20.2). In tropical rainforests, large seeds (> 30 mm in length) are seldom buried by dung beetles, although this may vary greatly depending on the local beetle assemblage. On the other hand, nearly all small seeds (≤ 3 mm in length) are buried. However, it is methodologically difficult to follow the fate of such small seeds in the field, and thus not much is known about the overall effect of secondary seed dispersal by dung beetles on the regeneration of plants with seeds of this size (see Dalling, Chapter 3, this volume, for general information on the fate of small seeds).

Variation in the relationship between seed size and burial probability (Fig. 20.2) has two components: inter-site and intra-site variation. Inter-site variability is most probably due to differences in the dung beetle communities (most importantly abundance and size of beetles), brought about by differences in habitat and climate among these five forests. The variability within sites reflects, among others, methodological differences in the experimental setup for each of the seed species or beads tested in each site (e.g. Andresen, 2002a), the great micro-spatial and -temporal variability

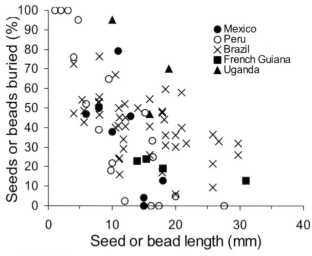

Fig. 20.2. Negative relationship between the percentage of seeds or beads (surrogate seeds) buried by dung beetles and seed or bead size in five tropical rainforests. Beads and seeds in these studies were naturally buried by dung beetles on the forest floor. Data from: Mexico (seeds), Estrada and Coates-Estrada (1991); Peru (beads), Andresen (1999); Brazil (beads and seeds), Andresen (2002a) and Andresen and Levey (2004); French Guiana (seeds), Feer (1999); Uganda (seeds), Shepherd and Chapman (1998).

displayed by dung beetle assemblages (Doube, 1991; Hanski and Cambefort, 1991c), and individual variation in dung beetle behaviour.

One study in Mexico (Estrada and Coates-Estrada, 1991; Fig. 20.3B) and one study in Brazil (Andresen, 2002a) have also shown a negative relationship between seed size and burial depth. Large seeds, being larger contaminants, are more likely to be excluded by dung beetles early during dung processing than small seeds. Small seeds will often be incorporated into the final brood- or feeding-ball, and thus buried more

deeply. Vulinec (2002) also found that on average seeds < 5 mm long were buried more deeply than seeds > 10 mm long. Feer (1999) found a negative relationship between burial depth and seed size for only one dung beetle species out of three tested.

Regarding the depths at which dung beetles bury seeds, much variability has been reported, with some seeds being buried only 0.5 cm deep, and others buried > 20 cm deep (e.g. Feer, 1999; Andresen, 2002a). However, several studies have found that most seeds are buried < 5 cm deep (Estrada and Coates-Estrada, 1991; Shepherd and

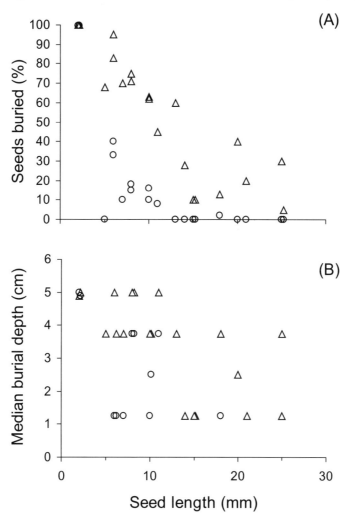

Fig. 20.3.   The effect of seed length on percentage of seeds buried (A) and burial depth (B) for roller (circles) and tunneller (triangles) dung beetles in controlled experiments with soil-filled cylinders, in Mexico. Data from Estrada and Coates-Estrada (1991).

Chapman, 1998; Andresen, 1999, 2002a; Vulinec, 2000).

Beetle size

Dung beetles in tropical forests often show a wide range of sizes among species (e.g. 2.5–37 mm in Brazil and French Guiana; Feer, 2000; Andresen, 2002a). Since seeds are dung contaminants from the beetles' perspective, a given seed will be a relatively smaller contaminant to a large beetle than to a small beetle. Additionally, larger beetles need more resources and thus process larger amounts of dung than smaller beetles (Doube, 1990; Hanski and Cambefort, 1991d). Thus, it is not surprising that several studies have found that larger beetles bury a higher percentage of seeds present in dung than smaller beetles (Feer, 1999; Vulinec, 2000, 2002; Andresen, 2002a). These studies also showed that larger beetles not only bury more seeds, but they also bury larger seeds than smaller beetles. Finally, larger species of beetles tend to bury seeds more deeply than smaller species with the same dung-processing behaviour (Feer, 1999; Vulinec, 2000, 2002).

Dung-processing behaviour

Regarding the ways in which dung beetles process dung, beetles can be classified as rollers, tunnellers (also known as burrowers) and dwellers (Halffter and Edmonds, 1982; Cambefort and Hanski, 1991). Rollers collect dung to form a ball, which they then roll some horizontal distance. Average horizontal distances of 1.0 and 1.2 m have been reported for some Peruvian (Andresen, 1999) and Mexican (Estrada and Coates-Estrada, 1991) rainforest roller species, respectively. However, maximum distances in these sites range up to 5 m, and distances up to 10 m have been reported in an African rainforest (Engel, 2000). In contrast, tunnellers make a relatively deep burrow right underneath or very close to the dung source, and then start provisioning their tunnel by pulling and pushing portions of dung into it. They occasionally move larger pieces of dung by pushing them short distances away from the source. Finally, dwellers process the dung inside the pat or immediately below the dung pat, without making deep underground tunnels (Halffter and Edmonds, 1982). Consequently, only tunneller and roller dung beetles are likely to act as secondary dispersers, moving seeds away from the initial site of deposition.

Tunnellers generally bury a higher proportion of seeds compared to rollers (Fig. 20.3A). In French Guiana, Hingrat and Feer (2002) found that seed burial rates were positively correlated to the weighted abundance (as an approximation of biomass) of tunneller dung beetles (see also Estrada and Coates-Estrada, 1991). In most Neotropical forests, tunneller dung beetles have been found to be more speciose and abundant than roller beetles (Peck and Howden, 1984; Feer, 2000; Andresen, 2002a). Also, on average, tunnellers are represented by larger species than rollers (Estrada and Coates-Estrada, 1991). This difference in size could account in part for the greater amount of seed burial; however, tunnellers also process larger amounts of dung than rollers of similar size (Doube, 1990) and probably move more seeds embedded in the dung. Finally, it seems that rollers need to be more selective than tunnellers when they gather dung. By excluding seeds at the moment of ball-formation, rollers maximize the quality of the resources collected, given the constraints on the weight and the size of the ball. In contrast to tunnellers, which build a tunnel close to the dung source, rollers cannot adjust the volume of the resources gathered if they exclude seeds just before burial. On the other hand, tunnellers probably exclude many of the seeds from the dung while moving it through the tunnel. Thus, seed burial is not as 'expensive' for tunnellers, relative to the net amount of dung gathered, as it is for rollers.

Tunnellers bury seeds at greater depths than roller dung beetles, and tunnellers bury larger seeds than rollers (Fig. 20.3). These results could be caused by the larger relative size of tunnellers (Estrada and Coates-Estrada, 1991). Whether rollers bury seeds less deeply and bury smaller seeds than tunnellers of equivalent size warrants further study.

Because tunnellers bury more seeds and more species of seeds present in dung, it may be assumed that these beetles play a more important role as secondary seed dispersers than rollers. However, since tunnellers concentrate their activity close to the dung source, they maintain a relatively clumped distribution of seeds, which could be detrimental for plant demography (Howe, 1989). Roller dung beetles, on the other hand, may perform higher-quality dispersal (*sensu* Schupp, 1993) by removing seeds from seed clumps, thus decreasing the probability of seedling competition and/or density-dependent seed predation (Andresen, 1999, 2002b). The effect of this aspect of secondary seed dispersal on plant fitness still remains to be quantified.

The plant's perspective: seed predation, seedling emergence and seedling establishment

From a plant's perspective, secondary seed dispersal by dung beetles will be advantageous only if it increases the probability of seedling establishment over seeds not moved by dung beetles. In the case of seeds buried by dung beetles, enhanced seedling establishment could be achieved through increased probabilities in seed survival, germination and/or seedling emergence.

Germination is sometimes inhibited under conditions that would be unfavourable for seedling emergence, establishment or survival. For example, in some species germination is inhibited when seeds are on the soil surface, to avoid conditions of reduced moisture availability (Pons, 2000). Similarly, in other species germination is inhibited by lack of light when buried, to avoid germination too deep in the soil for the seedling to reach the surface (Pons, 2000). However, many seed species, in particular larger ones, will germinate even when buried too deeply to successfully emerge as seedlings (Fenner, 1987; Vander Wall, 1993; Dalling *et al.*, 1994; Chen and Maun, 1999). Studies focusing specifically on the effect of seed burial by dung beetles, on

germination, are still needed. Studies exist, however, showing that burial as well as depth of burial affect both the probability that a seed will survive and the probability that a seedling will successfully emerge. This evidence will be discussed in the following sections.

Effects of seed burial and burial depth on seed survival

Buried seeds have much higher probabilities of escaping seed predation than seeds on the surface (Chambers and MacMahon, 1994; Crawley, 2000). A study in Brazil, using seeds of 11 tree species buried by dung beetles, showed that burial decreased by threefold the probability of seed predation by rodents (Andresen and Levey, 2004). Moreover, studies in tropical forests in Mexico, Peru and Uganda have shown that the probability of predation decreases as burial depth increases (Estrada and Coates-Estrada, 1991; Shepherd and Chapman, 1998; Andresen, 1999) (Fig. 20.4). The study in Mexico used an experimental set-up in which seeds were buried inside a large cage which contained captive rodents (Estrada and Coates-Estrada, 1991), while the ones in Uganda and Peru assessed predation on seeds buried along forest transects (Shepherd and Chapman, 1998; Andresen, 1999). The results from Mexico indicate that seeds buried 5.1–8 cm deep still suffered 17% predation rates, while results from the other two sites indicate that seeds buried 1–5 cm deep only suffered 1–4% predation rates. It is likely that experiments with captive rodents would overestimate natural predation rates. Still, we can generalize that seeds buried > 3 cm deep by dung beetles will most likely escape predation by rodents.

Other factors, besides burial depth, also affect the probability that a given seed predator will remove a buried seed. These factors include soil texture, soil moisture, seed species and seed size (Johnson and Jorgensen, 1981; Vander Wall, 1998; Crawley, 2000). Regarding seed size, beetles bury small seeds deeper than large seeds. Further, large seeds present more conspicuous olfactory

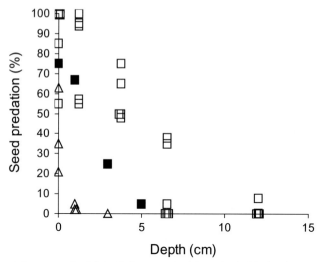

Fig. 20.4. Seed predation versus burial depth for several seed species in Mexico (open squares), Peru (filled squares) and Uganda (triangles). Data from: Mexico, Estrada and Coates-Estrada (1991); Peru, Andresen (1999); Uganda, Shepherd and Chapman (1998).

cues for seed predators, and consequently predation of buried seeds increases with seed size (Hulme and Borelli, 1999; Crawley, 2000). However, in some tropical rainforests large seeds are often secondarily dispersed by seed-caching rodents, rather than eaten by them (e.g. Forget et al., 1998). Thus, for plants that produce seeds scatterhoarded by rodents, studies on the fate of seeds buried by dung beetles should include a quantitative assessment of the role of rodents both as seed predators and as secondary dispersers.

Effects of seed burial and burial depth on seedling emergence

Seed burial can be advantageous for plant regeneration because it decreases the probability of predation, but it can also have a negative effect if shoots of elongating seedlings are unable to reach the soil surface (Fenner, 1987). Further, as burial depth increases, the probability of non-emergence also increases, and most species have their emergence rates greatly diminished with burial depths > 3 cm (Fig. 20.5). This relationship has been shown clearly in some tropical forest species for which seeds were experimentally sown at different depths (Fenner, 1987; Dalling et al., 1994; Shepherd

and Chapman, 1998; Feer, 1999; Engel, 2000; Hingrat and Feer, 2002; Pearson et al., 2002; Andresen and Levey, 2004). Deep burial of seeds by dung beetles can be a serious impediment to seedling emergence in some plant species in which > 90% of buried seeds fail to emerge (e.g. Hingrat and Feer, 2002; Andresen and Levey, 2004).

Studies on seed dispersal by dung beetles have speculated that plant species with larger seeds might be better able to emerge from greater depths than those with smaller seeds (Estrada and Coates-Estrada, 1991; Shepherd and Chapman, 1998; Andresen, 1999). Large seeds have large energy reserves (Foster, 1986; Westoby et al., 1992) and produce large seedlings. Several studies have shown that seedlings from such seeds are better able to emerge from greater soil depths than small seeds (Leishman et al., 2000, and references therein). However, in a comparison of ten seed species ranging in size from 11 to 26 mm in length, Andresen and Levey (2004) found no evidence of a relationship between seed size and probability of seedling emergence.

Along with seed size, other seed and seedling characteristics such as those related to functional morphology play an important role in determining the capacity

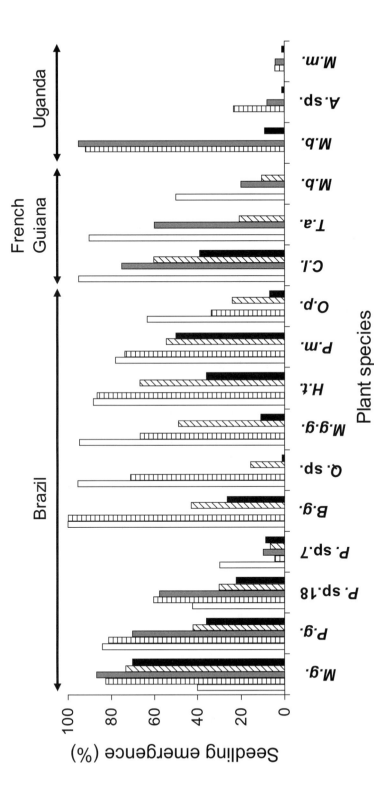

Fig. 20.5. Percentage of seedlings establishing from seeds experimentally sown at different depths: 0 cm (white bars), 1 cm (horizontally hatched bars), 3 cm (grey bars), 5 cm (diagonally hatched bars), 10 cm (black bars). Plant species in order from left to right are: *M.g., Minquartia guianensis* (Olacaceae); *P.g., Pourouma guianensis* (Moraceae); *P. sp. 18, Pouteria* sp. 18 (Sapotaceae); *P. sp. 7, Pouteria* sp. 7 (Sapotaceae); *B.g., Buchenavia grandis* (Combretaceae); *Q.* sp., Quiinaceae sp.; *M.g.g., Micropholis guyanensis guyanensis* (Sapotaceae); *H.t., Helicostylis tomentosa* (Moraceae); *P.m., Pourouma minor* (Moraceae); *O.p., Ocotea percurrens* (Lauraceae); *C.l., Chrysophyllum lucentifolium* (Sapotaceae); *T.a., Tetragastris altissima* (Burseraceae); *M.b., Manilkara bidentata* (Sapotaceae); *M.b., Mimusops bagshawei* (Annonaceae); *A.* sp., *Aframomum* sp. (Zingiberaceae); *M.m., Monodora myristica* (Annonaceae). When bars are not present at a particular depth treatment, it means that such a treatment was not tested, except for *Tetragastris altissima* and *Manilkara bidentata* at 10 cm, for which no seedlings established from seeds sown at this depth. Data from Andresen and Levey (2004) for Brazil; Feer (1999; unpublished data) and Hingrat and Feer (2002) for French Guiana; Shepherd and Chapman (1998) for Uganda.

of a seedling to emerge from a buried seed, and to survive throughout the establishment period (Garwood, 1996; Kitajima, 1996; Kitajima and Fenner, 2000; Ibarra-Manríquez *et al.*, 2001). Seedlings can be classified using a combination of three dichotomous traits that describe cotyledon exposure, position and texture (Garwood, 1996). First, seedlings are phanerocotylar if cotyledons emerge from the seed coat and become totally exposed, and cryptocotylar if cotyledons remain hidden within the seed coat. Second, seedlings are epigeal when cotyledons are raised above ground, and hypogeal if cotyledons are below ground or rest on the surface. Finally, seedlings can have foliaceous cotyledons that become photosynthetic, or fleshy cotyledons that store reserves. Andresen (2000a) found that the ability of a seedling to emerge was related to the functional morphology of seedlings. Except for one species, seedlings with cryptocotylar hypogeal reserve cotyledons were more successful at emerging from 10 cm depths than seedlings with phanerocotylar epigeal foliaceous cotyledons. Seedlings with phanerocotylar epigeal reserve cotyledons had intermediate success (Fig. 20.6).

Although it seems that in general seed burial diminishes the probability of seedling emergence and thus seedling establishment of rainforest plants, it is important to mention that some species, under controlled conditions in which predation is prevented, show higher seedling establishment for shallowly buried seeds than for surface seeds (e.g. *Minquartia guianensis* and *Pouteria* sp. 18 in Fig. 20.5).

In summary, we have seen that most seeds are buried by dung beetles between 1 and 5 cm. Within this range, the highest probability of seed predation, but also the highest probability of seedling emergence, occurs between 1 and 3 cm. In turn, the lowest probability of seed predation, but also the lowest probability of seedling emergence, occurs between 3 and 5 cm. Thus it seems that an average burial depth of 3 cm would be optimum for seedling establishment of large-seeded (5–25 mm) tropical plants whose seeds are buried by dung beetles.

Seedling establishment

Seed burial has a positive impact, through increased seed survival, and a negative

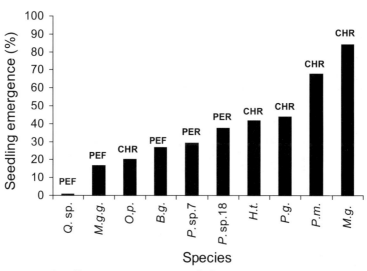

Fig. 20.6.   Percentage of seedling emergence in a controlled germination experiment at 10 cm depth, relative to baseline establishment at 0 cm (1 cm for *Minquartia guianensis* and *Pouteria* sp. 18). Letters above bars indicate functional morphology of seedlings: P, phanaerocotylar; C, cryptocotylar; E, epigeal cotyledons; H, hypogeal cotyledons; F, foliaceous cotyledons; R, reserve cotyledons; see text for further explanation. Plant species same as Fig. 20.5.

impact, through decreased seedling emergence (Vander Wall, 1993; Shepherd and Chapman, 1998). But what is the net effect of seed burial, i.e. the effect on seedling establishment? Only one study has followed the fate of seeds placed in dung on the forest floor, which were subsequently buried or not buried by dung beetles, until seedling establishment (Andresen, 2001; Andresen and Levey, 2004). In that study, seven species, of 11 seed species tested, showed a significant positive net effect of seed burial by dung beetles, while the other four species showed no effect of burial on seedling establishment (Fig. 20.7). The overall effect of seed burial by dung beetles, for all the species considered together, was a twofold increase in seedling establishment (Andresen and Levey, 2004). More studies are necessary to ascertain if the overall positive effect of seed burial by dung beetles also holds for other forests and other ecosystems.

Several interesting avenues for future research focusing on the effect of dung

beetles on seed survival and seedling establishment need to be pointed out. First, as already mentioned, virtually nothing is known about the final fate of small seeds (≤ 3 mm) secondarily dispersed by dung beetles. Because of their small size, most of these seeds are buried by dung beetles, and they also tend to be buried more deeply than larger seeds. It has in general been assumed that the effect of burial by beetles on such seeds will be negative due to very low seedling emergence from deeply buried seeds. However, this effect has not been quantified. Also, since some small-seeded species show long-term dormancy (Leishman et al., 2000), it may be that small seeds incorporated deeply into the soil seed bank by dung beetles are eventually brought back to the surface by abiotic (Putz, 1983) or biotic factors (Grant, 1983; Willems and Huijsmans, 1994). Such seeds may present pulses of germination and establishment, when seeds are brought back to the surface, without suffering a substantial negative effect due to initial deep burial by beetles. Although it

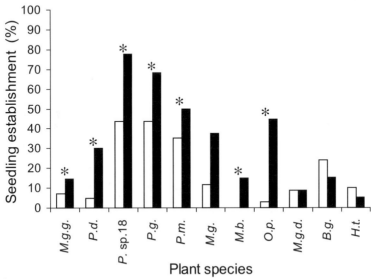

Fig. 20.7. Seedling establishment in Brazil for seeds placed on the forest floor surrounded by dung that were subsequently buried by dung beetles (black bars) or not buried by dung beetles (white bars). Seed species, in order from left to right, are as follows: *M.g.g*, *Micropholis guyanensis guyanensis*; *P.d.*, *Pouteria durlandii*; *P.* sp. 18, *Pouteria* sp. 18; *P.g.*, *Pourouma guianensis*; *P.m.*, *Pourouma minor*; *M.g.*, *Minquartia guianensis*; *M.b.*, *Manilkara bidentata*; *O.p.*, *Ocotea percurrens*; *M.g.d.*, *Micropholis guyanensis duckeana*; *B.g.*, *Buchenavia grandis*; and *H.t.*, *Helicostylis tomentosa*. *Indicates a statistically significant difference (P < 0.05).

presents methodological difficulties, this is an interesting topic for future research (see also Dalling, Chapter 3, this volume).

Second, while it is known that adult dung beetles do not eat seeds, it is not well known how much fibre the larvae consume. Larvae have unspecialized mouthparts that allow them to process the fibre content of brood balls (Halffter and Edmonds, 1982), and thus the possibility exists that they consume some of the seeds or germinants embedded in the dung. Thus it remains to be elucidated whether dung beetle larvae could be acting as predators of seeds dispersed by the adults.

Finally, two additional under-studied potentially beneficial effects of seed dispersal by dung beetles on seedling establishment are the effect of horizontal dispersal of seeds and the effect of seed cleaning. Since most seeds defecated by vertebrates, in particular mammals, are deposited in clumps of several to many seeds (Howe, 1989), scattering of seeds by dung beetles, regardless of seed burial, may greatly diminish seed clumping and consequently reduce seedling competition. On the other hand, seed cleaning caused by the removal of dung surrounding seeds, independent of seed movement, can have important consequences for seedling establishment. It is known that rodent and insect seed predators are attracted by dung odour (Janzen, 1982; Levey and Byrne, 1993; Andresen, 1999); thus it is possible that seeds which are cleaned of the surrounding dung through dung beetle activity will be less detectable to predators. Similarly, seeds cleaned by dung beetles have been shown to suffer lower rates of fungal attack than seeds in dung from which beetles were excluded (Jones, 1994).

Extrinsic Factors Affecting the Beetle–Seed Interaction

Extrinsic factors are those that affect the composition of the dung beetle assemblage that is attracted to a given dung pat, and will thus have an indirect effect on the fate of seeds in dung. In this section we will

only focus on those extrinsic factors that have been investigated in the context of secondary seed dispersal by dung beetles. However, many other extrinsic factors affect the composition of local dung beetle assemblages, but their effects on seed dispersal and seed fate remain unknown. Some of these are: type of dung (including the dung of different mammal seed dispersers, as well as the dung of bird, reptile and amphibian seed dispersers; e.g. Young, 1981; Estrada et al., 1993), soil characteristics (hardness, humidity, texture; e.g. Osberg et al., 1994), forest type (e.g. Escobar, 2000; Vulinec, 2002) and composition of the vertebrate community (e.g. Estrada et al., 1999).

Before we discuss in some detail a few of the extrinsic factors, it is important to mention that many of these not only affect the composition of the dung beetle assemblages but also the composition and behaviour of seed predator and pathogen assemblages. We must remember that, while the seed–beetle interaction is taking place, a seed–predator and/or seed–pathogen interaction is also occurring, and the challenge becomes to understand how these multiple interactions themselves interact to ultimately determine the fate of a seed (Andresen and Levey, 2004).

Defecation patterns

How and where frugivorous animals deposit dung vary greatly among species, depending on features such as body size, social behaviour, movement patterns, and digestive physiology. These characteristics influence the primary dispersal event, determining the initial spatial distribution of dung and seeds on the forest floor (Andresen, 1999, 2002b; Zhang and Wang, 1995; Julliot, 1997; McConkey, 2000).

The amount of dung that surrounds seeds is known to affect their short-term fate. Larger amounts of dung attract more species and individuals of dung beetles, and in some cases also larger beetles (Peck and Howden, 1984; Andresen, 2002a). Using seeds of several tree species, as well as plastic beads, surrounded by 5, 10 or 25 g of monkey dung, Andresen (2002a) and Andresen and Levey (2004) showed that seeds and beads surrounded by more dung were buried more often and more deeply by dung beetles, than seeds surrounded by less dung. Thus, seeds embedded in larger amounts of dung are likely to suffer high non-emergence but low predation rates, while seeds embedded in less dung are likely to suffer low non-emergence but high predation rates. These opposite effects might often compensate each other, and dung amount will have no significant effect on the long-term fate of seeds, with the same number of seedlings establishing from seeds embedded in more dung as from seeds embedded in less dung (e.g. Andresen, 2001).

Two other factors related to defecation patterns that could potentially affect beetle behaviour and seed fate are spatial distribution (scattered versus clumped defecations) and seed density. Data gathered thus far have found no effect of these factors (Feer, 1999; Andresen, 2002b), but more experiments should be conducted before a generalization can be made.

Time of day

The composition and structure of diurnal and nocturnal dung beetle assemblages in most habitats are quite distinct. For example, in tropical forests the nocturnal dung beetle assemblages consist of more individuals (although not necessarily more species), larger average size of species and a higher relative proportion of tunneller beetles than the diurnal assemblage (Hanski and Cambefort, 1991d; Halffter et al., 1992; Feer, 2000; Andresen, 2002a). Thus, it is not surprising that in four independent experiments Andresen (2002a) found that more seeds and beads were buried, and at greater depths, when they were handled by the nocturnal assemblage of dung beetles than when they were handled by the diurnal assemblage. That study also reported a significant effect of time of day on dung removal rates, with more dung removed

during the nocturnal period than during the diurnal period. Similarly, Hingrat and Feer (2002) found a positive correlation between seed burial rate and the weighted abundance of the large nocturnal tunnellers.

Season

Andresen (2002b) found a significant difference in short-term seed fate between dry and wet seasons in a Brazilian rainforest: more seeds and beads were buried during the wet season (39%) than during the dry season (32%). Burial depths, however, were the same in both seasons. That study also found that significantly fewer dung piles were completely removed after 24 h during the dry season than during the wet season. These differences were detected even though in this forest, as in other forests with weakly pronounced seasons, the composition and abundance of the dung beetle assemblages change little between seasons (Peck and Forsyth, 1982; Andresen, 2002b).

Habitat disturbance

The structure of dung beetle communities is greatly affected by environmental characteristics such as vegetation cover, soil characteristics and climatic conditions (Janzen, 1983; Hanski and Cambefort, 1991d; Halffter et al., 1992; Osberg et al., 1994; Davis, 1996; Escobar, 2000). Also, because dung beetles depend on other animals for dung, their populations are affected by changes in the populations of other animals, in particular mammals (Lumaret and Kirk, 1991; Lumaret et al., 1992; Estrada et al., 1999). Consequently, dung beetles respond to biotic and abiotic changes in their habitat, be these natural or human-induced disturbances (Janzen, 1983; Klein, 1989; Amezquita et al., 1999; Vulinec, 2002). Changes in a dung beetle community due to habitat disturbance can occur in species richness, species composition, temporal structure (diurnal versus nocturnal species), and functional structure (rollers versus tunnellers versus dwellers). In fact, dung beetles have been used as bio-indicators of forest disturbance in several tropical forests (Nummelin and Hanski, 1989; Halffter *et al.*, 1992; Halffter and Favila, 1993; Estrada *et al.*, 1998; Davis, 2000; Davis *et al.*, 2000, 2001).

If dung beetle assemblages are affected by habitat disturbances, their ecological importance as secondary seed dispersers will probably also be affected. However, functional or second-order effects of habitat disturbance are rarely measured directly; rather they are inferred based on data of first-order effects on species richness and abundance (Didham, 1996). Thus, while several studies have measured the effects of habitat disturbances such as forest frag-mentation on the structure of dung beetle communities, and indirectly inferred detri-mental effects for secondary seed dispersal and plant regeneration (e.g. Estrada *et al.*, 1998, 1999; Vulinec, 2000, 2002), only two studies have directly measured the effects of disturbance on secondary dispersal by dung beetles.

In a study in Brazil, in which seed fate was compared in continuous forest, 1 ha forest fragments, and 10 ha forest fragments, Andresen (2003) found that a significantly higher proportion of seeds were buried by dung beetles in continuous forest than in forest fragments for two of the three species. However, no significant differences were found among sites in the total number of seedlings establishing from all seeds (buried and not buried). None the less, for one of the plant species, the difference between seed-ling establishment from buried seeds com-pared to surface seeds was much greater in the forest fragments than in the continuous forest. Thus, seed burial seems to be even more advantageous in fragments than in continuous forest, most likely due to very high predation pressures on surface seeds in fragments (Andresen, 2003), but more experiments need to be conducted to confirm this. In a second study in French Guiana, Hingrat and Feer (2002) found no difference in the percentage of seeds of one tree species buried by beetles in continuous forest versus islands in an artificially flooded landscape, even though the lowest seed-burial rates were recorded on some of the smallest islands. Similarly, they did not find differences in seed predation by rodents among the habitats.

Thus, even though is seems reasonable to conclude that the impact of dung beetles on seed dispersal will be diminished if the dung beetle community is affected negatively by forest disturbance, this needs to be corroborated directly via observational or experimental studies. Further, as shown by the study in Brazil (Andresen, 2003), it is not enough to simply quantify secondary seed dispersal by beetles, but one must also follow the fate of dispersed and undispersed seeds until seedling establishment.

Concluding Remarks

How prevalent is the dung-beetle–seed interaction in tropical rainforests? In these forests, the biomass of frugivorous animals constitutes a large proportion of the total vertebrate biomass. Also, most tree species produce fleshy fruits adapted for animal consumption and primary seed dispersal through defecation. Dung beetle communi-ties in these forests can be tremendously rich both in terms of species and indivi-duals. And, finally, given that the dung of vertebrates, particularly that of herbivores (*sensu lato*), is the preferred resource for most dung beetle species, it is safe to say that the dung-beetle–seed interaction is widespread in most undisturbed tropical rainforests.

How important is secondary dispersal of seeds by beetles, relative to primary dis-persal? Primary dispersal through defeca-tion by mammals if often characterized by seeds being deposited well away from the parent tree, but also by seeds being depos-ited in clumps of several to many seeds surrounded by faecal material (Howe, 1989; Andresen, 2000b). Although deposition away from the parent tree is beneficial for many plant species due to higher per capita seed and/or seedling survival at some distance away from the parent tree (Harms *et al.*, 2000), both the clumping of seeds and the presence of dung are known to have potentially negative effects on plant fitness. Seed clumping can cause reduced seed or seedling survival due to the activity of density-dependent predators or pathogens, and it can cause reduced seedling survival due to intense competition. Dung presence can cause reduced seed survival because it attracts seed predators, and because it can favour fungal attack. Thus, although most primary dispersal by mammals will successfully deposit seeds in a site with higher survival and establishment probabilities, relative to the immediate vicinity of the parent tree, dung beetles will further increase survival and establishment probabilities by scattering and burying seeds and dung.

Finally, what role do dung beetles play in the evolution of plant traits? Dung beetles have probably interacted with seeds present in dung for a very long time. Just in the way beetles move and bury seeds embedded in the dung of present-day frugivorous animals, millions of years ago they moved and buried seeds embedded in the dung of frugivorous dinosaurs (Chin and Gill, 1996). Thus, it is tempting to propose that, among the many evolutionary pressures shaping plant traits, secondary seed dispersal by dung beetles may also have played a role. For example, Shepherd and Chapman (1998) suggested that, if dung beetles consistently reduced the variability in seed survival by always depositing seeds in safer sites, then they may be favouring directional selection of seed size, towards a size that increases the probability of burial by dung beetles, i.e. towards smaller seed sizes. However, if larger seeds are better able to establish seedlings from seeds buried by dung beetles (because they have larger reserves or because they are buried less deeply) than smaller seeds, this should also act as a directional selection force, but in the opposite direction: towards larger seed sizes. Alternatively, these opposing effects of burial by dung beetles, increased seed survival and decreased seedling emergence could select in favour of increased seed size variation, which has been proposed as an adaptive

strategy in plants (e.g. Moegenburg, 1996; Geritz, 1998).

Even though the evolutionary effect of dung beetles on seeds is not yet clear, the ecological effects have began to be demonstrated. Many aspects of this plant–animal interaction still have to be elucidated, but it is clear that it is not a simple one. Not only do many biotic and abiotic factors play a role in determining the outcome of the beetle–seed interaction, but other organisms may have synergistic or antagonistic effects, forming a complex web of interactions (Andresen and Levey, 2004).

Acknowledgements

We would like to thank Britta Kunz, Trond Larsen and two anonymous reviewers for their insightful comments on the manuscript.

References

Amezquita, S.J.M., Forsyth, A., Lopera, A.T. and Camacho, A.M. (1999) Comparison of the composition and species richness of dung beetles (Coleoptera: Scarabaeidae) in forest remnants in the Colombian Orinoquia. *Acta Zoologica Mexicana Nueva Serie* 76, 113–126.

Andresen, E. (1999) Seed dispersal by monkeys and the fate of dispersed seeds in a Peruvian rainforest. *Biotropica* 31, 145–158.

Andresen, E. (2000a) The role of dung beetles in the regeneration of rainforest plants in Central Amazonia. PhD thesis, University of Florida, Gainesville, Florida.

Andresen, E. (2000b) Ecological roles of mammals: the case of seed dispersal. In: Entwistle, A. and Dunstone, N. (eds) *Future Priorities for the Conservation of Mammalian Diversity: Has the Panda had its Day?* Cambridge University Press, Cambridge, pp. 11–25.

Andresen, E. (2001) Effects of dung presence, dung amount, and secondary dispersal by dung beetles on the fate of *Micropholis guyanensis* (Sapotaceae) seeds in Central Amazonia. *Journal of Tropical Ecology* 17, 61–78.

Andresen, E. (2002a) Dung beetles in a Central Amazonian rainforest and their ecological role as secondary seed dispersers. *Ecological Entomology* 27, 257–270.

Andresen, E. (2002b) Primary seed dispersal by red howler monkeys and the effect of defecation pattern on the fate of dispersed seeds. *Biotropica* 34, 261–272.

Andresen, E. (2003) Effect of forest fragmentation on dung beetle communities and functional consequences for plant regeneration. *Ecography* 26, 87–97.

Andresen, E. and Levey, D.J. (2004) Effects of dung and seed size on secondary dispersal, seed predation, and seedling establishment of rain forest trees. *Oecologia* 139, 145–154.

Bergstrom, B.C., Maki, L.R. and Werner, B.A. (1976) Small dung beetles as biological control agents: laboratory studies of beetle action on trichostrongylid eggs in sheep and cattle feces. *Proceedings of the Helminthology Society of Washington* 43, 171–174.

Böhning-Gaese, K., Gaese, B.H. and Rabemanantsoa, S.B. (1999) Importance of primary and secondary seed dispersal in the Malagasy tree *Commiphora guillaumini*. *Ecology* 80, 821–832.

Byrne, M.M. and Levey, D.J. (1993) Removal of seeds from frugivore defecations by ants in a Costa Rican rain forest. *Vegetatio* 107/108, 363–374.

Cambefort, Y. and Hanski, I. (1991) Dung beetle population biology. In: Hanski, I. and Cambefort, Y. (eds) *Dung Beetle Ecology*. Princeton University Press, Princeton, New Jersey, pp. 36–50.

Chambers, J.C. and MacMahon, J.A. (1994) A day in the life of a seed: movements and fates of seeds an their implications for natural and managed systems. *Annual Review of Ecology and Systematics* 25, 263–292.

Chen, H. and Maun, M.A. (1999) Effects of sand burial depth on seed germination and seedling emergence of *Cirsium pitcheri*. *Plant Ecology* 140, 53–60.

Chin, K. and Gill, G.D. (1996) Dinosaurs, dung beetles, and conifers: participants in a Cretaceous food web. *Palaios* 11, 280–285.

Crawley, M.J. (2000) Seed predators and plant population dynamics. In: Fenner, M. (ed.) *Seeds. The Ecology of Regeneration in Plant Communities*, 2nd edn. CAB International, Wallingford, UK, pp. 167–182.

Dalling, J.W., Swaine, M.D. and Garwood, N.C. (1994) Effect of soil depth on seedling emergence in tropical soil seed-bank investigations. *Functional Ecology* 9, 119–121.

Davis, A.J. (2000) Does reduced-impact logging help preserve biodiversity in tropical rainforests? A case study from Borneo using dung beetles (Coleoptera: Scarabaeoidea) as indicators. *Environmental Entomology* 29, 467–475.

Davis, A.J., Huijbregts, H. and Krikken, J. (2000) The role of local and regional processes in shaping dung beetle communities in tropical forest plantations in Borneo. *Global Ecology and Biogeography* 9, 281–292.

Davis, A.J., Holloway, J.D., Huijbregts, H., Kirk-Spriggs, A.H. and Sutton, S.L. (2001) Dung beetles as indicators of change in the forests of northern Borneo. *Journal of Applied Ecology* 38, 593–616.

Davis, A.L. (1996) Community organization of dung beetles (Coleoptera: Scarabaeidae): differences in body size and functional group structure between habitats. *African Journal of Ecology* 34, 258–275.

Didham, R.K. (1996) Insects in fragmented forests: a functional approach. *Trends in Ecology and Evolution* 11, 255–260.

Doube, B.M. (1990) A functional classification analysis of the structure of dung beetle assemblages. *Ecological Entomology* 15, 371–383.

Doube, B.M. (1991) Dung beetles of Southern Africa. In: Hanski, I. and Cambefort, Y. (eds) *Dung Beetle Ecology*. Princeton University Press, Princeton, New Jersey, pp. 133–155.

Engel, T.R. (2000) *Seed Dispersal and Forest Regeneration in a Tropical Lowland Biocenosis (Shimba Hills, Kenya)*. Logos Verlag, Berlin, 344 pp.

Escobar, S.F. (2000) The diversity of dung beetles (Scarabaeidae, Scarabaeinae) in different habitats in the Nukak Natural Reserve, Guaviare department, Colombia. *Acta Zoologica Mexicana Nueva Serie* 79, 103–121.

Estrada, A. and Coates-Estrada, R. (1991) Howler monkeys (*Alouatta palliata*), dung beetles (Scarabaeidae) and seed dispersal: ecological interactions in the tropical rain forest of Los Tuxtlas, Mexico. *Journal of Tropical Ecology* 7, 459–474.

Estrada, A., Halffter, G., Coates-Estrada, R. and Meritt, D.A. Jr (1993) Dung beetles attracted to mammalian herbivore (*Alouatta palliata*) and omnivore (*Nasua narica*) dung in the tropical rain forest of Los Tuxtlas, Mexico. *Journal of Tropical Ecology* 9, 45–54.

Estrada, A., Coates-Estrada, R., Anzures Dadda, A. and Cammarano, P. (1998) Dung and carrion beetles in tropical rain forest fragments and agricultural habitats at Los Tuxtlas, Mexico. *Journal of Tropical Ecology* 14, 577–593.

Estrada, A., Anzures, D.A. and Coates-Estrada, R. (1999) Tropical rain forest fragmentation, howler monkeys (*Alouatta palliata*), and dung beetles at Los Tuxtlas, Mexico. *American Journal of Primatology* 48, 253–262.

Feer, F. (1999) Effects of dung beetles (Scarabaeidae) on seeds dispersed by howler monkeys (*Alouatta seniculus*) in the French Guianan rain forest. *Journal of Tropical Ecology* 15, 129–142.

Feer, F. (2000) Les coléoptères coprophages et nécrophages (Scarabaeidae s. str. et Aphodidae) de la forêt de Guyane française: composition spécifique et structure des peuplements. *Annales de la Société Entomologique de France* 36, 119–145.

Fenner, M. (1987) Seedlings. *New Phytologist* 106 (Supplement), 35–47.

Fenner, M. (2000) *Seeds. The Ecology of Regeneration in Plant Communities*, 2nd edn. CAB International, Wallingford, UK, 410 pp.

Forget, P.-M. and Milleron, T. (1991) Evidence for secondary seed dispersal by rodents in Panama. *Oecologia* 87, 596–599.

Forget, P.-M., Milleron, T. and Feer, F. (1998) Patterns in post-dispersal seed removal by Neotropical rodents and seed fate in relation to seed size. In: Newbery, D.M., Prins, H.H.T. and Brown, N.D. (eds) *Dynamics of Tropical Communities*. Blackwell Science Ltd, Oxford, pp. 25–49.

Foster, S.A. (1986) On the adaptive value of large seeds for tropical moist forest trees: a review and synthesis. *The Botanical Review* 52, 260–299.

Garwood, N.C. (1996) Functional morphology of tropical tree seedlings. In: Swaine, M.D. (ed.) *The Ecology of Tropical Forest Tree Seedlings*. UNESCO and the Parthenon Publishing Group, Paris, pp. 59–130.

Geritz, S.A.H. (1998) Co-evolution of seed size and seed predation. *Evolutionary Ecology* 12, 891–911.

Gill, B. (1991) Dung beetles in tropical American forests. In: Hanski, I. and Cambefort, Y. (eds) *Dung Beetle Ecology*. Princeton University Press, Princeton, New Jersey, pp. 211–229.

Grant, J.D. (1983) The activities of earthworms and the fates of seeds. In: Satchell, J.E. (ed.) *Earthworm Ecology*. Chapman & Hall, London, pp. 107–122.

Halffter, G. and Edmonds, W.D. (1982) *The Nesting Behavior of Dung Beetles (Scarabaeinae). An Ecological and Evolutive Approach.* Instituto de Ecologia, Mexico DF, 176 pp.

Halffter, G. and Favila, M.E. (1993) The Scarabaeinae (Insecta: Coleoptera) an animal group

for analysing, inventoring and monitoring biodiversity in tropical rainforest and modified landscapes. *Biology International* 27, 15–21.

Halffter, G., Favila, M.E. and Halffter, V. (1992) A comparative study of the structure of the scarab guild in Mexican tropical rain forest and derived ecosystems. *Folia Entomologica Mexicana* 84, 131–156.

Hanski, I. (1989) Dung beetles. In: Lieth, H. and Werger, M.J.A. (eds) *Tropical Rain Forest Ecosystems*. Elsevier, Amsterdam, pp. 489–511.

Hanski, I. and Cambefort, Y. (1991a) *Dung Beetle Ecology*. Princeton University Press, Princeton, New Jersey, 463 pp.

Hanski, I. and Cambefort, Y. (1991b) Competition in dung beetles. In: Hanski, I. and Cambefort, Y. (eds) *Dung Beetle Ecology*. Princeton University Press, Princeton, New Jersey, pp. 305–329.

Hanski, I. and Cambefort, Y. (1991c) Spatial processes. In: Hanski, I. and Cambefort, Y. (eds) *Dung Beetle Ecology*. Princeton University Press, Princeton, New Jersey, pp. 283–304.

Hanski, I. and Cambefort, Y. (1991d) Resource partitioning. In: Hanski, I. and Cambefort, Y. (eds) *Dung Beetle Ecology*. Princeton University Press, Princeton, New Jersey, pp. 330–349.

Harms, K.E., Wright, S.J., Calderon, O., Hernandez, A. and Herre, E.A. (2000) Pervasive density-dependent recruitment enhances seedling diversity in a tropical forest. *Nature* 404, 493–495.

Hingrat, Y. and Feer, F. (2002) Effets de la fragmentation forestière sur l'activité des coléoptères coprophages: dispersion secondaire des graines en Guyane Française. *Revue d'Ecologie (la Terre et la Vie)* 57, 165–179.

Howe, H.F. (1989) Scatter- and clump-dispersal and seedling demography: hypothesis and implications. *Oecologia* 79, 417–426.

Hulme, P.E. and Borelli, T. (1999) Variability in post-dispersal seed predation in deciduous woodland: relative importance of location, seed species, burial and density. *Plant Ecology* 145, 149–156.

Ibarra-Manríquez, G., Martínez Ramos, M. and Oyama, K. (2001) Seedling functional types in a lowland rain forest in Mexico. *American Journal of Botany* 88, 1801–1812.

Janzen, D.H. (1971) Seed predation by animals. *Annual Review of Ecology and Systematics* 2, 465–492.

Janzen, D.H. (1982) Removal of seeds from horse dung by tropical rodents: influence of habitat and amount of dung. *Ecology* 63, 1887–1900.

Janzen, D.H. (1983) Seasonal change in abundance of large nocturnal dung beetles (Scarabaeidae) in a Costa Rican deciduous forest and adjacent horse pasture. *Oikos* 41, 274–283.

Janzen, D.H. (1986) Mice, big mammals and seeds: it matters who defecates what where. In: Estrada, A. and Fleming, T.H. (eds) *Frugivores and Seed Dispersal.* Dr W. Junk Publishers, Dordrecht, The Netherlands, pp. 314–338.

Johnson, T. and Jorgensen, C.D. (1981) Ability of desert rodents to find buried seeds. *Journal of Range Management* 34, 312–314.

Jones, M.B. (1994) Secondary seed removal by ants, beetles, and rodents in a Neotropical moist forest. MSc thesis, University of Florida, Gainesville, Florida.

Julliot, C. (1997) Impact of seed dispersal by red howler monkeys *Alouatta seniculus* on the seedling population in the understory of tropical rain forest. *Journal of Ecology* 85, 431–440.

Kitajima, K. (1996) Cotyledon functional morphology, patterns of seed reserve utilization and regeneration niches of tropical tree seedlings. In: Swaine, M.D. (ed.) *The Ecology of Tropical Forest Tree Seedlings.* UNESCO and the Parthenon Publishing Group, Paris, pp. 193–210.

Kitajima, K. and Fenner, M. (2000) Ecology of seedling regeneration. In: Fenner, M. (ed.) *Seeds. The Ecology of Regeneration in Plant Communities,* 2nd edn. CAB International, Wallingford, UK, pp. 331–359.

Klein, B.C. (1989) Effect of forest fragmentation on dung and carrion beetle communities in central Amazonia. *Ecology* 70, 1715–1725.

Leishman, M.R., Wright, I.J., Moles, A.T. and Westoby, M. (2000) Ecology of seedling regeneration. In: Fenner, M. (ed.) *Seeds. The Ecology of Regeneration in Plant Communities,* 2nd edn. CAB International, Wallingford, UK, pp. 31–57.

Levey, D.J. and Byrne, M.M. (1993) Complex ant–plant interactions: rain forest ants as secondary dispersers and post-dispersal seed predators. *Ecology* 74, 1802–1812.

Lumaret, J.-P. and Kirk, A.A. (1991) South temperate dung beetles. In: Hanski, I. and Cambefort, Y. (eds) *Dung Beetle Ecology.* Princeton University Press, Princeton, New Jersey, pp. 97–115.

Lumaret, J.P., Kadiri, N. and Bertrand, M. (1992) Changes in resources: consequences for the dynamics of dung beetle communities. *Journal of Applied Ecology* 29, 349–356.

McConkey, K.R. (2000) Primary seed shadow generated by gibbons in the rain forests of Barito Ulu, Central Borneo. *American Journal of Primatology* 52, 13–29.

McKinney, G.T. and Morley, F.H.W. (1975) The agronomic role of introduced dung beetles in grazing systems. *Journal of Applied Ecology* 12, 831–837.

Miranda, C.H.B., dos Santos, J.C.C. and Bianchin, I. (1998) Contribution of *Onthophagus gazella* to soil fertility improvement by bovine fecal mass incorporation into the soil. 1. Greenhouse studies. *Revista Brasileira de Zootecnia* 27, 681–685.

Mittal, I.C. (1993) Natural manuring and soil conditioning by dung beetles. *Tropical Ecology* 34, 150–159.

Moegenburg, S.M. (1996) *Sabal palmetto* seed size: causes of variation, choices of predators, and consequences for seedlings. *Oecologia* 106, 539–543.

Nealis, V.G. (1977) Habitat associations and community analysis of south Texas dung beetles (Coleoptera: Scarabaeinae). *Canadian Journal of Zoology* 55, 138–147.

Nummelin, M. and Hanski, I. (1989) Dung beetles of the Kibale forest, Uganda; comparison between virgin and managed forests. *Journal of Tropical Ecology* 5, 349–352.

Osberg, D.C., Doube, B.M. and Hanrahan, S.A. (1994) Habitat specificity in African dung beetles: the effect of soil type on the survival of dung beetle immatures (Coleoptera Scarabaeidae). *Tropical Zoology* 7, 1–10.

Pearson, T.R.H., Burslem, D.F.R.P., Mullins, C.E. and Dalling, J.W. (2002) Germination ecology of Neotropical pioneers: interacting effects of environmental conditions and seed size. *Ecology* 83, 2798–2807.

Peck, S.B. and Forsyth, A. (1982) Composition, structure, and competitive behaviour in a guild of Ecuadorian rain forest dung beetles (Coleoptera; Scarabaeidae). *Canadian Journal of Zoology* 60, 1624–1634.

Peck, S.B. and Howden, H.F. (1984) Response of a dung beetle guild to different sizes of dung bait in a Panamanian rainforest. *Biotropica* 16, 235–238.

Pons, T.L. (2000) Seed responses to light. In: Fenner, M. (ed.) *Seeds. The Ecology of Regeneration in Plant Communities.* CAB International, Wallingford, UK, pp. 237–260.

Price, M.V. and Jenkins, S.H. (1986) Rodents as seed consumers and dispersers. In: Murray, D.R. (ed.) *Seed Dispersal*. Academic Press, Sydney, pp. 191–235.

Putz, F.E. (1983) Treefall pits and mounds, buried seeds, and the importance of soil disturbance to pioneer trees on Barro Colorado Island, Panama. *Ecology* 64, 1069–1074.

Schupp, E.W. (1993) Quantity, quality and the effectiveness of seed dispersal by animals. *Vegetatio* 107/108, 15–29.

Shepherd, V.E. and Chapman, C.A. (1998) Dung beetles as secondary seed dispersers: impact on seed predation and germination. *Journal of Tropical Ecology* 14, 199–215.

Terborgh, J. (1986) Community aspects of frugivory in tropical forests. In: Estrada, A. and Fleming, T.H. (eds) *Frugivores and Seed Dispersal*. Dr W. Junk Publishers, Dordrecht, The Netherlands, pp. 371–384.

Vander Wall, S.B. (1993) A model of caching depth: implications for scatter hoarders and plant dispersal. *The American Naturalist* 141, 217–232.

Vander Wall, S.B. (1998) Foraging success of granivorous rodents: effects of variation in seed and soil water on olfaction. *Ecology* 79, 233–241.

Vulinec, K. (1999) Dung beetles, monkeys, and seed dispersal in the Brazilian Amazon. PhD thesis, University of Florida, Gainesville, Florida.

Vulinec, K. (2000) Dung beetles (Coleoptera: Scarabeidae), monkeys, and conservation in Amazonia. *Florida Entomologist* 83, 229–241.

Vulinec, K. (2002) Dung beetle communities and seed dispersal in primary forest and disturbed land in Amazonia. *Biotropica* 34, 297–309.

Westoby, M., Jurado, E. and Leishman, M. (1992) Comparative evolutionary ecology of seed size. *Trends in Ecology and Evolution* 7, 368–372.

Willems, J.H. and Huijsmans, K.G.A. (1994) Vertical seed dispersal by earthworms: a quantitative approach. *Ecography* 17, 124–130.

Young, O.P. (1981) The attraction of Neotropical scarabaeinae (Coleoptera: Scarabaeidae) to reptile and amphibian fecal material. *The Coleopterists Bulletin* 35, 345–348.

Zhang, S.Y. and Wang, L.-X. (1995) Fruit consumption and seed dispersal of *Ziziphus cinnamomum* (Rhamnaceae) by two sympatric primates (*Cebus apella* and *Ateles paniscus*) in French Guiana. *Biotropica* 27, 397–401.

21 Post-dispersal Seed Fate of Some Cloud Forest Tree Species in Costa Rica

Daniel G. Wenny

Illinois Natural History Survey, Lost Mound Field Station, 3159 Crim Drive, Savanna, IL 61074, USA

Introduction

Research on post-dispersal seed fates has expanded rapidly in the past 20 years. The finding that rodents worldwide scatter-hoard many types of seeds and that such scatterhoarding is a second stage of dispersal for some plants has led to the realization that one cannot assume that seed removal is equivalent to seed predation (Forget *et al.*, 1998; Jansen, 2003). Several studies indicate that secondary dispersal can be an important phase of plant recruitment (Forget and Milleron, 1991; Vander Wall, 1992; Forget, 1993; Levey and Byrne, 1993; Bohning-Gaese *et al.*, 1999; Brewer and Rejmanek, 1999; Hoshizaki *et al.*, 1999). Therefore, simply observing post-dispersal seed removal is insufficient evidence to assess the importance of primary dispersal patterns unless the fate of removed seeds is also assessed. A variety of organisms, ranging from rodents to beetles to micro-organisms, interact with dispersed seeds but the roles they play are poorly known (Steele *et al.*, 1996).

Despite several well-documented examples of secondary dispersal (Vander Wall and Longland, 2004), the extent to which most vertebrate-dispersed plant species benefit from secondary dispersal is unknown. In a given habitat with dozens to hundreds of plant species, it is difficult to predict which species might benefit from secondary dispersal because we lack analyses of plant characteristics that may be associated with secondary dispersal syndromes. Perhaps the best example of a secondary dispersal syndrome is the presence of ant-attracting elaiosomes on seeds with ballistic primary dispersal (Passos and Ferreira, 1996; Gomez and Espadaler, 1998). For trees and other woody species, the occurrence of secondary dispersal syndromes is much less clear. Dung beetles act as secondary dispersers for seeds in mammalian dung but apparently not for seeds regurgitated or defecated by birds (Andresen and Feer, Chapter 20, this volume). As dung beetles are selecting dung rather than seeds, a syndrome for this type of secondary dispersal is unlikely. Rodents are important seed predators and dispersers worldwide, and defining a syndrome for rodent dispersal may be possible (Vander Wall and Longland, Chapter 18, this volume). Caviomorph rodents that act as both primary and secondary dispersers tend to scatterhoard large seeds (> 1 g) that have some physical defence such as a thick or hard seed coat. Smaller rodents eat a wide variety of smaller seeds but the extent of scatterhoarding in these species is poorly known (Price and Jenkins, 1986; Brewer and Rejmanek, 1999; Vander Wall *et al.*, 2001).

Although secondary dispersal syndromes are beginning to be defined, more information is needed on the fate of different kinds of seeds in a variety of habitats.

In this study, I examined seed fate in several vertebrate-dispersed plant species in a montane cloud forest in Costa Rica. I focused on seeds that receive primary dispersal from frugivorous birds and arboreal mammals with the expectation that some seed species would be secondarily dispersed by terrestrial seed-caching rodents. Most studies on post-dispersal seed removal use artificially dispersed seeds placed in a specific arrangement in the field (Forget and Wenny, Chapter 23, this volume). In this study, however, I used naturally dispersed seeds at their original dispersal locations to monitor subsequent seed fate. My intent was to characterize post-dispersal fate of a variety of large-seeded, vertebrate-dispersed plant species, particularly lauraceous trees, for comparison with an ongoing study on *Ocotea endresiana* (Wenny, 2000b). The specific goals were to determine which plant species benefited from secondary dispersal, which animals provided it and which, if any, seed characteristics could be used to predict the likelihood of secondary dispersal by seed-caching rodents.

Methods

Study site

This study was conducted from April 1994 to November 1995 in the Monteverde Cloud Forest Preserve (10°12′N, 84°42′W) in the Cordillera de Tilaran in north-western Costa Rica. The Tropical Science Centre of San José, Costa Rica, administers this reserve. The average annual rainfall is about 2500 mm, with most occurring between May and November (Clark *et al.*, 2000). Additional precipitation from mist and cloud interception is probably substantial (Clark and Nadkarni, 2000). The study area is in relatively undisturbed lower montane rainforest (Haber, 2000a) along the continental divide at 1600 m elevation. A

5-ha area 500 m from the beginning of the Valley Trail (Sendero El Valle) was mapped and marked into a 10 m × 10 m grid with corners marked with PVC pipes. The vegetation and ecology of the area is described in more detail elsewhere (Lawton and Dryer, 1980; Nadkarni and Wheelwright, 2000).

Seed fate

I located naturally dispersed seeds in April–August 1994 and *Beilschmeidia costaricensis* seeds from February to March 1995 by systematically searching the ground for freshly regurgitated, dropped, or defecated seeds. The searches proceeded within 10-m wide belt transects delineated by the PVC markers. It was impossible to search the entire site with equal intensity, but an effort was made to cover the entire site at least once every 2–3 weeks. Seeds with a damaged seed coat (by gnawing of squirrels or other animals) directly under fruiting trees were not included, nor were fallen or dropped fruits because the emphasis of the study was determining the occurrence of secondary dispersal after endozoochory by birds and arboreal mammals. Typically, recently deposited seeds could be distinguished from older seeds, and only recently dispersed seeds were used in this study. Each seed was weighed, length and diameter measured, and any signs of insect attack (presumably pre-dispersal oviposition) were noted.

I marked seeds by gluing 50–75 cm of unwaxed dental floss to them, and tying about 50 cm of flagging tape to the floss. Flagging tape was pink, orange, or white and yellow striped depending on the species. Seeds were allowed to air-dry for 1–3 h in a lab room before gluing. Each marked seed was returned to its original location the next morning. A pilot study at this study site indicated no difference in seed removal rates for marked and unmarked *O. endresiana* seeds (Wenny, 2000b). In addition, much of the seed removal at this site takes place at night when visual cues are less likely to bias seed removal (Wenny, 2000b). Similar marking procedures have been used in several other Neotropical sites with no evidence of effects on seed removal (Schupp, 1988; Forget, 1996; Peres *et al.*, 1997; Brewer and Rejmanek, 1999). The influence of marking systems on retrieval of cached seeds, however, remains unexplored.

We conducted censuses weekly for 12–16 weeks. Because new seeds were found throughout the study period, the total census period for each seed varied but was at least 12 weeks for seeds that remained intact for that long. If a marked seed was removed, the surrounding area was searched until the flagging tape–dental floss assembly was found. The distance from the original location was measured and the fate of the seed was characterized by examining any seed remains, the condition of the line glued to the seed, and the surrounding area. Seed fate was classified as seed predation if a seed was removed from the line and pieces of the seed or seed coat remained. Seeds that had been moved from the original location but remained intact were considered removed. If a seed was buried in the soil it was classified as cached by rodents. Seeds covered by leaves were not considered cached because natural leaf fall was responsible for such situations (D. Wenny, unpublished data). At the end of the study, all remaining seeds were examined and classified as either viable if the seed appeared healthy, or not viable if it was infested with insects or was hollow, mealy, discoloured or rotten.

Rodent trapping

I used Sherman live traps (23.5 cm × 8.5 cm × 9 cm) to capture and identify small rodents. Nine sets of three traps were set for 3 nights each month for 8 months between September 1993 and June 1994. Each set of three traps had one trap at the base of a tree, one beside a fallen log and one in a treefall gap. Rodents were identified, measured, marked and released. Traps were baited with seeds or peanut butter and oatmeal in alternate trapping sessions. Traps were opened in the afternoon and

checked the next morning. Traps were washed and rebaited each day during a trapping session so that rodent odours did not influence capture rates.

Tracking stations

I used muddy sections of trails as tracking stations to record presence of animals too large to capture in the live traps and too difficult to observe otherwise. In five different places, an area of mud ≈ 2 m × 1 m was levelled and checked daily. The tracking stations were smoothed at least once each month. This procedure was not intended to estimate population sizes but only to note the occurrence of species not observed or captured during the course of this study.

Results

Seed and fruit characteristics

Plant species for which I found at least 20 seeds and with seeds large enough to mark were included in this analysis. Overall, 1175 seeds from 14 species in nine families were studied (Table 21.1). Seed size ranged from 0.1 g to 31.5 g. In this study the five species weighing < 1 g are considered small, six species weighing 1–5 g are considered intermediate size and three species weighing over 5 g are considered large. Although the mass of species in this study span a wide range, they represent the upper range of the seed sizes in the Monteverde area. Over 75% of 114 mainly bird-dispersed species for which data are available weigh less than 0.5 g (Wheelwright *et al.*, 1984). In this study seed size of the 14 species averaged 5.1 ± 8.3 g (mean $\pm$ 1 SD) with a median of 2.1 g.

Fruits of most of the focal species are single-seeded drupes or berries. *Guarea* spp. fruits are dehiscent capsules typically with four arillate seeds, and fruits of *Salacia* sp. are large indehiscent, odoriferous berries with three or more seeds. *Meliosma vernicosa* and *Chione sylvicola* seeds had hard

Table 21.1. Plant species included in this study. Dispersers were assigned according to Haber (2000b) and based on observations during this study ('mammals' indicates arboreal non-volant mammals). Physical defence refers to the type of seed coat (testa) or endocarp and is relative to the other species in this study.

Species (Family)	Disperser	Seed mass (g) mean ± SD	Physical defence
Ardisia palmana (Myrsinaceae)	Birds	0.1 ± 0.03	Hard testa
Geonoma edulis (Aracaceae)	Birds	0.2 ± 0.06	Hard testa
Chione sylvicola (Rubiaceae)	Birds	0.4 ± 0.10	Two-seeded stone
Ocotea endresiana (Lauraceae)	Birds	0.8 ± 0.13	None
Guarea glabra (Meliaceae)	Birds	0.8 ± 0.21	Hard testa
Prunus annularis (Rosaceae)	Birds	1.3 ± 0.30	None
Meliosma vernicosa (Sabiaceae)	Bats, birds	1.8 ± 0.31	Thick testa
Eugenia guatemalensis (Myrtaceae)	Bats, birds	2.3 ± 0.62	None
Ocotea meziana (Lauraceae)	Birds	3.2 ± 0.65	None
Guarea kunthiana (Meliaceae)	Birds	4.5 ± 2.1	Hard testa
Pleurothyrium palmanum (Lauraceae)	Birds	4.5 ± 1.4	None
Salacia sp. (Hippocrateaceae)	Mammals	7.6 ± 1.6	None
Beilschmiedia costaricensis (Lauraceae)	Birds	12.6 ± 3.2	Thick endocarp
Persea sp. (Lauraceae)	Mammals	31.5 ± 6.4	None

and thick seeds that could not be cut with a pocket knife. *Ardisia palmana*, *Geonoma edulis* and *Guarea kunthiana* seeds were fairly hard but could be cut with difficulty. *Guarea glabra* seeds were easier to cut than *G. kunthiana*. *Beilschmiedia costaricensis* fruits have a thickened endocarp (unlike all other Lauraceae in this study) that remains intact after removal of the pulp by frugivores. The remaining eight species of seeds have essentially no physical defence.

Primary dispersal

Most of the focal species are eaten and dispersed by birds (Haber, 2000b). Birds tend to regurgitate intermediate and large seeds. Regurgitated seeds are usually free of pulp and typically found singly or in loose aggregations depending on the foraging and perching behaviour of the dispersers (Wenny and Levey, 1998). Small seeds in this study were regurgitated by some bird species and defecated by others. Black guans (*Chamaepetes unicolor*) were not observed to regurgitate any seeds. Other important avian seed dispersers at this study site include resplendent quetzal

(*Pharomacrus mocinno*), orange-bellied trogon (*Trogon aurantiiventris*), emerald toucanet (*Aulacorhynchus prasinus*), three-wattled bellbird (*Procnias tricarunculata*), black-faced solitaire (*Myadestes melanops*) and mountain robin (*Turdus plebejus*).

Bats and arboreal mammals often eat the pulp and drop seeds containing remnants of pulp. *Eugenia guatemalensis* and *M. vernicosa* are eaten mainly by bats. *Salacia* sp. seeds are probably dispersed by primates and nocturnal arboreal mammals (Haber et al., 1996). Fruits of *Persea* sp., the largest seed by far in this study, are eaten by arboreal mammals but many fruits fall to the ground and are eaten or damaged by terrestrial rodents. Of the three species of primates in the area, spider monkeys (*Ateles geoffroyi*) are more frugivorous and more important seed dispersers for fleshy-fruited species than white-face capuchins (*Cebus capucinus*) or mantled howlers (*Alouatta palliata*). Numerous species of bats occur in the area (Timm and LaVal, 2000), but no sampling was done in conjunction with this study.

Minimum dispersal distances were not estimated because the parent was not known in most cases. Data for *O. endresiana*, *B. costaricensis*, *G. glabra* and *G. kunthiana* indicate that nearly all seeds were less than 100 m from the nearest conspecific tree and most seeds were within 30 m (Wenny, 1999, 2000a,b).

Post-dispersal seed fate

I discerned three patterns of seed removal: (i) animals seldom removed seeds with thick, hard seed coats regardless of size (*M. vernicosa* and *C. sylvicola*); (ii) animals removed 50–100% of small and intermediate seeds without thick, hard seed coats; and (iii) animals rarely removed large seeds regardless of seed coat characteristics (Fig. 21.1). Seed size alone was not a good predictor of removal rate ($r^2 = 0.10$) or distance moved ($r^2 = 0.06$).

Removal rates varied considerably among seed species (Fig. 21.1). Only nine of 815 seeds could not be relocated. For most species, 100% of the flagging tape used to mark seeds was found after removal (Table 21.2, column 9). Most seed species did not have a second stage of dispersal by seed-caching rodents (Table 21.2, column 3). Therefore, seed discovery by animals resulted in seed predation for most seeds. Only *Guarea kunthiana* (27%) and *G. glabra* (46%) had substantial numbers of seeds cached (Wenny, 1999). At the end of this study 8% of *G. kunthiana* and 30% of *G. glabra* remained cached. *Geonoma edulis*, an understorey palm with a hard seed coat, had a few seeds cached (4%), with 2% remaining cached after 12 weeks (Table 21.2). *Panopsis suaveolens* (Proteaceae) and *Pouteria fossicola* (Sapotaceae), for which I marked < 20 seeds, also had several seeds cached. For the nine species for which all removed seeds were found, no seeds were cached, but, for the three species with some caching, not all marked seeds were found after removal (Table 21.2). Thus caching cannot be ruled out for *A. palmana* and *C. sylvicola*. Although one *Ocotea meziana* seed was never found after removal, results for other Lauraceae suggest no secondary dispersal by rodents (Table 21.2; Wenny, 2000a,b).

Based on tooth marks on seeds, the patterns of seed removal and the places to

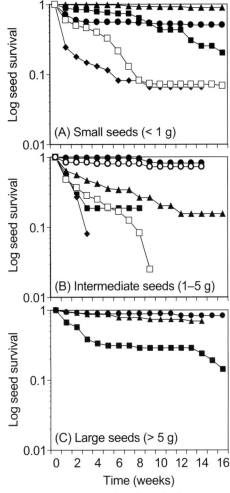

Fig. 21.1. Proportion of seeds (and subsequent seedlings) surviving (log scale) based on the Kaplan–Meier survival function. Standard error not shown for clarity. Species are listed in order of increasing seed size: (A) species with small seeds (< 1 g); *Ardisia palmana*, solid squares; *Geonoma edulis*, circles; *Chione sylvicola*, triangles; *Ocotea endresiana*, diamonds; *Guarea glabra*, open squares; (B) species with intermediate-sized seeds (1–5 g); *Prunus annularis*, solid squares; *Meliosma vernicosa*, solid circles; *Eugenia guatemalensis*, triangles; *Ocotea meziana*, diamonds; *Guarea kunthiana*, open squares; *Pleurothyrium palmanum*, open circles; (C) species with large seeds (> 5 g); *Salacia* sp., squares; *Beilschmiedia costaricensis*, circles; *Persea* sp., triangles.

Table 21.2. Seed movement and mortality after 12–16 weeks. Species listed in order of increasing seed size. All proportions calculated from the total sample size (n). Caching and movement includes the proportions of seeds that were cached whether or not they were later retrieved (cache), the proportion of seeds that were moved but not eaten or cached (moved) and the average distance seeds were moved by all causes. Seed mortality includes the proportions of seeds killed by vertebrates (vert.), insects (insect) and fungal pathogens (fungal). Final fate includes the proportions of seeds that were removed by animals but not relocated (miss.), seeds that were not killed or germinated by the end of the study (intact), seeds that remained cached (cached) and seeds that germinated and established seedlings (seedling). The sum of the last seven columns equals 1.00.

Species	n	Caching and movement		Distance (m) mean ± SD	Mortality and final fate						
		Cache	Moved		Vert.	Insect	Fungal	Miss.	Intact	Cached	Seedling
Ardisia palmana	39	0.00	0.26	0.8 ± 1.6	0.43	0.00	0.31	0.03	0.10	0.00	0.13
Geonoma edulis	57	0.04	0.23	0.9 ± 1.8	0.47	0.02	0.00	0.04	0.08	0.02	0.37
Chione sylvicola	82	0.00	0.09	1.1 ± 1.6	0.20	0.00	0.00	0.02	0.67	0.00	0.11
Ocotea endresiana	61	0.00	0.03	0.6 ± 0.8	0.91	0.02	0.00	0.00	0.00	0.00	0.07
Guarea glabra	404	0.46	0.25	1.2 ± 1.3	0.48	0.11	0.01	0.00	0.09	0.30	0.01
Prunus annularis	48	0.00	0.15	0.9 ± 1.2	0.81	0.00	0.00	0.00	0.02	0.00	0.17
Meliosma vernicosa	41	0.00	0.15	0.2 ± 0.3	0.19	0.00	0.00	0.00	0.76	0.00	0.05
Eugenia guatemalensis	64	0.00	0.23	1.0 ± 1.2	0.67	0.00	0.17	0.00	0.16	0.00	0.00
Ocotea meziana	37	0.00	0.16	0.3 ± 0.9	0.97	0.00	0.00	0.03	0.00	0.00	0.00
Guarea kunthiana	120	0.27	0.16	6.1 ± 1.3	0.78	0.02	0.10	0.02	0.00	0.08	0.00
Pleurothyrium palmanum	28	0.00	0.36	0.6 ± 1.3	0.21	0.04	0.02	0.00	0.15	0.00	0.58
Salacia sp.	42	0.00	0.29	1.7 ± 2.3	0.86	0.00	0.00	0.00	0.00	0.00	0.14
Beilschmiedia costaricensis	129	0.00	0.04	1.2 ± 1.9	0.07	0.10	0.00	0.00	0.05	0.00	0.78
Persea sp.	23	0.00	0.65	1.6 ± 2.1	0.30	0.00	0.00	0.00	0.00	0.00	0.70

which seeds were taken, small rodents were responsible for most post-dispersal seed removal and predation of small and intermediate seeds. The exception to this pattern is caching and consumption of *G. glabra* and *G. kunthiana* by agoutis with additional consumption of *G. kunthiana* by peccaries. Excluding *Guarea* spp., small rodents were responsible for at least 70% of vertebrate-caused seed mortality.

Most seed movement away from the original location was clearly associated with animals. Nearly 20% of all seeds, however, were moved short distances (< 1 m) by heavy rainfall or other abiotic factors. The number of seeds that experienced such abiotic movement was not highly correlated with seed size (Pearson correlation, $r = 0.42$, $P = 0.17$) as might be expected if rainfall was the only major cause. In addition, 12% of all seeds were chewed but not completely consumed. Some partially eaten seeds were eventually consumed over the course of several days. This fate was particularly common for *Salacia* sp., for which 60% were eaten piecemeal by black-breasted wood-quail (*Odontophorus leucolaemus*). Nearly 91% of *Persea* sp. seeds were damaged but only 30% suffered lethal damage (none was entirely consumed). All intermediate and large-seeded species except *M. vernicosa* had some seeds partially eaten (range 2–91%). The proportion of seeds partially eaten was positively correlated with seed size ($r = 0.82$, $P < 0.001$). Such partial consumption did not prevent germination and seedling establishment for most species with intermediate and large seeds. In contrast, seed mortality caused by fungal pathogens was often preceded by gnawing or partial seed consumption. For two species with the highest levels of mortality from fungal pathogens, *Ardisa palmana* (31%) and *Eugenia guatemalensis* (17%), fungal pathogen attack followed partial consumption in 82% of cases.

Beilschmiedia costaricensis seeds were occasionally buried (10%) by unidentified beetles (Nitidulae). The beetles loosened and churned the soil to inter a seed, leaving a slight mound, which compacted after several days. These seeds were not moved before burial, none of them germinated, and all were filled with frasse when examined at the end of the study (Wenny, 2000a), suggesting the beetles buried seeds to protect the resource in a manner similar to burying beetles (Silphidae) and dung beetles (Scarabaeidae). This behaviour is a rare example of beetles hoarding food (Vander Wall, 1990) and deserves further study. For other species, the combination of blemishes on newly dispersed seeds and the eventual determination that blemished seeds were filled with frasse suggests pre-dispersal infestation rather than post-dispersal seed consumption by adult insects.

Rodent trapping

Although trapping effort was relatively low, trapping success was high enough (18–27%) to characterize the small terrestrial mammal community. At least five species of small rodents were captured in live traps (Table 21.3). The most common species by far was *Peromyscus nudipes*, accounting for 81% of individuals captured. *Reithrodontomys* (probably two species) was captured in traps baited with peanut butter but never in traps baited with Lauraceae seeds. *Scontinomys teguina* was captured in traps with seeds but never chewed or ate the seeds. I captured more rodents than expected in gaps and fewer than expected near logs, but these trends were not statistically significant ($\chi^2 = 3.22$, df = 2, $P = 0.2$). Sample sizes were too small to examine microhabitat differences for each species. *Heteromys desmatestianus*, the only species with cheek pouches, usually carried seeds smaller than those in this study. We observed but did not capture pygmy squirrels (*Microsciurus alfari*) and a larger squirrel, presumably *Sciurus granatensis* (Reid, 1997).

Tracking stations

The tracking stations confirmed the presence of tapir (*Tapirus bairdii*), brocket

Table 21.3. Trapping effort (trap-nights) and number of small mammals captured in three microhabitats: near fallen logs, at the base trees > 20 cm diameter and in treefall gaps.

Species	Microsites			
	Log	Tree	Gap	All microsites
Heteromys desmarestianus	1	2	1	4
Oryzomys albigularis	3	3	6	12
Peromyscus nudipes	28	38	40	106
Reithrodontomys spp.	2	1	4	7
Scotinomys teguina	0	2	3	5
Total captures	34	46	54	134
Trap-nights	183	211	198	592

deer (*Mazama americana*), collared peccary (*Tayassu tajacu*), jaguar (*Panthera onca*), puma (*Puma concolor*), coati (*Nasua narica*), agouti (*Dasyprocta punctata*) and paca (*Agouti paca*) (nomenclature follows Reid, 1997). Peccary, brocket deer and agouti were also observed directly. Tapir, peccary and jaguar occurred irregularly, leaving tracks on several occasions over the course of a week or more, followed by long periods of apparent absence from the study site. All other species were present throughout the year. Tracks of several other smaller species were recorded, but I was not able to identify them.

Discussion

For most plant species in this study, post-dispersal seed removal resulted in predation. Terrestrial vertebrates, and in particular small rodents, were responsible for at least 70% of seed predation. Agoutis, collared peccaries and black-breasted wood-quail were also important seed predators but ate fewer seed species regularly (three, one and one, respectively) than the small rodents. Only two species experienced significant secondary dispersal by seed-caching rodents. One clear advantage of such secondary dispersal is protection from predation by small rodents. Caching may also protect seeds from insects and fungal pathogens, but this study was not designed to assess the full impact of these sources of mortality. In addition, some seeds with evidence of insect activity were eaten by rodents. Thus, the relatively high level of seed predation by rodents observed in this study probably underestimates the level of infestation by insects and pathogens. On the other hand, most insect seed predation in this study was probably pre-dispersal infestation (based on marks on newly dispersed seeds suggesting oviposition), and fungal pathogens appear to be more important mortality factors for seedlings than for seeds in this study. The roles of insects and fungal pathogens in seed fate need further study.

Another finding of this study is that almost 20% of seeds were moved but not eaten or cached. Such non-lethal seed movement has both biotic and abiotic causes. Heavy rainfall moves smaller seeds while animal activity, such as partial seed consumption, affects the larger seeded species. Because the study area was relatively flat, seed movement by rainfall was probably less than would occur in an area with steeper slopes. Most abiotic movement was less than 2 m and the extent to which such movements benefit the parent plants is unknown. It is unlikely that such short movements would affect the level of seed predation by rodents, but such movement could serve to scatter seeds that were dispersed together and lead to lower density-dependent seedling mortality. Such movement could also cause a seed to fall into a crack or depression in the soil, get buried under leaf litter, or become lodged against a log where accessibility to seed predators may be lower and subsequent survival may be higher.

Whether or not a seed is moved, it is clear that seed mortality continues after germination. Germination most probably is not a major limitation on recruitment for relatively large-seeded species (Wenny, 2000a,b). This and other studies indicate that post-dispersal seed predation is a major bottleneck between seed production and seedling recruitment. However, the links between the agents of post-dispersal seed mortality and seedling recruitment have rarely been noted. In this study, species with larger seeds had higher germination and post-germination survival rates. But larger seeds were also more likely to be gnawed, chewed or moved even after germination. It is likely that some species that experience low rates of seed predation are exposed to increased mortality once germination begins. This possibility was most evident for *M. vernicosa*, which has a very hard seed coat and very low levels of post-dispersal seed removal. Upon germination, the seed coat splits apart leaving the developing seedling with no physical defence. The slight increase in mortality after 6 weeks (Fig. 21.1) was a result primarily of seedling herbivory rather than seed predation *per se*. The fate of *C. sylvicola*, which also has a thick seed coat, was similar to that of *M. vernicosa* except that the seed does not split upon germination, but the seedling emerges from one end of the seed. Several newly germinated seeds appeared to have been gnawed open to extract the remaining seed reserves. Although this type of mortality blurs the distinction between seed predation and seedling herbivory, it suggests that rodents can track certain resources and attack different plant species at different stages of seedling development (e.g. Forget and Milleron, 1991; Jansen, 2003).

The tracking stations showed that large predators and large herbivores were present in the study area. Thus, this part of the Monteverde region (the continental divide and Atlantic slope) appears to contain a relatively intact ecosystem. The presence of large predators may explain the apparently lower abundance of agoutis in this area compared to sites on the Pacific slope (personal observation). The spiny pocket mouse (*Heteromys desmarestianus*) is also less common along the continental divide than at lower elevations on the Pacific slope (Anderson, 1982; G. Murray, personal communication). As both agoutis and pocket mice are known to scatterhoard seeds, their relatively low abundance in the study area may be partially responsible for the low incidence of secondary dispersal recorded in this study. The higher abundance of these species on the Pacific slope is likely to influence seed fate and forest succession (e.g. Guariguata *et al.*, 2000). Climatic changes in the region will complicate this relationship further (Pounds *et al.*, 1999; Lawton *et al.*, 2001). Plant species from the drier zones on the Pacific slope are expected to expand distributions into higher elevations and it would be interesting to study seed fate of species (or among related species with similar seed characteristics) spanning this elevational range as plant distributions shift.

Members of the Lauraceae seemed to fall into one of two seed fate categories. Two species of *Ocotea* experienced very high seed predation while one species each of *Beilschmiedia*, *Persea* and *Pleurothyrium* experienced low seed predation. In other studies in the Monteverde area, 95–100% post-dispersal seed removal has been noted for *O. monteverdensis* (Wheelwright, 1988) and *O. tonduzii* (D. Wenny, unpublished data). In a Colombian cloud forest, Samper (1992) recorded 100% removal of *Nectandra* sp. (*Ocotea* and *Nectandra* have very similar seeds). Although the seeds of lauraceous species are thought to be chemically defended from seed consumers (Castro, 1993), it appears that rodents can tolerate consuming *Ocotea* species better than seeds of some other genera. Part of the explanation for the lack of secondary dispersal in some Lauraceae may be large seed size and rapid germination rates, making these species poor candidates for scatterhoarding and consumption later in the year.

Results of this study suggest that seed fate can be explained, in part, by seed characteristics but that secondary dispersal syndromes may be more difficult to define.

While primary dispersal by scatterhoarding may represent a predictable syndrome with associated plant characteristics (Vander Wall, 2001), secondary dispersal via scatterhoarding involves much more variable fruit and seed traits (Vander Wall and Longland, Chapter 18, this volume). The classic examples of scatterhoarding involve plant species that have great annual variation in crop size with cascading effects on seed predator/disperser populations (Kelly and Sork, 2002). These species tend to have large, hard seeds and are fairly common in some habitats (e.g. *Quercus* spp. and other nut-bearing trees in some temperate habitats). Plant species secondarily dispersed by large rodents may fit some aspects of this syndrome, particularly the characteristics of the seed coat or physical defence against seed predators. In contrast, many bird-dispersed trees with high post-dispersal seed removal also exhibit high annual variation in crop size (e.g. Wheelwright, 1986), but a link between variable seed crop size and post-dispersal predator satiation has not been established. In *O. endresiana*, for example, crop size was at least an order of magnitude lower in 1994 than in 1993 or 1995 but seed removal was uniformly high in all 3 years and secondary dispersal did not occur (this study; Wenny, 2000b). Furthermore, the small rodents that appear to be responsible for most post-dispersal seed predation have shorter lifespans (1–2 years) than the much larger caviomorph rodents, and populations of small rodents may fluctuate in response to overall resource or moisture levels. Scatterhoarding and retrieval of seeds may not be an efficient foraging strategy for most of these species. Nevertheless, species such as *Heteromys desmarestianus* are known to be important scatterhoarders in some lowland tropical habitats (Brewer and Rejmanek, 1999). The roles of squirrels as scatterhoarders in tropical forests is largely unexplored. Studies on a wider variety of plant species over several years are needed to further examine the possibility of a secondary dispersal syndrome by scatterhoarding rodents in tropical forests.

References

Anderson, S.D. (1982) Comparative population ecology of *Peromyscus mexicanus* in a Costa Rican wet forest. PhD, University of Southern California, Los Angeles, California.

Bohning-Gaese, K., Gaese, B.H. and Rabemanantsoa, S.B. (1999) Importance of primary and secondary seed dispersal in the Malagasy tree *Commiphora guillaumini*. *Ecology* 80, 821–832.

Brewer, S.W. and Rejmanek, M. (1999) Small rodents as significant dispersers of tree seeds in a Neotropical forest. *Journal of Vegetation Science* 10, 165–174.

Castro, O.C. (1993) Chemical and biological extractives of Lauraceae species in Costa Rican tropical forests. *Recent Advances in Phytochemistry* 27, 65–87.

Clark, K.L. and Nadkarni, N.L. (2000) Microclimate variability. In: Nadkarni, N.M. and Wheelwright, N.T. (eds) *Monteverde: Ecology and Conservation of a Tropical Cloud Forest.* Oxford University Press, New York, pp. 33–34.

Clark, K.L., Lawton, R.O. and Butler, P.R. (2000) The physical environment. In: Nadkarni, N.M. and Wheelwright, N.T. (eds) *Monteverde: Ecology and Conservation of a Tropical Cloud Forest.* Oxford University Press, New York, pp. 33–37.

Forget, P.-M. (1993) Post-dispersal predation and scatterhoarding of *Dipteryx panamensis* (Papilionaceae) seeds by rodents in Panama. *Oecologia* 94, 255–261.

Forget, P.-M. (1996) Removal of seeds of *Carapa procera* (Meliaceae) by rodents and their fate in rainforest in French Guiana. *Journal of Tropical Ecology* 12, 751–761.

Forget, P.-M. and Milleron, T. (1991) Evidence for secondary dispersal by rodents in Panama. *Oecologia* 87, 596–599.

Forget, P.-M., Milleron, T. and Feer, F. (1998) Patterns in post-dispersal seed removal by neotropical rodents and seed fate in relation to seed size. In: Newbery, D.M., Prins, H.H.T. and Brown, N.D. (eds) *Dynamics of Tropical Communities.* British Ecological Society, London, pp. 25–49.

Gomez, C. and Espadaler, X. (1998) Seed dispersal curve of a Mediterranean myrmecochore: influence of ant size and the distance to nests. *Ecological Research* 13, 347–354.

Guariguata, M.R., Adame, J.J.R. and Finegan, B. (2000) Seed removal and fate in two selectively logged lowland forests with contrasting protection levels. *Conservation Biology* 14, 1046–1054.

Haber, W.A. (2000a) Plants and vegetation. In: Nadkarni, N.M. and Wheelwright, N.T. (eds) *Monteverde: Ecology and Conservation of a Tropical Cloud Forest.* Oxford University Press, New York, pp. 39–69.

Haber, W.A. (2000b) Vascular plants of Monteverde. In: Nadkarni, N.M. and Wheelwright, N.T. (eds) *Monteverde: Ecology and Conservation of a Tropical Cloud Forest.* Oxford University Press, New York, pp. 457–518.

Haber, W.A., Zuchowski, W. and Bello, E. (1996) *An Introduction to Cloud Forest Trees: Monteverde, Costa Rica.* La Nación, San José, Costa Rica, 197 pp.

Hoshizaki, K., Suzuki, W. and Nakashizuka, T. (1999) Evaluation of secondary dispersal in a large-seeded tree *Aesculus turbinata*: a test of directed dispersal. *Plant Ecology* 144, 167–176.

Jansen, P.A. (2003) Scatterhoarding and tree regeneration: ecology of nut dispersal in a Neotropical forest. PhD thesis, Wageningen University, Wageningen, The Netherlands.

Kelly, D. and Sork, V.L. (2002) Mast seeding in plants: why, how, where? *Annual Review of Ecology and Systematics* 33, 427–447.

Lawton, R. and Dryer, V. (1980) The vegetation of the Monteverde Cloud Forest Preserve. *Brenesia* 18, 101–116.

Lawton, R.O., Nair, U.S., Pielke, R.A. and Welch, R.M. (2001) Climatic impact of tropical lowland deforestation on nearby montane cloud forests. *Science* 294, 584–587.

Levey, D.J. and Byrne, M.M. (1993) Complex ant–plant interactions: rain forest ants as secondary dispersers and post-dispersal seed predators. *Ecology* 74, 1802–1819.

Nadkarni, N.M. and Wheelwright, N.T. (2000) *Monteverde: Ecology and Conservation of a Tropical Cloud Forest.* Oxford University Press, New York, 573 pp.

Passos, L. and Ferreira, S. (1996) Ant dispersal of *Croton priscus* (Euphorbiaceae) seeds in a tropical semideciduous forest in southeastern Brazil. *Biotropica* 28, 697–700.

Peres, C.A., Schiesari, L.C. and Dias-Leme, C.L. (1997) Vertebrate predation of Brazil-nuts (*Bertholletia excelsa*, Lecythidaceae), an agouti-dispersed Amazonian seed crop: a test of the escape hypothesis. *Journal of Tropical Ecology* 13, 69–79.

Pounds, J.A., Fogden, M.P.L. and Campbell, J.H. (1999) Biological response to climate change on a tropical mountain. *Nature* 398, 611–615.

Price, M.V. and Jenkins, S.H. (1986) Rodents as seed consumers and dispersers. In: Murray, D.R. (ed.) *Seed Dispersal.* Academic Press, Sydney, pp. 191–235.

Reid, F.A. (1997) *A Field Guide to the Mammals of Central America and Southeast Mexico.* Oxford University Press, New York, 334 pp.

Samper, C.T. (1992) Natural disturbance and plant establishment in an Andean cloud forest. PhD thesis, Harvard University, Cambridge, Massachusetts.

Schupp, E.W. (1988) Factors affecting post-dispersal seed survival in a tropical forest. *Oecologia* 76, 525–530.

Steele, M.A., Hadj-Chikh, L.Z. and Hazeltine, J. (1996) Caching and feeding decisions by *Sciurus carolinensis*: responses to weevil-infested acorns. *Journal of Mammalogy* 77, 305–314.

Timm, R.M. and LaVal, R.K. (2000) Mammals of Monteverde. In: Nadkarni, N.M. and Wheelwright, N.T. (eds) *Monteverde: Ecology and Conservation of a Tropical Cloud Forest.* Oxford University Press, New York, pp. 553–558.

Vander Wall, S.B. (1990) *Food Hoarding in Animals.* University of Chicago Press, Chicago, Illinois, 445 pp.

Vander Wall, S.B. (1992) The role of animals in dispersing a 'wind-dispersed' pine. *Ecology* 73, 614–621.

Vander Wall, S.B. (2001) The evolutionary ecology of nut dispersal. *Botanical Review* 67, 74–117.

Vander Wall, S.B. and Longland, W.S. (2004) Diplochory: are two seed dispersers better than one? *Trends in Ecology and Evolution* 19, 155–161.

Vander Wall, S.B., Thayer, T.C., Hodge, J.S., Beck, M.J. and Roth, J.K. (2001) Scatter-hoarding behavior of deer mice (*Peromyscus maniculatus*). *Western North American Naturalist* 61, 109–113.

Wenny, D.G. (1999) Two-stage dispersal of *Guarea glabra* and *G. kunthiana* (Meliaceae) in Costa Rica. *Journal of Tropical Ecology* 15, 481–496.

Wenny, D.G. (2000a) Seed dispersal of a high-quality fruit by specialized frugivores: high-quality dispersal? *Biotropica* 32, 327–337.

Wenny, D.G. (2000b) Seed dispersal, seed predation, and seedling recruitment of a neotropical montane tree. *Ecological Monographs* 70, 331–351.

Wenny, D.G. and Levey, D.J. (1998) Directed seed dispersal by bellbirds in a tropical cloud forest. *Proceedings of the National Academy of Sciences, USA* 95, 6204–6207.

Wheelwright, N.T. (1986) A seven-year study of individual variation in fruit production in tropical bird-dispersed tree species in the family Lauraceae. In: Estrada, A. and Fleming, T.H. (eds) *Frugivores and Seed Dispersal.* Dr W. Junk Publishers, Dordrecht, The Netherlands, pp. 19–35.

Wheelwright, N.T. (1988) Four constraints on coevolution between fruit-eating birds and fruiting plants: a tropical case study. *Proceedings of the International Ornithological Congress* XIX. University of Ottawa Press, Ottawa, pp. 827–845.

Wheelwright, N.T., Haber, W.A., Murray, K.G. and Guindon, C.A. (1984) Tropical fruit-eating birds and their food plants: a survey of a Costa Rican lower montane forest. *Biotropica* 16, 173–192.

22 Observing Seed Removal: Remote Video Monitoring of Seed Selection, Predation and Dispersal

Patrick A. Jansen and Jan den Ouden

Centre for Ecosystem Studies, Wageningen University, PO Box 47, 6700 AA, Wageningen, The Netherlands

---

Introduction

What happens to seeds once they have reached a surface – here referred to as post-dispersal seed fate (cf. Chambers and MacMahon, 1994) – is becoming an increasingly important question. A large body of literature is available evaluating post-dispersal seed fate (reviewed in Hulme, 1998, 2002; Crawley, 2000; Moles and Westoby, 2003; see Hulme and Kollmann, Chapter 2, this volume; Vander Wall and Longland, Chapter 18, this volume). Many studies investigating post-dispersal seed fate, however, measure only seed removal from experimental plots, and implicitly or explicitly assume that removed seeds were killed by seed-eaters (see Hulme and Kollmann, Chapter 2, this volume; Forget and Wenny, Chapter 23, this volume; Wenny, Chapter 21, this volume). This approach may be inadequate, as several recent studies in which removed seeds were tracked have shown that many of these seeds are secondarily dispersed rather than consumed (e.g. Vander Wall, 1992, 1997; Forget, 1993; Levey and Byrne, 1993; Brewer and Rejmanek, 1999; Hoshizaki et al., 1999; Jansen et al., 2002). Scatter-hoarding animals eat some of the seeds they gather but bury most seeds as food reserves in scattered caches in the soil (e.g. Jansen et al., 2002, 2004).

Scatterhoarding may enhance seed survival and seedling establishment because it moves seeds away from the parent and siblings, secures seeds from seed predators such as insects and wild pigs, and puts seeds in situations that often enhance germination and establishment (Vander Wall, 1990). Cached seeds can only establish seedlings, however, if animals leave some proportion of their food reserves untouched. Several studies of secondary dispersal (e.g. Vander Wall, 1994; Jansen, 2003) have shown that scatterhoarded seeds may indeed escape consumption and develop into seedlings. Thus, seed survival and seedling establishment can occur not only right after primary dispersal, but also after one or more subsequent bouts of secondary dispersal (Price and Jenkins, 1986). Moreover, some studies of large-seeded species have observed seeds dying from desiccation or seed predation if they were not removed and buried by scatterhoarding animals (Shaw, 1968; Jansen et al., 2004). Hence, traditional studies of post-dispersal seed survival that simply measure seed removal are insufficient for estimating seed mortality (see

Hulme and Kollmann, Chapter 2, this volume; Wenny, Chapter 21, this volume).

An important aspect of studying post-removal seed fate is finding out which animal species are responsible for seed removal and how they treat seeds. Most studies rely on circumstantial evidence for inferences on agents of seed removal, such as footprints, observations or captures near experimental seed plots, dental traces on seeds and caching profiles. However, there are several techniques by which actors of seed removal can be identified with greater accuracy and certainty. The simplest are direct observations. Forget (1990), for example, observed from a hide to record removal of *Voucapoua americana* seeds from an experimental plot in French Guiana. He observed that all seeds were removed by a single red acouchy (*Myoprocta exilis*), and he later retrieved most of these seeds in soil surface caches. A technique that allows for monitoring seed removal during longer time spans and at numerous locations simultaneously uses remote cameras (Kucera and Barrett, 1993), in which an infrared sensor detects movement by warm-bodied animals and triggers a camera. Remote cameras have been used successfully to identify terrestrial animals in several studies on terrestrial seed or fruit removal (Miura *et al.*, 1997; Yasuda *et al.*, 2000; Page *et al.*, 2001; Takada *et al.*, 2001; Beck and Terborgh, 2002).

The aim of this chapter is to demonstrate the potential of a third technique, automated remote video monitoring (e.g. Hughes and Shorrock, 1998), for studying seed removal and seed fate. To do so, we test the following hypotheses using data collected with this technique on animal behaviour and seed removal in large-seeded tree species:

1. Most seed removal is by scatterhoarding animals.
2. Scatterhoarding animals detect seeds more rapidly than do non-dispersing animals.
3. Scatterhoarding animals adapt their foraging activity rhythm to seed abundance.
4. Bigger, more nutritious seeds are removed faster than smaller ones.

5. Bigger, more nutritious seeds are more likely to be scatterhoarded than smaller ones.
6. Higher removal rates are correlated with higher rates of scatterhoarding rather than immediate seed predation.
7. Seed handling time is proportional to the effort scatterhoarders spend on caching, and correlated with dispersal distance.
8. Larger-bodied scatterhoarders disperse large seeds further than do smaller-bodied scatterhoarders.

Note that these hypotheses and tests serve to examine the utility of remote video monitoring rather than as presentations of experimental results.

Sites and Equipment

Most of the data presented in this chapter were collected in 1996–2000 at the Nouragues Biological Station, French Guiana (4°02′N, 52°42′W), an undisturbed tropical lowland rainforest site with an intact fauna (see Bongers *et al.*, 2001, for an extensive description of the site). Terrestrial granivores common in this area include two scatterhoarding rodents: red acouchy (*Myoprocta acouchy*) and red-rumped agouti (*Dasyprocta leporina*). These are important seed dispersers that harvest and cache seeds during the wet fruiting season for use during the lean dry season (Forget, 1990, 1996; Jansen and Forget, 2001; Jansen *et al.*, 2002). Other rodent species in the area, i.e. Guianan red squirrel (*Sciurus aestuans*), Cuvier's terrestrial spiny rat (*Proechimys cuvieri*), bristle mouse (*Neacomys guianae*) and rice rat (*Oryzomys megacephalus*), are primarily seed predators (Guillotin, 1982; Guillotin *et al.*, 1994; Henry, 1997). Species in these genera may hoard seeds, but probably put most seeds in places that are unfavourable for seedling establishment, such as in larders, burrows or up in trees (Forget, 1993; Brewer and Rejmanek, 1999; P.A. Jansen, Nouragues, 1998, personal observation). White-lipped peccary (*Tajassu pecari*) and collared peccary (*Pecari tajacu*) are potential dispersers of small seeds (see Hulme and Kollmann, Chapter 2, this volume; Wenny, Chapter 21, this volume), but they destroy most of the large seeds they ingest (Kiltie, 1981, 1982; Bodmer, 1991; see Beck, Chapter 6, this volume). The ecological role of the two *Mazama* deer species in the area (see Gayot *et al.*, 2004), finally, is probably comparable to that of peccaries. With the exception of rats and mice, all species are (primarily) diurnal at Nouragues.

In addition, we collected data in 1998 at Speulderbos, a mixed oak–beech forest in The Netherlands (52°15′N, 5°40′E; see den Ouden, 2000, for further site description). Terrestrial granivores in this area include two rodent species, bank vole (*Clethriono-mys glareolus*) and wood mouse (*Apodemus sylvaticus*), and two ungulates, Eurasian wild boar (*Sus scrofa scrofa*) and red deer (*Cervus elaphus*). Here, wood mice are the only mammals that scatterhoard seeds (see den Ouden *et al.*, Chapter 13, this volume).

At Nouragues, we established over 250 experimental plots with more than 6500 seeds in the understorey throughout the area. Each plot measured 0.5 to 1.0 m² where we removed litter and placed 9, 25, 49 or occasionally 100 fresh seeds. We arranged seeds in regular grids in order to identify them individually by their position. Overall, we used a dozen different seed species, but most plots contained seeds of either *Carapa procera* (Meliaceae; $n = 109$ plots, 3500 seeds) or *Licania alba* (Chrysobalanaceae; $n = 59$ plots, 1300 seeds), two large-seeded tree species (henceforth *Carapa* and *Licania*). These seeds were weighed, thread-marked and individually numbered. Marking enabled retrieval and identification of seeds after removal, as the thread and tag protrude from the soil even when seeds are cached (Forget, 1990; see Forget and Wenny, Chapter 23, this volume). Our experimental period included 3 years of abundant fruiting (1996, 1998 and 2000) and 2 lean years (1997 and 1999). At Speulderbos, we established five experimental seed plots, each including 49 acorns of *Quercus robur* (Fagaceae) arranged in a 7×7 grid. At Speulderbos and in most experiments at Nouragues, we manipulated seed fresh mass within

seed plots to study size discrimination by scatterhoarding rodents: each plot included as wide a range of (intraspecific) seed masses as possible (cf. Jansen *et al.*, 2002, 2004).

We monitored animal activity and seed removal by remote video recording at plots for several hours up to several days, depending on the seed removal rate. Then, from the playback, we recorded the identity of animals and the time of removal for each individual seed, as well as some behavioural observations. Our equipment consisted of a monochrome CCD video surveillance camera (Philips VCM 6250/00T) in a weather-proof housing (Eneo VHL-2EC) and a time-lapse video recorder (Panasonic AG-1070 DC). Power came from 12V 18Ah car batteries that we replaced daily and recharged at the field lab using solar panels. Much smaller, more energy-efficient and more advanced systems are now available (see Discussion). The camera was mounted on a tree next to the plot at ~1.5 m height using an aluminium wall bracket (Eneo WD-14/MK). The recorder was attached to a tree 2–3 m away from the plot, in a waterproof bag with silicagel. Depending on ambient humidity, video recording was either triggered by a passive infrared detector (ASIM IR 207, not waterproof enough for rainforest conditions) or continuous in time-lapse mode (four frames per second). Recordings were on ordinary VHS tapes, on which date and time in seconds were also visible. At Speulderbos, we only recorded seed removal overnight, using an infrared (IR) lamp (Dennard 880M20 microlight). In Nouragues, in contrast, nocturnal seed removal in a first series of trials with this IR lamp was so rare that we decided to use all power for recording diurnal animal activity and seed removal.

Results

The most basic purpose of video monitoring seed removal is to identify the animal species that visit seed sources and remove seeds. At our seed plots in Nouragues, we recorded a suite of different animals, including most of the ungulate and terrestrial rodent species occurring in the study area, and many other terrestrial animals. The recordings allowed separating animals that actually took seeds from those that did not. The only animals taking seeds by day in Nouragues were red acouchy, red-rumped agouti, collared peccary (Fig. 22.1A–C, respectively) and Guianan red squirrel. Most, but not all, visits by these species resulted in seed removal (here including *in situ* consumption of seeds). If we had inventoried seed removers by occurrence of footprints, a frequently used method in studies of seed removal, we would have produced a longer, less accurate list of potential seed dispersers and predators. Likewise, we found that only wood mice removed acorns from our seed plots at Speulderbos, whereas live rodent trapping prior to the experiments had shown that bank voles were more common than wood mice.

Discovery of seeds

More beneficial to plants than simply removing seeds is that scatterhoarders remove seeds before non-dispersing animals get a chance to find them. We studied how much time it took different granivorous mammals to discover the artificial seed sources in Nouragues, using the time indication on our video recordings. We measured the time until discovery for each animal species at 102 plots, and compared the rates using Kaplan Meier survival analysis. Times were right-censored at the end of our recordings if a species was not seen visiting a plot within the video observation period. We found that the two scatterhoarding rodents detected plots significantly more rapidly than did the other granivores (log-rank test: $\chi^2_1 = 111$, $P < 0.001$; Fig. 22.2A). Acouchies were the first to discover seed plots in 96% of the cases, and were always earlier than agoutis. Their higher abundance and smaller home range compared to agoutis and peccaries (Dubost, 1988) increased the probability of acouchies

Fig. 22.1. Video stills of granivorous mammals, viewed from above, visiting cafeteria plots with different-sized *Carapa procera* seeds in a French Guianan rainforest. (A) Red acouchy (*Myoprocta acouchy*), (B) agouti (*Dasyprocta leporina*) and (C) collared peccary (*Pecari tajacu*). The time indication (upper left corners) and organization of the seeds in a grid allowed for measurement of time and sequence of removal of individual seeds by different species or even individuals. The black dots visible on photo B are black-painted heads of roofing nails, pressed in the ground below each seed, that appear upon seed removal.

discovering and rapidly sequestering seeds before larger and stronger competitors could monopolize the food.

An important determinant of the ability of species to obtain seeds is their activity rhythm. In Nouragues, seed removal occurs primarily during the day, and the early morning should therefore be a good time for granivores to monopolize fallen seeds that have accumulated overnight. We used our recordings to investigate the foraging activity of different granivores during the day, counted how many seeds each species took for each hour of the day, and calculated rates of seed removal for each species. We did this separately for years of high and low seed abundance to see how food availability influenced activity patterns. We found that the overall foraging activity of acouchies

peaked in the early morning and the late afternoon (Fig. 22.3). Both agoutis and acouchies showed reduced activity at midday. This 'siesta' may serve to avoid the high midday temperatures as well as the rain showers that frequently occurred around noon, but we also observed that acouchies got increasingly bothered by deer flies and other insects over the course of the morning.

Acouchies showed a strong apparent response to food abundance. As predicted, their morning activity was greater in low seed years than in high seed years, and also continued for longer. Peccaries, in contrast, were only recorded around noon and only in low seed years, but this pattern may have been an artefact of the low number of encounters. Discovery times that we recorded for *Carapa* seed plots in years of

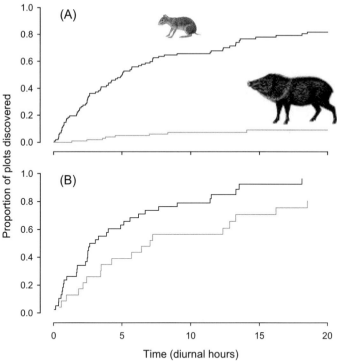

Fig. 22.2. Time to detection of experimental seed plots by granivorous mammals in a French Guianan rainforest, based on video surveillance data. (A) Agoutis and acouchies (upper line) were much more rapid in detecting plots than were non-scatterhoarders such as collared peccaries (lower line). Data are from 102 seed plots ranging in seed species, size, site and date. Seeds were at least 1 g fresh mass and included monkey-dispersed species, bird-dispersed species and nut-bearing species. Artwork by Fiona A. Reid (Engstrom *et al.*, 1999). (B) Acouchies detected *Carapa procera* seed plots more rapidly in low seed years (upper line) than in high seed years (lower line).

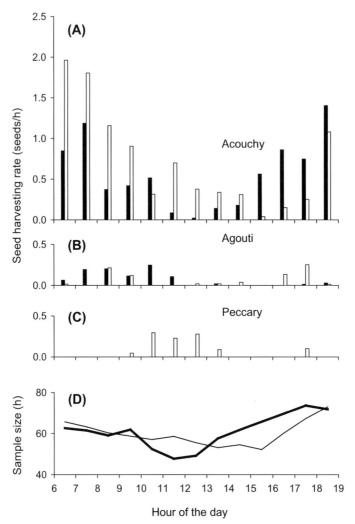

Fig. 22.3. Activity patterns of three granivorous mammals at 152 experimental seed plots in French Guianan rainforest, based on video recordings. Plotted is the average number of seeds removed per hour, for different hours of the day, by acouchies (A), agoutis (B) and peccaries (C), respectively. The bottom graph (D) shows the sampling intensity in terms of total hours of video per time interval. White bars and thin lines indicate low seed years, black bars and bold lines represent high seed years. Plots included *Carapa procera* (68) and *Licania alba* (49) or one of ten other species (45 plots). 1670 seed removal events were recorded.

contrasting seed abundance also suggest that acouchies increase foraging activity under food scarcity. Acouchies discovered the plots significantly more rapidly in low seed years (median time 171 min, $n = 38$ plots) than in high seed years (422 min, $n = 23$) (log rank test: $\chi^2_1 = 5.4$, $P = 0.02$; Fig. 22.2B). Food scarcity apparently increased foraging effort, which increases the probability of acouchies finding a given seed source.

Predators or dispersers?

Most plant species have several animal species that potentially eat their seeds, but these animals often differ in their impact on seed fate. Some species ignore the seeds, others act primarily as seed predators that immediately consume the seeds they find, and yet others act as potential seed dispersers by transporting seeds away from the parent plant and caching them. Members of the latter group, in turn, will differ in their performance as seed dispersers. Determining the impact of animal species therefore requires detailed seed fate data. We compared the contributions of different granivore species as predators and dispersers of two seed species, *Carapa* and *Licania*. Seeds of *Licania* are larger and have a greater nutritional content and longer storage life than *Carapa* (Jansen and Forget, 2001) and thus should be more valuable for storage purposes. We used video monitoring to identify the agent of removal for each individual seed in our experimental plots. We then searched the surrounding area to retrieve removed seeds by their numbered thread marks and record their fate. This made it possible to link individual seed fates to granivorous species or even individuals, and to assess the contribution of each animal species to predation or dispersal.

Figure 22.4 shows the proportions of seeds removed by four species of granivores for each seed species. Peccaries and squirrels removed few of the seeds (peccaries removed 5% of *Carapa*, squirrels 2% of *Licania*), yet they accounted for an important part of seed mortality. Peccaries crushed and consumed all *Carapa* seeds they handled, and accounted for 35% of all *Carapa* seeds found eaten. Squirrels take *Licania* seeds into trees and liana tangles (P.A. Jansen, Nouragues, 1998, personal observation). Because these sites are unsuitable for seedling establishment, arboreal caching is equivalent to killing seeds even if the seeds are not eaten. Acouchies accounted for the majority of recorded removal in both *Carapa* (62%) and *Licania* (60%). By far, most of the removed seeds that we retrieved were scatterhoarded (86% in *Carapa* and 99% in *Licania*). Agoutis, finally, were less important seed removers than acouchies, even though agoutis are especially well-known for scatterhoarding seeds in the Neotropics (e.g. Smythe, 1970; Forget and Milleron, 1991; Forget, 1992, 1993; Hallwachs, 1994; Peres and Baider, 1997; Silvius and Fragoso, 2003). They removed few *Carapa* seeds (2% of removal) but quite a number of *Licania* seeds (13%). We retrieved a low proportion of agouti-dispersed seeds, and, while most of these in *Carapa* were eaten (71%), all were scatterhoarded in *Licania*. For acouchy and agouti, removed seeds that we did not find were probably carried beyond the radius of our search area (see below). Hence, the smaller percentage consumed in *Licania* suggests that acouchies stored more seeds of this high-value species, while the smaller proportion of seeds we located suggests that dispersal of this preferred species was also further.

Body size and dispersal distance

Agoutis are two to six times heavier than acouchies (Emmons and Feer, 1997). Hence, we may expect that handling and transporting large nuts is easier for agoutis than for acouchies, and that agoutis disperse a given large seed species further than do acouchies. To test this prediction, we compared plot-to-cache distances of 415 individually thread-marked *Licania* seeds removed by acouchies to 105 seeds removed by agoutis. Contrary to our expectations, we found no significant difference in dispersal distance between agoutis and acouchies (log rank test: $\chi^2_1 = 0.44$, $P = 0.508$; Fig. 22.5), at least not for the first 25 m. This suggests that agoutis and acouchies have a similar impact with regard to dispersal distance, even though they differ in their seed preference.

Seed removal rates

The rate of seed removal varies widely among plant species, animal species, year and habitat. Many published studies treat removal rate as predation rate, with rapid removal treated as equivalent to low post-dispersal seed survival. If, however, the removal rate represents the motivation of animals to sequester and scatterhoard seeds, removal rate might rather be indicative for secondary seed dispersal, with rapid removal by scatterhoarders

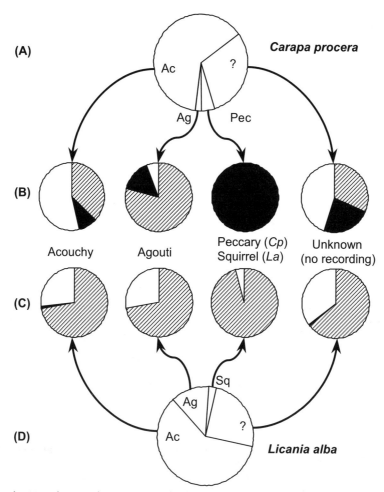

Fig. 22.4. Identities of mammals removing seeds of two large-seeded tree species in French Guianan rainforest, determined from remote video recordings, and the subsequent fate of seeds. (A) Proportions of rapidly germinating *Carapa procera* seeds ($n = 1537$) removed by different granivorous mammal species, and (B) the primary seed fate for each agent, with white slices indicating cached, black eaten, and hatched unknown (not retrieved on soil within the search radius). (D) Proportions of the slowly germinating *Licania alba* seeds ($n = 963$) removed by different granivorous mammal species, and (C) primary seed fate for each agent (slice patterns as in B). Question marks in (A) and (D) indicate the proportion of seed removal for which we had no video recordings. *L. alba* has a much slower germination rate and, thus, longer seed storage life than *C. procera*, and was therefore favoured as food. For acouchy and agouti, many of the missing seeds (B and C) were presumably dispersed beyond the search radius, especially in *L. alba*.

shortening the seed's exposure to seed predators. To investigate whether immediate predation of seeds in Nouragues increased with removal speed, we correlated the proportion of *Carapa* seeds consumed by mammals to the median time-to-removal of individual plots. We used video surveillance to record the exact times that seeds were removed during 1 or a few days. If it took longer for plots to be depleted, we switched to the standard method of recording seed presence according to a regular scheme of censuses. This approach allowed us to capture subtler differences in removal speed than with the standard method alone. Mean removal times of *Carapa* seed plots in French Guiana ranged from several hours to several weeks (Fig. 22.6). Plots with high

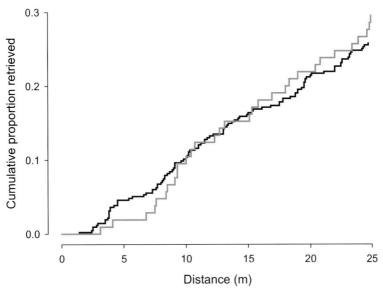

Fig. 22.5. Comparison of dispersal distances between different-sized scatterhoarders in French Guiana. Plotted is the cumulative retrieval of *Licania alba* seed caches made by red acouchies (*Myoprocta acouchy*; black line) and considerably larger red-rumped agoutis (*Dasyprocta leporina*; grey line), as a function of an increasing search radius from seed plots. Remote video recordings of seed removal were used to assign seeds and caches to animal species. Estimated cache distances were not significantly different (see text). Note that less than 30% of the seeds were found cached within 25 m; most of the remaining 70% were assumed to be dispersed beyond that distance, not actually found by us.

removal rate tended to have lower proportions of removed seeds eaten, rather than scatterhoarded, than did plots with slow removal rates (linear regression with time log-transformed: $\beta = 0.124$, $F_{1,43} = 7.9$, $P = 0.007$, $R^2 = 0.16$). Rapid seed removal in Nouragues is also correlated with larger dispersal distances (Jansen *et al.*, 2004). Both rapid removal and greater dispersal distance indicate a higher motivation of rodents to carefully store the seeds. If this pattern holds for other species and habitats, rapid seed removal could be a sign of higher-quality seed dispersal rather than higher predation.

Selectivity

Agoutis and acouchies in French Guiana and wood mice in The Netherlands remove and disperse only one seed at a time. We hypothesized that rodents would give

priority to removing the biggest seeds, because those have the greatest nutritional value, and yield the highest reward per caching effort and per cache (Smith and Reichman, 1984). The alternative strategy of removing seeds randomly, regardless of food value, would lengthen the exposure time of high-value seeds, and thus increase the risk of such seeds being monopolized by competitors. To test whether animals discriminated among seed sizes during removal, we determined the order of seed removal from *Carapa* experimental plots in which seed fresh mass varied widely (cf. Jansen *et al.*, 2004). We used video recordings to determine the exact sequence of seed removal. Then, we standardized and lumped the data from all plots, and used least absolute deviation (LAD) regression for different quantiles to test whether heavier seeds were given any priority. The significance values for fitted quantile regressions were calculated from 10,000 permutations (see Jansen *et al.*, 2002, for

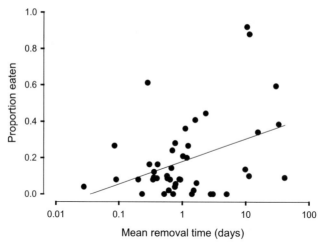

Fig. 22.6. Rate of *Carapa procera* seed removal from experimental plots in French Guiana and the proportion of seeds subsequently found consumed. Rate was calculated as (mean removal time)$^{-1}$, and mean removal time was estimated as the total waiting time of all seeds divided by the number of seeds removed. Rapid removal was correlated with low rates of immediate seed consumption (see text).

analysis details). There was indeed a tendency for large seeds to be removed before small seeds (Fig. 22.7): regressions at all quantiles had negative slopes, although significant only at the quantiles ≥ 0.25. The coefficient values, however, were modest (average $\beta = -0.003$), indicating that the effect was rather weak. The position of seeds within plots can have a greater effect on seed removal than seed size (Jansen *et al.*, 2002; see den Ouden *et al.*, Chapter 13, this volume).

Seed handling time

Seed fate studies generally suffer from incomplete retrieval of cached seeds (see Forget and Wenny, Chapter 23, this volume). Missing seeds are difficult to account for because they can be either overlooked within the search area, taken into burrows or into trees where they die, or dispersed outside the search area. Not accounting for missing seeds can cause serious bias in dispersal distances and survival rates, but so can assumptions on the fate of such seeds. We therefore investigated whether the handling time could serve as an alternative estimate for dispersal. In our experiments at

Speulderbos, we used video recordings to calculate the exact time that elapsed between the removal of numbered and magnet-tagged acorns by individual wood mice and their return to the plot for removing the next seed. Subsequently, we searched for scatterhoarded seeds with a magnetometer and measured their distance to the experimental plots as well as cache depth. Indeed, the time interval between visits correlated significantly with both plot-to-cache distance (linear regression, time and distance log-transformed: $R^2 = 0.42$, $F_{1,185} = 138$, $P < 0.001$; Fig. 22.8A) and cache depth ($R^2 = 0.10$, $F_{1,185} = 19.4$, $P < 0.001$; Fig. 22.8B), two elements of dispersal positively correlated with seed survival (e.g. Vander Wall, 1993; Hammond and Brown, 1998). The combined model $\log(t) = 1.57 + 0.73 \times \log(\text{distance in m}) + 0.37 \times \log(\text{depth in cm})$ explained more than half of the variation in interval duration (multiple regression: $R^2 = 0.51$, $F_{2,185} = 96$, $P < 0.001$). Cache distance and depth were not correlated ($r = 0.02$, $P = 0.78$). We conclude that interval time between visits was proportional to the investment of rodents in scatterhoarding and their performance as seed dispersers, and can be used as a potential estimate of dispersal under incomplete seed retrieval.

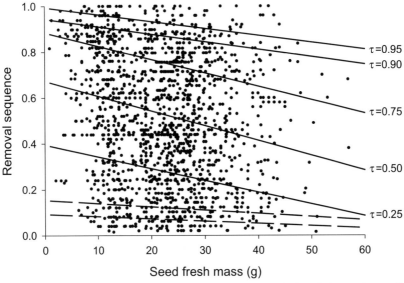

Fig. 22.7. Size discrimination by granivorous mammals during removal of seeds from cafeteria plots in a French Guianan rainforest. Plotted is the standardized removal sequence for 1638 *Carapa procera* seeds from 44 plots. Seeds that were among the first to be removed from a plot have sequence values near 0, whereas seeds among the last to be removed have values near 1. Lines are least absolute deviation regressions for different quantiles τ. The continuous lines indicate quantile regression slopes that were significantly different from zero ($P < 0.0002$), indicating that large seeds were removed faster than smaller ones. The quantile regressions at $\tau = 0.05$ and 0.10 (broken lines) were not significant ($P > 0.05$). P values were determined from 10,000 permutations.

Discussion

We have demonstrated that remote video monitoring of post-dispersal seed removal makes it possible to address a variety of questions that cannot be answered using traditional field methods. Video recordings allow for distinguishing actual dispersers and predators, and for linking agents to the fate of individual seeds. Both in French Guiana and The Netherlands, we found that most or all seed removal was by a single rodent species which scatterhoarded most or all of the seeds it took. We compared the performance of acouchies and the larger agoutis as seed dispersers and found that they produced indistinguishable seed shadows, at least at shorter distances. Video surveillance can also be used to measure the exact times at which events occur. Time recordings revealed that scatterhoarding rodents discovered seed plots much more rapidly than non-scatterhoarding mammals. Acouchies had the highest probability

of encountering seed sources, and were able to further increase this probability by shifting their activity towards the early morning in low seed years.

We have also shown how time recordings are an improvement compared to the standard method of recording seed presence in intervals, because they yield continuous data that can be more effectively analysed using survival analysis (e.g. Fox, 1993; Zens and Peart, 2003), without problems of interval-censoring, and, if necessary, with detection time and removal time separated. The ability to capture subtle differences in removal rate allows for comparisons even if seed removal is rapid (e.g. for preferred seed species or under seed scarcity). Thus, we found that rapid removal of nuts in French Guiana was indicative of high secondary dispersal rather than high immediate predation. Finally, remote video monitoring yields data on the behaviour of free-ranging animals that discover and explore seed plots. We showed, for example, that

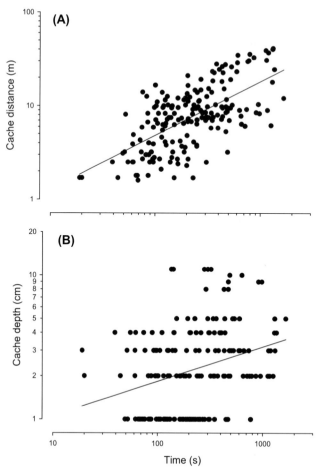

Fig. 22.8. Scatterhoarding of acorns (*Quercus robur*) by wood mice (*Apodemus sylvaticus*) in a Dutch oak–beech (*Fagus sylvatica*) forest. Time elapsed between an individual removing an acorn and that same individual returning for the next acorn was correlated with the subsequent (A) cache distance and (B) cache depth of that same acorn. The interval between subsequent removal events was proportional to caching effort and dispersal distance.

granivores in French Guiana tended to remove large seeds before smaller ones, supporting the idea that large seeds are preferred to small ones (Smith and Reichman, 1984). And we showed that time intervals between successive visits of wood mice to seed plots in The Netherlands were proportional to the dispersal distance of scatterhoarded acorns. Interactions between food competitors at seed sources are another example of this type of behavioural data (e.g. den Ouden *et al.*, Chapter 13, this volume).

Remote video monitoring has already been applied in a wide range of animal field studies, including provisioning rates and predation at bird nests (e.g. Ouchley *et al.*, 1994; Franzreb and Hanula, 1995; Delaney and Grubb, 1999; Keedwell and Sanders, 2002; Stake and Cimprich, 2003), behaviour of badgers (*Meles meles*) at middens (Stewart *et al.*, 1997), activity of overwintering bats in caves (Sedgeley, 2001), predation of reef sponges (Dunlap and Pawlik, 1996) and the use of fauna passages (Mathiasen and Madsen, 2000). Video cameras have even been mounted on the back of seals to study their hunting behaviour (Davis *et al.*, 1999). However, this technology has not been applied to seed fate studies (but see Jansen *et al.*, 2002, 2004; see den Ouden *et al.*, Chapter 13, this volume). Complications for use in field studies in the past have been, amongst others, the equipment's bulkiness, sensitivity to weather and high cost.

Recent technological and methodological innovations have made remote monitoring of seed removal in the field increasingly feasible. Compact, high quality systems can now be assembled largely from consumer electronics readily available at reasonable cost. Digital techniques permit computer storage and analysis of video fragments, which can be copied into presentations or other media. Digital video cameras can record at low light levels, and have built-in features for nocturnal recording using an infrared beam to illuminate study areas. Most models can record date and time in seconds. The cameras come with compact lithium batteries that allow for up to 10 or more hours of recording and up to several weeks of stand-by, and higher-capacity batteries are also available. Waterproof housing, required for operating these cameras outdoors, is available from camera brands as well as specialist diving companies. The typically yellow cases can be painted in camouflage colours and mounted on a tree or post using a wall bracket. Passive infrared sensors allow recording of events only, switching the camera to stand-by if there is no vertebrate activity. A more up-to-date system that we are now using is a Sony DCR-VTR18E mini DV handycam in Sony ® SPK-TRV33 weatherproof housing in combination with a TrailMaster ® 700v passive infrared sensor. A good but cheaper alternative for electronic control of digital cameras, 35 mm cameras and video camcorders is PixController ® control boards that use an integrated Passive Infrared (PIR) Motion Control circuit to trigger devices.

Monitoring seed removal with remote video cameras is especially efficient if seed removal is rapid. Then, it takes only 1 or a few days to record near-complete depletion of a seed plot, after which the camera can readily be used for the next replicate. Realizing ten replicates per month with a single camera unit is then feasible, allowing sufficient replication within a limited time span (fieldwork period or fruiting season) to capture variation in removal rates and seed fates between sites and individuals with only one or few cameras. However, remote video cameras are less practical for full monitoring of plots with slow seed removal rates, because one would need to operate many parallel systems. In our experiments, we usually stopped video monitoring after 2 or 3 days if seed removal was slow, because we found replication more important. Photo-camera traps are cheaper and seem a more suitable way for monitoring slow seed removal and other relatively rare events simultaneously at many places.

Remote video monitoring cannot replace studies of post-removal seed fate, in which seeds are tracked until they either die or establish seedlings. Without data on the ultimate fate of cached seeds, video recordings still leave plenty of room for the argument that scatterhoarding is only putting off the evil hour – cached seeds are food reserves – and is of negligible net benefit (e.g. Larson and Howe, 1987; Pena-Claros and De Boo, 2002). What video surveillance can do, however, is strongly reinforce seed fate tracking experiments and thereby make such high-effort studies more rewarding.

Acknowledgements

This paper was presented at the 2002 annual meeting of the Association for Tropical Biology and Conservation in Panama City, Panama. We thank Pierre-Michel Forget and Joe Wright for the invitation. We are grateful to Martijn Bartholomeus, Marion Diemel, Jelmer Elzinga, Pascalle Jacobs, Sander Koenraadt, Willem Loonen, Melchert Meijer zu Schlochtern, Anne Rutten, Almira Siepel and Jaco van Altena for their help with the fieldwork. We thank the Centre National de Recherche Scientifique/Pierre Charles-Dominique and the Dutch State Forest Service/Harry Heer for permission to work in Nouragues and Speulderbos, respectively. Frans Bongers, Steve Vander Wall, Harald Beck and Steven Brewer provided useful comments on the manuscript. Fieldwork was supported by the Netherlands Foundation for the Advancement of Tropical Research (NWO-WOTRO grant W84–407) and Wageningen University.

References

Beck, H. and Terborgh, J. (2002) Groves versus isolates: how spatial aggregation of *Astrocaryum murumuru* palms affects seed removal. *Journal of Tropical Ecology* 18, 275–288.

Bodmer, R.E. (1991) Strategies of seed dispersal and seed predation in Amazonian ungulates. *Biotropica* 23, 255–261.

Bongers, F., Charles-Dominique, P., Forget, P.-M. and Théry, M. (2001) *Nouragues: Dynamics and Plant–Animal Interactions in a Neotropical Rainforest*. Kluwer Academic Publishers, Dordrecht, The Netherlands, 421 pp.

Brewer, S.W. and Rejmanek, M. (1999) Small rodents as significant dispersers of tree seeds in a Neotropical forest. *Journal of Vegetation Science* 10, 165–174.

Chambers, J.C. and Macmahon, J.A. (1994) A day in the life of a seed: movements and fates of seeds and their implications for natural and managed systems. *Annual Review of Ecology and Systematics* 25, 263–292.

Crawley, M.J. (2000) Seed predators and plant population dynamics. In: Fenner, M. (ed.) *Seeds. The Ecology of Regeneration in Plant Communities*. CAB International, Wallingford, UK, pp. 157–191.

Davis, R.W., Fuiman, L.A., Williams, T.M., Collier, S.O., Hagey, W.P., Kanatous, S.B., Kohin, S. and Horning, M. (1999) Hunting behavior of a marine mammal beneath the Antarctic fast ice. *Science* 283, 993–996.

Delaney, D.K. and Grubb, T.G. (1999) Activity patterns of nesting Mexican spotted owls. *Condor* 101, 42–49.

den Ouden, J. (2000) The role of bracken (*Pteridium aquilinum*) in forest succession. PhD thesis, Wageningen University, Wageningen, The Netherlands.

Dubost, G. (1988) Ecology and social life of the red acouchy, *Myoprocta exilis*; comparisons with the orange-rumped agouti, *Dasyprocta leporina*. *Journal of Zoology* 214, 107–113.

Dunlap, M. and Pawlik, J.R. (1996) Video-monitored predation by Caribbean reef fishes on an array of mangrove and reef sponges. *Marine Biology Berlin* 126, 117–123.

Emmons, L.H. and Feer, F. (1997) *Neotropical Rainforest Mammals: A Field Guide*, 2nd edn. University of Chicago Press, Chicago, Illinois, 307 pp.

Engstrom, M.D., Lim, B.K. and Reid, F.A. (1999) *A Guide to the Mammals of the Iwotrama Forest*. Iwokrama International Centre, Georgetown, Guyana, 32 pp. Available at: http://www.iwokrama.org/mammals/guide.pdf

Forget, P.-M. (1990) Seed dispersal of *Vouacapoua americana* (Caesalpiniaceae) by caviomorph rodents in French Guiana. *Journal of Tropical Ecology* 6, 459–468.

Forget, P.-M. (1992) Seed removal and seed fate in *Gustavia superba* (Lecythidaceae). *Biotropica* 24, 408–414.

Forget, P.-M. (1993) Post-dispersal predation and scatterhoarding of *Dipteryx panamensis* (Papilionaceae) seeds by rodents in Panama. *Oecologia* 94, 255–261.

Forget, P.-M. (1996) Removal of seeds of *Carapa procera* (Meliaceae) by rodents and their fate in rainforest in French Guiana. *Journal of Tropical Ecology* 12, 751–761.

Forget, P.-M. and Milleron, T. (1991) Evidence for secondary seed dispersal by rodents in Panama. *Oecologia* 87, 596–599.

Fox, G.A. (1993) Failure-time analysis: emergence, flowering, survivorship, and other waiting times. In: Scheiner, S.M. and Gurevitch, J. (eds) *Design and Analysis of Ecological Experiments*. Chapman & Hall, New York, pp. 253–289.

Franzreb, K.E. and Hanula, J.L. (1995) Evaluation of photographic devices to determine nestling diet of the endangered Red-cockaded Woodpecker. *Journal of Field Ornithology* 66, 253–259.

Gayot, M., Henry, O., Dubost, G. and Sabatier, D. (2004) Comparative diet of the two forest cervids of the genus *Mazama* in French Guiana. *Journal of Tropical Ecology* 20, 31–43.

Guillotin, M. (1982) Activités et régimes alimentaires: leurs interactions chez *Proechimys cuvieri* et *Oryzomys capito velutinus* (Rodentia) en forêt guyanaise. *Revue d'Ecologie (La Terre et la Vie)* 36, 337–371.

Guillotin, M., Dubost, G. and Sabatier, D. (1994) Food choice and food competition among the three major primate species of French Guiana. *Journal of Zoology* 233, 551–579.

Hallwachs, W. (1994) The clumsy dance between agoutis and plants: scatterhoarding by Costa Rican dry forest agoutis (*Dasyprocta punctata*: Dasyproctidae: Rodentia). PhD thesis, Cornell University, Ithaca, New York.

Hammond, D.S. and Brown, V.K. (1998) Disturbance, phenology and life-history characteristics: factors influencing frequency-dependent attack on tropical seeds and seedlings. In: Newbery, D.M., Brown, N. and Prins, H.H.T. (eds) *Dynamics of Tropical Communities*. Blackwell Science, Oxford, pp. 51–78.

Henry, O. (1997) The influence of sex and reproductive state on diet preference in four terrestrial mammals of the French Guianan rain forest. *Canadian Journal of Zoology* 75, 929–935.

Hoshizaki, K., Suzuki, W. and Nakashizuka, T. (1999) Evaluation of secondary dispersal in a large-seeded tree *Aesculus turbinata*: a test of directed dispersal. *Plant Ecology* 144, 167–176.

Hughes, A.G. and Shorrock, G. (1998) Design of a durable event detector and automated video surveillance unit. *Journal of Field Ornithology* 69, 549–556.

Hulme, P.E. (1998) Post-dispersal seed predation: consequences for plant demography and evolution. *Perspectives in Plant Ecology, Evolution and Systematics* 1, 32–46.

Hulme, P.E. (2002) Seed eaters: seed dispersal, destruction and demography. In: Levey, D.J., Silva, W.R. and Galetti, M. (eds) *Seed Dispersal and Frugivory: Ecology, Evolution and Conservation*. CAB International, Wallingford, UK, pp. 257–273.

Jansen, P.A. (2003) Scatterhoarding and tree regeneration: ecology of nut dispersal in a Neotropical rainforest. PhD thesis, Wageningen University, Wageningen, The Netherlands.

Jansen, P.A. and Forget, P.M. (2001) Scatterhoarding and tree regeneration. In: Bongers, F., Charles-Dominique, P., Forget, P.-M. and Théry, M. (eds) *Nouragues: Dynamics and Plant–Animal Interactions in a Neotropical Rainforest*. Kluwer Academic Publishers, Dordrecht, The Netherlands, pp. 275–288.

Jansen, P.A., Bartholomeus, M., Bongers, F., Elzinga, J.A., Den Ouden, J. and Van Wieren, S.E. (2002) The role of seed size in dispersal by a scatterhoarding rodent. In: Levey, D.J., Silva, W.R. and Galetti, M. (eds) *Seed Dispersal and Frugivory: Ecology, Evolution and Conservation*. CAB International, Wallingford, UK, pp. 209–225.

Jansen, P.A., Hemerik, L. and Bongers, F. (2004) Seed mass and mast seeding enhance dispersal by a Neotropical scatter-hoarding rodent. *Ecological Monographs* 74, 569–589.

Keedwell, R.J. and Sanders, M.D. (2002) Nest monitoring and predator visitation at nests of Banded Dotterels. *Condor* 104, 899–902.

Kiltie, R.A. (1981) Stomach contents of rain forest peccaries (*Tayassu tajacu* and *T. pecari*). *Biotropica* 13, 234–236.

Kiltie, R.A. (1982) Bite force as a basis for niche differentiation between rain forest peccaries (*Tayassu tajacu* and *T. pecari*). *Biotropica* 14, 188–195.

Kucera, T.E. and Barrett, R.H. (1993) The Trailmaster camera system for detecting wildlife. *Wildlife Society Bulletin* 21, 505–508.

Larson, D. and Howe, H.F. (1987) Dispersal and destruction of *Virola surinamenis* seeds by agoutis: appearance and reality. *Journal of Mammalogy* 68, 859–860.

Levey, D.J. and Byrne, M.M. (1993) Complex ant–plant interactions: rain forest ants as secondary dispersers and post-dispersal seed predators. *Ecology* 74, 1802–1812.

Mathiasen, R. and Madsen, A.B. (2000) Infrared video-monitoring of mammals at a fauna underpass. *Zeitschrift für Saeugetierkunde* 65, 59–61.

Miura, S., Yasuda, M. and Ratnam, L.C. (1997) Who steals the fruits? Monitoring frugivory of mammals in a tropical rain forest. *Malayan Nature Journal* 50, 183–193.

Moles, A.T. and Westoby, M. (2003) Latitude, seed predation and seed mass. *Journal of Biogeography* 30, 105–128.

Ouchley, K., Hamilton, R.B. and Wilson, S.D. (1994) Nest monitoring using a micro-video camera. *Journal of Field Ornithology* 65, 410–412.

Page, L.K., Swihart, R.K. and Kazacos, K.R. (2001) Seed preferences and foraging by granivores at raccoon latrines in the transmission dynamics of the raccoon roundworm (*Baylisascaris procyonis*). *Canadian Journal of Zoology* 79, 616–622.

Pena-Claros, M. and De Boo, H. (2002) The effect of forest successional stage on seed removal of tropical rain forest tree species. *Journal of Tropical Ecology* 18, 261–274.

Peres, C.A. and Baider, C. (1997) Seed dispersal, spatial distribution and population structure of Brazilnut trees (*Bertholletia excelsa*) in southeastern Amazonia. *Journal of Tropical Ecology* 13, 595–616.

Price, M.V. and Jenkins, S.H. (1986) Rodents as seed consumers and dispersers. In: Murray, D.R. (ed.) *Seed Dispersal*. Academic Press, Sydney, pp. 191–235.

Sedgeley, J.A. (2001) Winter activity in the tree-roosting lesser short-tailed bat, *Mystacina tuberculata*, in a cold-temperate climate in New Zealand. *Acta Chiropterologica* 3, 179–195.

Shaw, M.W. (1968) Factors affecting the natural regeneration of sessile oak (*Quercus petraea*) in north Wales. I. A preliminary study of acorn production, viability and losses. *Journal of Ecology* 56, 565–583.

Silvius, K.M. and Fragoso, J.M.V. (2003) Red-rumped Agouti (*Dasyprocta leporina*) home range use in an Amazonian forest: implications for the aggregated distribution of forest trees. *Biotropica* 35, 74–83.

Smith, C.C. and Reichman, O.J. (1984) The evolution of food caching by birds and mammals. *Annual Review of Ecology and Systematics* 15, 329–351.

Smythe, N. (1970) Ecology and behavior of the agouti (*Dasyprocta punctata*) and related species on Barro Colorado Island, Panama. PhD thesis, University of Maryland, Baltimore, Maryland.

Stake, M.M. and Cimprich, D.A. (2003) Using video to monitor predation at black-capped vireo nests. *Condor* 105, 348–357.

Stewart, P.D., Ellwood, S.A. and MacDonald, D.W. (1997) Remote video-surveillance of wildlife: an introduction from experience with the European badger *Meles meles*. *Mammal Review* 27, 185–204.

Takada, M., Yasuda, M., Kanzaki, M. and Yamakura, T. (2001) Seed dispersal of *Sapindus mukorossi* identified by an automatic camera system at the Kasugayama forest, Central Japan. *Journal of Kansai Organization of Nature Conservation* 23, 3–12.

Vander Wall, S.B. (1990) *Food Hoarding in Animals*. University of Chicago Press, Chicago, Illinois, 445 pp.

Vander Wall, S.B. (1992) The role of animals in dispersing a 'wind-dispersed' pine. *Ecology* 73, 614–621.

Vander Wall, S.B. (1993) A model of caching depth: implications for scatter hoarders and plant dispersal. *American Naturalist* 141, 217–232.

Vander Wall, S.B. (1994) Seed fate pathways of antelope bitterbrush: dispersal by seed-caching yellow pine chipmunks. *Ecology* 75, 1911–1926.

Vander Wall, S.B. (1997) Dispersal of singleleaf pinon pine (*Pinus monophylla*) by seed-caching rodents. *Journal of Mammalogy* 78, 181–191.

Yasuda, M., Miura, S. and Hussein, N.A. (2000) Evidence for food hoarding behaviour in terrestrial rodents in Pasoh forest reserve, a Malaysian lowland rain forest. *Journal of Tropical Forest Science* 12, 164–173.

Zens, M.S. and Peart, D.R. (2003) Dealing with death data: individual hazards, mortality and bias. *Trends in Ecology and Evolution* 18, 366–373.

23 How to Elucidate Seed Fate? A Review of Methods Used to Study Seed Removal and Secondary Seed Dispersal

Pierre-Michel Forget[1] and Dan Wenny[2]

[1]Muséum National d'Histoire Naturelle, Département Ecologie et Gestion de la Biodiversité, UMR 5176 CNRS-MNHN, 4 avenue du Petit Château, F-91800 Brunoy, France; [2]Illinois Natural History Survey, Lost Mound Field Station, 3159 Crim Drive, Savanna, IL 61074, USA

Introduction

The role of frugivorous vertebrates as seed dispersers in tropical rainforests has received much attention over the past 30 years (Estrada and Fleming, 1986; Fleming and Estrada, 1993; Levey *et al.*, 2002). Most studies have focused on endozoochory by birds, primates and bats. In contrast, the scatterhoarding activities of animals, shown to be effective for seedling recruitment in temperate regions, were poorly known in the tropics. Such caching or hoarding of seeds by invertebrates (ants, dung beetles) or vertebrates (birds, rodents) is often a second stage of dispersal that leads to rearrangement of the seed shadow and thus is a vital link in the recruitment stage of plants. Despite recurrent references to hoarding behaviour by animals (Vander Wall, 1990), their role as efficient seed dispersers in the tropics only started to be documented in the last decade when a variety of new techniques were developed (Levey *et al.*, 2002).

The aim of this chapter is to review the methods used to track seed fate during and after removal with a special emphasis on secondary seed dispersal and diplochory (see Vander Wall and Longland, Chapter 18, this volume). Our objectives are to review the benefits and limitations of marking methods with regard to seed size, animal size, habitat and duration of study. Then, in light of the recent literature, we present recommendations on how to document the fate of removed seeds with greater precision. To prepare this review, we compiled literature first from available peer-reviewed journals, and from references cited in these papers, and finally from online keyword searches with Current Contents and Biosis. This review is certainly incomplete in the sense that we did not attempt to review all seed removal studies but focused on those that were designed to determine final seed fate. We believe that the large sample of studies we consulted is probably representative of the methods used and information available.

The term 'seed fate' has become widely used in recent years to describe what happens to seeds between production and seedling establishment, though an exact definition should include pre-dispersal mortality, removal and dispersal away from the parent tree. In the 1970s and 1980s, studies typically used the term seed removal synonymously with seed predation even though ultimate fate often was not studied. Unfortunately, this is all too often still true. The possibility of further dispersal after primary dispersal generally was not considered in most early studies. Although seed caching by animals in the temperate zone was known to be common in many communities (Vander Wall, 1990), it was typically the primary stage of dispersal (e.g. squirrels caching nuts). Seed caching by agoutis was known in the early 1970s (Smythe, 1970, 1978), but such behaviour was not studied as an important stage of dispersal until the late 1980s (Forget, 1990, 1991; Forget and Milleron, 1991). During this period, it became apparent that several groups of tropical rodents played important roles in forest dynamics as seed predators and primary and secondary dispersers. At the same time, it became clear that seed removal is not synonymous with seed predation, and studies of seed movement beyond seed removal are needed to fully evaluate the role of dispersers. Studies throughout the tropics in the 1990s led to the realization that secondary dispersal is a diverse, widespread phenomenon. Although some authors continue to assume that seed removal is equivalent to seed predation, most researchers have become more careful to clearly state the elements of seed fate included in their study.

Seed Fate Methods

Our survey of the literature yielded 16 methods to track seeds or fruits. These methods range from simple and inexpensive to sophisticated and expensive. Most studies have been done in the field but several have been conducted in a laboratory or aviary. We grouped the studies into three categories based on the methodology and results sought. First, studies that determined seed fate by direct visual observation of animals in the wild or in captivity. Many of these focused on animal behaviour rather than the fate of individual seeds. Second, studies in which seeds were attached to a

fixed point such as a line that is tied to a stake or plant stem. These studies typically involve monitoring seed removal over time to determine if seeds are removed by animals and at what rates. Variations of this method (e.g. spool-and-line) provide information on seed dispersal and fate. This method allows seeds to be moved beyond the immediate proximity, although the seeds remain attached to a point. Third, studies that involved marked seeds or fruits that were monitored to record removal and to determine ultimate destination and fate.

Visual observations

Seed dispersal behaviour can be observed directly in open habitats, with small animals that travel short distances, or with animals in captivity. Few observation-based studies, however, include data on final post-dispersal fate of seeds. Some studies estimate dispersal distances and cache locations with the help of landmarks. For example, Darley-Hill and Johnson (1981) studied seed dispersal of *Quercus palustris* acorns by blue jays (*Cyanocitta cristata*) in Virginia, USA. They colour-marked birds and followed them to determine whether they consumed or cached acorns. Based on the time jays travelled before caching acorns, they estimated dispersal distances. Similarly, Moreno and Carrascal (1995) studied European nuthatches (*Sitta europea*) and sunflowers in Spain. Birds were marked with coloured rings and observed. Distances to caches were estimated based on flight times. Hutchins *et al.* (1996) studied caching behaviour of Eurasian nutcrackers (*Nucifraga caryocatactes*) and red squirrels (*Sciurus vulgaris*) in a pine plantation in China. Nutcrackers transported seeds in sublingual pouches, caching up to nine Korean pine (*Pinus koraiensis*) seeds (mean 2.6 seeds) per site as far as 1 km from the plantation. Nuthatches (*S. europea*) and squirrels were observed taking seeds 10–50 m and 5–10 m, respectively. Chipmunks (*Tamias sibiricus*) gathered seeds from the ground, loaded them into their cheek pouches, and transported them at least 20 m to cache sites. Several avian post-dispersal seed predators were noticed during these observations. In all instances, long-distance flights of birds prevented a detailed examination of the seed dispersal process, and distances given in these studies remain rough estimates of seed dispersal by birds (see den Ouden, et al., Chapter 13, this volume). More precise data on cache locations are possible when diurnal rodents are the focal species, especially when the field of view is unobstructed. For instance, Cahalane (1942) studied fox squirrels (*S. niger rufiventer*) in the Ann Arbor cemetery in Michigan, USA. He was able to locate and mark caches that were made by habituated squirrels, and follow cache fate through winter and early spring. Similarly, Steele *et al.* (1996) observed habituated grey squirrels (*Sciurus carolinensis*) in an urban park in northeastern Pennsylvania to test the effect of acorn (*Quercus* spp.) infestation by weevils on seed removal, and documented the distance acorns were cached.

The ability to observe naturally foraging squirrels and birds and their propensity to make caches close to the food source has then been used intensively to test hypotheses regarding the evolution of scatterhoarding behaviour and optimal-density models (Stapanian and Smith, 1978; Kraus, 1983; Benkman *et al.*, 1984; Clarkson *et al.*, 1986). Hurly and Roberston (1987) observed red squirrels (*Tamiasciurus hudsonicus*) equipped with coloured collars caching seeds of scotch pine (*Pinus sylvestris*) and red pine (*P. resinosa*) 5.8 and 10.1 m, respectively, from the seed source. The larger and more valued items were taken farther than the smaller and less valuable items in a pine plantation. Similarly, Jokinen and Suhonen (1995) compared scatterhoarding by colour-marked tits (*Parus montanus* and *P. cristatus*) and found that, on average, caching distances were greater for large hulled sunflower (*Helianthus* sp.) seeds (36 m) than for smaller spruce (*Picea abies*) seeds (7 m).

In the Tropics, visual observation of animals removing seeds is more difficult.

Remote camera (Yasuda *et al.*, 2000; Beck and Terborgh, 2002; see Yasuda *et al.*, Chapter 9, this volume) and video surveillance systems (Jansen, 2003; Jansen and den Ouden, Chapter 22, this volume) have been used to record seed removal in both temperate and tropical habitats. In most instances only records of removal or predation are feasible, as an animal taking a seed will usually walk out of range of cameras. However, by individually marking seeds and animals, one may be able to identify with accuracy which individuals consume seeds at the source, or remove to consume or to cache them elsewhere. Such a method is also very profitable as one may carry out studies on selective seed choice to measure the effect of nutrient availability and seed size on seed removal and fate.

Direct observations of caching behaviour of spiny rats (*Proechimys* spp.) and acouchies (*Myoprocta acouchy*) in French Guiana were possible from a blind placed near feeding sites in the forest (Forget, 1990, 1991). Visual observations of other large neotropical rodents such as agoutis (*Dasyprocta* spp.) are facilitated in forests with open understorey (Hallwachs, 1986, 1994). Although time consuming, such opportunities to study caching behaviour should be used more often, particularly when it is possible to observe individually marked animals. However, marking seeds is essential in the wild when seeds are taken by vertebrates and transported out of sight, especially when seed fate is not solely dependent on the first, visible, short-distance movement (Chambers and MacMahon, 1994).

Determination of seed fate in the Tropics via observation is also regularly used in studies focusing on seed removal by invertebrates such as ants and dung beetles. Seeds in these studies are rarely marked because of their small size. However, the short distances travelled by dispersers eliminates the need for marking. Kaspari (1993) recorded foraging behaviour of soil-nesting ants in Costa Rica. He observed ants removing seeds from experimental bird droppings and carrying them to nest entrances an average distance of 45 cm (see also Levey and Byrne, 1993; Levey and Sargent, 2000).

Böhning-Gaese *et al.* (1995, 1999) in Madagascar observed large *Aphaenogaster swammerdami* ants carrying *Comniphora guillaumini* seeds an average of 4.4 m. In contrast, smaller *Pheidole* sp. ants carried seeds an average of 0.9 m. Similarly, while directly observing dung beetle rolling behaviour, Andresen (1999) determined that dispersal distance was related to beetle size. The larger ($\approx$ 13 mm long) and smaller ($\approx$ 8 mm) *Canthon* beetle species moved dung (which contained seeds) mean distances of 118 and 82 cm, respectively (see also Feer, 1999; Andresen and Feer, Chapter 20, this volume).

An alternative to direct observation of seed-handling animals is to develop indirect methods of recording animal movements. One such option is to mark trapped rodents with a fluorescent powder and retrace the travel route with an ultraviolet lamp (Lemen and Freeman, 1985; Miyaki and Kikuzawa, 1988; Longland and Clements, 1995). After release the animal will leave some powder at each cache location. This method works best in dry habitats or seasons, and is likely to be less effective in either tropical or temperate rainforests. In wet habitats, most powder rubs off and sticks to leaves in the first few metres (P. Jansen, Wageningen, 2003, personal communication). It also requires working at night to find caches, which may be difficult, dangerous, or disruptive in some habitats. Researchers should consider the impact of persistent pigments on the environment and on future studies in the same site (Halfpenny, 1992).

Fixed Marking Methods

Tethered line

Attaching seeds with a thread to a fixed point, such as a small tree or twig, is useful for distinguishing biotic and abiotic causes of seed movement. Fixed-attachment methods ensure that seeds or propagules are not lost and are useful for determining which types of animals remove seeds (e.g. rodents, birds or insects). This method avoids seeds

being blown away by wind or washed away by rain, and allows the experimenter to identify seed fate on the spot of release by eventually retrieving the seed (if not removed) or seed remains. Tethered lines have been used extensively in temperate forests (Kikusawa, 1988; Whelan *et al.*, 1991; Okamoto, 1994; Vander Wall, 1994; Moegenburg, 1996), tropical rainforests (Schupp, 1988a,b, 1990; Theimer and Gehring, 1999) and mangroves (Smith, 1987; Smith *et al.*, 1989; Sousa and Mitchell, 1999). Small seeds can also be glued to a 2.5-cm Petri dish that is glued to the head of a 10-cm nail (Mittelbach and Gross, 1984; Whelan *et al.*, 1991; Kollmann and Bassin, 2001). Alternatively, the initial location of seeds is sometimes marked with a wire-stake flag (Schupp and Frost, 1989) or a plastic toothpick (Sork and Boucher, 1977) or other man-made objects. This method documents seed removal or consumption on site, but the ultimate fate of removed seeds remains unknown.

A common practice when studying seed removal and survival is to mark locations with coloured stakes, flagging tape or toothpicks, or to place seeds in plastic dishes to limit seed losses from rain or other abiotic factors. While using wooden stakes 3 cm from caches made by squirrels, Cahalane (1942) acknowledged that there was a risk that such landmarks might be used by rodents to find buried items. To reduce this possibility, he added more stakes away from the nuts, and reported that his wooden markers did not seem to affect squirrel behaviour, but he did not explicitly test this possibility. The fact remains, however, that because virtually 100% of the nuts were retrieved by squirrels, the stakes may well have had an effect that has not been ruled out by adding more stakes. Thus, this method can have some serious drawbacks because these markers can serve as visual cues for some visually oriented seed predators (Vander Wall and Peterson, 1996; see Pyare and Longland, 2000). Most researchers assume that the behaviour of animals will be unaffected by such marks, but this is not true, or at least is rarely tested by experimenters prior to the study. Using white plastic labels, for instance, it is clear that animals may be attracted to them (P.-M. Forget, personal observation), thus potentially biasing the results of experiments. Animals can also learn to associate marking systems with food sources very quickly. If auxiliary markers are used, preliminary studies should be conducted to demonstrate that they do not influence seed removal rates. Markers may simulate germinating seedlings with white young sprouting leaves indicating the location of a potential resource, i.e. the buried cotyledons (N. Smythe, Gamboa, Panama, 1990, personal communication; P.-M. Forget, personal observation). Comparative studies need to be done to document which colours, sizes and types of artificial marks are more reliable and less likely to inform animals on the location of potential resources. Such experiments may be helpful with nocturnal rodents that may not be able to see the markers, but for diurnal animals this is likely to be a waste of time. Vander Wall's experience with yellow pine chipmunks, which may be true of rodents and birds in general, is that they will learn quickly whatever man-made objects one uses to mark cache sites if they are used consistently (S.B. Vander Wall, 2004, personal communication). The only solution he and colleagues found is to use natural objects (cones, twigs, pebbles, etc.) in unnatural patterns. However, even these can be learned if used in the same area repeatedly.

Spool-and-line

The spool-and-line protocol consists of a thread-filled bobbin from which the line is dispensed as the seed attached to the end of the line is carried away. The animal's travel route and final seed location and fate can be determined by following the line. The spool-and-line method is an extension of the fixed attachment methods. In the former, the line length varies and, in theory, accommodates the full range of distances seeds could be taken. Depending on the size of the item removed, the bobbin may be attached to the fixed point (Yasuda et al., 1991; Soné and Kohno, 1996), to the surface of the seed (Theimer, 2001, 2003), or inserted inside the seed (Hallwachs, 1986; Dennis, 2003). For example, Hallwachs (1986) replaced one seed in multi-seeded Hymenaea courbaril pods with a small bobbin to track the fate of the entire fruit. In this case, however, when the fruit is followed one must also mark the enclosed seeds as these will be taken, consumed or eventually secondarily dispersed after the fruit has been carried some distance and opened. Similarly, Dennis (2003) simulated fruit of Entiandra sankeyana (Lauraceae) by using a pitted Chinese date (Ziziphus jujube) surrounding a cotton bobbin representing the seed. The spool-and-line method allows tracking over fairly large distances, i.e. 55 m in Beilschmiedia bancroftii (Theimer, 2001), 68 m in the E. sankeyana simulated fruit (Dennis, 2003) and 225 m in Hymenaea courbaril (Hallwachs, 1986).

Compared to the tethered method, the spool-and-line method offers an advantage for the determination of seed fate, but is not without drawbacks. On one hand, tracing the line allows a rapid recovery of removed seeds that may otherwise only be found by chance. On the other hand, if the line is broken, a common event (Soné and Kohno, 1996), one gains little information other than removal. Animals sometimes abandon or consume marked seeds, when the line gets entangled in vegetation. Line entanglement is more likely when the spool is attached to the source rather than to the seed (P.A. Jansen, personal communication; D. Wenny, personal observation). Preparing spools can be time consuming when one needs hundreds or thousands of seeds for one experiment. The spool-and-line method is more efficient at describing seed shadows of ground-dwelling rather than arboreal animals. For this reason, once secondary seed dispersal has been demonstrated using the spool-and-line method, it might be better to individually mark seeds with a free-ranging labelling method (see next section) that would allow researchers to relocate cached and gnawed seeds over a large area. Such a two-part study could be used to test for biases in the spool-and-line method.

Free Marking Methods

Most seed fate studies utilize a free marking method to relocate the removed seeds, and to discriminate between seed predation and hoarding. Free marking methods may be either internal (radioisotopes, radio-transmitter, magnets or metal objects) or external (thread, flagging tape or other light conspicuous object). Internal marking methods necessarily involve some device to detect the removed seeds whereas external marking usually involves visual observations. Whatever type of free marking method is used, it is important to note that when seeds are large enough, it is useful to number and individually mark them, allowing one to follow seed fates at the level of the individual. Individual labelling of seeds can facilitate the understanding of complex seed fate pathways with primary and repeated recaching of seeds (Jansen et al., 2002; Vander Wall, 2002). This can be done with indelible ink (e.g. extra fine tip Sharpie®) directly on the seed or, when the line system is used, the line can be equipped with a numbered tag at the end (e.g. Zhang and Wang, 2001) or a bar code can be printed directly on the line (e.g. Forget, 1992, 1993).

Internal cue

Internal labelling has been used in a wide range of studies. These methods are appropriate when small seed size prevents addition of an external label without interfering with seed handling, significantly increasing seed weight, or when large seeds can accommodate a implanted device. Three methods of internal labelling have been developed. First, the insertion of a metal object (Sork, 1984; Mack, 1995; den Ouden, 2000) or magnet (Alverson and Diaz, 1989; Iida, 1996; Cintra and Horna, 1997; den Ouden, 2000) into seeds, and their

relocation with a magnetic locator or metal detector, proved to be a very useful method allowing recovery of seeds buried up to 40 cm deep. Depending on the time spent searching and the area surveyed, one can recover most, if not all, labelled seeds (Alverson and Diaz, 1989). As search time will limit recovery rates, a team of searchers is recommended for such experiments. In addition, the weight added by the nail or magnet may increase seed weight substantially and may thus affect dispersal distances or removal rates (Stapanian and Smith, 1978; Jansen et al., 2002). Such biases may be important for small and medium seeds (< 1 g). In the case of large seeds, the added weight is likely to be unimportant because it is included in the natural range of seed sizes observed in many species (Jansen et al., 2002). Another disadvantage is that the preparation of the seeds takes a considerable amount of time (Iida, 1996). As with other methods in which detection distances are relatively short (< 1 m), one may not detect seeds that are buried deep underground, in trees, or taken outside the search area. Consequently, few authors have used this method.

Fate of cached seeds can also be documented through the labelling of the cached item itself (as opposed to marking the location of the item). For example, Stapanian and Smith (1978) and Sork (1983) buried walnuts (Juglans nigra) and hickory nuts (Carya glabra), respectively, wrapped in aluminium foil to mimic scatterhoarding by rodents. Removal rates were documented by checking buried seeds with a metal detector. Although this method is useful for documenting post-dispersal predation and removal, it is inadequate for studying hoarding as rodents remove the foil before recaching seeds (Sork, 1983).

A second option possible only for large seeds is miniaturized radio-transmitters (Tamura, 1994). The few studies that have used this method have been carried out in temperate forests of Japan (Soné and Kohno, 1996; Tamura and Shibasaki, 1996; Tamura et al., 1999; Tamura, 2001). Tamura (1994) equipped walnuts (Juglans sp.; 10 g average), with a radio-transmitter weighing from

0.8 to 7 g fixed outside or inside the nut. Marked seeds weighed 13.1 g (range 5.8–20 g) with external radio-transmitters and 13.5 g (range 6.5–17.5 g) with internal transmitters. He observed that squirrels did not reject nuts because of their outer shape, but that seeds with transmitters weighing more than 5 g were never taken, and nuts weighing more than 15 g had lower removal rates. This method, more than any other, allows researchers to relocate seeds over long distances or in hard to reach areas. Tamura (1994) and Tamura and Shibasaki (1996), for example, were able to find seeds up to 19 m high in trees where squirrels often cache their seeds.

Labelling seeds using a radioisotope solution and relocating them after removal by animals with a portable Geiger counter is a very effective method, especially in dry environments (Lawrence and Rediske, 1959; Radvanyi, 1966; Abbott and Quink, 1970; Quink et al., 1970; Jensen, 1985; Jensen and Nielsen, 1986). Probably because radio-isotopes are perceived to be dangerous, this method was not widely applied until it was extensively used in the 1990s by Vander Wall and colleagues (Vander Wall, 1992a,b, 1993, 1994, 1995a,b,c, 1997, 2002; Vander Wall and Joyner, 1998; Longland et al., 2001). This method has many advantages: (i) a wide range of seed sizes can be studied; (ii) it is easy to prepare seeds; (iii) no marks are apparent on released seeds; (iv) the rate of recovery is high (50–100%); and (v) detection of seeds in burrows is possible up to 50 cm deep (Vander Wall, 1997). Overall, the radioisotope method may be the most efficient way to determine fates of small-seeded species. While the risk of using radionuclides is perceived to be great, the actual risk, if precautions are taken, is negligible.

The most commonly used radioisotope is scandium-46, which is a gamma-emitting radionuclide. Gamma rays penetrate soil and can be detected from about 50 cm away or 25 cm deep (depending on the strength of the label). Scandium-46 has a half-life of 83.8 days, which allows researchers to follow seed fate for 6–9 months, but after 1 year (about four half-lives) the signal is so weak that it is unlikely to interfere with experiments initiated the following year. Technetium-99 has also been used but has a half-life of 6 h so is only suitable for short-term studies, such as following seeds taken by ants. The decay product of scandium-46 is titanium, a non-radioactive and non-toxic substance. Scandium-46 is not absorbed or used by animals or plants so it is not seques-tered in the body and it does not travel through the food chain. Instead, any contamination passes through the animal's body within 2 days. Thus, animals are not affected and, at least at low concentrations, neither is seed germination. The fate of seeds can be determined because one can find both intact seeds and remains of eaten seeds. One millicurie (cost about $300) can label about 1000 seeds, depending on seed size and strength of label. Seeds are labelled by moistening them in a solution of scandium-46 in distilled water in a sealed container, then air-drying the seeds for 24–48 h. The main cost limitation is the expense of Geiger counters. Research with radioactive substances requires regulatory permits as well as attention to safety precautions and security concerns. Researchers interested in this method may contact S. Vander Wall for more details.

Entire plants can be injected with radio-nuclides, thus making seeds and seedlings from that plant detectable (Primack and Levy, 1988). This method can be used with small plants to assess dispersal but has rarely been used and never in relation to hoarding. Similarly, genetic and molecular studies (Gibson and Wheelwright, 1995; Cain et al., 2000; Degen et al., 2001; Naga-mitsu et al., 2001) offer powerful tools to study dispersal.

External cue

In humid tropical rainforests, often with little logistic support or resources, researchers need a simple, inexpensive method to mark and relocate seeds removed by rodents (Forget et al., 1998). A preliminary study carried out in April–May 1987 with thread-marked seeds of the large-seeded tree *Vouacapoua americana* (Caesalpiniaceae) showed that, as in the temperate regions, it is possible to study seed fates in the tropical environments (Forget, 1990). In comparison to some of the other methods discussed above, the use of thread or nylon line to mark seeds appears unsophisticated. However, it has proven to be very successful in dense tropical forest undergrowth. At least 20 different authors, mostly in Neotropical rainforests, have used thread-marking methods.

Depending on the author, the line might be either passed through the endosperm (Forget, 1990, 1991, 1992; Forget *et al.*, 2000; Jansen and Forget, 2001; Jansen *et al.*, 2002), passed through the tip of the hard seed coat after drilling a small hole through the seed (Forget, 1991, 1993; Forget *et al.*, 1994, 1998, 2002; Brewer and Rejmanek, 1999; Brewer, 2001; Brewer and Webb, 2001; Zhang and Wang, 2001; Li and Zhang, 2002; Charles-Dominique *et al.*, 2003; Chauvet *et al.*, 2004) or attached to fruit via a small metal screw (Adler and Kestell, 1998). When the focus of the study is seed removal and detecting caches or immediate predation, the damage to the seeds caused by drilling is unimportant. Longer-term studies designed to document seedling establishment following scatterhoarding, however, should avoid drilling seeds. Indeed, some species are attacked rapidly by ants or beetles, some dehydrate (Forget *et al.*, 2000), and some germinate rapidly after drilling (e.g. *H. courbaril*, P.-M. Forget, personal observation). For these reasons, many authors now prefer to super-glue or epoxy (e.g. Soldi-Mix®) the line directly to the seed tegument (Peres and Baider, 1997; Wenny, 1999, 2000a,b; Notman and Gorchov, 2001; Silva and Tabarelli, 2001; Midgley *et al.*, 2002; Russo, 2003), which does not damage the embryo. Nevertheless, use of glue may affect removal rates. For example, some rodents, such as agouti and acouchy, often peel fruits and seeds, cleaning all remaining parts of the pulp and any abnormal outgrowth visible on the seed surface. While manipulating such glued-marked seeds, rodents sometimes detach the line (Silva and Tabarelli, 2001; C. Baider, São Pedro, 2000, personal

communication). Therefore, the glue needs to be inconspicuous, and the line should be durable nylon instead of Dacron® polyester fibre or copper filament. Fine, steel wire, used in a few studies (Zhang and Wang, 2001; Li and Zhang, 2002; Xiao *et al.*, 2002), is difficult for rodents to cut.

Line lengths of 30–200 cm have been used. Generally, shorter lengths should be used to avoid entanglement, but, if retrieval rates are low, longer lines may show that seeds are being taken deep into burrows (Brewer and Rejmanek, 1999). Some authors have used brightly coloured line and/or fluorescent paint in order to better see the line in the dark forest understorey. Others added a label at the end of the line to number each seed individually. Variations on thread-marking methods include brightly coloured Dacron® fishing line (Midgley *et al.*, 2002), fluorescent paint sprayed on thread (Guariguata *et al.*, 2000, 2002), fluorescent green fishing line with 8 cm of fluorescent Day-Glo pink flagging at the end (Notman and Gorchov, 2001; Jansen *et al.*, 2002), brightly coloured flagging tape or plastic tag tied to the distal end of the thread (Yasuda *et al.*, 1991; Wenny, 1999, 2000a,b; Silva and Tabarelli, 2001; Wenny, Chapter 21, this volume), and a light tin tag at the end of a 3 cm long metal strand (Zhang and Wang, 2001; Li and Zhang, 2002; Xiao *et al.*, 2002). Even though such marking systems apparently do not influence the rate of seed removal (Wenny, 2000b), the potential biases associated with such methods have not been examined in detail. According to Peres and Baider (1997), such flagging tape and coloured marks do not deter hoarding animals.

In many instances, humans, especially males, have difficulty perceiving red-pink colours in the dark green understorey environment. Therefore, we recommend neutral bright colours such as white, yellow or bright blue that are highly visible and discernible on the leaf litter, to increase the chance that all seeds will be recovered by all experimenters with no bias. On the other hand, in grassland habitats, especially during the dormant season, yellow and white thread can be difficult to see (D. Wenny, personal observation). Studies should be done to verify that coloured lines, tags and other visual cues do not attract rodents or other seed removal agents or influence cache retrieval.

Other methods

Before the techniques discussed above were developed to track seed movement, it was fairly common to mark seeds or fruits with coloured, water-soluble paints (Gurnell, 1984), alcohol-based, indelible and water-proof ink (Ballardie and Whelan, 1986; Bossard, 1990; Kollmann and Schill, 1996; Hoshizaki et al., 1997, 1999; Hoshizaki and Hulme, 2002; Passos and Oliveira, 2002), typewriter correction fluid (Auld and Denham, 1999), or simply with a physical mark (Lee, 1985). The objectives of such marking were to distinguish tested fruits or seeds from older or newly fallen ones, to record removal rates, to determine if seed removal is biotic or abiotic, and to identify removed items when recovered later. This marking method involves visually searching the study area for marked seeds. Examples of these methods include documenting removal of Grevillea spp. seeds by various species of large ants, rodents and marsupials (Auld and Denham, 1999), dispersal of seeds by ants up to 9.6 m from the source (Ballardie and Whelan, 1986), transportation of Pandanus seeds by crabs to their sheltering areas at an average distance of 7.3 m (Lee, 1985), documenting caching of Pinus contorta cones in middens by red squirrels (Gurnell, 1984), and locating seedlings originating from previously marked seed caches (Kollmann and Schill, 1996; Hoshizaki et al., 1999). In Aesculus turbinata, for instance, repeated use of coloured ink allowed a description of the ultimate seed–seedling shadow by locating the seedling and their associated previously numbered seed. Mean distances of seedlings from the seed source or to the nearest parent tree varied from 14.8 to 114 m among 3 different years (Hoshizaki et al., 1999). Given the high propensity for successive

recaching of seeds of some pilfering rodents (Vander Wall and Joyner, 1998; Vander Wall, 2002), the use of this labelling method only allows description of the ultimate seed distribution, but not all hoarding sites or the seed pathway per se.

Plastic beads embedded in dung have been used to estimate dispersal distances and burial depth by dung beetles (Vulinec, 2000; Andresen and Feer, Chapter 20, this volume) although this method cannot be used to estimate post-dispersal seed fate. Similarly, plastic beads have been used to distinguish biotic from abiotic removal (Hughes and Westoby, 1992). Levey and Sargent (2000) described a method for tracking seeds dispersed by vertebrates using fluorescent microspheres. They sprayed microspheres on to fruits and examined faecal samples collected in seed traps under perches to document patterns of seed dispersal (Tewksbury et al., 2002). In order to detect the presence of microspheres, faecal samples must be examined under an epifluorescent microscope (Levey and Sargent, 2000).

Concluding Recommendations

Labelling methods contrast between regions, as well as between authors. The simplicity of free line methods has led to its use in temperate habitats (Hoshizaki et al., 1999; Li and Zhang, 2002; Midgley et al., 2002). Total cost is another crucial factor in determining which method to use. The low cost of free line methods most probably explains its widespread use in the Tropics compared to other more expensive methods such as magnets or radio-transmitters. These latter methods are, indeed, much more efficient to retrieve seeds, especially over long distances, but few students can afford losing a large number of them. Indeed, the proportion of seeds lost (i.e. not retrieved by the experimenter) may range up to 100% (E.W. Schupp, 2002, personal communication). Sample sizes can be much larger for radiolabelling and free line methods than for radio-transmitters and spool-and-line systems, which are expensive or time-intensive to prepare.

Which marking method to study seed fate?

Both the radioisotope in the temperate region and the line in the Tropics have advantages and disadvantages that might encourage authors to test the other methods listed in this chapter. In the temperate zone, although radioisotopes have proven effective in tracking a variety of seed types, the cost of instrumentation, concern about radioactivity, and licensing requirements might be barriers to their general use. When dispersal distances are short (< 50 m) and seeds are small, the most effective marking method is radioisotopes. Use of radio-transmitters is very efficient for describing long-distance dispersal patterns, but would be expensive to incorporate in a complex multi-factorial study on seed fate. Note that the vast majority of plant species have seeds < 1 g. Use of radio-transmitters is impractical for small or medium-sized seeds and is expensive, but it has been proved to be effective for large (> 5–10 g) seeds. For large seeds that weigh 10–100 g, adding a small 2 g radio-transmitter would probably not affect final dispersal distances. For dung beetles, use of lines is efficient to recover seeds taken into galleries. For larger animals, the type of line method chosen is a trade-off between visibility (and higher recovery rates) and entanglement (and shorter apparent dispersal distances or lower removal rates). Because the methods of marking developed this far, with the exception of radioisotopes, work best for larger seeds, we know the least about the occurrence of seed fates in small-seeded plants.

In conclusion, we emphasize a need for studies examining the biases of various methods and to standardize experimental protocols across different regions and species so that studies are more comparable. We recognize that most studies focus on common short-distance events while rare long-distance events are probably important in plant community dynamics. Therefore,

studies that combine techniques will be invaluable in advancing our understanding of seed fate.

Acknowledgements

This chapter greatly benefited from comments by Stephen Vander Wall and Patrick Jansen, and unpublished materials were also kindly provided by Katrin Böhning, Johannes Kollmann, Evan Notman, Kazuhiko Hoshizaki, Marco Pizo, Sabrina Russo, Mike Steele, Steven Vander Wall and Masatoshi Yasuda. Thanks to all of them.

References

Abbott, H.G. and Quink, T.F. (1970) Ecology of eastern white pine seed caches made by small forest mammals. *Ecology* 51, 271–278.

Adler, G.H. and Kestell, D.W. (1998) Fates of neotropical tree seeds influenced by spiny rats (*Proechimys semispinosus*). *Biotropica* 30, 677–681.

Alverson, W.S. and Diaz, A.G. (1989) Measurements of the dispersal of large seeds and fruits with a magnetic locator. *Biotropica* 21, 61–63.

Andresen, E. (1999) Seed dispersal by monkeys and the fate of dispersed seeds in a Peruvian rain forest. *Biotropica* 31, 145–158.

Auld, T.D. and Denham, A.J. (1999) The role of ants and mammals in dispersal and post-dispersal seed predation of the shrubs *Grevillea* (Proteaceae). *Plant Ecology* 144, 201–213.

Ballardie, R.T. and Whelan, R.J. (1986) Masting, seed dispersal, and seed predation in the cycad *Macrozamia communis*. *Oecologia* 70, 100–105.

Beck, H. and Terborgh, J. (2002) Groves versus isolates: how spatial aggregation of *Astrocaryum murumuru* palms affects seed removal. *Journal of Tropical Ecology* 18, 275–288.

Benkman, C.W., Balda, R.P. and Smith, C.C. (1984) Adaptations for seed dispersal and the compromises due to seed predation in limber pine. *Ecology* 65, 632–642.

Böhning-Gaese, K., Gaese, B.H. and Rabemantsoa, S.B. (1995) Seed dispersal by frugivorous tree visitors in the Malagasy tree species

Commiphora guillaumini. Ecotropica 1, 41–50.

Böhning-Gaese, K., Gaese, B.H. and Rabemanantsoa, S.B. (1999) Importance of primary and secondary seed dispersal in the Malagasy tree *Commiphora guillaumini. Ecology* 80, 821–832.

Bossard, C.C. (1990) Tracing of ant-dispersed seeds: a new technique. *Ecology* 71, 2370–2371.

Brewer, S.W. (2001) Predation and dispersal of large and small seeds of a tropical palm. *Oikos* 92, 245–255.

Brewer, S.W. and Rejmanek, M. (1999) Small rodents as significant dispersers of tree seeds in a neotropical forest. *Journal of Vegetation Science* 10, 165–174.

Brewer, S.W. and Webb, M.A.H. (2001) Ignorant seed predators and factors affecting the seed survival of a tropical palm. *Oikos* 93, 32–41.

Cahalane, V.H. (1942) Caching and recovery of food by the western fox squirrel. *Journal of Wildlife Management* 6, 338–352.

Cain, M.L., Milligan, B.G. and Strand, A.E. (2000) Long-distance seed dispersal in plant populations. *American Journal of Botany* 87, 1217–1227.

Chambers, J.C. and MacMahon, J.A. (1994) A day in the life of a seed: movements and fates of seeds and their implications for natural and managed systems. *Annual Review of Ecology and Systematics* 25, 263–292.

Charles-Dominique, P., Chave, J., Dubois, M.-A., de Granville, J.-J., Riera, B. and Vezzoli, C. (2003) Colonization front of the understorey palm *Astrocaryum sciophilum* in a pristine rain forest of French Guiana. *Global Ecology and Biogeography* 12, 237–248.

Chauvet, S., Feer, F. and Forget, P.-M. (2004) Seed fate of two Sapotaceae species in a Guianan rain forest in the context of escape and satiation hypotheses. *Journal of Tropical Ecology* 20, 1–9.

Cintra, R. and Horna, V. (1997) Seed and seedling survival of the palm *Astrocaryum murumuru* and the legume tree *Dipteryx micrantha* in gaps in Amazonian forest. *Journal of Tropical Ecology* 13, 257–277.

Clarkson, K., Eden, S.F., Sutherland, W.J. and Houston, A.I. (1986) Density dependence and magpie food hoarding. *Journal of Animal Ecology* 55, 111–121.

Darley-Hill, S. and Johnson, W.C. (1981) Acorn dispersal by the blue jay (*Cyanocitta cristata*). *Oecologia* 50, 231–232.

Degen, B., Caron, H., Bandou, E., Maggia, L., Chevallier, M.H., Leveau, A. and Kremer, A.

(2001) Fine-scale spatial genetic structure of eight tropical tree species as analysed by RAPDs. *Heredity* 87, 497–507.

den Ouden, J. (2000) The role of bracken (*Pteridium aquilinum*) in forest dynamics. PhD thesis, Wageningen University, Wageningen, The Netherlands.

Dennis, A.J. (2003) Scatter-hoarding by musky rat-kangaroos, *Hypsiprymnodon moschatus*, a tropical rain-forest marsupial from Australia: implications for seed dispersal. *Journal of Tropical Ecology* 19, 619–627.

Estrada, R. and Fleming, T.H. (1986) *Frugivores and Seed Dispersal*. Dr W. Junk Publishers, Dordrecht, The Netherlands.

Feer, F. (1999) Effects of dung beetles (Scarabaeidae) on seeds dispersed by howler monkeys (*Alouatta seniculus*) in the French Guiana rainforest. *Journal of Tropical Ecology* 15, 1–14.

Fleming, T.H. and Estrada, A. (1993) *Frugivory and Seed Dispersal: Ecological and Evolutionary Aspects*. Kluwer Academic Publishers, Dordrecht, The Netherlands, 392 pp.

Forget, P.-M. (1990) Seed-dispersal of *Vouacapoua americana* (Caesalpiniaceae) by caviomorph rodents in French Guiana. *Journal of Tropical Ecology* 6, 459–468.

Forget, P.-M. (1991) Scatterhoarding of *Astrocaryum paramaca* by *Proechimys* in French Guiana: comparison with *Myoprocta exilis. Tropical Ecology* 32, 155–167.

Forget, P.-M. (1992) Seed removal and seed fate in *Gustavia superba* (Lecythidaceae). *Biotropica* 24, 408–414.

Forget, P.-M. (1993) Post-dispersal predation and scatterhoarding of *Dipteryx panamensis* (Papilionaceae) seeds by rodents in Panama. *Oecologia* 94, 255–261.

Forget, P.-M. and Milleron, T. (1991) Evidence for secondary seed dispersal in Panama. *Oecologia* 87, 596–599.

Forget, P.-M., Munoz, E. and Leigh, E.G. Jr (1994) Predation by rodents and bruchid beetles on seeds of *Scheelea* palms on Barro Colorado Island, Panama. *Biotropica* 26, 420–426.

Forget, P.-M., Milleron, T. and Feer, F. (1998) Patterns in post-dispersal seed removal by neotropical rodents and seed fate in relation to seed size. In: Newbery, D.M., Prins, H.T.T. and Brown, N.D. (eds) *Dynamics of Tropical Communities*. Blackwell Science, Oxford, pp. 25–49.

Forget, P.-M., Milleron, T., Feer, F., Henry, O. and Dubost, G. (2000) Effects of dispersal pattern and mammalian herbivores on seedling recruitment for *Virola michelii*

(Myristicaceae) in French Guiana. *Biotropica* 32, 452–462.

Forget, P.-M., Hammond, D., Milleron, T. and Thomas, R. (2002) Seasonality of fruiting and food hoarding by rodents in Neotropical forests: consequences for seed dispersal and seedling recruitment. In: Levey, D.J., Silva, W.R. and Galetti, M. (eds) *Seed Dispersal and Frugivory: Ecology, Evolution and Conservation.* CAB International, Wallingford, UK, pp. 241–253.

Gibson, J.P. and Wheelwright, N.T. (1995) Genetic structure in a population of a tropical tree *Ocotea tenera* (Lauraceae): influence of avian seed dispersal. *Oecologia* 103, 49–54.

Guariguata, M.R., Adame, J.J.R. and Finegan, B. (2000) Seed removal and fate in two selectively logged lowland forests with contrasting protection levels. *Conservation Biology* 14, 1046–1054.

Guariguata, M.R., AriasLeClaire, H. and Jones, G. (2002) Tree seed fate in a logged and fragmented forest landscape, northeastern Costa Rica. *Biotropica* 34, 405–415.

Gurnell, J. (1984) Home range, territoriality, caching behaviour and food supply of the red squirrel (*Tamasciurus hudsonicus fremonti*) in a subalpine lodgepole pine forest. *Animal Behavior* 32, 1119–1131.

Halfpenny, J.C. (1992) Environmental impacts of powdertracking using fluorescent pigments. *Journal of Mammalogy* 73, 680–682.

Hallwachs, W. (1986) Agoutis *Dasyprocta punctata*: the inheritors of guapinol *Hymenaea courbaril* (Leguminosae). In: Estrada, R. and Fleming, T.H. (eds) *Frugivores and Seed Dispersal.* Dr W. Junk Publishers, The Hague, pp. 119–135.

Hallwachs, W. (1994) The clumsy dance between agoutis and plants: scatterhoarding by Costa Rican dry forest agoutis (*Dasyprocta punctata*: Dasyproctidae: Rodentia). PhD thesis, Cornell University, Ithaca, New York.

Hoshizaki, K. and Hulme, P.E. (2002) Mast seedling and predator-mediated indirect interactions in a forest community: evidence from post-dispersal fate of rodent-generated caches. In: Levey, D.J., Silva, W.R. and Galetti, M. (eds) *Seed Dispersal and Frugivory: Ecology, Evolution and Conservation.* CAB International, Wallingford, UK, pp. 227–239.

Hoshizaki, K., Suzuki, W. and Sasaki, S. (1997) Impacts of secondary seed dispersal and herbivory on seedling survival in *Aesculus turbinata. Journal of Vegetation Science* 8, 735–742.

Hoshizaki, K., Suzuki, W. and Nakashizuka, T. (1999) Evaluation of secondary dispersal in a large-seeded tree *Aesculus turbinata*: a test of directed dispersal. *Plant Ecology* 144, 167–176.

Hughes, L. and Westoby, M. (1992) Fate of seeds adapted for dispersal by ants in Australian sclerophyll vegetation. *Ecology* 73, 1285–1299.

Hurly, T.A. and Robertson, R.J. (1987) Scatter-hoarding by territorial red squirrels: a test of the optimal density model. *Canadian Journal of Zoology – Revue Canadienne de Zoologie* 65, 1247–1252.

Hutchins, H.E., Hutchins, S.A. and Liu, B. (1996) The role of birds and mammals in Korean pine (*Pinus koraiensis*) regeneration dynamics. *Oecologia* 107, 120–130.

Iida, S. (1996) Quantitative analysis of acorn transportation by rodents using magnetic locator. *Vegetatio* 124, 39–43.

Jansen, P.A. (2003) Scatterhoarding and tree regeneration. Ecology of nut dispersal in a Neotropical forest. PhD thesis, Wageningen University, Wageningen, The Netherlands.

Jansen, P.A. and Forget, P.-M. (2001) Scatterhoarding by rodents and tree regeneration in French Guiana. In: Bongers, F., Charles-Dominique, P., Forget, P.-M. and Théry, M. (eds) *Nouragues: Dynamics and Plant–Animal Interactions in a Neotropical Rainforest.* Kluwer Academic Publishers, Dordrecht, The Netherlands, pp. 275–288.

Jansen, P.A., Bartholomeus, M., Bongers, F., Elzinga, J.A., den Ouden, J. and Van Wieren, S.E. (2002) The role of seed size in dispersal by a scatter-hoarding rodent. In: Levey, D.J., Silva, W.R. and Galetti, M. (eds) *Seed Dispersal and Frugivory: Ecology, Evolution and Conservation.* CAB International, Wallingford, UK, pp. 209–225.

Jensen, T.S. (1985) Seed–seed predator interactions of European beech, *Fagus sylvatica* and forest rodents, *Clethrionomys glareolus* and *Apodemus flavicollis. Oikos* 44, 149–156.

Jensen, T.S. and Nielsen, O.F. (1986) Rodents as seed dispersers in a heath–oak wood succession. *Oecologia (Berlin)* 70, 214–221.

Jokinen, S. and Suhonen, J. (1995) Food caching by willow and crested tits: a test of the scatter-hoarding models. *Ecology* 76, 892–898.

Kaspari, M. (1993) Removal of seeds from Neotropical frugivore droppings. *Oecologia* 95, 81–88.

Kikusawa, K. (1988) Dispersal of *Quercus mongolica* acorns in a broadleaved deciduous forest, 1. Disappearance. *Forest Ecology and Management* 25, 168.

Kollmann, J. and Bassin, S. (2001) Effects of management on seed predation in wildflower strips in northern Switzerland. *Agriculture Ecosystems and Environment* 83, 285–296.

Kollmann, J. and Schill, H.-P. (1996) Spatial patterns of dispersal, seed predation and germination during colonization of abandoned grassland by *Quercus petraea* and *Coryllus avellana*. *Vegetatio* 125, 193–205.

Kraus, B. (1983) A test of the optimal-density model for seed scatterhoarding. *Ecology* 64, 608–610.

Lawrence, W.H. and Rediske, J.H. (1959) Radio-tracer technique for determining the fate of broadcast Douglas-fir seed. *Proceedings of the American Society of Foresters* 1959, 99–101.

Lee, M.A.B. (1985) The dispersal of *Pandanus tectorius* by the land crab *Cardisoma carnifex*. *Oikos* 45, 169–173.

Lemen, C.A. and Freeman, P.W. (1985) Tracking mammals with fluorescent pigments: a new technique. *Journal of Mammalogy* 66, 134–136.

Levey, D.J. and Byrne, M.M. (1993) Complex ant–plant interactions: rain forest ants as secondary dispersers and post-dispersal seed predators. *Ecology* 74, 1802–1812.

Levey, D.J. and Sargent, S. (2000) A simple method for tracking vertebrate-dispersed seeds. *Ecology* 81, 267–274.

Levey, D.J., Silva, W.R. and Galetti, M. (eds) (2002) *Seed Dispersal and Frugivory: Ecology, Evolution and Conservation*. CAB International, Wallingford, UK, 511 pp.

Li, H.-J. and Zhang, Z.-B. (2002) Effects of rodents on acorn dispersal and survival of the Liaodong oak (*Quercus liaotungensis* Koidz.). *Forest Ecology and Management* 6026, 1–10.

Longland, W.S. and Clements, C. (1995) Use of fluorescent pigments in studies of seed caching by rodents. *Journal of Mammalogy* 76, 1260–1266.

Longland, W.S., Jenkins, S.H., Vander Wall, S.B., Veech, J.A. and Pyare, S. (2001) Seedling recruitment in *Oryzopsis hymenoides*: are desert granivores mutualists or predators? *Ecology* 82, 3131–3148.

Mack, A.L. (1995) Distance and non-randomness of seed dispersal by the dwarf cassowary (*Casuarius bennetti*). *Ecography* 18, 286–295.

Midgley, J., Anderson, B., Bok, A. and Fleming, T. (2002) Scatter-hoarding of Cape Proteaceae nuts by rodents. *Evolutionary Ecology Research* 4, 623–626.

Mittelbach, G.G. and Gross, K.L. (1984) Experimental studies of seed predation in old-fields. *Oecologia* 65, 7–13.

Miyaki, M. and Kikuzawa, K. (1988) Dispersal of *Quercus mongolica* acorns in a broadleaved deciduous forest. 2. Scatterhoarding by mice. *Forest Ecology and Management* 25, 9–16.

Moegenburg, S.M. (1996) *Sabal palmetto* seed size: causes of variation, choices of predators, and consequences for seedlings. *Oecologia* 106, 539–543.

Moreno, E. and Carrascal, L.M. (1995) Hoarding nuthatches spend more time hiding a husked seed than an unhusked seed. *Ardea* 83, 391–395.

Nagamitsu, T., Ichikawa, S., Ozawa, M., Shimamura, R., Kachi, N., Tsumura, Y. and Muhammad, N. (2001) Microsatellite analysis of the breeding system and seed dispersal in *Shorea leprosula* (Dipterocarpaceae). *International Journal of Plant Sciences* 162, 155–159.

Notman, E. and Gorchov, D.L. (2001) Variation in post-dispersal seed predation in mature Peruvian lowland tropical forest and fallow agricultural sites. *Biotropica* 33, 621–636.

Okamoto, M. (1994) Postdispersal seed predation by eastern turtle dove: an experimental study in an urban botanical garden. *Bulletin of the Osaka Museum of Natural History* 48, 1–8.

Passos, L. and Oliveira, P.S. (2002) Ants affect the distribution and performance of seedlings of *Clusia criuva*, a primarily bird-dispersed rain forest tree. *Journal of Ecology* 90, 517–528.

Peres, C.A. and Baider, C. (1997) Seed dispersal, spatial distribution and population structure of Brazilnut trees (*Bertholletia excelsa*) in southeastern Amazonia. *Journal of Tropical Ecology* 13, 595–616.

Primack, R.B. and Levy, C.K. (1988) A method to label seeds and seedlings using gamma-emitting radionuclides. *Ecology* 69, 796–800.

Pyare, S. and Longland, W.S. (2000) Seedling-aided cache detection by heteromyid rodents. *Oecologia* 122, 66–71.

Quink, T.F., Abbott, H.G. and Mellen, W.J. (1970) Locating tree seed caches of small mammals with a radioisotope. *Forest Science* 16, 147–148.

Radvanyi, A. (1966) Destruction of radio-tagged seeds of white spruce by small mammals during summer months. *Forest Science* 12, 307–314.

Russo, S.E. (2003) Linking spatial patterns of seed dispersal and plant recruitment in a neotropical tree, *Virola calophylla* (Myristicaceae). PhD thesis, University of Illinois, Urbana, Illinois.

Schupp, E.W. (1988a) Factors affecting post-dispersal seed survival in a tropical forest. *Oecologia* 76, 525–530.

Schupp, E.W. (1988b) Seed and early seedling predation in the forest understory and in treefall gaps. *Oikos* 51, 71–78.

Schupp, E.W. (1990) Annual variation in seedfall, postdispersal predation, and recruitment of a neotropical tree. *Ecology* 71, 504–515.

Schupp, E.W. and Frost, E.J. (1989) Differential predation of *Welfia georgii* seeds in treefall gaps and the forest understory. *Biotropica* 21, 200–203.

Silva, M.G. and Tabarelli, M. (2001) Seed dispersal, plant recruitment and spatial distribution of *Bactris acanthocarpa* Martius (Arecaceae) in a remnant of Atlantic forest in northeast Brazil. *Acta Oecologica – International Journal of Ecology* 22, 259–268.

Smith, T.J. III (1987) Seed predation in relation to tree dominance and distribution in mangrove forests. *Ecology* 68, 266–273.

Smith, T.J. III, Chan, H.T., McIvor, C.C. and Robblee, M.B. (1989) Comparison of seed predation in tropical tidal forests from three continents. *Ecology* 70, 146–151.

Smythe, N. (1970) Relationships between fruiting seasons and seed dispersal methods in a neotropical forest. *The American Naturalist* 104, 25–35.

Smythe, N. (1978) The natural history of the Central American agouti (*Dasyprocta punctata*). *Smithsonian Contribution to Zoology* 257, 1–52.

Soné, K. and Kohno, A. (1996) Application of radiotelemetry to the survey of acorn dispersal by *Apodemus* mice. *Ecological Research* 11, 187–192.

Sork, V.L. (1983) Distribution of pignut hickory (*Carya glabra*) along a forest to edge transect, and factors affecting seedling recruitment. *Bulletin of the Torrey Botanical Club* 110, 494–506.

Sork, V.L. (1984) Examination of seed dispersal and survival in red oak, *Quercus rubra* (Fagaceae), using metal-tagged acorns. *Ecology* 65, 1020–1022.

Sork, V.L. and Boucher, D.H. (1977) Dispersal of sweet pignut hickory in a year of low fruit production, and the influence of predation by a curculionid beetle. *Oecologia (Berlin)* 28, 289–299.

Sousa, W.P. and Mitchell, B.J. (1999) The effect of seed predators on plant distributions: is there a general pattern in mangroves? *Oikos* 86, 55–66.

Stapanian, M.A. and Smith, C.C. (1978) A model for seed scatterhoarding: coevolution of fox squirrels and black walnuts. *Ecology* 59, 884–896.

Steele, M.A., Hadj-Chikh, L.Z. and Hazeltine, J. (1996) Caching and feeding decisison by *Sciurus carolinensis*: responses to weevil-infested acorns. *Journal of Mammalogy* 77, 305–314.

Tamura, N. (1994) Application of a radio-transmitter for studying seed dispersion by animals. *Journal of the Japanese Forestry Society* 76, 607–610.

Tamura, N. (2001) Walnut hoarding by the Japanese wood mouse, *Apodemus speciosus* Temminck. *Journal of Forestry Research* 6, 187–190.

Tamura, N. and Shibasaki, E. (1996) Fate of walnut seeds, *Juglans ailanthifolia*, hoarded by Japanese squirrels, *Sciurus lis*. *Journal of Forestry Research* 1, 219–222.

Tamura, N., Hashimoto, Y. and Hayashi, F. (1999) Optimal distances for squirrels to transport and hoard walnuts. *Animal Behavior* 58, 635–642.

Tewkesbury, J.J., Levey, D.J., Haddad, N.M., Sergent, S., Orrock, J.L., Weldon, A., Danielson, B.J., Brinkerhoo, J., Danshen, E.L. and Townsend, P. (2002) Corridors affect plants, animals and their interactions in fragmented landscapes. *Proceedings of the National Academy of Sciences USA* 99, 12923–12926.

Theimer, T.C. (2001) Seed scatterhoarding by white-tailed rats: consequences for seedling recruitment by an Australian rain forest tree. *Journal of Tropical Ecology* 17, 177–189.

Theimer, T.C. (2003) Intraspecific variation in seed size affects scatterhoarding behaviour of an Australian tropical rain-forest rodent. *Journal of Tropical Ecology* 19, 95–98.

Theimer, T.C. and Gehring, C.A. (1999) Effects of a litter-disturbing bird species on tree seedling germination and survival in an Australian tropical rain forest. *Journal of Tropical Ecology* 15, 737–749.

Vander Wall, S.B. (1990) *Food Hoarding in Animals*. University of Chicago Press, Chicago, Illinois, 45 pp.

Vander Wall, S.B. (1992a) Establishment of Jeffrey pine seedlings from animal caches. *Western Journal of Applied Forestry* 7, 14–20.

Vander Wall, S.B. (1992b) The role of animals in dispersing a 'wind-dispersed' pine. *Ecology* 73, 614–621.

Vander Wall, S.B. (1993) Cache site selection by chipmunk (*Tamias* spp.) and its influence on the effectiveness of seed dispersal in Jeffrey pine (*Pinus jeffreyi*). *Oecologia* 96, 246–252.

Vander Wall, S.B. (1994) Removal of wind-dispersed pine seeds by ground-foraging vertebrates. *Oikos* 69, 125–132.

Vander Wall, S.B. (1995a) Dynamics of yellow pine chipmunk (*Tamias amoenus*) seed caches: underground traffic in bitterbrush seeds. *Ecoscience* 2, 216–266.

Vander Wall, S.B. (1995b) The effects of seed value on the caching behavior of yellow pine chipmunks. *Oikos* 74, 533–537.

Vander Wall, S.B. (1995c) Influence of substrate water on the ability of rodents to find buried seeds. *Journal of Mammalogy* 76, 851–856.

Vander Wall, S.B. (1997) Dispersal of singleleaf piñon pine (*Pinus monophylla*) by seed-caching rodents. *Journal of Mammalogy* 78, 181–191.

Vander Wall, S.B. (2002) Secondary dispersal of Jeffrey pine seeds by rodent scatter hoarders: the roles of pilfering, recaching, and a variable environment. In: Levey, D.J., Silva, W.R. and Galetti, M. (eds) *Seed Dispersal and Frugivory: Ecology, Evolution and Conservation*. CAB International, Wallingford, UK, pp. 193–208.

Vander Wall, S.B. and Joyner, J.W. (1998) Recaching of Jeffrey pine (*Pinus jeffreyi*) seeds by yellow pine chipmunks (*Tamias amoenus*): potential effects on plant reproductive success. *Canadian Journal of Botany* 76, 154–162.

Vander Wall, S.B. and Peterson, E. (1996) Associative learning and the use of cache markers by yellow pine chipmunks (*Tamias amoenus*). *Southwestern Naturalist* 41, 88–90.

Vulinec, K. (2000) Dung beetles (Coleoptera: Scarabaeidae), monkeys, and conservation in Amazonia. *Florida Entomologist* 83, 229–241.

Wenny, D.G. (1999) Two-stage dispersal of *Guarea glabra* and *G. kunthiana* (Meliaceae) in Monteverde, Costa Rica. *Journal of Tropical Ecology* 15, 481–496.

Wenny, D.G. (2000a) Seed dispersal of a high quality fruit by specialized frugivores: high quality dispersal? *Biotropica* 32, 327–337.

Wenny, D.G. (2000b) Seed dispersal, seed predation, and seedling recruitment of a neotropical montane tree. *Ecological Monographs* 70, 331–351.

Whelan, C.J., Willson, M.F., Tuma, C.A. and Souza-Pinto, I. (1991) Spatial and temporal patterns of post-dispersal seed predation. *Canadian Journal of Botany* 69, 428–436.

Xiao, Z.-S., Wang, Y.-S., Zhang, Z.-B. and Ma, Y. (2002) Preliminary studies on the relationships between communities of small mammals and types of habitats in Dujiangyan Region. *Chinese Biodiversity*, 163–169. (In Chinese with English summary.)

Yasuda, M., Nagagoshi, N. and Takahashi, F. (1991) Examination of the spool-and-line method as a quantitative technique to investigate seed dispersal by rodents. *Japanese Journal of Ecology* 41, 257–262.

Yasuda, M., Miura, S. and Hussein, N.A. (2000) Evidence for hoarding behaviour in terrestrial rodents in Pasoh Forest reserve, a Malaysian lowland rain forest. *Journal of Tropical Forest Science* 12, 164–173.

Zhang, Z.-B. and Wang, F.-S. (2001) Effects of rodents on seed dispersal and survival of wild apricot (*Prunus armeniaca*). *Acta Ecologica Sinica* 21, 839–846.

Index

Note: page numbers in *italics* refer to figures and tables

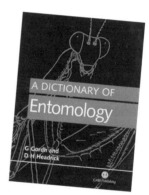
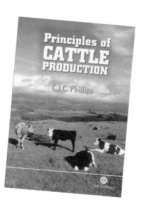